Weather Dynamics
An Introduction

.

Weather Dynamics
An Introduction

Thomas Frisius
Climate Service Center Germany (GERICS), Germany

Klaus Fraedrich
Max Planck Institute of Meteorology, Germany

World Scientific

NEW JERSEY · LONDON · SINGAPORE · BEIJING · SHANGHAI · HONG KONG · TAIPEI · CHENNAI · TOKYO

Published by

World Scientific Publishing Co. Pte. Ltd.

5 Toh Tuck Link, Singapore 596224

USA office: 27 Warren Street, Suite 401-402, Hackensack, NJ 07601

UK office: 57 Shelton Street, Covent Garden, London WC2H 9HE

Library of Congress Control Number: 2024035245

British Library Cataloguing-in-Publication Data
A catalogue record for this book is available from the British Library.

WEATHER DYNAMICS
An Introduction

ISBN 978-981-12-7628-6 (hardcover)
ISBN 978-981-12-7723-8 (paperback)
ISBN 978-981-12-7629-3 (ebook for institutions)
ISBN 978-981-12-7630-9 (ebook for individuals)

For any available supplementary material, please visit
https://www.worldscientific.com/worldscibooks/10.1142/13410#t=suppl

Desk Editor: Kannan Krishnan/Amanda Yun

Typeset by Stallion Press
Email: enquiries@stallionpress.com

Preface

This book provides an introduction to the dynamics of weather in the atmosphere. It is designed as a textbook addressed to students. The material is based on a three-semester course on theoretical meteorology held at the University of Hamburg by the authors. Parts of the book were inspired by the lecture "Theoretical Meteorology" given by Prof. F. Herbert at the Goethe University Frankfurt am Main. The book is primarily meant for undergraduate and graduate students of meteorology or atmospheric sciences. However, it is also suitable for students studying other natural science disciplines or interested laypersons, if they have solid basic knowledge in mathematics and physics.

The chapters of the book are built upon each other and the reader should understand the material presented here without using additional literature. An appendix assists readers without a solid background in mathematics to get into the subject. Often, we cite additional literature that can be used for further advancing the knowledge about a particular topic. The cited literature also provides an overview of relevant historical contributions to dynamic meteorology. To further advance the understanding of weather dynamics, we also employ a range of nonlinear numerical models developed in the working environments of the authors. These include the global circulation model Planet Simulator (PlaSim) and the cloud model HURMOD, which have also been used in research. The authors highly appreciate comments for future improvements of this text.

Thomas Frisius and Klaus Fraedrich

Notation

In this book, we use the following mathematical notation:

1. Real and complex variables and constants are denoted by italic letters, e.g. a or A.
2. Scales and associated dimensionless quantities are written with Roman letters, e.g. L for the length scale and Ro for the Rossby number.
3. Vectors and tensors in three-dimensional space are represented by boldface letters, e.g. $\mathbf{a}$ or $\mathbf{A}$.
4. A complex variable a is split into real and imaginary parts by the subscripts r and i, respectively, i.e. $a = a_r + ia_i$.
5. The superscript C symbolizes a complex conjugation, i.e. $a^C = a_r - ia_i$.
6. The functions $\Re(..)$ and $\Im(..)$ yield real and imaginary parts of the argument, respectively, e.g. $\Re(a) = a_r$ and $\Im(a) = a_i$.
7. Vectors and matrices based on phase space coordinates (amplitudes of selected field functions) are indicated by $\vec{}$ and $\overset{\leftrightarrow}{}$, respectively, e.g. $\vec{a}$ and $\overset{\leftrightarrow}{a}$.
8. The superscript T indicates the transpose of a tensor or matrix, e.g. $\mathbf{A}^{\mathrm{T}}$ or $\overset{\leftrightarrow}{a}{}^{\mathrm{T}}$.

About the Authors

Thomas Frisius is currently employed at the Climate Service Center Germany (GERICS) of the Helmholtz-Zentrum Hereon, Germany, which he joined in 2018. He studied Meteorology at the University of Hamburg and received a PhD in 1999 on the topic of "The impact of symmetry breaking instabilities on the nonlinear development of baroclinic waves". From 2000 until 2007 he was a lecturer at the Institute for Meteorology and Geophysics of the Goethe-University Frankfurt am Main. Afterward, he worked at the Max Planck Institute for Meteorology in the Max Planck Fellow Group "Modeling Continuum Climate Variability". In 2009, he was appointed as head of the Junior Research Group "Dynamical Systems" within CLiSAP, the "Integrated Climate System Analysis and Prediction" Cluster of Excellence, at the University of Hamburg where he lectured on "Theoretical Meteorology". Dr Frisius' research interests include baroclinic waves, tropical cyclones, dynamical systems, global and regional atmospheric modeling and ocean dynamics.

Klaus Fraedrich is a guest scientist at the Max-Planck Institute for Meteorology, Hamburg, Germany. His prior affiliations include Professor at the University of Hamburg, Freie Universität Berlin, Rheinische Friedrich-Wilhelms-Universität Bonn (all in Germany); visiting Professor at the Universities of New South Wales (Sydney, Australia), Washington (Seattle, USA), Wisconsin (Madison, USA), Nanjing University of Information Science and Technology (China), Head of Long-Range Weather Forecasting Research, Bureau of Meteorology Research Centre (Melbourne, Australia); and Postdoc at Colorado State University (Fort Collins, USA), visiting researcher in South America (Venezuela) and in Africa (East African Meteorology Department in Nairobi, Kenya). He has received awards including the Lewis Frey Richardson Medal (European Geophysical Union), Alfred Wegener Medal (Germany), Gay-Lussac-Humboldt Prize (France-Germany), Max Planck Research Prize (Germany), Max Planck Fellowship (Germany), Academy of Sciences and Humanities (Hamburg). His research interests span across regional climates, nonlinear systems analysis and predictability, weather dynamics of tropics and extra-tropics, building of a user-friendly and open-access GCM hierarchy, and climate variability and attribution.

Contents

Introduction

Weather events are part of everyday life and in this book, we will learn the physical mechanisms that are responsible for the development of these phenomena. Weather results from the motion of **air** and takes place in the **Earth's atmosphere** that can be divided into several layers. These are the **troposphere, stratosphere, mesosphere** and **thermosphere**. Figure 1 shows the vertical arrangement of these layers together with the vertical temperature profile of the **international standard atmosphere**. The troposphere represents the lowest atmospheric layer and is characterized by a temperature decrease in the vertical direction. Most of the weather activity happens here and, in this book, we will focus on this layer. The **tropopause** at 11 km height forms the boundary between the troposphere and stratosphere. The temperature does not change with height in the lower part of the stratosphere and increases in its upper part. The stratosphere and mesosphere are separated by the **stratopause** at 47 km height. In the mesosphere, the temperature decreases again with height up to the **mesopause** at about 85 km height. The thermosphere lies above the mesopause and here the temperature increases strongly with height. However, the air density is very low, so that the usual understanding of temperature is not helpful. The **exosphere** that describes the transition to outer space begins above the **thermopause** at about 500 km. The mass density and pressure decrease strongly with height and approach zero in the thermosphere. The troposphere has already 75% of the total atmospheric mass. The actual vertical temperature profiles differ in time and space from that of the international standard atmosphere

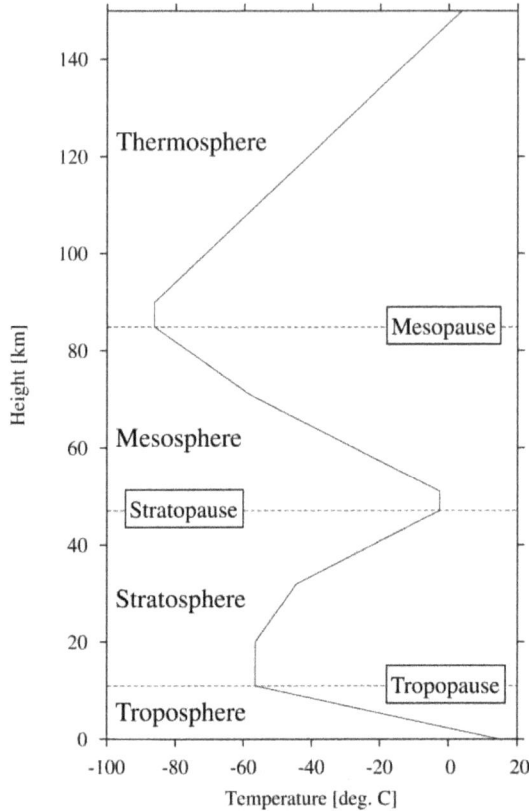

Fig. 1.   Vertical layers in the atmosphere. The dashed lines mark the boundaries of the layer and the solid line shows temperature in °C of the international
standard atmosphere.

but the layer classification remains valid. In this book, we will not
deal with the formation of these layers but state that it is linked to
electromagnetic radiation.

The gaseous constituents of air have except for **water vapor** constant volume shares in the atmosphere up to an altitude of 100 km.
Dry air is mainly composed of **nitrogen**(∼78%), **oxygen** (∼21%),
argon (∼0.9%) and **carbon dioxide** (∼0.04%). The latter is slowly
increasing over time due to anthropogenic emissions. There are further **trace gases** in the atmosphere which have a low volume share
but can be important for the **greenhouse effect** due to the **absorption** of **infrared radiation**. Water vapor, cloud drops, ice particles

and aerosols have variable concentrations in the air because of significant sinks and sources.

The long-term time average of atmospheric flow describes the **general circulation of the atmosphere**. An average over one season is already sufficient to see the characteristic pattern. Figure 2 shows a **meridional** (south–north) cross-section through the atmosphere which displays observed temperature, eastward air velocity and streamlines of the **meridional overturning circulation**. The associated fields have been averaged over the year 2008 and in **zonal** (west–east) direction along the latitude circles. Note that the vertical coordinate is not height but the ratio of pressure and surface pressure. The reason is that the surface of the earth is not flat which leads to difficulties in averaging along latitude circles at constant height. On the other hand, the pressure ratio is one at the surface and, therefore, perfectly adapted to the mountainous surface. In general, pressure decreases with height so that it represents a suitable vertical coordinate. We see that temperature decreases in a vertical direction until it reaches the tropopause where the change reverses. Temperature has

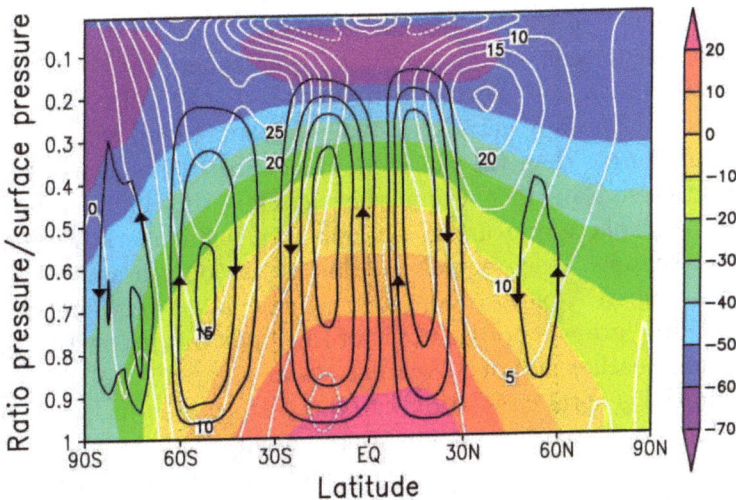

Fig. 2.   Annual and zonal average of temperature (colored shadings, °C), zonal wind (white isolines, contour interval 5 m/s), and streamlines of the meridional overturning circulation (black lines) as a function of latitude and the ratio pressure divided by surface pressure. The displayed fields for the year 2008 are based on NCEP Reanalysis (see Kalnay *et al.*, 1996 for more details).

a maximum near the equator and decreases monotonically towards
the poles. This meridional temperature gradient establishes because
more solar radiation reaches the earth at low latitudes than at high
latitudes. The eastward wind velocity is known as the **zonal wind**
and it has maxima close to the tropopause in the mid-latitudes of the
Southern and Northern Hemispheres. These maxima describe the **jet
streams** that form because of the meridional temperature gradients.
We will explain this relation later in this book. The Southern Hemisphere has two jet streams while in the Northern Hemisphere, only
one jet stream is observed. Five cells of the meridional overturning
circulation can be identified. The density of the plotted streamlines
is related to the strength of the cell. The two strong cells north and
south of the equator are the **Hadley cells**. They have an upward
motion around the equator and a downward motion in the subtrop-
ics. They form because heating at low latitudes is much higher than
at high latitudes. The overturning in the mid-latitudes is described
by the **Ferrel cells**. They have a reversed sense of rotation compared
to adjacent Hadley cells. The Ferrel cell in the Northern Hemisphere
has less strength than the one in the Southern Hemisphere. South of
60S another cell appears which constitutes the **polar cell** and has
the same sense of rotation as the southern Hadley cell. Obviously,
no such cell is seen in the Northern Hemisphere. The visible differences between both hemispheres stem from the different land-sea
distribution. The Northern Hemisphere has more land area and more
mountain ranges than the Southern Hemisphere. The general circulation has also a significant annual cycle. Only the Hadley cell of the
winter hemisphere appears in the northern hemispheric winter and
summer seasons. The general circulation has only a small interan-
nual variability. A picture from another year looks very similar. The
general circulation is not the topic of this book but it constitutes the
frame for weather activity. Weather systems experience also a feedback on the general circulation, i.e. the pattern seen in Fig. 2 would
look different if weather systems were absent.

Figure 3 shows sea level pressure and the horizontal wind vector
close to the ground for two consecutive days. The region displayed
encompasses large parts of North America, the North Atlantic and
Europe. We can detect several local pressure minima and maxima
which are marked by L and H, respectively. These **lows** and **highs**
steer the direction of the wind. Here in the northern Hemisphere
air tends to flow anticlockwise around lows and clockwise around

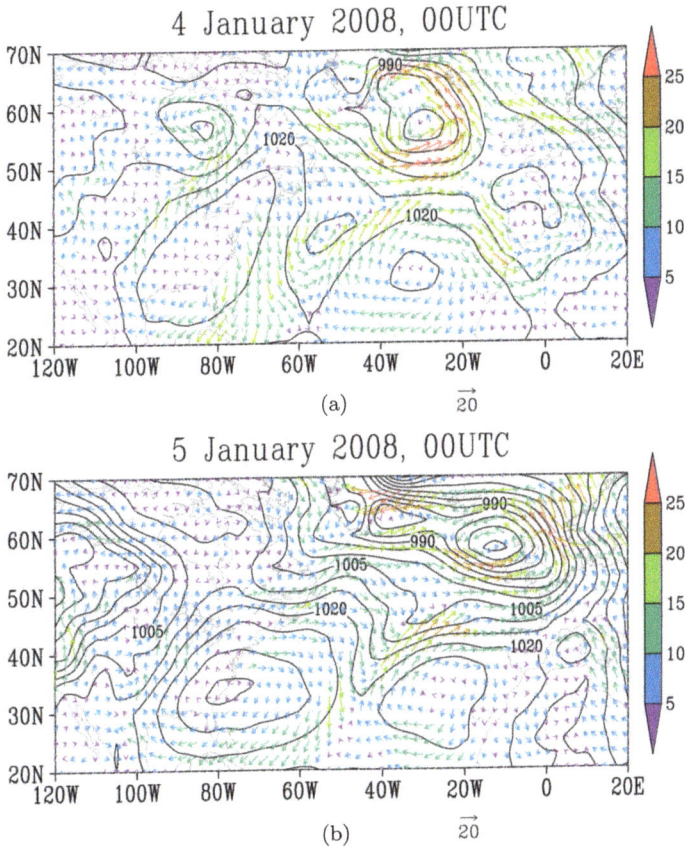

Fig. 3. Sea level pressure (isolines, contour interval 5 hPa) and near-surface wind vector (colored arrows, m/s) on (a) January 4, 2008, 00 UTC and (b) January 5, 2008, 00 UTC.* The displayed fields are based on NCEP Reanalysis.

Note: *UTC is the abbreviation for Coordinated Universal Time and refers to the time at 0° longitude.

highs. For this reason, these pressure anomalies are also known as **cyclones** and **anticyclones**. We can also see in the figure that the wind speed is related to the density of the pressure isolines, i.e. a large difference in pressure between nearby locations yields a strong wind. The wind speed is also influenced by the underlying surface with higher values above the oceans and lower values above land. The pressure anomalies migrate in the course of time and usually with an eastward component. Furthermore, the magnitude and shape of the anomalies change. Therefore, the pressure fields of two consecutive days differ noticeably as seen in the figure. It is not easy to predict the time evolution of these fields but nowadays this is possible for a number of days by using a **numerical weather forecast model**. Such developments describe the weather dynamics in the atmosphere.

Many different weather phenomena appear in the atmosphere. In addition to their individual characteristics, we can distinguish them by their typical length and time scales. Figure 4 shows a **scale**

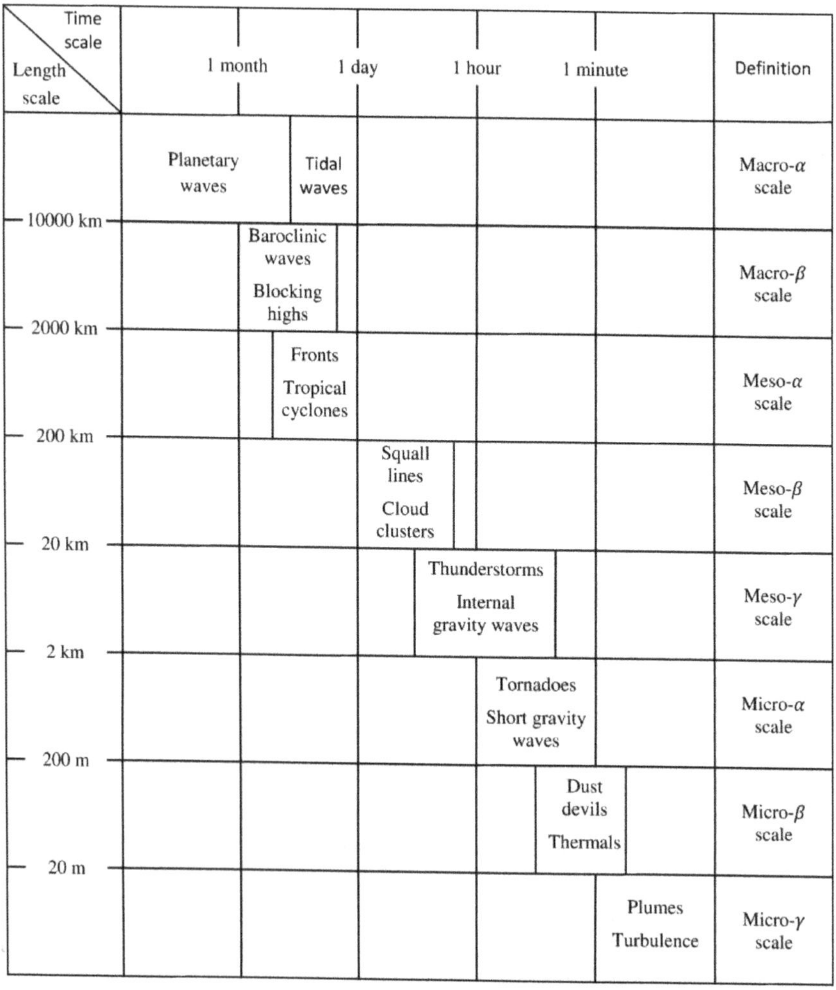

Length scale \ Time scale	1 month	1 day	1 hour	1 minute	Definition
— 10000 km	Planetary waves	Tidal waves			Macro-α scale
— 2000 km		Baroclinic waves / Blocking highs			Macro-β scale
— 200 km		Fronts / Tropical cyclones			Meso-α scale
— 20 km			Squall lines / Cloud clusters		Meso-β scale
— 2 km			Thunderstorms / Internal gravity waves		Meso-γ scale
— 200 m			Tornadoes / Short gravity waves		Micro-α scale
— 20 m				Dust devils / Thermals	Micro-β scale
				Plumes / Turbulence	Micro-γ scale

Fig. 4.   Scale diagram for atmospheric weather phenomena.
Source: Adapted from Orlanski (1973).

diagram in which a selection of weather phenomena is inserted. Length scales range from a few meters to more than 10,000 km while time scales vary from one minute to one month. It is easy to imagine that a simultaneous prediction of all these phenomena is virtually impossible. Therefore, it is useful to group the scales into several categories. The shown classification in **microscale, mesoscale** and **macroscale** was introduced by Orlanski (1975). A further subdivision is defined by the additional Greek letters α, β and γ. Different approximations of the physical laws are used for different scales. For example, for the microscale, the assumption of an incompressible fluid is valid while for the macroscale we can assume a force balance in vertical direction because we have in this case very shallow systems due to the relatively small height of the troposphere. We see in the diagram that the time scales of the phenomena increase with their length scales, i.e. larger weather systems have a larger lifetime. In this book, we will mainly consider the dynamics of macroscale weather systems but the last chapter is exclusively devoted to the smaller scales. Instead of macroscale, we will use the more common term **synoptic scale**.

This book provides a physical explanation for the formation and behavior of weather systems. The first chapter introduces the basis for atmospheric fluid dynamics. In the second chapter, the governing equations for synoptic-scale weather systems are derived by a scale analysis and their energetics are discussed. The third chapter deals with macroscale wave phenomena in the barotropic atmosphere and the quasigeostrophic approximation is derived. **Barotropic** means essentially that horizontal wind does not depend on height and temperature does not change in horizontal direction. This limiting assumption is relaxed in chapter four in which **baroclinic** (non-barotropic) weather dynamics is tackled. This chapter also explains baroclinic instability that is responsible for the formation of many synoptic-scale weather systems. Chapter five addresses nonlinear processes that can lead to chaotic and nonpredictable behavior. This chapter also provides an introduction to dynamical system theory. The last chapter treats a selection of sub-synoptic weather systems, namely convection, fronts and tropical cyclones.

Chapter 1

Foundations of Atmospheric Dynamics

1.1 Introduction

The atmosphere is essentially composed of gas. The fractions of most gas types do not vary in the lowermost 20 km of the atmosphere. Exceptions are water vapor and some trace gases. The latter have little direct impact on atmospheric dynamics and, therefore, we can neglect their variation. However, the variation of water vapor can indeed influence the flow, e.g., by higher buoyancy. In the context of atmospheric dynamics, one can treat all gases as ideal. We will explain later what characterizes an ideal gas. In the atmosphere also cloud particles (**hydrometeors**) and **aerosols** occur beside the various gases. However, in this textbook, we restrict our scope to synoptic and sub-synoptic weather systems. The scales of these phenomena allow the neglect of the impact of cloud particles and aerosols on the dynamics although such atmospheric elements play crucial roles in cloud microphysics which will not be treated here. However, condensation and evaporation of water vapor may have an effect on the flows to be considered here. We will account for these processes and, for consistency of the equations, we also include sedimenting cloud hydrometeors but in most cases, we assume a dry atmosphere in which no water vapor exists and no condensation or evaporation takes place.

1.2   Preliminaries

First of all, we postulate that everything happens in an infinite three-dimensional **Euclidean space** having the coordinates x, y and z. This is a **Cartesian coordinate system** and forms the inertial frame of reference and the position in space is expressed in terms of the position vector

$$\mathbf{r} = x\,\mathbf{e}_x + y\,\mathbf{e}_y + z\,\mathbf{e}_z \qquad (1.1)$$

where $\mathbf{e}_x$, $\mathbf{e}_y$ and $\mathbf{e}_z$ are the unit vectors in the direction of the coordinate axes.

Furthermore, there is a time coordinate t that also extends over an infinite range. The position vector of an individual mass element is uniquely determined by time t. All laws are invariant with respect to a **Galilean transformation** which is given by

$$\mathbf{r}' = \mathbf{r} + \mathbf{r}_0 + \mathbf{u}(t - t_0), \quad t' = t \qquad (1.2)$$

Relativistic effects are ignored throughout the entire book.

1.3   Continuum Hypothesis

The mixture of ideal gases forms a **fluid**. The hydrodynamic equations determine the fluid dynamics and can be justified on the basis of statistical mechanics. The reason is that a volume of air contains a huge number of molecules ($>10^{20}$ per liter). Due to frequent collisions, the molecules describe a very irregular movement in space. In this **microscopic** view (Fig. 1.1(a)) it is impossible to see a **macroscopic** phenomenon like a vortex (Fig. 1.1(b)). For practical purposes, we must, therefore, derive macroscopic equations. The **statistical mechanics** take full account of the quantum nature of matter and make some assumptions to derive the hydrodynamic equations. Here, we treat a simpler path by ignoring the quantum nature and considering the air as a **continuum**, that is, the air forms a mass that fills the atmosphere completely.

Then, we can define a **volumetric mass density**[a] ρ:

$$\rho = \rho(\mathbf{r}, t) \qquad (1.3)$$

---

[a]In the following, we will loosely use the term "density" instead of volumetric mass density.

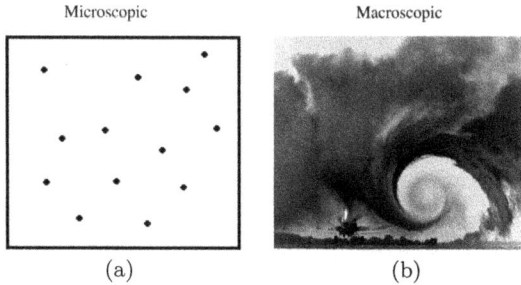

Microscopic   Macroscopic

(a)   (b)

Fig. 1.1. (a) Microscopic and (b) macroscopic picture of a gas.
Source: (b) https://upload.wikimedia.org/wikipedia/commons/f/fe/Airplane_vortex_edit.jpg.

In most cases, it is assumed that the **field function** $\rho(\mathbf{r}, t)$ is continuous and differentiable but sometimes we will introduce **discontinuities** in idealized models to describe meteorological phenomena like fronts. Then, the discontinuity forms the boundary of the considered fluid volumes. However, in the real atmosphere discontinuities in density or any other hydrodynamic field do not exist. Therefore, we can apply the common differential and integral theorems for finding a proper equation set to describe atmospheric dynamics. The **mass** M of any arbitrary and even arbitrarily small **fluid volume** can be deduced from the mass density by the volume integral

$$M = \iiint_V \rho\, dV \tag{1.4}$$

We can also define a fluid velocity vector field $\mathbf{v}(\mathbf{r}, t)$. It yields the velocity by which infinitesimal fluid volumes move at position $\mathbf{r}$ and time t. The vector field $\mathbf{v}(\mathbf{r}, t)$ is like $\rho(\mathbf{r}, t)$ continuous and differentiable. In hydrodynamics, it is called **flow velocity** field while in a meteorological context, one usually uses the term **wind** field. With the wind field, one can calculate **streamlines** $\mathbf{r}_S(l)$ and **trajectories** $\mathbf{r}_T(t)$ by solving the differential equations

$$\frac{d\mathbf{r}_S}{dl} = \mathbf{v}(\mathbf{r}_S, t_0), \quad \frac{d\mathbf{r}_T}{dt} = \mathbf{v}(\mathbf{r}_T, t) \tag{1.5}$$

for initial conditions $\mathbf{r}_S(0) = \mathbf{r}_{S0}$ and $\mathbf{r}_T(t_0) = \mathbf{r}_{T0}$, respectively. A streamline is always parallel to wind field vectors while a trajectory includes the time development of the wind field and yields a line along which the barycenter of an infinitesimal fluid volume moves. Streamlines coincide with trajectories in time-independent wind fields.

1.4 Budget Equation

To derive the hydrodynamic equations, we must introduce the **fluid volume** $V_F(t)$. It is defined by a specified volume of a fluid that changes in the course of time due to the flow, that is, the surface elements of the volume $V_F(t)$ move with the flow velocity field $\mathbf{v}(\mathbf{r}, t)$ which implicates a deformation of the volume in the course of time.

We distinguish between **intensive** and **extensive properties** of a fluid. These are related to each other by

$$B(t) = \iiint_{V_F(t)} \rho(\mathbf{r}, t)b(\mathbf{r}, t)dV \qquad (1.6)$$

where $B(t)$ is the extensive and $b(\mathbf{r}, t)$ the intensive property. The latter can be interpreted as the local amount of the property per unit mass. For finding solutions to hydrodynamic equations, it is necessary to derive equations for the intensive properties since the extensive property $B(t)$ cannot be predicted without knowing the local processes that change $b(\mathbf{r}, t)$. On the other hand, hydrodynamic laws refer to extensive properties like the mass or momentum of the fluid volume. Often, a hydrodynamic law is constituted like the following budget equation:

$$\frac{dB}{dt} = \iiint_{V_F(t)} \rho \dot{b} dV - \oiint_{\partial V_F(t)} \mathbf{j}_b \cdot \mathbf{n} da \qquad (1.7)$$

In this equation, $\dot{b} = \dot{b}(\mathbf{r}, t)$ denotes the local **source** of the property[b], $\mathbf{j}_b = \mathbf{j}_b(\mathbf{r}, t)$ the **non-advective flux** of the property, $\partial V_F(t)$ the surface of the fluid volume V_F and $\mathbf{n}$ the unit vector perpendicular to the differential surface element da. To derive an equation for the intensive property, it is necessary to replace the time derivative in front of the volume integral by an operator that applies to the

---

[b]The dot above the letter indicates a time tendency of the respective quantity.

integrand. This can be done by **Reynolds transport theorem:**

$$\frac{d}{dt}\left(\iiint_{V_F(t)} b_V dV\right) = \iiint_{V_F(t)} \frac{\partial b_V}{\partial t} dV + \oiint_{\partial V_F(t)} b_V \mathbf{v} \cdot \mathbf{n} da$$

$$(1.8)$$

where $b_V = \rho b$ denotes the **volumetric density** of the property.

To proof this theorem, we consider the fluid volume at the two points in time t and $t + \Delta t$. We find that

$$\frac{d}{dt}\left(\iiint_{V_F(t)} b_V dV\right)$$

$$= \lim_{\Delta t \to 0} \frac{1}{\Delta t}\left[\iiint_{V_F(t+\Delta t)} b_V(\mathbf{r}, t+\Delta t) dV - \iiint_{V_F(t)} b_V(\mathbf{r}, t) dV\right]$$

$$= \lim_{\Delta t \to 0} \iiint_{V_F(t+\Delta t)} \frac{b_V(\mathbf{r}, t+\Delta t) - b_V(\mathbf{r}, t)}{\Delta t} dV$$

$$+ \lim_{\Delta t \to 0}\left[\iiint_{V_F(t+\Delta t)} \frac{b_V(\mathbf{r}, t)}{\Delta t} dV - \iiint_{V_F(t)} \frac{b_V(\mathbf{r}, t)}{\Delta t} dV\right]$$

$$= \lim_{\Delta t \to 0} \iiint_{V_F(t+\Delta t)} \frac{b_V(\mathbf{r}, t+\Delta t) - b_V(\mathbf{r}, t)}{\Delta t} dV$$

$$+ \lim_{\Delta t \to 0}\left[\iiint_{\Delta V_F^+} \frac{b_V(\mathbf{r}, t)}{\Delta t} dV - \iiint_{\Delta V_F^-} \frac{b_V(\mathbf{r}, t)}{\Delta t} dV\right] \quad (1.9)$$

Here, ΔV_F^+ denotes the volume at the time $t + \Delta t$ that adds to the original one at time t while ΔV_F^- denotes the volume that vanishes in the time period between t and $t + \Delta t$. The deformation of the volume in this time period is illustrated in Fig. 1.2. The last term on the right-hand side of Eq. (1.9) contains two volume integrals that convert into the following surface integrals in the limit $\Delta t \to 0$:

$$\lim_{\Delta t \to 0}\left[\iiint_{\Delta V_F^+} \frac{b_V(\mathbf{r}, t)}{\Delta t} dV - \iiint_{\Delta V_F^-} \frac{b_V(\mathbf{r}, t)}{\Delta t} dV\right]$$

$$= \iint_{a_F^+} b_V \mathbf{v} \cdot \mathbf{n} da + \iint_{a_F^-} b_V \mathbf{v} \cdot \mathbf{n} da = \oiint_{\partial V_F} b_V \mathbf{v} \cdot \mathbf{n} da \quad (1.10)$$

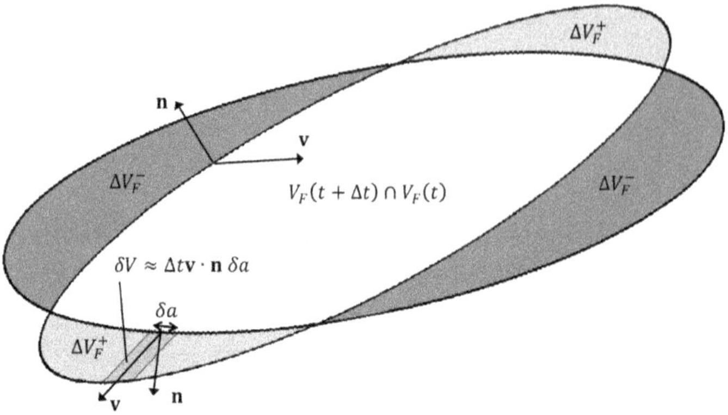

Fig. 1.2.   Sketch of the volume deformation by the flow. For simplicity, the sketch displays the volumes as two-dimensional areas. The area in light gray shows the added volume ΔV_F^+ and the area in dark gray the vanishing volume ΔV_F^-. The volumes ΔV_F^+ and ΔV_F^+ can be assembled by small rhombi.

This calculation becomes evident by considering a small rhombohedron volume element of the different volumes ΔV_F^+ and ΔV_F^- like that sketched in Fig. 1.2. The volume content of this rhombohedron is approximately $\Delta t\, \mathbf{v} \cdot \mathbf{n} \delta a$. Taking the limit $\Delta t \to 0$ and $\delta a \to da$ leads to exact fulfillment of the approximation and, therefore, we obtain $dV/\Delta t = \mathbf{v} \cdot \mathbf{n}\, da$ which proves the identities (1.10) and thereby (1.8).

Reynolds transport theorem (1.8) states that the time tendency of the extensive property splits into the volume integral of the local time tendencies of the volumetric property density and the surface integral of the property flux across the boundaries of the fluid volume. With **Gauss's divergence theorem** (see Appendix A.4.1) we find

$$\frac{dB}{dt} = \iiint_{V_F(t)} \frac{\partial b_V}{\partial t} dV + \oiint_{\partial V_F(t)} b_V \mathbf{v} \cdot \mathbf{n} da$$

$$= \iiint_{V_F(t)} \left[\frac{\partial b_V}{\partial t} + \nabla \cdot (b_V \mathbf{v}) \right] dV \qquad (1.11)$$

A useful relation results by setting $b_V = 1$:

$$\frac{dV_F}{dt} = \frac{d}{dt} \left(\iiint_{V_F(t)} dV \right) = \iiint_{V_F(t)} \nabla \cdot \mathbf{v} dV \qquad (1.12)$$

Therefore, the time tendency of the fluid volume content equals the volume integral of the flow divergence. With Eq. (1.11) we obtain for the budget equation

$$\iiint_{V_F(t)} \frac{\partial}{\partial t}(\rho b) + \nabla \cdot (\rho b \mathbf{v}) dV = \iiint_{V_F(t)} \rho \dot{b} - \nabla \cdot \mathbf{j}_b dV \qquad (1.13)$$

Since the integral theorems are valid for arbitrary volumes they also apply to arbitrarily small volumes. Then, the integrand varies only with a negligible degree and we can move it in front of the integral resulting in the local budget equation

$$\frac{\partial}{\partial t}(\rho b) + \nabla \cdot (\rho b \mathbf{v}) = \rho \dot{b} - \nabla \cdot \mathbf{j}_b \qquad (1.14)$$

This equation constitutes the **local budget equation** in **flux form**. It is a prognostic equation for the volumetric density $b_V = \rho b$. Later we can also derive the budget equation in advection and Lagrangian form but we must first introduce the law of mass conservation.

1.5   Law of Mass Conservation

With the **law of mass conservation**, it is postulated that the mass of a fluid volume is conserved. Therefore, we can write

$$\frac{dM}{dt} = \frac{d}{dt}\left(\iiint_{V_F} \rho dV\right) = 0 \qquad (1.15)$$

With Eq. (1.11) we can derive the local flux form of the mass conservation law

$$\frac{\partial \rho}{\partial t} + \nabla \cdot (\rho \mathbf{v}) = 0 \qquad (1.16)$$

This equation constitutes the **mass continuity equation.**

The mass continuity equation can be employed to write the budget equation in different forms. Applying the product rule of differentiation to the left-hand side of the local budget equation (1.14) leads to

$$b\left[\frac{\partial \rho}{\partial t} + \nabla \cdot (\rho \mathbf{v})\right] + \rho\left[\frac{\partial b}{\partial t} + \mathbf{v} \cdot \nabla b\right] = \rho \dot{b} - \nabla \cdot \mathbf{j}_b \qquad (1.17)$$

By substituting the mass continuity equation (1.16) we obtain the local budget equation in **advection form**:

$$\frac{\partial b}{\partial t} + \mathbf{v} \cdot \nabla b = \dot{b} - \frac{1}{\rho}\nabla \cdot \mathbf{j}_b \qquad (1.18)$$

With the advection form we get a prognostic equation for the intensive property b. Another form of the budget equation results when we introduce the **material derivative**. To explain the material derivative, we assume that the **barycenter** (center of mass) of a **fluid parcel** (differential fluid volume) follows the **trajectory** $\mathbf{r}(t) = \mathbf{r}_T(t)$. Then, the time tendency of b observed within the fluid parcel becomes

$$\frac{d}{dt}b[\mathbf{r}_T(t), t] = \frac{\partial b}{\partial t} + \frac{d\mathbf{r}_T}{dt} \cdot \nabla b = \frac{\partial b}{\partial t} + \mathbf{v} \cdot \nabla b \qquad (1.19)$$

The second identity results because the fluid parcel moves with the flow velocity $\mathbf{v}$. This time derivative is called material derivative since it determines the time tendency of the property of an individual fluid parcel. To distinguish the material derivative from the ordinary time derivative we use the notation

$$\frac{Db}{Dt} \equiv \frac{\partial b}{\partial t} + \mathbf{v} \cdot \nabla b \qquad (1.20)$$

With this notation, we can also write the local budget equation as follows:

$$\frac{Db}{Dt} = \dot{b} - \frac{1}{\rho}\nabla \cdot \mathbf{j}_b \qquad (1.21)$$

The budget equations (1.18) and (1.21) refer to the **Eulerian** and **Lagrangian frames of reference**, respectively. The difference between the local time derivative and the material derivative yields the **advection**

$$\frac{\partial b}{\partial t} - \frac{Db}{Dt} = -\mathbf{v} \cdot \nabla b \qquad (1.22)$$

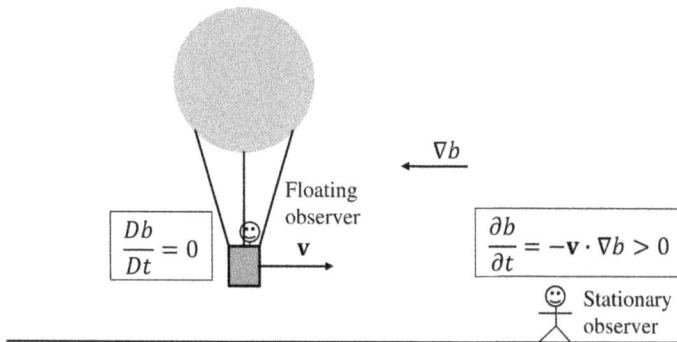

Fig. 1.3. Sketch of the difference between the Eulerian frame of reference (stationary observer) and the Lagrangian frame of reference (floating observer). The floating observer does not experience a change of property b while the stationary observer realizes an increase of b due to advection.

Therefore, an observer at a fixed location detects a time tendency of b that differs in the advection from the time tendency of an observer following the fluid flow. The property b becomes an **individually conserved quantity** if the right-hand side of (1.21) vanishes. Then, the material derivative vanishes and the local time derivative is identical to the advection. This situation is sketched in Fig. 1.3. The observer in the balloon does not experience any change of b. However, the observer at the ground detects an increase of b with time since b decreases to the right.

By revisiting the mass continuity equation, we can now write it in the Lagrangian frame of reference

$$\frac{D\rho}{Dt} = -\rho \nabla \cdot \mathbf{v} \tag{1.23}$$

We see that the material change of mass density occurs when the flow is **divergent** ($\nabla \cdot \mathbf{v} > 0$) or **convergent** ($\nabla \cdot \mathbf{v} < 0$). This result can be understood on the basis of Reynolds transport theorem (cf. Eq. (1.13)).

1.6 Newton's Laws of Motion

Newton's laws of motion form the basis of mechanics and can also be postulated for fluids. Then, we obtain the following three laws:

Newton's first law of motion makes the following statement: *The center of gravity of a fluid volume will move at a constant velocity if no forces are acting on the volume.* The center of gravity $\mathbf{r}_g(t)$ is defined by

$$\mathbf{r}_g(t) = \frac{1}{M} \iiint_{V_F(t)} \rho \mathbf{r} dV \tag{1.24}$$

Therefore, in the case that no forces are applied, we have

$$\mathbf{r}_g(t) = \mathbf{r}_{g0} + \mathbf{v}_0(t - t_0) \tag{1.25}$$

The consequence of this law can be seen by taking the time derivative of Eq. (1.24) and applying the Reynolds transport theorem[c] so that

$$\frac{1}{M} \iiint_{V_F} \frac{\partial}{\partial t}(\rho \mathbf{r}) + \nabla \cdot (\rho \mathbf{v} \mathbf{r}) dV = \mathbf{v}_0 \tag{1.26}$$

With the continuity equation (1.16) we find that

$$\mathbf{P} = \iiint_{V_F} \rho \mathbf{v} dV = M \mathbf{v}_0 = \text{const.} \tag{1.27}$$

where $\mathbf{P}$ is the **momentum** of the fluid volume. Consequently, Newton's first law of motion becomes the law of momentum conservation. However, in the general case momentum is not conserved.

Newton's second law of motion declares the following: *The time tendency of momentum $\mathbf{P}(t)$ is identical to the sum of the applied forces.* Therefore, Newton's second law can be expressed mathematically as

$$\frac{d\mathbf{P}}{dt} = \sum_{j=1}^{J} \mathbf{F}_j \tag{1.28}$$

where the $\mathbf{F}_j$ are the **forces** acting on the fluid volume V_F.

Newton's third law of motion states: *If the fluid volume V_1 applies a force $\mathbf{F}_{12}$ on fluid volume V_2, then fluid volume V_2 exerts a*

---

[c]This can be seen most easily by applying this theorem for each vector component and adding the results afterwards.

counterforce $\mathbf{F}_{21}$ *on fluid volume* V_1 *such that* $\mathbf{F}_{21} = -\mathbf{F}_{12}$ (actio = reactio). Newton's third law of motion is not valid for **Lorentz forces** (see, e.g., McCormmach, 1970). However, these are not of relevance in our context.

The forces split up into **body** and **surface forces** so that

$$\sum_{j=1}^{J} \mathbf{F}_j = \sum_{j=1}^{J_B} \iiint_{V_F} \rho \mathbf{f}_j dV + \oiint_{\partial V_F} \boldsymbol{\tau} da \qquad (1.29)$$

where $\mathbf{f}_j(\mathbf{r}, t)$ is the force density of the jth body force and $\boldsymbol{\tau}$ the **stress** acting on the surface of the volume.

Surface forces result from collisions between molecules. The average distance between two collisions of a specified molecule (**mean free path**) is very small relative to the scales of atmospheric flow phenomena. Therefore, in the interior of the volume V_F, only collisions between molecules that are already part of the volume take place. Due to Newton's third law of motion, a change of momentum $\mathbf{P}$ requires the interaction with matter from outside of the volume. Hence, only collisions very close to the surface of the fluid volume V_F can contribute to a change of momentum $\mathbf{P}$. It is justifiable to assume that the resulting forces only act on the surface of the volume. Body forces stem from other interactions with external matter. The **gravity force** represents the only body force in most applications of atmospheric dynamics. Furthermore, body forces are usually **conservative forces**, that is, the force density results from the gradient of a **potential**. Hence, we have

$$\mathbf{f}_j = -\nabla \phi_j \qquad (1.30)$$

It is convenient to sum up all potentials in one, i.e. $\phi = \sum_{j=1}^{J_B} \phi_j$. Then, Newton's second law of motion becomes

$$\frac{d\mathbf{P}}{dt} = -\iiint_{V_F} \rho \nabla \phi dV + \oiint_{\partial V_F} \boldsymbol{\tau} da \qquad (1.31)$$

It is not possible to derive a local form from this equation by applying the Reynolds transport theorem because the surface integral on the right-hand side of Eq. (1.31) cannot be converted into a volume integral by Gauss's integral theorem. To derive a suitable form of this integral we assert that the stress $\boldsymbol{\tau}$ does not only depend upon the location and time but also on the orientation of the surface element da. Therefore, we have

$$\boldsymbol{\tau} = \boldsymbol{\tau}(\mathbf{r}, \mathbf{n}, t) \qquad (1.32)$$

That $\boldsymbol{\tau}$ depends on $\mathbf{n}$ becomes clear when we apply Newton's third law of motion to two fluid volumes which share a part of their boundary surface (Fig. 1.4). There, the stress yields an interaction force between the two volumes and it is evident that $\boldsymbol{\tau}(\mathbf{r}, \mathbf{n}, t) = -\boldsymbol{\tau}(\mathbf{r}, -\mathbf{n}, t)$ must hold to ensure that the interaction forces offset against each other.

To reveal the dependence on the surface normal $\mathbf{n}$ we consider a little tetrahedron displayed in Fig. 1.5. It has three edges parallel to the three coordinate axes and the vertices of the tetrahedron are located at the locations $(x_T, 0, 0)$, $(0, y_T, 0)$, $(0, 0, z_T)$ and $(0, 0, 0)$ The orientation of the front surface can be varied by the choice of

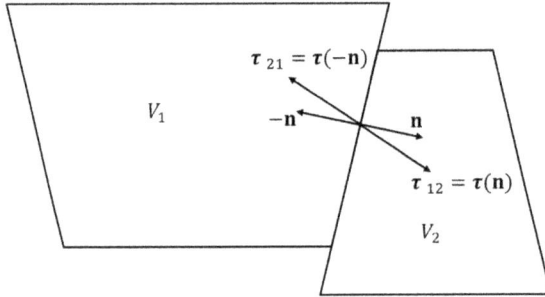

Fig. 1.4.   The stress $\boldsymbol{\tau}$ at the surface between two interacting fluid volumes.

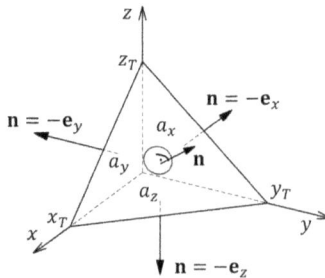

Fig. 1.5.   Sketch of the tetrahedron used to derive the dependency of the stress $\boldsymbol{\tau}$ on the surface normal $\mathbf{n}$.

x_T, y_T and z_T. Obviously, the normal $\mathbf{n}$ of this surface becomes

$$
\begin{aligned}
\mathbf{n} &= \frac{(x_T\mathbf{e}_x - z_T\mathbf{e}_z) \times (y_T\mathbf{e}_y - z_T\mathbf{e}_z)}{|(x_T\mathbf{e}_x - z_T\mathbf{e}_z) \times (y_T\mathbf{e}_y - z_T\mathbf{e}_z)|} \\
&= \frac{y_T z_T\mathbf{e}_x + x_T z_T\mathbf{e}_y + x_T y_T\mathbf{e}_z}{\sqrt{y_T^2 z_T^2 + x_T^2 z_T^2 + x_T^2 y_T^2}}
\end{aligned}
\tag{1.33}
$$

The lateral faces parallel to the coordinate axes have the surface normals $\mathbf{n} = -\mathbf{e}_x$, $\mathbf{n} = -\mathbf{e}_y$ and $\mathbf{n} = -\mathbf{e}_z$ and the associated surface areas are

$$
a_x = \frac{1}{2}y_T z_T, \quad a_y = \frac{1}{2}x_T z_T, \quad a_z = \frac{1}{2}x_T y_T
\tag{1.34}
$$

Therefore, the normal of the front surface can also be written as

$$
\mathbf{n} = \frac{a_x\mathbf{e}_x + a_y\mathbf{e}_y + a_z\mathbf{e}_z}{\sqrt{a_x^2 + a_y^2 + a_z^2}}
\tag{1.35}
$$

while the area of the front surface becomes

$$
\begin{aligned}
a &= \frac{1}{2}|(x_T\mathbf{e}_x - z_T\mathbf{e}_z) \times (y_T\mathbf{e}_y - z_T\mathbf{e}_z)| \\
&= \sqrt{a_x^2 + a_y^2 + a_z^2}
\end{aligned}
\tag{1.36}
$$

From these two equations, we find that

$$
a_x = \mathbf{n} \cdot \mathbf{e}_x a = n_x a, \quad a_y = \mathbf{n} \cdot \mathbf{e}_y a = n_y a, \quad a_z = \mathbf{n} \cdot \mathbf{e}_z a = n_z a
\tag{1.37}
$$

The tetrahedron has such a small size that we can neglect the dependency of stress $\boldsymbol{\tau}$ on the location in space. Therefore, we obtain for the applied surface force on the tetrahedron the following result:

$$
\begin{aligned}
\oiint_{\partial V_F} \boldsymbol{\tau}\,da &= \boldsymbol{\tau}(-\mathbf{e}_x)a_x + \boldsymbol{\tau}(-\mathbf{e}_y)a_y + \boldsymbol{\tau}(-\mathbf{e}_z)a_z + \boldsymbol{\tau}(\mathbf{n})a \\
&= [\boldsymbol{\tau}(-\mathbf{e}_x)n_x + \boldsymbol{\tau}(-\mathbf{e}_y)n_y + \boldsymbol{\tau}(-\mathbf{e}_z)n_z + \boldsymbol{\tau}(\mathbf{n})]a
\end{aligned}
\tag{1.38}
$$

In the case that no other forces act on the volume the time tendency of momentum becomes approximately

$$\frac{d\mathbf{P}}{dt} \approx \rho V_F \frac{D\mathbf{v}}{Dt}$$
$$\approx [\boldsymbol{\tau}(-\mathbf{e}_x)n_x + \boldsymbol{\tau}(-\mathbf{e}_y)n_y + \boldsymbol{\tau}(-\mathbf{e}_z)n_z + \boldsymbol{\tau}(\mathbf{n})]a \quad (1.39)$$

The approximation translates into an identity in the limit of an infinitesimally small volume. To fulfill this equation the terms in brackets on the right-hand side must add to zero for a finite acceleration because the fraction a/V_F approaches infinity in this limit. Therefore, we obtain for the dependence of stress on surface normal orientation:

$$\boldsymbol{\tau}(\mathbf{n}) = -\boldsymbol{\tau}(-\mathbf{e}_x)n_x - \boldsymbol{\tau}(-\mathbf{e}_y)n_y - \boldsymbol{\tau}(-\mathbf{e}_z)n_z \quad (1.40)$$

Due to Newton's third law, we have the relation $\boldsymbol{\tau}(-\mathbf{n}) = \boldsymbol{\tau}(\mathbf{n})$ Therefore, we can write

$$\boldsymbol{\tau}(\mathbf{n}) = \boldsymbol{\tau}(\mathbf{e}_x)n_x + \boldsymbol{\tau}(\mathbf{e}_y)n_y + \boldsymbol{\tau}(\mathbf{e}_z)n_z \quad (1.41)$$

Obviously, the stress depends on nine components, namely the components of the three vectors on the right-hand side of Eq. (1.41) Consequently, it is appropriate to introduce the stress tensor $\mathbf{T}$ so that

$$\boldsymbol{\tau}(\mathbf{n}, \mathbf{r}, t) = \mathbf{n} \cdot \mathbf{T}(\mathbf{r}, t) \quad (1.42)$$

In Cartesian coordinates we obtain

$$\mathbf{T} = \mathbf{e}_x \boldsymbol{\tau}_x + \mathbf{e}_y \boldsymbol{\tau}_y + \mathbf{e}_z \boldsymbol{\tau}_z$$
$$= \mathbf{e}_x \mathbf{e}_x \tau_{xx} + \mathbf{e}_x \mathbf{e}_y \tau_{xy} + \mathbf{e}_x \mathbf{e}_z \tau_{xz} + \mathbf{e}_y \mathbf{e}_x \tau_{yx} + \mathbf{e}_y \mathbf{e}_y \tau_{yy}$$
$$+ \mathbf{e}_y \mathbf{e}_z \tau_{yz} + \mathbf{e}_z \mathbf{e}_x \tau_{zx} + \mathbf{e}_z \mathbf{e}_y \tau_{zy} + \mathbf{e}_z \mathbf{e}_z \tau_{zz} \quad (1.43)$$

where $\boldsymbol{\tau}_x$, $\boldsymbol{\tau}_y$ and $\boldsymbol{\tau}_z$ are the stresses when the surface normal is directed in x, y and z direction, respectively (Appendix A.2 gives a brief introduction to tensors). Now, the correspondence to Eq. (1.41) becomes clear.

With the representation of the stress by a stress tensor we can convert the surface integral in Eq. (1.31) into a volume integral because

$$\oiint_{\partial V_F} \tau da = \oiint_{\partial V_F} \mathbf{n} \cdot \mathbf{T} da = \iiint_{V_F} \nabla \cdot \mathbf{T} dV \qquad (1.44)$$

The application of Gauss's integral theorem to a tensor can be derived by the decomposition (1.43). It allows the representation of the stress tensor in the form

$$\mathbf{T} = \tau_x^r \mathbf{e}_x + \tau_y^r \mathbf{e}_y + \tau_z^r \mathbf{e}_z \qquad (1.45)$$

Then, we can easily see that the surface integral splits up as follows:

$$\oiint_{\partial V_F} \tau da = \oiint_{\partial V_F} \mathbf{n} \cdot \tau_x^r da \mathbf{e}_x + \oiint_{\partial V_F} \mathbf{n} \cdot \tau_y^r da \mathbf{e}_y$$
$$+ \oiint_{\partial V_F} \mathbf{n} \cdot \tau_z^r da \mathbf{e}_z \qquad (1.46)$$

Now, Gauss's integral theorem can be applied to each of these integrals and it becomes evident that

$$\nabla \cdot \mathbf{T} = \nabla \cdot (\tau_x^r) \mathbf{e}_x + \nabla \cdot (\tau_y^r) \mathbf{e}_y + \nabla \cdot (\tau_z^r) \mathbf{e}_z \qquad (1.47)$$

Therefore, the divergence of a tensor is just the left dot product between the Nabla operator and the tensor itself.

Using (1.46) we obtain for Newton's second law of motion

$$\frac{d\mathbf{P}}{dt} = -\iiint_{V_F} \rho \nabla \phi dV + \iiint_{V_F} \nabla \cdot \mathbf{T} dV \qquad (1.48)$$

The right-hand side includes only volume integrals and with Reynolds transport theorem (Eq. (1.11)), we can derive the local budget equation for momentum in flux form:

$$\frac{\partial}{\partial t}(\rho \mathbf{v}) + \nabla \cdot (\rho \mathbf{v v}) = -\rho \nabla \phi + \nabla \cdot \mathbf{T} \qquad (1.49)$$

With the continuity equation (1.16) we can also convert this equation into advection form

$$\frac{\partial \mathbf{v}}{\partial t} + \mathbf{v} \cdot \nabla \mathbf{v} = -\nabla \phi + \frac{1}{\rho} \nabla \cdot \mathbf{T} \qquad (1.50)$$

and into Lagrangian form

$$\frac{D\mathbf{v}}{Dt} = -\nabla\phi + \frac{1}{\rho}\nabla\cdot\mathbf{T} \tag{1.51}$$

These equations are known as the **equations of motion**. In these forms, they are valid for every type of continuous matter including solids. For a fluid, the stress tensor must fulfill an additional condition, namely that in a **hydrostatic** fluid $[\mathbf{v}(\mathbf{r},t) = 0]$ the stresses are always directed perpendicular and toward the surface element of the fluid volume. Therefore,

$$\boldsymbol{\tau}(\mathbf{n},\mathbf{r},t) = -p(\mathbf{r},t)\mathbf{n} \quad \text{if} \quad \mathbf{v}(\mathbf{r},t) = 0 \quad \text{for all } \mathbf{r},t \tag{1.52}$$

where $p(\mathbf{r},t)$ constitutes the **pressure**. With this property, it is not possible to have shear stresses (parallel to the surface) in a hydrostatic fluid and this means that the fluid cannot resist shear stresses. The pressure does not depend on the surface orientation as this would violate the linear relationship given by Eq. (1.42). In the hydrostatic case, the stress tensor becomes

$$\mathbf{T} = -p(\mathbf{e}_x\mathbf{e}_x + \mathbf{e}_y\mathbf{e}_y + \mathbf{e}_z\mathbf{e}_z) = -p\mathbf{I} \tag{1.53}$$

where $\mathbf{I}$ denotes the unit tensor. For the divergence of $\mathbf{T}$ we obtain

$$\nabla\cdot\mathbf{T} = -\frac{\partial p}{\partial x}\mathbf{e}_x - \frac{\partial p}{\partial y}\mathbf{e}_y - \frac{\partial p}{\partial z}\mathbf{e}_z = -\nabla p \tag{1.54}$$

Obviously, the divergence of the stress tensor translates into the pressure gradient.

In a hydrostatic fluid, we have no accelerations and therefore, we get

$$\frac{1}{\rho}\nabla p = -\nabla\phi \tag{1.55}$$

This is the **hydrostatic balance equation**. It replaces the equation of motion in the static case and it describes the balance between the applied body force and the pressure gradient force. In a meteorological context, the gravity of Earth forms the only relevant body force. Therefore, the pressure gradient is almost exactly directed towards the Earth's center in a hydrostatic atmosphere.

In the general case, the stress tensor is separated as follows:

$$\mathbf{T} = -p\mathbf{I} + \mathbf{J} \tag{1.56}$$

where $\mathbf{J}$ is the **viscous stress tensor**. $\mathbf{J}$ depends on the flow field $\mathbf{v}(\mathbf{r}, t)$ and describes the momentum exchange by **molecular diffusion** which acts to reduce velocity gradients. The molecular diffusion does not have any direct relevance for meso- and synoptic-scale weather systems. Therefore, we can neglect $\mathbf{J}$ in the equation of motion. However, $\mathbf{J}$ becomes important in microscale turbulence which in turn impacts on larger-scale flows. How to incorporate these impacts will be explained later. With Eq. (1.56) the equation of motion takes the form

$$\frac{\partial \mathbf{v}}{\partial t} + \mathbf{v} \cdot \nabla \mathbf{v} = -\frac{1}{\rho}\nabla p - \nabla \phi + \frac{1}{\rho}\nabla \cdot \mathbf{J} \tag{1.57}$$

which yields together with the continuity equation a sufficient basis to predict flows with constant density ρ and known potential ϕ. While ϕ can indeed be taken as a time-independent known field in meteorology it is not appropriate to assume constant density. Therefore, we must consider more laws which enable us to determine the density.

1.7 First Law of Thermodynamics

The **first law of thermodynamics** is essentially an energy budget equation. It includes in addition to the mechanical energies the so-called **inner energy**:

$$E_I(t) = \iiint_{V_F(t)} \rho e_I(\mathbf{r}, t)\, dV \tag{1.58}$$

where $e_I(\mathbf{r}, t)$ denotes the **specific inner energy** which is a thermodynamic state variable. Inner energy comprises the kinetic energy of the random molecule motion and potential energy due to intermolecular forces. On the other hand, the **kinetic energy** of the fluid volume is given by

$$E_K(t) = \iiint_{V_F(t)} \frac{\rho}{2}\mathbf{v}^2(\mathbf{r}, t)\, dV = \iiint_{V_F(t)} \rho\, e_K(\mathbf{r}, t)\, dV \tag{1.59}$$

where $e_K(\mathbf{r}, t)$ denotes the **specific kinetic energy**. The molecular motion is subdivided into a random and uniform part by the distinction of these two energies. The energies of these motions are assigned to **inner kinetic energy** and (conventional) kinetic energy, respectively. In general, **inner potential energy** also contributes to inner energy.

We postulate the following energy equation:

$$\frac{d}{dt} \left(\iiint_{V_F} \rho e_K dV + \iiint_{V_F} \rho e_I dV \right) = - \iiint_{V_F} \rho \mathbf{v} \cdot \nabla \phi dV$$

$$+ \oiint_{\partial V_F} \mathbf{v} \cdot \boldsymbol{\tau} da - \oiint_{\partial V_F} \mathbf{n} \cdot \mathbf{j}_R da - \oiint_{\partial V_F} \mathbf{n} \cdot \mathbf{j}_{e_I} da \qquad (1.60)$$

where $\mathbf{j}_R(\mathbf{r}, t)$ denotes the energy flux by **electromagnetic radiation**, and $\mathbf{j}_{e_I}(\mathbf{r}, t)$ the **diffusive flux** of inner energy. On the left-hand side, we find the tendencies of kinetic energy and inner energy. The first and second terms on the right-hand side yield the **work** done on the fluid volume by body and surface forces, respectively, while the third term results from the energy conversion due to electromagnetic radiation.[d] The last term on the right-hand side describes the diffusive flux of inner energy across the boundary of the fluid volume. The effect of diffusive fluxes of kinetic energy is already part of the work done by stresses.

We must convert the surface integrals to volume integrals to derive a local budget equation. For the work done by stresses we obtain by using Gauss's integral theorem

$$\oiint_{\partial V_F} \mathbf{v} \cdot \boldsymbol{\tau} da = \oiint_{\partial V_F} (\mathbf{n} \cdot \mathbf{T}) \cdot \mathbf{v} da = \oiint_{\partial V_F} \mathbf{n} \cdot (\mathbf{T} \cdot \mathbf{v}) da$$

$$= \iiint_{V_F(t)} \nabla \cdot (\mathbf{T} \cdot \mathbf{v}) dV \qquad (1.61)$$

Therefore, the local budget equation for specific energy $e_K + e_I$ becomes

$$\frac{\partial}{\partial t}[\rho(e_K + e_I)] + \nabla \cdot [\rho \mathbf{v}(e_K + e_I)]$$

$$= -\rho \mathbf{v} \cdot \nabla \phi + \nabla \cdot (\mathbf{T} \cdot \mathbf{v}) - \nabla \cdot \mathbf{j}_R - \nabla \cdot \mathbf{j}_{e_I} \qquad (1.62)$$

---

[d]The radiation pressure can be neglected in the meteorological context.

The first law of thermodynamics does not include kinetic energy. Therefore, we must subtract a prognostic equation for specific kinetic energy to remove its tendency on the left-hand side of Eq. (1.62). It is possible to derive such an equation by taking the dot product of velocity $\mathbf{v}$ with the equation of motion (1.50). We get

$$\frac{\partial e_K}{\partial t} + \mathbf{v} \cdot \nabla e_K = -\mathbf{v} \cdot \nabla \phi + \frac{1}{\rho} \mathbf{v} \cdot \nabla \cdot \mathbf{T} \qquad (1.63)$$

With the continuity equation (1.16) we can convert this equation into flux form:

$$\frac{\partial}{\partial t}(\rho e_K) + \nabla \cdot (\mathbf{v} \rho e_K) = -\rho \mathbf{v} \cdot \nabla \phi + \mathbf{v} \cdot \nabla \cdot \mathbf{T} \qquad (1.64)$$

Subtracting this equation from Eq. (1.62) leads to a budget equation for inner energy in flux form

$$\frac{\partial}{\partial t}(\rho e_I) + \nabla \cdot (\rho \mathbf{v} e_I) = \nabla \cdot (\mathbf{T} \cdot \mathbf{v}) - \mathbf{v} \cdot \nabla \cdot \mathbf{T} - \nabla \cdot \mathbf{j}_R - \nabla \cdot \mathbf{j}_{e_I} \quad (1.65)$$

With this equation, we see an interesting result, namely that the work by surface forces contributes both to kinetic energy and inner energy while the work by body forces only leads to a change of kinetic energy. This becomes understandable by considering the fact that the origin of the surface forces is rooted in molecular collisions. Then, only a part of the triggered molecular motion can be assigned to the flow velocity. The body force on the other hand appears uniform at the molecular scale. Hence such a force only triggers uniform acceleration that will be entirely assigned to the increase of flow $\mathbf{v}$ in time.

It is worthwhile to split the work assigned to inner energy into the parts stemming from pressure and viscosity, respectively. We obtain by inserting (1.56)

$$\nabla \cdot (\mathbf{T} \cdot \mathbf{v}) - \mathbf{v} \cdot \nabla \cdot \mathbf{T} = -\nabla \cdot (p\mathbf{v}) + \nabla \cdot (\mathbf{J} \cdot \mathbf{v}) + \mathbf{v} \cdot \nabla p - \mathbf{v} \cdot \nabla \cdot \mathbf{J}$$

$$= -p\nabla \cdot \mathbf{v} + \mathbf{J} : \nabla \mathbf{v} \qquad (1.66)$$

where we have used Eq. (A.60)[e] of the appendix and $\mathbf{J} : \nabla \mathbf{v}$ constitutes the **dissipative heating**. With this subdivision the budget

---

[e]All equations beginning with A refer to the appendix that introduces some mathematical foundation.

equation for inner energy becomes

$$\frac{\partial}{\partial t}(\rho e_I) + \nabla \cdot (\rho \mathbf{v} e_I) = -p\nabla \cdot \mathbf{v} + \mathbf{J} : \nabla \mathbf{v} - \nabla \cdot \mathbf{j}_R - \nabla \cdot \mathbf{j}_{e_I} \quad (1.67)$$

The first term on the right-hand side forms the **pressure-volume work**. Using the continuity equation (1.23) we can write this term in the form

$$-p\nabla \cdot \mathbf{v} = \frac{p}{\rho}\frac{D\rho}{Dt} = \frac{p}{\rho}\frac{D}{Dt}\left(\frac{1}{v}\right) = -\frac{p}{\rho v^2}\frac{Dv}{Dt} = -p\rho\frac{Dv}{Dt} \quad (1.68)$$

where $v = 1/\rho$ denotes the **specific volume**. Therefore, we get for the Lagrangian form of the inner energy equation

$$\frac{De_I}{Dt} = -p\frac{Dv}{Dt} + \frac{1}{\rho}\mathbf{J} : \nabla \mathbf{v} - \frac{1}{\rho}\nabla \cdot \mathbf{j}_R - \frac{1}{\rho}\nabla \cdot \mathbf{j}_{e_I} \quad (1.69)$$

This equation represents the first law of thermodynamics formulated in local form for a fluid parcel.

An alternative form of the first law of thermodynamics results by substitution of **enthalpy**

$$h = e_I + pv \quad (1.70)$$

Then, we obtain

$$\frac{Dh}{Dt} = \frac{1}{\rho}\frac{Dp}{Dt} + \frac{1}{\rho}\mathbf{J} : \nabla \mathbf{v} - \frac{1}{\rho}\nabla \cdot \mathbf{j}_R - \frac{1}{\rho}\nabla \cdot \mathbf{j}_{e_I} \quad (1.71)$$

This form is more appropriate in the meteorological context since it is easier to measure changes in pressure than in specific volume.

1.8   Equations of State and Moisture Variables

A closed set of equations needs one equation that determines the density as a function of other fluid properties. This is the **thermal equation of state**. We assume that the air in the atmosphere without condensate constitutes an **ideal gas**. Then, the thermal equation of state becomes

$$\rho = \frac{p}{R_a T} \quad (1.72)$$

where R_a is the specific gas constant of the air parcel and T its temperature. The temperature T is a measure of the inner kinetic energy of the air. We see that the density of the air rises with increasing pressure when temperature is held constant. This is **Boyle's law**. On the other hand, pressure increases with increasing temperature for a fixed density which is known as **Gay-Lussac's law**. The combination of both laws yields the thermal equation of state (1.72).

We already know that air comprises several gases. The gases which make up 99% of the volume do not vary in its concentration in the lower 100 km of the atmosphere (**homosphere**) except for water vapor. It is known that the mixture of ideal gases also forms an ideal gas but the specific gas constants of the various gases vary. Due to **Dalton's law**, we can add the partial pressures of the various gases to obtain the pressure of the mixture, that is

$$p = \sum_{j=1}^{J_G} p_j = \sum_{j=1}^{J_G} R_j \rho_j T = \sum_{j=1}^{J_G} R_j m_j \rho T \qquad (1.73)$$

where the index j denotes the particular gas, $m_j = M_j/M$ the associated mass fraction and J_G the number of the various gases. Since only water vapor variations matter in atmospheric dynamics, we can write this equation as

$$p = (R_d m_d + R_v m_v)\rho T \qquad (1.74)$$

where the indices d and v refer to dry air (all gases except for water vapor) and water vapor, respectively. Comparison with Eq. (1.73) yields

$$R_d = \frac{\sum_{j=1}^{J_d} R_j m_j}{m_d}, \quad m_d = \sum_{j=1}^{J_d} m_j = 1 - m_v \qquad (1.75)$$

where the summation in this equation includes all gases except for water vapor. The specific gas constant for dry air and water vapor amounts to $R_d = 278.1$ J/kg/K and $R_v = 461.4$ J/kg/K, respectively. Therefore, from the equation of state (1.72) we see that an air parcel with **moist air** (dry air plus water vapor) has a lower density than a dry air parcel having the same pressure and temperature. The mass fraction for water vapor m_v is called **specific humidity** and represents besides vapor pressure p_v an important **moisture variable**.

In many applications the **virtual temperature** T_v is used to consider the impact of specific humidity on density. It is given by

$$T_v = [1 + (R_v/R_d - 1)m_v]T \qquad (1.76)$$

With virtual temperature T_v we can write the thermal equation of state in the form

$$\rho = \frac{p}{R_d T_v} \qquad (1.77)$$

A nonzero amount of condensate characterizes **cloud air**. Since the condensate has also a mass, we must consider it in the equation of state. We restrict condensate to cloud droplets and denote its mass fraction by m_l. Although an ensemble of cloud droplets can hardly be viewed as a continuous field, it is a widely used approach to consider m_l as continuous and we do it likewise. Then, the specific volume of cloud air becomes

$$v = m_d \frac{R_d T}{p} + m_v \frac{R_v T}{p} + m_l v_l \qquad (1.78)$$

where v_l denotes the specific volume of liquid water. Therefore, the equation of state takes the form

$$\rho = \frac{p}{m_d R_d T + m_v R_v T + m_l v_l} \qquad (1.79)$$

The volume occupied by liquid water is usually very small compared to that of gas. Therefore, we can neglect its contribution and with the relation $m_d = 1 - m_v - m_l$ we obtain

$$\rho = \frac{p}{R_d T \left[1 + \left(\frac{R_v}{R_d} - 1\right) m_v - m_l\right]} \qquad (1.80)$$

Consequently, the presence of condensate increases the density of air due to its additional mass and very small specific volume. To take this effect into account in the virtual temperature it is useful to introduce the generalized definition

$$T_v = [1 + (R_v/R_d - 1)m_v - m_l]T \qquad (1.81)$$

which includes the effect of condensate loading.

We need another equation for the determination of inner energy e_I as a function of other thermodynamic state variables. The **caloric equation of state** provides this function. The inner kinetic energy is proportional to the temperature and in a mixture, the inner kinetic energies of the various components can be added to obtain the total inner kinetic energy. Therefore, we can write for the inner kinetic energy of air

$$e_{IK} = [c_{vd}(1 - m_v - m_l) + c_{vv}m_v + c_{vl}m_l]T \qquad (1.82)$$

where $c_{vd} = 717.9\,\text{J/kg}$, $c_{vv} = 1408.6\,\text{J/kg}$ and $c_l = 4185.4\,\text{J/kg}$ denote the **specific heat capacities at constant volume** for dry air, water vapor and liquid water, respectively.

Besides inner kinetic energy, we must consider inner potential energy due to the intermolecular forces. These forces can be neglected in dry air. In water vapor, **van der Waals forces** exist which may initiate phase transitions. A proper consideration of these forces would make it necessary to replace the equation of state for an ideal gas by one for a real gas. In a meteorological context, it is, however, acceptable to treat water vapor as an ideal gas. On the other hand, a non-negligible amount of inner potential energy will be released when a **phase transition** takes place. This is indeed the most important impact of water vapor on atmospheric dynamics. To take this into account we must introduce the energy of condensation

$$\Delta E_c = L_{v0}\Delta M_v = \iiint_{V_F} L_{v0}\Delta\rho_v dV \qquad (1.83)$$

where L_{v0} is a constant and ΔM_v the condensed mass of water vapor and $\Delta\rho_v$ the associated change of water vapor density during the condensation process. Therefore, the specific inner potential energy difference between the vapor and liquid phases of water becomes

$$e_{IP} = L_{v0}m_v \qquad (1.84)$$

Addition of inner kinetic energy (1.82) and inner potential energy (1.84) yields the total inner energy of moist air

$$e_I = [c_{vd}(1 - m_v - m_l) + c_{vv}m_v + c_l m_l]T + L_{v0}m_v \qquad (1.85)$$

We see that the inner kinetic energy also changes when a phase transition takes place. This is because dry air, water vapor and liquid water have different heat capacities. This effect contributes to the change of temperature during a phase transition. With Eq. (1.78), (1.85) and the assumption $v_l \approx 0$ we can deduce the specific enthalpy

$$h = e_I + pv = [c_{pd}(1 - m_v - m_l) + c_{pv}m_v + c_l m_l]T + L_{v0}m_v \quad (1.86)$$

where $c_{pd} = c_{vd} + R_d = 1005\,\text{J/kg}$ and $c_{pv} = c_{vv} + R_v = 1870\,\text{J/kg}$ denote the specific heat capacities for dry air and water vapor at constant pressure, respectively. Note that the heat capacities for liquid water at constant volume and constant pressure are identical since we have neglected the volume. Anyhow, the actual difference is very small since liquid water is nearly incompressible. Furthermore, we see that the inner potential energy yields another contribution to enthalpy. **Sublimation** is a phase transition of water vapor that leads to the formation of ice crystals and it also can take place in clouds. The potential energy release due to sublimation is larger than that of condensation. Sublimation can be an important process in rain formation. However, we ignore sublimation throughout this book since we do not treat cloud microphysical processes in detail. Specific enthalpy can also be written as

$$h = h_d m_d + h_v m_v + h_l m_l \quad (1.87)$$

where $h_d = c_{pd}T$, $h_v = c_{pv}T + L_{v0}$ and $h_l = c_l T$ are the **partial specific enthalpies** of the respective constituent.

Specific humidity represents a mass fraction and it would be a conserved quantity if phase transitions and diffusion were absent. In general, the budget equation for specific humidity becomes

$$\frac{Dm_v}{Dt} = S_{l,v} - \frac{1}{\rho}\nabla \cdot \mathbf{j}_{m_v} \quad (1.88)$$

where $S_{l,v}$ denotes the evaporation rate of cloud droplets and $\mathbf{j}_{m_v}$ the diffusive flux of water vapor. Note that a negative evaporation rate describes the condensation of cloud droplets. In cloud air, we must consider an additional equation for the mass fraction of liquid water

$$\frac{Dm_l}{Dt} = -S_{l,v} - \frac{1}{\rho}\nabla \cdot \mathbf{j}_{m_l} \quad (1.89)$$

where $\mathbf{j}_{m_l}$ is the diffusive flux of liquid water which is of a different nature. The liquid water flux $\mathbf{j}_{m_l}$ mainly describes the sedimentation of cloud hydrometeors due to gravity. There is no molecular diffusion since the cloud droplet bonds liquid water. With Eqs. (1.88) and (1.89) it is taken for granted that $S_{l,v}$ yields a phase transition from liquid water to water vapor. We can also formulate a budget equation for dry air:

$$\frac{Dm_d}{Dt} = -\frac{1}{\rho}\nabla \cdot \mathbf{j}_{m_d} \tag{1.90}$$

Since we have the constraint $m_d + m_v + m_l = 1$ we do not need to solve this equation. However, we see that due to this constraint, the diffusive flux for dry air is $\mathbf{j}_{m_d} = -\mathbf{j}_{m_l} - \mathbf{j}_{m_v}$. Therefore, the diffusive flux of dry air compensates for the fluxes of water vapor and liquid water.

Now, we can derive a **temperature tendency equation**. Inserting Eq. (1.86) into the first law of thermodynamics in enthalpy form and using Eqs. (1.88) and (1.89) yield:

$$c_p\frac{DT}{Dt} = -L_vS_{l,v} + \frac{1}{\rho}\frac{Dp}{Dt} + \frac{1}{\rho}\mathbf{J}:\nabla\mathbf{v} - \frac{1}{\rho}\nabla\cdot\mathbf{j}_R - \frac{1}{\rho}\nabla\cdot\mathbf{j}_r$$
$$- \frac{1}{\rho}(c_{pv} - c_{pd})\mathbf{j}_{m_v}\cdot\nabla T - \frac{1}{\rho}(c_l - c_{pd})\mathbf{j}_{m_l}\cdot\nabla T \tag{1.91}$$

where c_p denote the **specific heat capacity at constant pressure** of the cloud air parcel, $\mathbf{j}_r$ the **reduced heat flux** and L_v the **latent heat of vaporization**. The heat capacity of cloud air at constant pressure, c_p, depends on the mass fractions and is given by

$$c_p = c_{pd}(1 - m_v - m_l) + c_{pv}m_v + c_{vl}m_l \tag{1.92}$$

The latent heat of vaporization, L_v, depends on temperature and becomes

$$L_v = h_v - h_l = L_{v0} - (c_l - c_{pv})T \tag{1.93}$$

At the freezing point ($0°C \equiv 273.15\,\text{K}$) L_v takes the value $L = 2.501 \times 10^6\,\text{J/kg}$. Therefore, $L_{v0} = 3.133 \times 10^6\,\text{J/kg}$. The latent heat of vaporization also takes the higher heat capacity of water in the liquid phase into account which must be considered in the energy budget of a phase transition. The reduced heat flux $\mathbf{j}_r$ is given by

$$\mathbf{j}_r = \mathbf{j}_{e_I} - (h_v - h_d)\mathbf{j}_{mv} - (h_l - h_d)\mathbf{j}_{m_l} \qquad (1.94)$$

This flux is reduced by the water vapor and liquid water fluxes which change enthalpy but not the temperature. The last two terms on the right-hand side of Eq. (1.91) correct the temperature advection by the contributions due to the diffusion fluxes. With these terms, it is taken into account that the barycenters of water vapor and liquid water move with a different velocity than that of dry air. A striking example of this effect is the sedimentation of raindrops. Raindrops usually fall downward from colder levels and cool the air since the heat capacity of liquid water is higher than that of dry air.[f]

In cloud air one needs a law which predicts the condensation or evaporation. The second law of thermodynamics to be presented in the next sections states that liquid water and water vapor tend toward an equilibrium state where both phases can coexist. Cloud air is usually very close to this phase equilibrium state in which the water vapor pressure results from the **Clausius–Clapeyron equation**

$$\frac{dp_{vs}}{dT} = \frac{L_v p_{vs}}{R_v T^2} \qquad (1.95)$$

where p_{vs} is referred to as the **saturation vapor pressure**. It is only a function of temperature. The Clausius–Clapeyron equation will be derived in the next section after introducing the second law of thermodynamics. Integration of Eq. (1.95) gives the saturation vapor pressure as a function of temperature

$$
\begin{aligned}
p_{vs}(T) &= p_{vs}(T_0) \left(\frac{T}{T_0}\right)^{\frac{c_{pv}-c_l}{R_v}} \exp\left[\frac{L_{v0}}{R_v}\left(\frac{1}{T_0} - \frac{1}{T}\right)\right] \\
&= p_{vs}(T_0) \left(\frac{T}{T_0}\right)^{\frac{c_{pv}-c_l}{R_v}} \exp\left(\frac{L_{v0}}{R_v T_0}\frac{T - T_0}{T}\right) \qquad (1.96)
\end{aligned}
$$

---

[f]Here, we made the assumption that the temperatures of moist air and liquid water in the cloud air parcel are identical. This assumption could be challenged but is a consequence of the local thermodynamic equilibrium which will be assumed in the next section.

Using the freezing point as the reference temperature T_0 and $p_{vs}(T_0) = 6.1094\,\text{hPa}$ we obtain the following formula:

$$p_{vs}(T) = 6.1094\,\text{hPa} \left(\frac{T}{T_0}\right)^{-5.018} \exp\left(24.86\frac{T - 273.15\,\text{K}}{T}\right) \quad (1.97)$$

Many studies in meteorology make use of the **Magnus Formula**:

$$p_{vs}(T) = 6.1094\,\text{hPa} \exp\left(17.625\frac{T - 273.15\,\text{K}}{T - 30\,\text{K}}\right) \quad (1.98)$$

which matches better the observations. Figure 1.6 compares both formulas and we see that the curves are nearly identical in the range of observed atmospheric temperatures ($<50°$C). Therefore, we can also apply the first formula resulting directly from the Clausius–Clapeyron equation. Note that at the **boiling point** (100°C) the saturation vapor pressure takes the value of the mean surface pressure. Above this point, the water vapor cannot be in equilibrium with liquid water with the consequence that all liquid water evaporates.

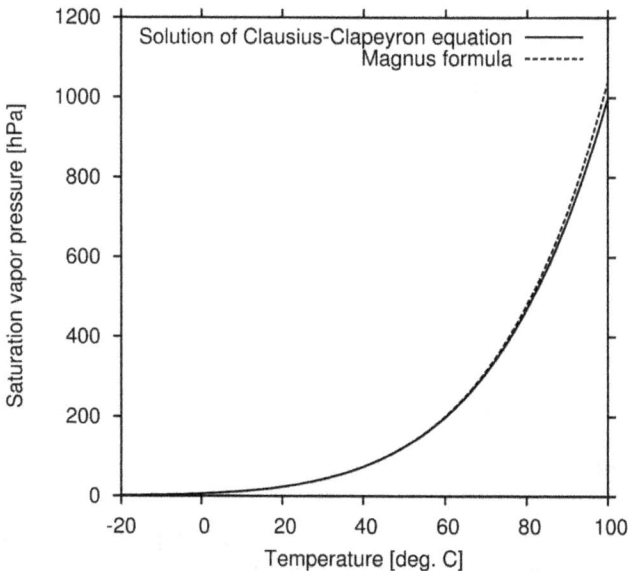

Fig. 1.6.   Saturation vapor pressure (hPa) as a function of temperature in degree Celsius. The solid curve displays the solution of the Clausius–Clapeyron equation while the dashed curve results from the Magnus Formula (1.98).

Condensation appears when the vapor pressure p_v exceeds the saturation vapor pressure p_{vs}. The usual cloud formation takes place when the temperature of a rising air parcel decreases until p_{vs} becomes small enough that condensation happens. Besides saturation vapor pressure p_{vs} one can also derive the **saturation-specific humidity**. It results from the ratio of vapor density to total density:

$$m_{vs} = \frac{\rho_{vs}}{\rho} = \frac{p_{vs}}{p}\frac{R_d T_v}{R_v T} = \frac{p_{vs}}{p}\frac{R_d}{R_v}\left[1 + \left(\frac{R_v}{R_d} - 1\right)m_{vs} - m_l\right] \quad (1.99)$$

Solving this equation for m_{vs} yields

$$m_{vs} = \frac{R_d}{R_v}\frac{p_{vs}(1 - m_l)}{p - (1 - R_d/R_v)p_{vs}} \quad (1.100)$$

Another important moisture variable is the **relative humidity** $\mathcal{H}$ which results from the fraction of vapor pressure and saturation vapor pressure. It is related to specific humidity in the following way:

$$\mathcal{H} = \frac{p_v}{p_{vs}} = \frac{m_v}{m_{vs}}\frac{1 + (R_v/R_d - 1)m_{vs} - m_l}{1 + (R_v/R_d - 1)m_v - m_l} \approx \frac{m_v}{m_{vs}} \quad (1.101)$$

Therefore, relative humidity is also approximately the ratio of specific humidity and saturation specific humidity. Other moisture variables are the **mixing ratio**

$$r_v = \frac{m_v}{m_d} \quad (1.102)$$

and the **dewpoint temperature** T_d which is a solution of the equation

$$p_{vs}(T_d) = p_v \quad (1.103)$$

That is, the dewpoint temperature arises when the saturation vapor pressure becomes identical to the vapor pressure, which happens when the air is cooled until it reaches the dewpoint temperature.

We still have to answer the question of how to determine the phase transition from liquid to vapor, $S_{l,v}$. Since the time scale of condensation is small compared to that of weather systems it is often assumed in atmospheric models that the adaptation to m_{vs} takes place immediately. In this case the condensation rate in saturated air $(m_{vs} = m_v)$ becomes

$$
S_{l,v} = \begin{cases} \dfrac{Dm_{vs}}{Dt} & \text{for } \dfrac{Dm_{vs}}{Dt} \leq 0 \\[2ex] 0 & \text{for } \dfrac{Dm_{vs}}{Dt} > 0 \quad \text{and} \quad m_l = 0 \\[2ex] \dfrac{Dm_{vs}}{Dt} & \text{for } \dfrac{Dm_{vs}}{Dt} > 0 \quad \text{and} \quad m_l > 0 \end{cases} \tag{1.104}
$$

Since m_{vs} is a function of temperature and pressure, condensation occurs only if at least one of these two quantities changes.

1.9   Second Law of Thermodynamics

The second law of thermodynamics makes the statement that macroscopic physical systems tend to a thermodynamic equilibrium state. To explain this, we must first introduce the term **exact differential**. Suppose that a fluid property b can be expressed as a function of two other fluid properties b_1 and b_2. Then, we obtain for the material derivative:

$$
\frac{Db}{Dt} = \frac{\partial b}{\partial b_1}\frac{Db_1}{Dt} + \frac{\partial b}{\partial b_2}\frac{Db_2}{Dt} \tag{1.105}
$$

We can also represent this relation in **differential form** as

$$
db = \frac{\partial b}{\partial b_1}db_1 + \frac{\partial b}{\partial b_2}db_2 \tag{1.106}
$$

The differential db means an infinitesimal change of b by any process affecting the fluid parcel. The differential (1.106) forms an exact differential since $b = b(b_1, b_2)$. However, differentials must not be exact. The first law of thermodynamics represents an example that can be written in the form of an **inexact differential**

$$
\delta Q = de_I + p\,dv \tag{1.107}
$$

Here, the heat differential δQ includes the processes of diffusion, dissipative heating and heating by radiation. While e_I and v represent fluid properties the heating does not. It depends on the properties of neighboring fluid parcels and even on processes in remote locations, as it would be the case when radiative absorption takes place. Therefore, δQ represents an inexact differential and Q cannot be written as a function of fluid properties. We define in our meteorological context a **thermodynamic state function** by a fluid property that can be written as a function of the thermodynamic state variables temperature, pressure and mass fractions of water vapor and liquid water. Therefore, specific inner energy e_I is obviously a thermodynamic state function since it results from the caloric equation of state (1.85). It can also be expanded in the form of an exact differential

$$de_I = \frac{\partial e_I}{\partial T} dT + \frac{\partial e_I}{\partial m_v} dm_v + \frac{\partial e_I}{\partial m_l} dm_l \qquad (1.108)$$

On the other hand, specific kinetic energy e_K is a fluid property but not a thermodynamic state function since it is a function of the velocity components and, therefore, not a function of T, p, m_v and m_l.

Here, we introduce the second law of thermodynamics in the framework of **classical irreversible thermodynamics**. This means that the **local thermodynamic equilibrium** holds. This has the consequence that the thermodynamic state of a fluid parcel can be uniquely described by the thermodynamic state variables. Then, the second law of thermodynamics is based on the following three postulates:

(1) The **specific entropy** s is a thermodynamic state function and it results from the exact differential

$$ds = \frac{de_I}{T} + \frac{p}{T} dv - \frac{\mu_d}{T} dm_d - \frac{\mu_v}{T} dm_v - \frac{\mu_l}{T} dm_l \qquad (1.109)$$

where μ_d, μ_v and μ_l constitute the **chemical potentials** for dry air, water vapor and liquid water, respectively. They are defined by

$$\mu_d = h_d - T s_d, \quad \mu_v = h_v - T s_v \quad \text{and} \quad \mu_l = h_l - T s_l \quad (1.110)$$

where s_d, s_v and s_l are the **partial specific entropies** such that

$$s = s_d m_d + s_v m_v + s_l m_l \qquad (1.111)$$

Eq. (1.109) forms the **Gibbs' fundamental equation**.

(2) The **entropy** S of a materially and energetically isolated fluid volume, constituting a **closed system**, increases in time or remains constant. Therefore, the following inequality equation is fulfilled

$$\frac{dS}{dt} = \frac{d}{dt}\left(\iiint_{V_F} \rho s\, dV\right) \geq 0 \qquad (1.112)$$

(3) For **open systems** in which no material and energetic isolation exists, the entropy change in time can be divided into an internal and external part so that

$$\frac{dS}{dt} = \frac{d_i S}{dt} + \frac{d_e S}{dt} = -\oiint_{\partial V_F} \mathbf{j}_s \cdot \mathbf{n}\, da + \iiint_{V_F} \rho \dot{s}_i\, dV \qquad (1.113)$$

where $\mathbf{j}_s$ denotes the **entropy flux** and $\dot{s}_i \geq 0$ the **entropy production**. The fluid volume is closed for $\mathbf{j}_s = 0$ on the volume boundary and the second statement can be verified in this case.

Practically, the inequality relation (1.112) means that a closed system tends toward a thermodynamic equilibrium in which the temperature and the mass fractions of the various fluid volume components are uniformly distributed. One consequence is that a volume cannot heat another warmer volume by heat conduction without investing work which is known as **Clausius statement**.

By using the enthalpy $h = e_I + pv$, we can write Gibbs' fundamental equation also in the form

$$ds = \frac{dh}{T} - \frac{v}{T}\, dp - \frac{\mu_d}{T}\, dm_d - \frac{\mu_v}{T}\, dm_v - \frac{\mu_l}{T}\, dm_l \qquad (1.114)$$

This equation is also valid when we replace the differentials by material time derivatives. Therefore, we can make use of the first law of thermodynamics (1.71) to eliminate Dh/Dt and obtain a prognostic equation for specific entropy

$$\frac{Ds}{Dt} = \frac{\mathbf{J} : \nabla \mathbf{v} - \nabla \cdot \mathbf{j}_R - \nabla \cdot \mathbf{j}_{e_I}}{\rho T} - \frac{\mu_d}{T}\frac{Dm_d}{Dt} - \frac{\mu_v}{T}\frac{Dm_v}{Dt} - \frac{\mu_l}{T}\frac{Dm_l}{Dt}$$
$$(1.115)$$

Substituting the material time derivatives of m_d, m_v and m_l yields

$$\frac{Ds}{Dt} = \frac{\mathbf{J}:\nabla\mathbf{v} - \nabla\cdot\mathbf{j}_R - \nabla\cdot\mathbf{j}_{e_I}}{\rho T} - \frac{a_{lv}}{T}S_{l,v}$$

$$+ \frac{\mu_v - \mu_d}{\rho T}\nabla\cdot\mathbf{j}_{m_v} + \frac{\mu_l - \mu_d}{\rho T}\nabla\cdot\mathbf{j}_{m_l} \qquad (1.116)$$

where $a_{lv} = \mu_l - \mu_v$ denotes the **affinity of evaporation**. This equation can be converted into the form (according to Eq. (1.113))

$$\frac{Ds}{Dt} = -\frac{1}{\rho}\nabla\cdot\mathbf{j}_s + \dot{s}_i \qquad (1.117)$$

where the entropy flux $\mathbf{j}_s$ and the entropy production $\dot{s}_i$ are given by

$$\mathbf{j}_s = \frac{\mathbf{j}_r}{T} + \frac{\mathbf{j}_R}{T} + (s_v - s_d)\mathbf{j}_{m_v} + (s_l - s_d)\mathbf{j}_{m_l} \qquad (1.118)$$

and

$$\dot{s}_i = \frac{\mathbf{J}:\nabla\mathbf{v}}{\rho T} - \frac{\mathbf{j}_R\cdot\nabla T}{\rho T^2} - \frac{\mathbf{j}_{e_I}\cdot\nabla T}{\rho T^2} - \frac{\mathbf{j}_{m_v}}{\rho}\cdot\nabla\left(\frac{\mu_v-\mu_d}{T}\right)$$

$$- \frac{\mathbf{j}_{m_l}}{\rho}\cdot\nabla\left(\frac{\mu_l-\mu_d}{T}\right) - \frac{a_{lv}}{T}S_{l,v} \qquad (1.119)$$

We see that the entropy flux $\mathbf{j}_s$ comprises the reduced diffusive heat flux, the radiative flux and the diffusive fluxes of the partial specific entropies which change the composition of the fluid parcel constituents. The expression for the entropy production can be converted into a more suitable form by noting that

$$\nabla\left(\frac{\mu_j}{T}\right) = \frac{1}{T}\nabla\mu_j - \frac{\mu_j}{T^2}\nabla T = \frac{1}{T}\frac{\partial\mu_j}{\partial T}\nabla T + \frac{1}{T}\nabla_T\mu_j - \frac{\mu_j}{T^2}\nabla T$$

$$= -\frac{s_j}{T}\nabla T + \frac{1}{T}\nabla_T\mu_j - \frac{\mu_j}{T^2}\nabla T = \frac{1}{T}\nabla_T\mu_j - \frac{h_j}{T^2}\nabla T$$

$$\text{for } j = d, v \quad \text{and} \quad l \qquad (1.120)$$

where ∇_T denotes that temperature is held fixed when the Nabla operator is applied to a thermodynamic state function. In Eq. (1.120) we made use of the relation $\partial \mu_j / \partial T = -s_j$ which results from the differential for the **Gibbs-function** $\mu = h - Ts$ by using (1.114):

$$d\mu = dh - Tds - sdT$$
$$= -sdT + vdp + \mu_d dm_d + \mu_v dm_v + \mu_l dm_l \qquad (1.121)$$

This is an exact differential and we therefore have $\partial \mu / \partial T = -s$. Then, $\partial \mu_j / \partial T = -s_j$ also holds since μ can be decomposed into partial specific quantities like enthalpy and entropy. Inserting the division (1.120) in the formula for entropy production (1.119) gives

$$\dot{s}_i = \frac{\mathbf{J} : \nabla \mathbf{v}}{\rho T} - \frac{\mathbf{j}_R \cdot \nabla T}{\rho T^2} - \frac{\mathbf{j}_r \cdot \nabla T}{\rho T^2}$$
$$- \frac{\mathbf{j}_{m_v}}{\rho T} \cdot \nabla_T (\mu_v - \mu_d) - \frac{\mathbf{j}_{m_l}}{\rho T} \cdot \nabla_T (\mu_l - \mu_d) - \frac{a_{lv}}{T} S_{l,v}$$
$$(1.122)$$

This formula shows that the entropy production comprises terms with a bilinear form. These are composed of an unknown "flux" and a known "thermodynamic force". The various fluxes impacting on a cloud air parcel are sketched in Fig. 1.7. They are given by the viscous stress tensor $\mathbf{J}$, the radiative flux $\mathbf{j}_R$, the reduced heat flux $\mathbf{j}_r$, the evaporation rate $S_{l,v}$, the diffusion fluxes $\mathbf{j}_{m_v}$ and $\mathbf{j}_{m_l}$ while

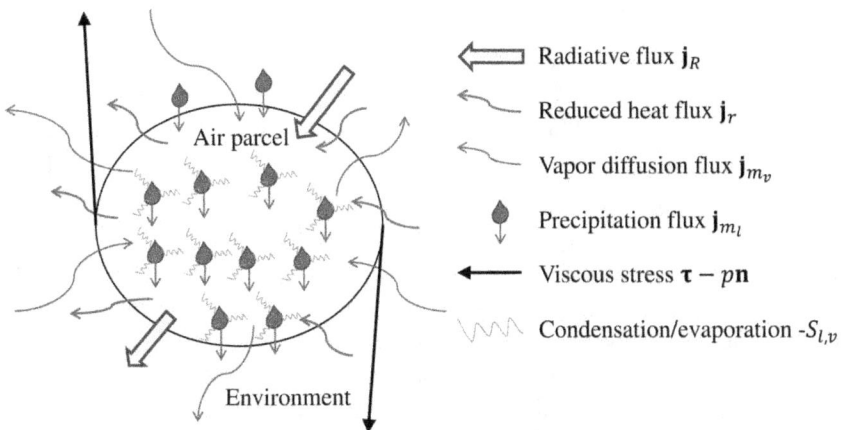

Fig. 1.7.   Sketch of irreversible processes in a cloud air parcel.

the corresponding forces are the velocity gradient $\nabla \mathbf{v}$, the temperature gradient ∇T, the affinity a_{lv} and the gradients of the chemical potential differences $\nabla_T(\mu_v - \mu_d)$ and $\nabla_T(\mu_l - \mu_d)$. To ensure that the entropy production is positive it is necessary that the fluxes must be pointed to a certain direction. A simple example is heat conduction in a one-component fluid at rest. Then, the right-hand side of Eq. (1.122) becomes simply $-\mathbf{j_r} \cdot \nabla T/(\rho T^2)$. It is obvious that a positive sign results from **Fourier's law of heat conduction** $\mathbf{j}_r = -K_T \nabla T$. Parameterization of other irreversible processes can be undertaken in a similar manner. We note that the **Navier–Stokes equations** and **Fick's law of diffusion** fulfill besides Fourier's law the requirement of a positive entropy production. Essentially, a positive entropy production holds if heat conduction is directed from cold to warm air, the diffusive momentum flux from high to low momentum and the diffusion flux of a constituent from high to low concentration. By these processes, the system approaches an equilibrium state in which the gradients of temperature, pressure, momentum and the concentrations of the constituents vanish. Special attention must be devoted to the last term in (1.122). Here, the affinity a_{lv} represents the thermodynamic force and the evaporation rate $S_{l,v}$ the flux. Evaporation ($S_{l,v} > 0$) and condensation ($S_{l,v} < 0$) can only occur for $a_{lv} \leq 0$ and $a_{lv} \geq 0$, respectively. By the corresponding phase transition, the vapor pressure approaches an equilibrium value, namely, the saturation vapor pressure. In this state, the affinity of evaporation vanishes. This shows that two phases can coexist in thermodynamic equilibrium. We note that a positive entropy production does not necessarily result in radiation and sedimentation of cloud droplets. Solar radiation is rarely directed antiparallel to the temperature gradient and cloud droplets usually fall downwards but not always toward regions with less liquid water density. These examples reveal the limitations of classical irreversible thermodynamics and the assumption of a local thermodynamic equilibrium. However, the specific entropy proved to be a relevant thermodynamic quantity in atmospheric dynamics and we will derive it as a function of the thermodynamic variables in the following.

To derive the specific entropy as a function of the thermodynamic state variables we use Gibbs' fundamental equation (1.114). Inserting the differential for enthalpy

$$dh = c_p dT + h_d dm_d + h_v dm_v + h_l dm_l \qquad (1.123)$$

gives

$$ds = \frac{c_p}{T}\,dT - \frac{v}{T}\,dp + s_d dm_d + s_v dm_v + s_l dm_l \qquad (1.124)$$

Since this differential is exact, the following differential equations result:

$$\frac{\partial s}{\partial T} = \frac{c_p}{T}, \quad \frac{\partial s}{\partial p} = -\frac{v}{T}, \quad \frac{\partial s}{\partial m_j} = s_j \quad \text{for } j = d, v \quad \text{and} \quad l \quad (1.125)$$

According to Eq. (1.111), it is useful to determine the partial specific entropies. They can be obtained by considering the components in isolation. We get

$$s_d = c_{pd}\ln\left(\frac{T}{T_0}\right) - R_d\ln\frac{p_d}{p_0} \qquad (1.126)$$

$$s_v = c_{pv}\ln\left(\frac{T}{T_0}\right) - R_d\ln\frac{p_v}{p_0} \qquad (1.127)$$

$$s_l = c_l\ln\left(\frac{T}{T_0}\right) \qquad (1.128)$$

Inserting these in Eq. (1.111) leads to the specific entropy of the moist air

$$s = c_p\ln\left(\frac{T}{T_0}\right) - R_d m_d\ln\left(\frac{p_d}{p_0}\right) - R_v m_v\ln\left(\frac{p_v}{p_0}\right) \qquad (1.129)$$

So far it is not clear that this expression fulfills $\partial s/\partial m_j = s_j$. The last equation in (1.125) is true when

$$\frac{\partial s_d}{\partial m_j}m_d + \frac{\partial s_v}{\partial m_j}m_v + \frac{\partial s_l}{\partial m_j}m_l = 0 \quad \text{for } j = d, v \quad \text{and} \quad l \quad (1.130)$$

To show this we prove that $p_d = n_d p$ and $p_v = n_v p$ where n_d and n_v are the **molar fractions** of dry air and water vapor, respectively. The molar fraction measures the number of molecules of the respective component in relation to the total amount of molecules in the gas. An ideal gas has the property that its volume at a given temperature, pressure and number of molecules is independent of the molecule type. Therefore, the gas constant per molecule, R^*, becomes

$$R^* = \bar{m}_d R_d = \bar{m}_v R_v \tag{1.131}$$

where $\bar{m}_j$ denotes the **molecular weight** of the component with the index j. Consequently, the molar fraction is related to the mass fraction by the formula

$$n_j = \frac{m_j/\bar{m}_j}{\frac{m_d}{\bar{m}_d} + \frac{m_v}{\bar{m}_v}} \quad \text{for } j = d \quad \text{and} \quad v \tag{1.132}$$

Due to Dalton's law (1.73) we have

$$p = p_d + p_v = R_d m_d \rho T + R_v m_v \rho T = R^* \frac{N_d}{M} \rho T + R^* \frac{N_v}{M} \rho T$$

$$= R^* \frac{N}{M} n_d \rho T + R^* \frac{N}{M} n_v \rho T = n_d p + n_v p \tag{1.133}$$

where N, N_d and N_v denote the numbers of molecules of the gas components, dry air and water vapor, respectively. Therefore, the relations $p_d = n_d p = (1 - n_v)p$ and $p_v = n_v p$ are valid. By inserting these in Eq. (1.129) one can verify using (1.132) that (1.125) is fulfilled.

Phase transitions are not considered in the entropy expression so far. They lead to positive or zero entropy production. Then,

$$\frac{a_{lv}}{T} = \frac{\mu_l - \mu_v}{T} = -\frac{L_v}{T} + s_v - s_l \begin{cases} \leq 0 & \text{for } S_{l,v} > 0 \\ \geq 0 & \text{for } S_{l,v} < 0 \end{cases} \tag{1.134}$$

In thermodynamic equilibrium, the affinity vanishes. Phase transitions happen fast so that the cloud droplets are nearly in phase equilibrium with the surrounding water vapor. Therefore, we have in cloud air ($m_l > 0$) the approximate relation

$$s_{vs} - s_l = \frac{L_v}{T} \tag{1.135}$$

where s_{vs} is the partial specific entropy at saturation. On the other hand, we find from Eqs. (1.127) and (1.128) that

$$s_{vs} - s_l = (c_{pv} - c_l)\ln\left(\frac{T}{T_0}\right) - R_v \ln\left(\frac{p_{vs}}{p_0}\right) = \frac{L_v}{T} \tag{1.136}$$

Taking the derivative with respect to temperature yields

$$\frac{c_{pv} - c_l}{T} - \frac{R_v}{p_{vs}}\frac{dp_{vs}}{dT} = \frac{1}{T}\frac{dL_v}{dT} - \frac{L_v}{T^2} \rightarrow \frac{dp_{vs}}{dT} = \frac{L_v p_{vs}}{R_v T^2} \qquad (1.137)$$

Due to the temperature dependence of L_v (cf. Eq. (1.93)), the first terms on the left- and right-hand sides of (1.137) cancel and we obtain the Clausius–Clapeyron equation (1.95). We note that it is only valid for the equilibrium of pure water vapor above a plane sheet of liquid water. Therefore, we have assumed that the presence of dry air has no impact on the partial pressure of water vapor at equilibrium. Finally, we can write for the specific entropy by inserting (1.136) in (1.111):

$$s = (c_{pd}m_d + c_l m_t)\ln\left(\frac{T}{T_0}\right) - R_d m_d \ln\left(\frac{p - p_v}{p_0}\right)$$

$$- R_v m_v \ln(\mathcal{H}) + \frac{L_v}{T}m_v \qquad (1.138)$$

where $m_t = m_v + m_l$ denotes the mass fraction of total water.

1.10 Reversible Adiabats and Potential Temperature

The determination of a **reversible adiabat** is a frequent application in atmospheric thermodynamics. It describes the temperature of an air parcel that undergoes an **adiabatic** and **reversible** process. Adiabatic means that the air parcel is materially and energetically isolated (closed system) while reversible states that no entropy production takes place. Therefore, the specific entropy of such an air parcel is conserved. A reversible adiabat of a dry air parcel represents the simplest case. Then, the mass fractions of water vapor and liquid water vanish ($m_v = m_l = 0$, $m_d = 1$) and we obtain for the entropy

$$s = c_{pd}\ln\left(\frac{T}{T_0}\right) - R_d\ln\left(\frac{p}{p_0}\right) \qquad (1.139)$$

In this context, it is useful to introduce the **potential temperature** θ of dry air which is related to specific entropy by

$$s = c_{pd}\ln\left(\frac{\theta}{T_0}\right) \qquad (1.140)$$

Substituting this expression in (1.139) yields potential temperature as a function of pressure and temperature

$$\theta = T\left(\frac{p_0}{p}\right)^{\frac{R_d}{c_{pd}}} = \frac{T}{\Pi} \tag{1.141}$$

where $\Pi = (p/p_0)^{R_d/c_{pd}}$ denotes the **Exner function** (Exner 1925). The isolated air parcel attains the potential temperature if it is moved on the pressure level where $p = p_0$. The temperature T of a reversible adiabat decreases with height since pressure also decreases due to gravity.

In unsaturated moist air ($m_l = 0$, $m_v \neq 0$, $m_d \neq 0$) phase transitions do not take place and the entropy production by mixing has no effect on the adiabat since it does not change the temperature or pressure. Furthermore, diffusion must be neglected when the air parcel undergoes a reversible process. Therefore, we set $dm_d = dm_v = 0$ and find from (1.124) the specific entropy

$$s = (c_{pd}m_d + c_{pv}m_v)\ln\left(\frac{T}{T_0}\right) - (R_d m_d + R_v m_v)\ln\left(\frac{p}{p_0}\right)$$

$$= [c_{pd} + (c_{pv} - c_{pd})m_v]\ln\left(\frac{T}{T_0}\right) - [R_d + (R_v - R_d)m_v]\ln\left(\frac{p}{p_0}\right) \tag{1.142}$$

The definition

$$s = [c_{pd} + (c_{pv} - c_{pd})m_v]\ln\left(\frac{\theta_m}{T_0}\right) \tag{1.143}$$

yields the potential temperature of moist air:

$$\theta_m = T\left(\frac{p_0}{p}\right)^{\frac{R_d + (R_v - R_d)m_v}{c_{pd} + (c_{pv} - c_{pd})m_v}} \tag{1.144}$$

The potential temperature θ_m is conserved in the case of adiabatic and reversible processes when phase transitions are absent. Usually, the potential temperatures θ and θ_m do not deviate much from each other so that simply the potential temperature θ can be used for the calculation of the adiabat of a moist unsaturated air parcel.

In cloud air ($m_l \neq 0$, $m_v \neq 0$, $m_d \neq 0$) phase transitions cannot be ignored. Then, the appropriate specific entropy is given in Eq. (1.138) for $\mathcal{H} = 1$. By setting

$$s = (c_{pd}m_d + c_l m_t)\ln\left(\frac{\theta_{es}}{T_0}\right) \qquad (1.145)$$

we obtain the **saturated equivalent potential temperature**

$$\theta_{es} = T\left(\frac{p_0}{p - p_{vs}}\right)^{\frac{R_d(1-m_t)}{\tilde{c}_p}} \exp\left[\frac{L_v m_{vs}}{\tilde{c}_p T}\right] \qquad (1.146)$$

where $\tilde{c}_p = c_{pd} + (c_l - c_{pd})m_t$. In the derivation of this expression, we made use of the relation $m_d = 1 - m_t$. Now, it is not possible to deduce the temperature of the corresponding adiabat as a function of pressure analytically but it can be properly estimated with a numerical method. The resulting temperature curve is called **reversible moist adiabat**. For practical purposes, the reversible moist adiabat includes the adiabat of the subcloud air since air often rises from the boundary layer close to the surface to larger heights where condensation takes place and a cloud forms. The rising of the parcel until this **lifting condensation level** follows approximately an adiabat resulting from Eq. (1.144) for a fixed θ_m. Afterwards, θ_{es} is conserved which yields another temperature curve due to the release of latent heat. Another quantity to characterize the reversible moist adiabat is the **equivalent potential temperature** θ_e. It is based on the entropy given in (1.138)

$$s = \tilde{c}_p \ln\left(\frac{T}{T_0}\right) - R_d(1 - m_t)\ln\left(\frac{p - p_v}{p_0}\right)$$

$$- R_v m_v \ln\mathcal{H} + \frac{L_v}{T}m_v \qquad (1.147)$$

The **equivalent potential temperature** results from the relation

$$s = \tilde{c}_p \ln\left(\frac{\theta_e}{T_0}\right) \qquad (1.148)$$

and reads

$$\theta_e = T\left(\frac{p_0}{p - p_v}\right)^{\frac{R_d(1-m_t)}{\tilde{c}_p}} (\mathcal{H})^{-\frac{R_v m_v}{\tilde{c}_p}} \exp\left[\frac{L_v m_v}{\tilde{c}_p T}\right] \qquad (1.149)$$

We see that it is identical to the saturated equivalent potential temperature in the saturated case $\mathcal{H} = 1$. It is conserved when the processes are adiabatic and reversible including phase transitions.

Often, only an approximation is used for equivalent potential temperature. Typically, we have $m_t \ll 1$, $R_v m_v \ll \tilde{c}_p$, $p_v \ll p$ and $L_v \approx 2.5 \times 10^6$ J/kg. Then, the approximation

$$\theta_e \approx \theta \exp\left(\frac{L_v m_v}{c_{pd}T}\right) \qquad (1.150)$$

holds. Figure 1.8 displays a reversible moist adiabat and a dry adiabat in a **T-log(P) diagram** that is also called **emagram**. This diagram has the temperature on the abscissa and the logarithm of pressure on the ordinate. This representation has the advantage that the area of a closed curve is proportional to the volume work of an ideal gas that undergoes a thermodynamic cycle along this closed curve. The ordinate is directed toward larger height because of the decreasing pressure. Therefore, the adiabats can be interpreted as the height-temperature profile of a rising or descending air parcel. The dry and reversible moist adiabat virtually coincide for pressures

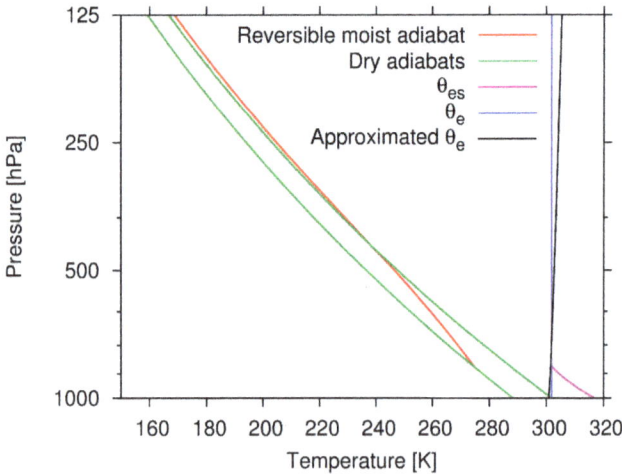

Fig. 1.8.   Temperature profiles in the T-log(P) diagram (emagram). Shown are the reversible moist adiabat (red curve), dry adiabats (green curves), the saturated equivalent potential temperature θ_{es} (blue curve), the equivalent potential temperature θ_{es} (violet curve) and the approximated potential temperature $\theta_e = \theta \exp\left[L_v m_v / (c_{pd}T)\right]$ (black curve).

above 844 hPa where the graph only displays the dry adiabat. There, the air is unsaturated since the partial vapor pressure falls below the saturation vapor pressure. The pressure 844 hPa marks the lifting condensation level above which the partial vapor pressure exceeds the saturation vapor pressure since the latter decreases with decreasing pressure. Consequently, condensation takes place and the associated latent heat release leads to a higher temperature than that of the dry adiabat. Nevertheless, temperature still decreases with decreasing pressure and increasing height so that condensation continues in the rising air parcel. Only the rate of temperature decrease has weakened by the latent heat release. The specific humidity approaches zero at large heights. There, the equivalent potential temperature is approximately given by

$$\theta_e = T \left(\frac{p_0}{p} \right)^{\frac{R_d(1-m_t)}{\bar{c}_p}} \tag{1.151}$$

Consequently, the reversible moist adiabat runs at large heights approximately parallel to the dry adiabat having the potential temperature $\theta = \theta_e$. However, we can see in Fig. 1.8 that the reversible moist adiabat has a slightly flatter slope since the heat capacity of the present liquid water is larger than that of the dry air. Figure 1.8 also displays the vertical profiles of the equivalent potential temperatures θ_{es} and θ_e. The saturated equivalent potential temperature θ_{es} decreases with height below the lifting condensation level because the saturation-specific humidity also decreases. On the other hand, the equivalent potential temperature θ_e remains constant below and above the lifting condensation level. Above this level θ_e and θ_{es} coincide since the air is saturated. We see in the diagram that the approximated θ_e as given by Eq. (1.150) slightly rises with height. Therefore, the approximated equivalent potential temperature is only crudely conserved along the reversible moist adiabat.

1.11 Transformation to a Rotating Reference Frame

The Earth rotates with the **angular velocity** $\Omega \approx \frac{2\pi}{24}\text{h} \approx 7.29 \times 10^{-5}\text{s}^{-1}$ with respect to Earth's axis. It is convenient to consider a coordinate frame that adheres to the rotating Earth. However, such a frame does not represent an inertial system and, therefore,

Newton's laws of motion are not valid. Nevertheless, it is possible to transform the equations from an inertial system to a rotating coordinate system. For this purpose, we define coordinates x', y', z' of a Cartesian system that has its origin in the barycenter of the rotating Earth and rotates with the angular velocity Ω. We assume that the barycenter of the Earth moves without acceleration and, therefore, we can also define coordinates x, y, z of a Cartesian system that forms an inertial system and has also its origin in the Earth's center. In both coordinate systems, the vertical coordinate axis (along z and z' direction) coincides with the Earth's axis. Before deriving the equation of motion in the rotating reference frame we address the transformation of the coordinates. Figure 1.9 displays the orientation of the horizontal axes. Let λ be the **azimuth angle** that specifies the angle to the x axis of the location vector projected onto the horizontal plane and $R = \sqrt{x^2 + y^2}$ denotes the distance of the projected vector $\mathbf{R}$ to the z-axis. Thus, R and λ constitute **polar coordinates** of the inertial system and we get

$$x = R \cos \lambda, \quad y = R \sin \lambda \tag{1.152}$$

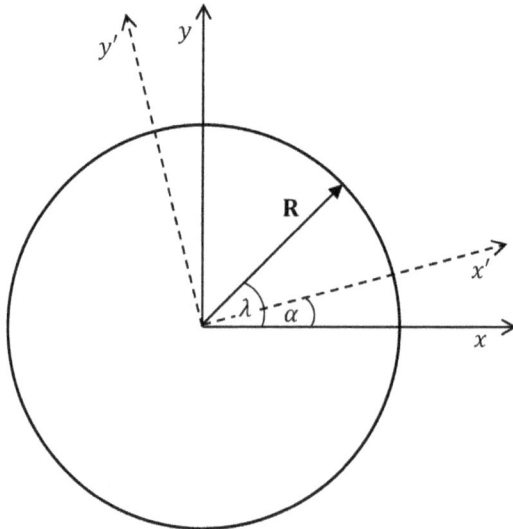

Fig. 1.9. Top view on the horizontal coordinate plane of the inertial system (x, y) and the rotating system (x', y').

Let α be the angle between the x and x' axes. Then, $\lambda - \alpha$ yields the angle of the position vector to the x' axis and we get

$$x' = R\cos(\lambda - \alpha), \quad y' = R\sin(\lambda - \alpha) \qquad (1.153)$$

With the angle addition theorem for trigonometrical functions, we find that

$$x' = R\cos(\lambda - \alpha) = R\cos\lambda\cos\alpha + R\sin\lambda\sin\alpha$$
$$= x\cos\alpha + y\sin\alpha \qquad (1.154)$$

and

$$y' = R\sin(\lambda - \alpha) = R\sin\lambda\cos\alpha - R\cos\lambda\sin\alpha$$
$$= y\cos\alpha - x\sin\alpha \qquad (1.155)$$

Since the x' and x' axes rotate with the angular velocity Ω the angle α becomes a function of time, namely $\alpha = \Omega t + \alpha_0$ where α_0 is a constant that can be set to zero without loss of generality. Consequently, the transformation equations become

$$x' = x\cos(\Omega t) + y\sin(\Omega t) \qquad (1.156)$$

$$y' = y\cos(\Omega t) - x\sin(\Omega t) \qquad (1.157)$$

$$z' = z \qquad (1.158)$$

By the time derivative, we can determine the velocity components in the rotating coordinate system. We obtain for the x' component

$$u' = \frac{dx'}{dt} = \frac{dx}{dt}\cos(\Omega t) + \frac{dy}{dt}\sin(\Omega t) - \Omega[x\sin(\Omega t) - y\cos(\Omega t)]$$
$$= u'_I + \Omega y' \qquad (1.159)$$

where u'_I denotes the x' component of the velocity in the inertial system without rotation. Note that identical vectors in different coordinate systems have different directions. Therefore, the velocity vector in the inertial system would have different components when evaluated in the rotating system. A corresponding relation results for the y' component:

$$v' = \frac{dy'}{dt} = v'_I - \Omega x' \qquad (1.160)$$

The coordinate does not change in z' direction so that

$$w' = \frac{dx'}{dt} = w'_I \tag{1.161}$$

This result can also be written in vector form. Let $\mathbf{e}_{x'}$, $\mathbf{e}_{y'}$ and $\mathbf{e}_{z'}$ be the unit vectors of the rotating coordinate system. Then, the velocities in the rotated and inertial system are linked by

$$\begin{aligned} \mathbf{v}' &= u'_I \mathbf{e}_{x'} + v'_I \mathbf{e}_{y'} + w'_I \mathbf{e}_{z'} + \Omega y' \mathbf{e}_{x'} - \Omega x' \mathbf{e}_{y'} \\ &= \mathbf{v}'_I - \mathbf{\Omega} \times \mathbf{r}' \end{aligned} \tag{1.162}$$

where $\mathbf{\Omega} = \Omega \mathbf{e}_{z'}$ and $\mathbf{r}'$ denotes the position vector in the rotating coordinate system. The difference vector between $\mathbf{v}'$ and $\mathbf{v}'_I$ constitutes the velocity field of Earth's rotation $\mathbf{v}'_E$ which is given by

$$\mathbf{v}'_E = \mathbf{\Omega} \times \mathbf{r}' \tag{1.163}$$

Therefore, we have a comprehensible result

$$\mathbf{v}'_I = \mathbf{v}' + \mathbf{v}'_E \tag{1.164}$$

The acceleration in the rotating coordinate system can also be determined by evaluating the time derivative. Differentiation of (1.159) yields

$$\begin{aligned} a'_x &= \frac{du'}{dt} = \frac{d^2 x}{dt^2} \cos(\Omega t) + \frac{d^2 y}{dt^2} \sin(\Omega t) \\ &\quad - 2\Omega \left[\frac{dx}{dt} \sin(\Omega t) - \frac{dy}{dt} \cos(\Omega t) \right] - \Omega^2 [x \cos(\Omega t) + y \sin(\Omega t)] \\ &= a'_{xI} + 2\Omega v'_I - \Omega^2 x' \end{aligned} \tag{1.165}$$

Here, a'_{xI} denotes the x' component of the acceleration in the inertial system. An observer in the rotating reference frame would notice two further accelerations which are called **inertial** or **pseudo accelerations**. They do not occur in the inertial system. For the other coordinate directions, we obtain a likewise derivation:

$$a'_y = a'_{yI} - 2\Omega u'_I - \Omega^2 y' \tag{1.166}$$

$$a'_z = a'_{zI} \tag{1.167}$$

and in vector form we get

$$\mathbf{a}' = \mathbf{a}'_I - 2\mathbf{\Omega} \times \mathbf{v}'_I + \mathbf{\Omega} \times (\mathbf{\Omega} \times \mathbf{r}')$$

$$= \mathbf{a}'_I - 2\mathbf{\Omega} \times \mathbf{v}'_I + \mathbf{\Omega} \times \mathbf{v}'_E \qquad (1.168)$$

Since $\mathbf{v}'_I = \mathbf{v}' + \mathbf{v}'_E$, Eq. (1.168) can also be written as

$$\mathbf{a}' = \mathbf{a}'_I - 2\mathbf{\Omega} \times \mathbf{v}' - \mathbf{\Omega} \times \mathbf{v}'_E \qquad (1.169)$$

Newton's second law of motion is only valid in an inertial system. Therefore, we can use (1.169) to obtain the acceleration in the rotating coordinate frame and we get for the equation of motion

$$\frac{D\mathbf{v}'}{Dt} + 2\mathbf{\Omega} \times \mathbf{v}' + \mathbf{\Omega} \times (\mathbf{\Omega} \times \mathbf{r}') = -\frac{1}{\rho}(\nabla p)' - (\nabla \phi)' + \frac{1}{\rho}(\nabla \cdot \mathbf{J})' \quad (1.170)$$

The scalar fluid properties got no prime in this equation because they are invariant with respect to a coordinate transform. For example, the temperature remains identical at the same location in the rotating coordinate system while the wind vector obtains a different orientation. Figure 1.10 sketches the orientation of the two inertial forces in the rotating system. These are the **Coriolis force** $-2\mathbf{\Omega} \times \mathbf{v}'$ and the **centrifugal force** $-\mathbf{\Omega} \times (\mathbf{\Omega} \times \mathbf{r}') = -\mathbf{\Omega} \times \mathbf{v}'_E$. The Coriolis force is always directed perpendicular to the wind vector since $\mathbf{v}' \cdot (\mathbf{\Omega} \times \mathbf{v}') = 0$. Therefore, the Coriolis force does not do work and does not contribute to any energy conversions in the rotating system. However, this is not the case for the centrifugal force. When, for example, a parcel at rest is moved to a location at a larger distance to the rotation axis it does not change its kinetic energy in the rotating system but in the inertial frame, it attains more kinetic energy due to the higher rotation speed of the earth. Therefore, the centrifugal force contributes to the energy budget. We can find a centrifugal potential that includes this effect and can be interpreted as potential energy in the rotating reference frame. With Eqs. (1.165)–(1.169) we see that:

$$\mathbf{\Omega} \times (\mathbf{\Omega} \times \mathbf{r}') = -\Omega^2 \mathbf{R}' = -\frac{\Omega^2}{2}(\nabla R^2)' \qquad (1.171)$$

Therefore, a modified potential can be defined as

$$\phi' = \phi - \frac{\Omega^2}{2} R^2 \qquad (1.172)$$

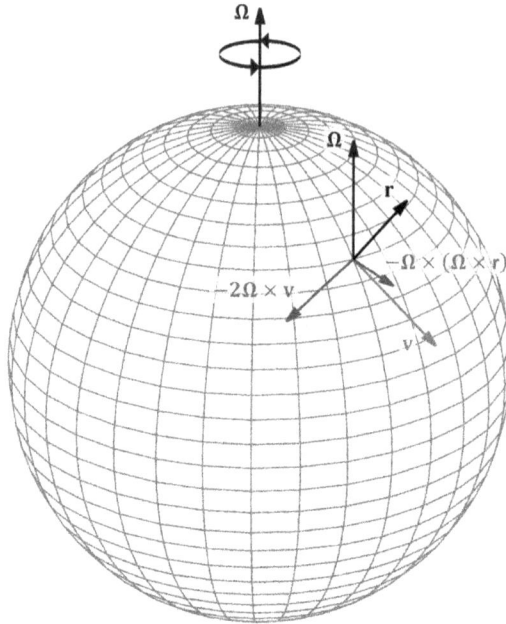

Fig. 1.10. Sketch of the Coriolis force $-2\mathbf{\Omega} \times \mathbf{v}$ and the centrifugal force $-\mathbf{\Omega} \times (\mathbf{\Omega} \times \mathbf{r}')$ in the rotating reference frame. Note that the position vector $\mathbf{r}$ is oriented perpendicular to the surface of the sphere.

It is the sum of the gravitational and centrifugal potential which is called **geopotential** in the meteorological context. In the rotating reference frame, a hydrostatic equilibrium refers to this frame and, therefore, the velocity $\mathbf{v}'_E$ is not interpreted as flow. Therefore, hydrostatic equilibrium means that the pressure surfaces run parallel to the surfaces of constant geopotential which deviate slightly from the surface of a sphere. Also, the Earth's surface adapts to the rotation which leads to a bigger Earth radius at the equatorial sea level compared to those at the poles. The difference between these radii is after all about 21 km but this value is very small compared to the mean Earth radius of 6371 km. In the remainder of the book, we refer to the rotating coordinate frame and we omit the prime for denoting transformed variables. Then, the equation of motion takes the form

$$\frac{D\mathbf{v}}{Dt} + 2\mathbf{\Omega} \times \mathbf{v} = -\frac{1}{\rho}\nabla p - \nabla\phi + \frac{1}{\rho}\nabla \cdot \mathbf{J} \qquad (1.173)$$

The continuity equation, the first law of thermodynamics, the budget equations for the various constituents, and the equations of state remain unaffected by the transformation to the non-inertial reference frame since they are equations for scalar properties.

1.12 Governing Equations and Energetics

In the previous sections, we compiled governing equations for atmospheric weather dynamics. We can now list a complete equation set:

(1) Equation of motion

$$\frac{\partial \mathbf{v}}{\partial t} + \mathbf{v} \cdot \nabla \mathbf{v} + 2\mathbf{\Omega} \times \mathbf{v} = -\frac{1}{\rho}\nabla p - \nabla \phi + \frac{1}{\rho}\nabla \cdot \mathbf{J} \qquad (1.174)$$

(2) Continuity equation

$$\frac{\partial \rho}{\partial t} + \nabla \cdot (\rho \mathbf{v}) = 0 \qquad (1.175)$$

(3) Temperature tendency equation

$$c_p \left(\frac{\partial T}{\partial t} + \mathbf{v} \cdot \nabla T \right) = -L_v S_{l,v} + \frac{1}{\rho}\frac{Dp}{Dt} + \frac{1}{\rho}\mathbf{J} : \nabla \mathbf{v}$$

$$- \frac{1}{\rho}\nabla \cdot \mathbf{j}_R - \frac{1}{\rho}\nabla \cdot \mathbf{j}_r$$

$$- (c_{pv} - c_{pd})\mathbf{j}_{m_v} \cdot \nabla T - (c_l - c_{pd})\mathbf{j}_{m_l} \cdot \nabla T$$

$$(1.176)$$

(4) Thermal equation of state

$$\rho = \frac{p}{R_d T \left[1 + \left(\frac{R_v}{R_d} - 1 \right) m_v - m_l \right]} \qquad (1.177)$$

(5) Budget equation for water vapor

$$\frac{\partial m_v}{\partial t} + \mathbf{v} \cdot \nabla m_v = S_{l,v} - \frac{1}{\rho}\nabla \cdot \mathbf{j}_{m_v} \qquad (1.178)$$

(6) Budget equation for liquid water

$$\frac{\partial m_l}{\partial t} + \mathbf{v} \cdot \nabla m_l = -S_{l,v} - \frac{1}{\rho}\nabla \cdot \mathbf{j}_{m_l} \qquad (1.179)$$

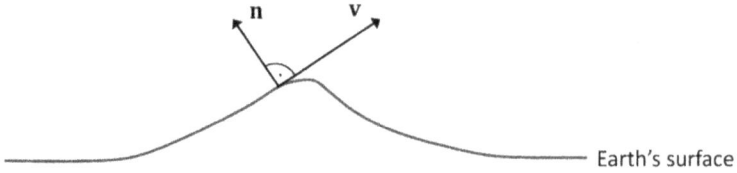

Fig. 1.11.   Sketch of the kinematic boundary condition at the Earth's surface.

The unknown variables of these equations are $\mathbf{v}$, p, ρ, T, m_v and m_l. On the other hand, we have six equations and, therefore, enough to solve these equations for the six unknowns by using an approximate numerical solution method. The formulation of the viscous stress tensor $\mathbf{J}$, the phase transition $S_{l,v}$, the radiative flux $\mathbf{j}_R$, the reduced heat flux $\mathbf{j}_r$, and the diffusive fluxes $\mathbf{j}_{m_v}$ and $\mathbf{j}_{m_l}$ have not been specified here. Suitable representation of these terms can be found on the basis of statistical mechanics, the theory of radiative transfer and cloud microphysics. In addition to these equations, we need boundary conditions for the determination of a solution. In the atmosphere, the lower boundary is the Earth's surface. The upper boundary is not clearly defined since the air smoothly approaches zero density at large heights. Boundary conditions must be formulated for $\mathbf{J}$, $\mathbf{j}_R$, $\mathbf{j}_r$, $\mathbf{j}_{m_v}$, $\mathbf{j}_{m_l}$, and $\mathbf{v}$. Usually, it is assumed that the velocity component normal to the boundary surface vanishes (see Fig. 1.11), i.e.

$$\mathbf{v} \cdot \mathbf{n} = 0 \text{ at the boundary surface} \qquad (1.180)$$

We note, however, that only the dry air velocity vanishes at the boundary surface since evaporation and precipitation yield nonzero velocities of water vapor and liquid water, respectively. Then, the barycenter of an air parcel moves across the boundary and its normal velocity component $\mathbf{v} \cdot \mathbf{n}$ does not vanish. So far, no study exists that the consideration of the correct boundary condition reveals a significant impact on meteorological dynamics. Therefore, we also ignore the existence of a nonzero normal velocity component at the boundary and apply Eq. (1.180).

The equation set (1.174)–(1.177) forms the **Navier–Stokes equations** for an ideal gas if the air is dry ($m_v = m_l = 0$) and radiative transfer is ignored ($\mathbf{j}_R = 0$). Since viscosity and diffusion result from molecular exchange, they do not have an appreciable impact on the flow of atmospheric weather systems. Therefore, it is legitimate to ignore these processes. Then, we obtain in the dry case without radiative transfer the **Euler equations**. These equations are reversible, that is, the transformation $\mathbf{v} \to -\mathbf{v}$, $\mathbf{\Omega} \to -\mathbf{\Omega}$ and $t \to -t$ yields identical equations. This has the consequence that the time-reversed development of a flow also underlies the hydrodynamic laws. On the other hand, this would not be true if viscosity and diffusion are taken into account. Therefore, we considered these as irreversible processes which contribute to entropy production (see Section 1.8). In extreme weather systems like hurricanes, the dissipative heating $\mathbf{J} : \nabla\mathbf{v}/\rho$ could play a certain role. In such contexts, one must find a suitable representation of this term. In Section 1.13, we will not neglect dissipative heating for energetic consistency in the parameterization of microturbulent exchange. However, we neglect all other processes connected to molecular exchange processes from now on.

Another simplification regards the treatment of liquid water. A large part of condensate eventually leaves the atmosphere in the form of precipitation. The time scale of this process is short in comparison to those of large-scale weather systems. Therefore, it is justified in this case to assume that condensed water precipitates immediately with the consequence that the mass fraction of liquid water, m_l vanishes everywhere. Then, the condensation rate determines the precipitation rate. The temperature profile that results in the absence of heat sources from phase transitions with immediate removal of the condensate is called **pseudoadiabat**. The temperature of a pseudoadiabat is similar to that of a reversible moist adiabat but the mass density can be appreciably smaller due to the high mass density of liquid water. Furthermore, we neglect the dependency of the heat capacities, latent heat and gas constant on specific humidity so that $c_p = \text{const.}$, $c_v = \text{const.}$, $L_v = \text{const.}$, and $\rho = p/(R_d T)$. These simplifications are suitable for the dynamics of synoptic weather systems but may lead to significant inaccuracies in cloud-scale flows that incorporate phase transitions.

With the aforementioned simplifications, the set of equations becomes

$$\frac{\partial \mathbf{v}}{\partial t} + \mathbf{v} \cdot \nabla \mathbf{v} + 2\mathbf{\Omega} \times \mathbf{v} = -\frac{1}{\rho}\nabla p - \nabla \phi \qquad (1.181)$$

$$\frac{\partial \rho}{\partial t} + \nabla \cdot (\rho \mathbf{v}) = 0 \qquad (1.182)$$

$$c_p \left(\frac{\partial T}{\partial t} + \mathbf{v} \cdot \nabla T \right) = -L_v S_{l,v} + \frac{1}{\rho} \frac{Dp}{Dt} - \frac{1}{\rho} \nabla \cdot \mathbf{j}_R \qquad (1.183)$$

$$\rho = \frac{p}{R_d T} \qquad (1.184)$$

$$\frac{\partial m_v}{\partial t} + \mathbf{v} \cdot \nabla m_v = S_{l,v} \qquad (1.185)$$

In many applications, the potential temperature represents a more suitable prognostic variable than temperature. Then, we can derive the prognostic equation for potential temperature from (1.183) using the definition (1.141):

$$\frac{\partial \theta}{\partial t} + \mathbf{v} \cdot \nabla \theta = -\frac{L_v}{c_p \Pi} S_{l,v} - \frac{1}{c_p \Pi \rho} \nabla \cdot \mathbf{j}_R \qquad (1.186)$$

Note that potential temperature is now conserved in the absence of phase transitions and radiation.

From the equation set (1.181)–(1.185), we can derive the following two energy budget equations for specific inner energy $e_I = c_v T + L_v m_v$ and specific kinetic energy $e_K = \mathbf{v}^2/2$:

$$\frac{De_I}{Dt} = -\frac{p}{\rho} \nabla \cdot \mathbf{v} - \frac{1}{\rho} \nabla \cdot \mathbf{j}_R \qquad (1.187)$$

$$\frac{De_K}{Dt} = -\mathbf{v} \cdot \nabla \phi - \frac{1}{\rho} \mathbf{v} \cdot \nabla p \qquad (1.188)$$

To refine the energetics further, we introduce **specific potential energy**:

$$e_P = \phi \qquad (1.189)$$

With the division of specific inner energy into specific inner kinetic energy $e_{IK} = c_v T$ and specific inner potential energy $e_{IP} = L_v m_v$ we get the following set of energy equations:

$$\frac{De_{IK}}{Dt} = -L_v S_{l,v} - \frac{p}{\rho} \nabla \cdot \mathbf{v} - \frac{1}{\rho} \nabla \cdot \mathbf{j}_R \qquad (1.190)$$

$$\frac{De_{IP}}{Dt} = L_v S_{l,v} \qquad (1.191)$$

$$\frac{De_K}{Dt} = -\mathbf{v} \cdot \nabla\phi - \frac{1}{\rho}\mathbf{v} \cdot \nabla p \qquad (1.192)$$

$$\frac{De_P}{Dt} = \mathbf{v} \cdot \nabla\phi \qquad (1.193)$$

The first term on the right-hand side of the equation for inner kinetic energy (1.190) is referred to as the **latent heat release** in the case of condensation ($S_{l,v} < 0$) and the specific inner potential energy e_{IP} can be regarded as specific **latent energy**. This is a reasonable term since latent energy does not play a role when the air is undersaturated but it can induce intense currents when saturation takes place. By adding all four equations we obtain the budget for specific total energy $e_T = e_I + e_K + e_P$

$$\frac{De_T}{Dt} = -\frac{1}{\rho}\nabla \cdot (p\mathbf{v}) - \frac{1}{\rho}\nabla \cdot \mathbf{j}_R \qquad (1.194)$$

Therefore, a change in total energy results when work is performed or electromagnetic radiation is absorbed. All other terms on the right-hand side form energy conversions which are illustrated in Fig. 1.12. However, we note that the energy conversions from e_{IK}

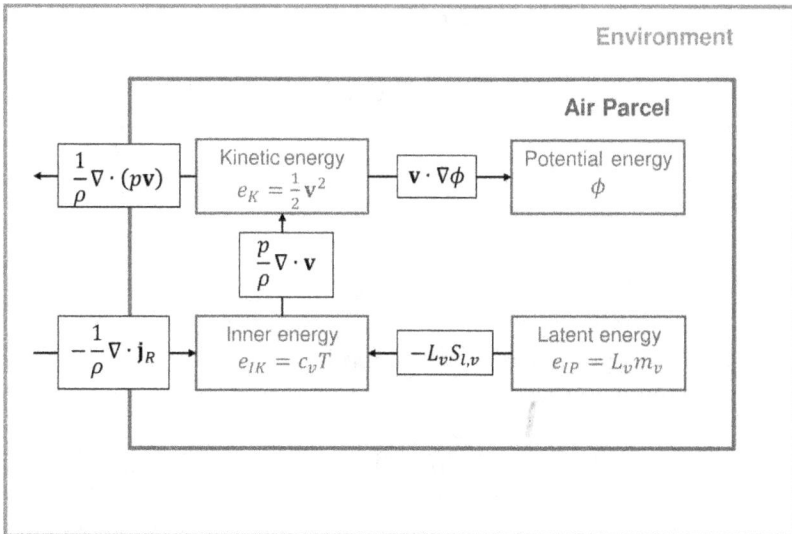

Fig. 1.12. Energy conversions in a fluid parcel and energy transfers to the surroundings.

to e_K is ambiguous since we can replace the conversion $(p\nabla \cdot \mathbf{v})/\rho$ by $-(\mathbf{v} \cdot \nabla p)/\rho$. Then, the transfer to the environment by volume work emanates from the inner kinetic energy instead of kinetic energy.

We can also derive equations for the volume integrated energies

$$\frac{dE_{IK}}{dt} = \frac{d}{dt} \iiint_{V_F} \rho c_v T dV = -\iiint_{V_F} \rho L_v S_{l,v} dV - \iiint_{V_F} p\nabla \cdot \mathbf{v} dV$$

$$- \iiint_{V_F} \nabla \cdot \mathbf{j}_R dV \qquad (1.195)$$

$$\frac{dE_{IP}}{dt} = \frac{d}{dt} \iiint_{V_F} \rho L_v m_v dV = \iiint_{V_F} \rho L_v S_{l,v} dV \qquad (1.196)$$

$$\frac{dE_K}{dt} = \frac{d}{dt} \iiint_{V_F} \frac{1}{2}\rho \mathbf{v}^2 dV = -\iiint_{V_F} \rho \mathbf{v} \cdot \nabla \phi dV$$

$$- \iiint_{V_F} \mathbf{v} \cdot \nabla p dV \qquad (1.197)$$

$$\frac{dE_P}{dt} = \frac{d}{dt} \iiint_{V_F} \rho \phi dV = \iiint_{V_F} \rho \mathbf{v} \cdot \nabla \phi dV \qquad (1.198)$$

From these equations, we obtain the following total energy equation for the fluid volume:

$$\frac{dE_T}{dt} = \frac{d}{dt} \iiint_{V_F} \rho \left(c_v T + L_v m_v + \phi + \frac{1}{2}\mathbf{v}^2 \right) dV$$

$$= -\oiint_{\partial V_F} p\mathbf{v} \cdot \mathbf{n} da - \oiint_{\partial V_F} \mathbf{j}_R \cdot \mathbf{n} da \qquad (1.199)$$

If this volume extends over the complete atmosphere the volume pressure work on the right hand side vanishes since at the boundaries the normal velocity component becomes zero. Then, the total energy only changes in time due to the absorption of radiation. Without radiation total energy becomes a **globally conserved quantity**, i.e. the extensive quantity does not change with time. Equation (1.7) reveals that individually conserved quantities are globally conserved quantities at the same time. However, globally conserved quantities do not necessarily constitute individually conserved quantities as is obvious for total energy.

1.13 Microturbulent Exchange Processes and Boundary Layer Physics

Although the equation set (1.181)–(1.185) appears acceptable for large-scale atmospheric dynamics, it lacks important processes. The reason is that processes on smaller scales also influence the dynamics at the large scale. However, it is often feasible to describe this feedback in an adequate way without explicitly calculating the small-scale phenomena. The technique is called **parameterization**. It is possible to parameterize the impact of **microturbulence** (spatial scales between 1 cm and 100 m). **Convection** represents another important subscale process which can be assigned to the meso-γ scale (100 m–10 km). It is also possible to describe the feedback of convection on large-scale flows but the procedure is less straightforward and still debatable. Here, we only focus on microturbulence and postpone the parameterization of convection to a subsequent chapter.

The first step is to derive filtered equations which do not resolve microturbulent flows. Applying a moving time average to the field variables is a suitable filter method that is referred to as **Reynolds-averaging**. Before applying the method, we introduce a second-time coordinate $\tilde{t}$ which runs from $-\tau/2$ to $\tau/2$ where τ denotes a timescale that is much larger than the timescale of microturbulent flows. The time coordinate describes the preceding and subsequent evolution of the field at each time t. With this new time coordinate the Reynolds-averaged variable becomes

$$\bar{b}(\mathbf{r}, t) = \frac{1}{\tau} \int_{-\tau/2}^{\tau/2} b(\mathbf{r}, t, \tilde{t}) \, d\tilde{t} \tag{1.200}$$

By this average, all microturbulent eddies are smoothed out so that the variable describes only the large scale. The **eddy perturbation** of this average is denoted by b'. Obviously, we have:

$$\bar{b'} = \overline{b - \bar{b}} = 0 \tag{1.201}$$

and also, the relation

$$\frac{\partial \bar{b}}{\partial t} = \overline{\left(\frac{\partial b}{\partial t}\right)} \tag{1.202}$$

holds. Another useful time filter is the **Hesselberg average** which is a moving time average of mass-weighted field quantities. It reads

$$\hat{b}(\mathbf{r}, t) = \frac{\frac{1}{\tau} \int_{-\tau/2}^{\tau/2} \rho(\mathbf{r}, t, \tilde{t}) b(\mathbf{r}, t, \tilde{t}) \, d\tilde{t}}{\frac{1}{\tau} \int_{-\tau/2}^{\tau/2} \rho(\mathbf{r}, t, \tilde{t}) \, d\tilde{t}} = \frac{\overline{\rho b}}{\bar{\rho}} \tag{1.203}$$

The eddy perturbation from the Hesselberg average is denoted by b''. The advantage of the Hesselberg average is that Reynolds-averaged mass fluxes can be written like in the unfiltered equation. Then, the continuity equation (1.16) keeps its original form. It becomes

$$\frac{\partial \bar{\rho}}{\partial t} + \nabla \cdot (\bar{\rho} \hat{\mathbf{v}}) = 0 \tag{1.204}$$

On the other hand, Reynolds-averaging of the budget equation in flux form gives

$$\frac{\partial}{\partial t} (\bar{\rho} \hat{b}) + \nabla \cdot (\widehat{\bar{\rho} b \mathbf{v}}) = \bar{\rho} \hat{b} - \nabla \cdot \overline{\mathbf{j}_b} \tag{1.205}$$

where the advective flux term can be decomposed as follows:

$$b\mathbf{v} = (\hat{b} + b'')(\hat{\mathbf{v}} + \mathbf{v}'') = \hat{b}\hat{\mathbf{v}} + b''\hat{\mathbf{v}} + \hat{b}\mathbf{v}'' + b''\mathbf{v}'' \tag{1.206}$$

Applying the Hesselberg average yields

$$\widehat{b\mathbf{v}} = \hat{b}\hat{\mathbf{v}} + \widehat{b''\mathbf{v}''} \tag{1.207}$$

Therefore, the budget equation (1.205) becomes

$$\frac{\partial}{\partial t} (\bar{\rho} \hat{b}) + \nabla \cdot (\bar{\rho} \hat{b}\hat{\mathbf{v}}) = \bar{\rho} \hat{b} - \nabla \cdot (\bar{\rho} \widehat{b''\mathbf{v}''}) - \nabla \cdot \overline{\mathbf{j}_b} \tag{1.208}$$

and by the use of the continuity equation, we obtain the Reynolds-averaged budget equation in advection form

$$\frac{\partial \hat{b}}{\partial t} + \hat{\mathbf{v}} \cdot \nabla \hat{b} = \hat{b} - \frac{1}{\bar{\rho}} \nabla \cdot (\bar{\rho} \widehat{b''\mathbf{v}''}) - \frac{1}{\bar{\rho}} \nabla \cdot \overline{\mathbf{j}_b} \tag{1.209}$$

In this way, we can average all governing tendency equations and we end up with the following set of equations:

$$\frac{\partial \hat{\mathbf{v}}}{\partial t} + \hat{\mathbf{v}} \cdot \nabla \hat{\mathbf{v}} + 2\mathbf{\Omega} \times \hat{\mathbf{v}} = -\frac{1}{\bar{\rho}} \nabla \bar{p} - \nabla \phi - \frac{1}{\bar{\rho}} \nabla \cdot (\widehat{\bar{\rho} \mathbf{v}'' \mathbf{v}''}) \quad (1.210)$$

$$\frac{\partial \bar{\rho}}{\partial t} + \nabla \cdot (\bar{\rho} \hat{\mathbf{v}}) = 0 \quad (1.211)$$

$$\frac{\partial \hat{\theta}}{\partial t} + \hat{\mathbf{v}} \cdot \nabla \hat{\theta} = -\frac{L_v}{c_p \bar{\bar{\Pi}}} \hat{S}_{l,v} - \frac{1}{c_p \bar{\bar{\Pi}} \bar{\rho}} \nabla \cdot \bar{\mathbf{j}_R} - \frac{1}{\bar{\rho}} \nabla \cdot (\widehat{\bar{\rho} \mathbf{v}'' \theta''}) \quad (1.212)$$

$$\bar{\rho} = \frac{\bar{p}}{R_d \hat{\theta} \bar{\bar{\Pi}}} \quad (1.213)$$

$$\frac{\partial \hat{m}_v}{\partial t} + \hat{\mathbf{v}} \cdot \nabla \hat{m}_v = \hat{S}_{l,v} - \frac{1}{\bar{\rho}} \nabla \cdot (\widehat{\bar{\rho} \mathbf{v}'' m_v''}) \quad (1.214)$$

After averaging we have neglected eddy perturbations of Π since these perturbations are small compared to the absolute value, i.e. $|\Pi'| \ll \Pi$. Therefore, $\overline{(b/\Pi)} \approx \bar{b}/\bar{\Pi}$. We see that we have obtained an equation set that resembles the non-averaged governing equations (1.181), (1.182), (1.184), (1.185) and (1.186). However, the Reynolds-averaged equations have additional eddy terms that stem from the advective fluxes.

The eddy terms become relevant in the **planetary boundary layer** which extends from the surface to a variable height having the scale of 1 km. The planetary boundary layer is, as sketched in Fig. 1.13, subdivided into the **viscous sublayer** (typically within 1 m above the surface), the **surface layer** (1–100 m height) and the **Ekman layer** (100–1000 m height). Eddy exchange plays also an important role in convective clouds but in most cases, the turbulent exchange is small in the **free atmosphere** above the boundary layer.

To parameterize microturbulence it is useful to understand why it occurs at all. A strong wind gradient in the form of **wind shear** can occur in the boundary layer and also in convective clouds. The flow becomes dynamically unstable at a certain magnitude of wind shear. Then, small-scale vortices evolve which tend to reduce the shear. In Section 3.11.2, we will treat such instability in more detail since it can also occur at synoptic scales. The vortices also modify other

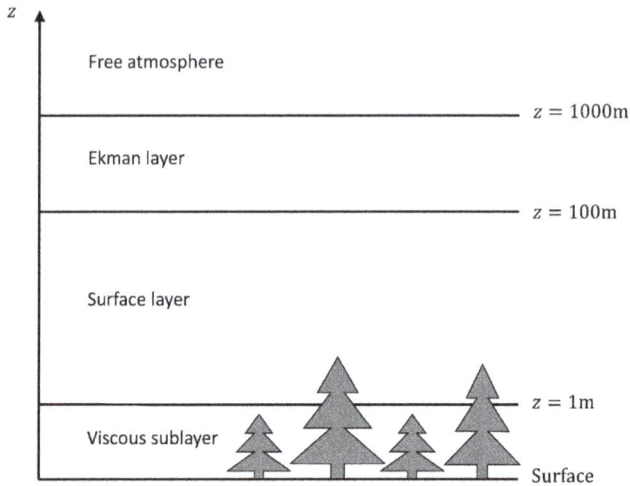

Fig. 1.13. Vertical structure of the planetary boundary layer. Note that the vertical axis is logarithmically scaled.

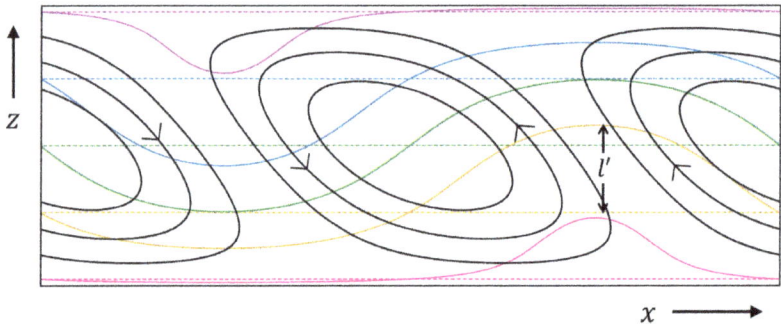

Fig. 1.14. Sketch of microturbulent eddies impacting on the field of a conserved fluid property. The dashed-colored lines are isolines of the Reynolds-averaged field while the solid-colored isolines refer to the perturbed field. The black lines show streamlines of the microturbulent eddy field.

hydrodynamic fields such that their gradients decline. A suitable parameterization method is based on the **mixing length theory** by Ludwig Prandtl (see Prandtl, 1942). In this theory, eddies have a typical length scale l' and cause deflections of fluid parcels of the same order. Figure 1.14 sketches an example for better understanding. Let b be a conserved field quantity that decreases in the coordinate direction perpendicular to the boundary, say the vertical direction z, at a constant rate $\partial \bar{b}/\partial z$. Then, we obtain for the perturbation after the deflection of the fluid parcel

$$b' = -l'\frac{\partial \bar{b}}{\partial z} \tag{1.215}$$

At the location where the parcel is deflected, we also have a velocity perturbation w' in the direction of the elongation. It is correlated with l' and, therefore, the Reynolds-averaged eddy flux becomes

$$\overline{w'b'} = -\overline{w'l'}\frac{\partial \bar{b}}{\partial z} = -K_b\frac{\partial \bar{b}}{\partial z} \quad \text{with} \quad K_b > 0 \tag{1.216}$$

The factor K_b constitutes the **eddy exchange coefficient** for the field property b. In this way, the microturbulent flux takes the form of a diffusive flux. Therefore, it is assumed that turbulence has a similar effect like diffusion induced by molecular motion. However, the exchange coefficient K_b is in contrast to molecular diffusion not a function of temperature only. It depends on the characteristics of the flow and the distance to the ground. One dependency can also be understood by the mixing length theory. Suppose the averaged flow is horizontal in coordinate direction x and increases at a constant rate $\partial \bar{u}/\partial z$ in coordinate direction z. In the case of horizontal homogeneity, the velocity component u is a conserved field quantity. Therefore, we find

$$u' = -l'\frac{\partial \bar{u}}{\partial z} \tag{1.217}$$

We see later in Section 3.11 that in shear flow instabilities the components u' and w' are anticorrelated for $\partial \bar{u}/\partial z > 0$ and correlated for $\partial \bar{u}/\partial z < 0$. Consequently, we obtain for the wind component w':

$$w' = r_v l' \left|\frac{\partial \bar{u}}{\partial z}\right| \tag{1.218}$$

where r_v denotes the amplitude ratio of vertical to horizontal wind perturbation. With this expression the eddy exchange coefficient becomes

$$K_b = l_m^2 \left|\frac{\partial \bar{u}}{\partial z}\right| \tag{1.219}$$

where $l_m = (r_v \overline{l'^2})^{1/2}$ is the **mixing length**. It is obvious that the eddy exchange coefficient is not a constant since the vertical shear $\partial \bar{u} / \partial z$ varies in space and time. Furthermore, the mixing length l_m also varies. It decreases close to solid boundaries and decreases in the case when the gravity force is exerted in a negative z direction and the vertical gradient of potential temperature, $\partial \bar{\theta} / \partial z$, increases. The latter results because the potential temperature of a rising air parcel remains approximately constant and, therefore, the buoyancy is negative due to the lower parcel temperature compared to the environment. This leads to a repelling force that inhibits vertical deflections.

The surface layer (see Fig. 1.13) is also known as the **Prandtl layer**. Here, we can assume that the frictional deceleration force at the surface (**surface drag**) becomes very important for the flow so that large wind shear arises. We can estimate the magnitude of the terms in the horizontal momentum equation for a horizontally uniform and steady boundary layer

$$\hat{w} \frac{\partial \hat{\mathbf{v}}_h}{\partial z} = \frac{1}{\bar{\rho}} \frac{\partial}{\partial z} (\widehat{\bar{\rho} w'' \mathbf{v}_h''}) \tag{1.220}$$

where $\hat{\mathbf{v}}_h$ denotes the horizontal wind vector which is parallel to the surface. In this equation we have neglected the Coriolis force since in the Prandtl layer it does not play an appreciable role. Furthermore, the density variations become very small so that we can assume a constant density. Then, the Hesselberg average coincides with the Reynolds average and we can write $\hat{\mathbf{v}}_h = \bar{\mathbf{v}}_h$. In the horizontally uniform steady-state boundary layer, we obtain for the continuity equation

$$\frac{\partial \bar{\rho}}{\partial t} + \frac{\partial}{\partial z} (\bar{\rho} \hat{w}) \approx \bar{\rho} \frac{\partial \bar{w}}{\partial z} = 0 \tag{1.221}$$

Therefore, the mean vertical velocity vanishes due to boundary condition (1.180) and we get from the momentum equation:

$$\frac{\partial}{\partial z} (\overline{w' \mathbf{v}_h'}) = 0 \tag{1.222}$$

This equation states that the vertical eddy momentum flux does not change in vertical direction. This is the reason why the Prandtl layer is often called the **constant flux layer**. Since the direction of $\bar{\mathbf{v}}_h$ cannot vary in the constant flux layer we can set $\bar{\mathbf{v}}_h = \bar{u}\mathbf{e}_x$. Then, the constant flux condition (1.222) becomes by using (1.216) and (1.219)

$$\frac{\partial}{\partial z}\left(l_m^2\left|\frac{\partial\bar{u}}{\partial z}\right|\frac{\partial\bar{u}}{\partial z}\right) = 0 \tag{1.223}$$

For a neutrally stratified boundary layer ($\partial\bar{\theta}/\partial z = 0$) the mixing length is approximately proportional to height so that

$$l_m = \kappa z \tag{1.224}$$

where κ denotes the **von Kármán constant** that has approximately the value 0.4. With this approach, we can solve Eq. (1.223) analytically. Integration yields

$$(\kappa z)^2\left|\frac{\partial\bar{u}}{\partial z}\right|\frac{\partial\bar{u}}{\partial z} = \text{const.} \tag{1.225}$$

This equation can be integrated again after taking the square root to yield the **logarithmic wind profile**

$$\bar{u} = \pm\frac{u_*}{\kappa}\ln\left(\frac{z}{z_0}\right) \tag{1.226}$$

where $u_* > 0$ denotes the **friction velocity**. The sign in front of u_* depends on the direction of the flow above the Prandtl layer. Figure 1.15 shows a graph of this profile. It can be seen that the wind vanishes at a finite height $z = z_0$ which is known as the **roughness length**. Near this height the surface layer turns into the **viscous sublayer** where the flow is highly influenced by the surface roughness and molecular exchange becomes non-negligible which causes fluxes of momentum and heat across the surface boundary. The roughness length depends highly on the surface properties. Table 1.1 lists some values revealing that the roughness length of a city center is more than 10^4 times larger than that of an open sea.

The horizontal viscous stress at the surface becomes $\boldsymbol{\tau}_h = \pm\bar{\rho}u_*^2\mathbf{e}_x$. It describes the **surface stress**. Often, the surface stress

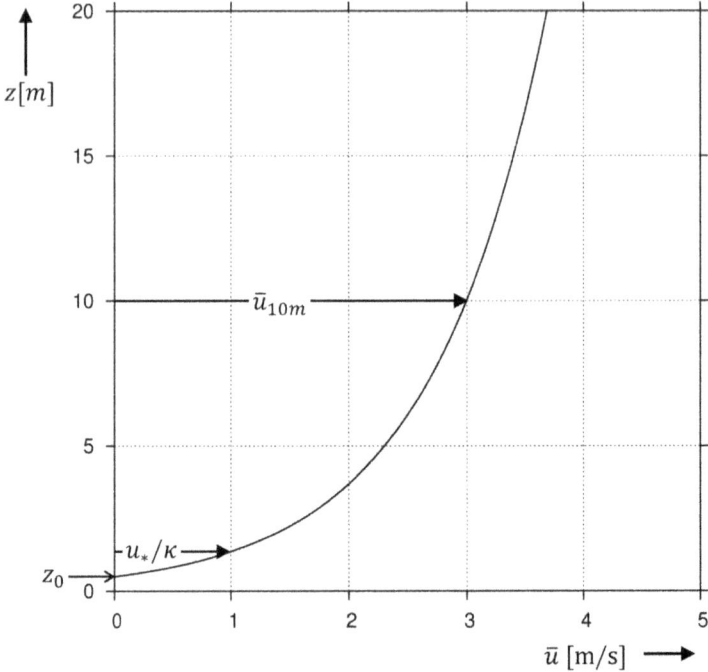

Fig. 1.15.   Logarithmic wind profile for $u_*/\kappa = 1\,\mathrm{m/s}$ and $z_0 = 0.5\,\mathrm{m}$.

Table 1.1.   Roughness length z_0 for various surface types.[g]

Surface type	Roughness length z_0 (m)
Open sea	0.0002
Open flat terrain (grass)	0.03
Low crops	0.1
High crops	0.25
Parkland, bushes, obstacles	0.5
Forest, regular large obstacles	1.0
City center	>2.0

is written as a function of wind at 10 m height in the form

$$\boldsymbol{\tau}_h = \bar{\rho} C_D |\bar{\mathbf{v}}_{h,10\,\mathrm{m}}| \bar{\mathbf{v}}_{h,10\,\mathrm{m}} \qquad (1.227)$$

---

[g] Adapted from WMO Guide to Meteorological Instruments and Methods of Observation WMO-No. 8, pp. I.5–12.

where C_D denotes the **drag coefficient** and the index 10 m refers to 10 m height above the surface. With the logarithmic wind profile (1.226) we find that the drag coefficient is related to the roughness length by

$$C_D = \left[\frac{\kappa}{\ln(10\,\mathrm{m}/z_0)} \right]^2 \tag{1.228}$$

We can calculate the vertical profile of any other conserved fluid property in the same way as for the horizontal wind u. For a conserved fluid property b we obtain

$$-\frac{\partial}{\partial z}(\overline{w'b'}) = \frac{\partial}{\partial z} \left(\kappa^2 z^2 \left| \frac{\partial \bar{u}}{\partial z} \right| \frac{\partial \bar{b}}{\partial z} \right) = 0 \tag{1.229}$$

and insertion of the solution for u gives

$$\frac{\partial}{\partial z} \left(u_* \kappa z \frac{\partial \bar{b}}{\partial z} \right) = 0 \tag{1.230}$$

Integration of this differential equation leads to the logarithmic profile:

$$\bar{b} = \frac{b_*}{\kappa} \ln \left(\frac{z}{z_0} \right) + b_s \tag{1.231}$$

where $b_* = (\overline{w'b'})/u_*$ and b_s are the characteristic and surface values of the property b, respectively.[h] The vertical diffusive flux can be expressed with values at 10 m height:

$$j_{b,z} = \bar{\rho}(\overline{w'b'}) = \bar{\rho} u_* b_* = \bar{\rho} C_b |\bar{\mathbf{v}}_{h,10\,\mathrm{m}}|(b_s - \bar{b}_{10\,\mathrm{m}}) \tag{1.232}$$

where C_b denotes the **surface transfer coefficient** for the property b. Comparison with (1.226) and (1.231) yields

$$C_b = C_D = \left[\frac{\kappa}{\ln(10\,\mathrm{m}/z_0)} \right]^2 \tag{1.233}$$

---

[h]Note that $\bar{b} = b_s$ is not fulfilled at the surface ($z = 0$) but at $z = z_0$. However, at and below $z = z_0$ the logarithmic profile is not appropriate since molecular exchange processes cannot be neglected anymore. Moreover, the viscous sublayer has in most cases a very small depth so that one can assume $\bar{b}(z_0) \approx b_s$.

Therefore, surface transfer coefficients and eddy exchange coefficients
do not depend upon the property in this constant flux boundary
layer model. However, observations reveal that this is not the case
in reality. Furthermore, the coefficients depend highly on the vertical gradient of potential temperature. A suitable description of this
dependency can be gained from the **Monin–Obukhov similarity
theory**. A relevant quantity is the **Monin–Obukhov** length:

$$\mathcal{L} = -\frac{\bar{\theta}}{\kappa g}\frac{u_*^3}{\overline{w'\theta'}} = \frac{\bar{\theta}}{\kappa g}\frac{u_*^3}{K_\theta \frac{\partial \bar{\theta}}{\partial z}} \tag{1.234}$$

where g denotes the vertical component of the gravity force which
is referred to as the **gravity acceleration**. The reciprocal value of
the Monin–Obukhov length measures the stability of stratification.
It influences the mixing length. For stable stratification ($\partial\bar{\theta}/\partial z > 0$)
deflections are hampered and the mixing length is reduced. In the
unstable case ($\partial\bar{\theta}/\partial z < 0$) the situation is vice versa. A suitable
description of surface layer profiles results from the following differential equations

$$\frac{\kappa z}{u_*}\frac{\partial |\bar{u}|}{\partial z} = \frac{\kappa z}{l_m} = \begin{cases} 1 + 4.7\dfrac{z}{\mathcal{L}} & \text{for } \mathcal{L} \geq 0 \\ \left(1 - 15\dfrac{z}{\mathcal{L}}\right)^{-1/4} & \text{for } \mathcal{L} < 0 \end{cases} \tag{1.235}$$

and

$$\frac{\kappa z}{\theta_*}\frac{\partial \bar{\theta}}{\partial z} = \frac{\kappa z}{l_\theta} = \begin{cases} 0.74 + 4.7\dfrac{z}{\mathcal{L}} & \text{for } \mathcal{L} \geq 0 \\ 0.74\left(1 - 9\dfrac{z}{\mathcal{L}}\right)^{-1/2} & \text{for } \mathcal{L} < 0 \end{cases} \tag{1.236}$$

where l_m and l_θ are the mixing lengths for momentum and potential
temperature, respectively. The validity of these differential equations
was verified in many field experiments (see Businger *et al.*, 1971). We
solve them for stable stratification ($\mathcal{L} > 0$). Then, we obtain for the
vertical profiles of horizontal wind and potential temperature

$$|\bar{u}| = \frac{u_*}{\kappa}\left[\ln\left(\frac{z}{z_0}\right) + 4.7\left(\frac{z}{\mathcal{L}} - \frac{z_0}{\mathcal{L}}\right)\right] \tag{1.237}$$

$$\bar{\theta} = \theta_s + \frac{\theta_*}{\kappa}\left[0.74\ln\left(\frac{z}{z_0}\right) + 4.7\left(\frac{z}{\mathcal{L}} - \frac{z_0}{\mathcal{L}}\right)\right] \tag{1.238}$$

From the profiles, we can deduce the surface transfer coefficients

$$C_D = \frac{\kappa^2}{\left[\ln\left(\frac{10\,\text{m}}{z_0}\right) + 4.7\frac{10\,\text{m}-z_0}{\mathcal{L}}\right]^2} \tag{1.239}$$

$$C_H = \frac{\kappa^2}{\left[\ln\left(\frac{10\,\text{m}}{z_0}\right) + 4.7\frac{10\,\text{m}-z_0}{\mathcal{L}}\right]\left[0.74\ln\left(\frac{10\,\text{m}}{z_0}\right) + 4.7\frac{10\,\text{m}-z_0}{\mathcal{L}}\right]} \tag{1.240}$$

where C_H is the surface transfer coefficient for **sensible heat** which is a more common denotation for inner kinetic energy. This coefficient regulates the turbulent potential temperature flux. For the eddy exchange coefficients, we get

$$K_m = l_m^2 \frac{\partial|\bar{u}|}{\partial z} = \frac{\kappa u_* z}{1 + 4.7z/\mathcal{L}} \tag{1.241}$$

$$K_\theta = l_m l_\theta \frac{\partial|\bar{u}|}{\partial z} = \frac{\kappa u_* z}{0.74 + 4.7\frac{z}{\mathcal{L}}} \tag{1.242}$$

where K_m denotes the eddy exchange coefficient for momentum. The profiles and exchange coefficients coincide in the neutral case ($1/\mathcal{L} = 0$) with those found in the previous consideration except for the factor 0.74 in Eqs. (1.236), (1.238) and (1.242). Therefore, the eddy exchange coefficient is higher for potential temperature than for horizontal wind.

The ratio of eddy exchange coefficients for momentum and temperature is called **turbulent Prandtl number** and it takes approximately the value Pr ≈ 0.74 in the presented example. For both quantities, we note a decrease in surface transfer with increasing static stability which reflects the fact that a stable stratification suppresses microturbulence. Figure 1.16 demonstrates the effect of stratification on vertical profiles of horizontal wind. Evidently, increasing stability leads to smaller vertical wind shears. This has the consequence that winds near the surface increase with decreasing stability of stratification.

Weather forecasts and general circulation models cannot resolve the surface layer since it has too small a thickness. One can only

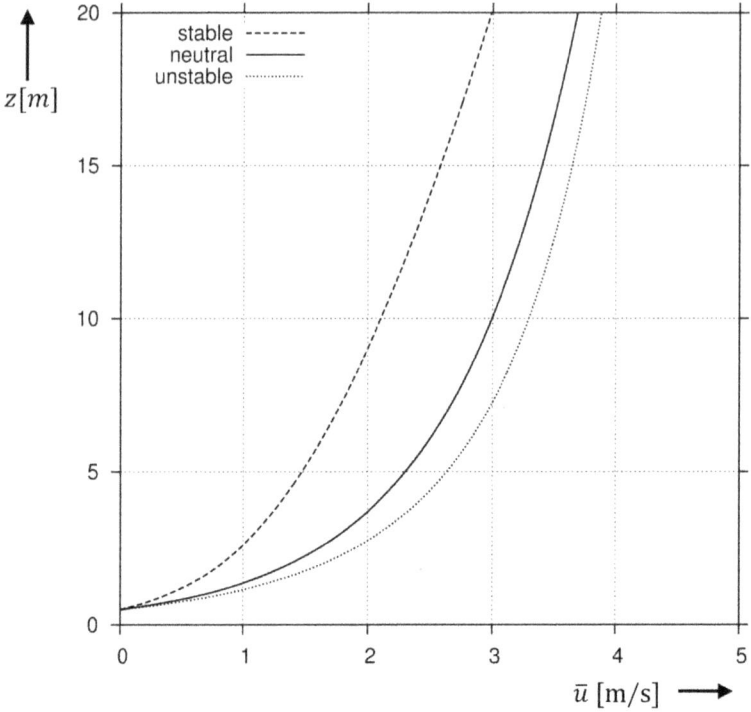

Fig. 1.16. Vertical profiles of horizontal wind in the surface layer for stable strat-ification ($\mathcal{L} = 50\,\text{m}$, dashed line), neutral stratification (solid line) and unstable stratification ($\mathcal{L} = -50\,\text{m}$, dotted line). All other parameters are as in Fig. 1.15 and the three profiles meet at $z = 40\,\text{m}$.

prescribe boundary conditions for the vertical turbulent fluxes at the surface according to the above-mentioned constant flux turbulence model. With the surface transfer coefficients, we can formulate the boundary conditions as follows:

$$(\widetilde{w''\mathbf{v}''}_h)_s = C_D|\hat{\mathbf{v}}_{h,10\,\text{m}}|\hat{\mathbf{v}}_{h,10\,\text{m}} \tag{1.243}$$

$$(\widetilde{w''\theta''})_s = C_H|\hat{\mathbf{v}}_{h,10\,\text{m}}|(\theta_s - \hat{\theta}_{10\,\text{m}}) \tag{1.244}$$

$$(\widetilde{w''m_v''})_s = C_E|\hat{\mathbf{v}}_{h,10\,\text{m}}|(m_{v,s} - \hat{m}_{v,10\,\text{m}}) \tag{1.245}$$

where the index s denotes evaluation at the surface and C_E the sur-face transfer coefficient for water vapor which regulates evaporation.

Note that these fluxes are strictly speaking not turbulent fluxes at the surface since only molecular exchange can take place there. However, they convert into turbulent fluxes above the viscous sublayer. For the determination of surface fluxes, one must know the fields at 10 m and the Monin–Obukhov length. In practical modeling, these values are estimated with numerical approximations.

The mixing length approach is also suitable above the surface layer where the so-called **Ekman layer** starts which extends vertically to a height of about 1 km and more. There, we cannot neglect the Coriolis force anymore and the mixing length is not proportional to height. The mixing length formula suggested by Blackadar (1962) describes a reasonable transition between the surface and the Ekman layers:

$$l_m = \frac{\kappa z}{1 + \frac{\kappa z}{100\,\text{m}}} \tag{1.246}$$

Therefore, the mixing length approaches the constant value $l_m = 100$ m with increasing height. In nonneutral boundary layers, the mixing length must be corrected by a factor that accommodates the effect of temperature stratification.

In the surface layer, the gradients of the hydrodynamic fields are virtually directed perpendicular to the surface. Therefore, turbulent fluxes depend only on one spatial coordinate. This may not be appropriate at larger heights. Especially, vertical motion in convective clouds provokes significant turbulent fluxes in all coordinate directions. A reasonable approach seems to set these fluxes proportional to the gradient of the corresponding fluid property. However, we note that the **Reynolds stress tensor** $\widehat{v''v''}$ is symmetric while the gradient $\nabla \hat{\mathbf{v}}$ is an asymmetric tensor, because $\partial \hat{v}_j / \partial x_k \neq \partial \hat{v}_k / \partial x_j$ holds in general where j and k indicate the Cartesian coordinate directions. Instead, a parameterization like for molecular viscosity in the Navier–Stokes momentum equation does not lead to this contradiction. It is given by

$$\widehat{v''v''} = -K_m \left(\nabla \hat{\mathbf{v}} + \nabla \hat{\mathbf{v}}^T - \frac{2}{3} \mathbf{I} \, \nabla \cdot \hat{\mathbf{v}} \right) \tag{1.247}$$

where $\mathbf{I}$ denotes the unit tensor. The tensor $\nabla \hat{\mathbf{v}} + \nabla \hat{\mathbf{v}}^T - 2/3 \, \mathbf{I} \, \nabla \cdot \hat{\mathbf{v}}$ is referred to as the deformation tensor since it attains nonzero elements when the fluid volume suffers deformation apart from isotropic compression. The latter results from the diagonal mean which vanishes in parameterization (1.247). Volume shrinking and expansion by these stresses are entirely caused by pressure.[i] The divergence term $2/3\,\mathbf{I}\,\nabla\cdot\hat{\mathbf{v}}$ can be neglected in most applications since the divergence of the three-dimensional flow is usually very small. We see that this approach reduces to (1.216) in the surface layer since $\widehat{w''u''} = -K_m(\partial\hat{w}/\partial x + \partial\hat{u}/\partial z) = -K_m\partial\hat{u}/\partial z$ in this case. However, in the general case, horizontal shear of vertical wind also contributes to the generation of turbulent fluxes. The fluxes of conserved scalar fluid properties are parameterized with a **gradient closure**. Then, we have

$$\widehat{\mathbf{v}''\theta''} = -K_\theta\nabla\hat{\theta} \tag{1.248}$$

$$\widehat{\mathbf{v}''m_v''} = -K_{m_v}\nabla\hat{m}_v \tag{1.249}$$

In this way, the microturbulent exchange terms resemble those for molecular viscosity, heat conduction and diffusion. Nevertheless, there are two crucial differences. First, the exchange coefficients depend highly on the flow field. Secondly, the temperature flux is proportional to the gradient of potential temperature and not to the temperature itself. This has the consequence that strong turbulent mixing generates an isentropic layer ($\hat{\theta}$ = const.) instead of an isothermal layer ($\hat{T}$ = const.) as it would be the case when only molecular heat conduction takes place. Therefore, turbulent heat exchange does not necessarily be in accordance with the second law of thermodynamics. On the other hand, one can formulate a microturbulent parameterization scheme that complies with energy conservation in a closed system. By multiplying the Reynolds-averaged momentum equation with $\hat{\mathbf{v}}$ and converting to the flux form we get

$$\frac{\partial}{\partial t}\left(\frac{\bar{\rho}}{2}\hat{\mathbf{v}}^2\right) + \nabla\cdot\left(\hat{\mathbf{v}}\frac{\bar{\rho}}{2}\hat{\mathbf{v}}^2 + \mathbf{J}_T\cdot\hat{\mathbf{v}}\right) = -\hat{\mathbf{v}}\cdot(\nabla\bar{p} + \bar{\rho}\nabla\phi) + \mathbf{J}_T : \nabla\hat{\mathbf{v}} \tag{1.250}$$

---

[i]The Navier–Stokes equations in general form also include a bulk viscosity that yields viscous isotropic normal stresses. However, the bulk viscosity does not play an important role in atmospheric fluid dynamics.

where $\mathbf{J}_T = \overline{\bar{\rho}\mathbf{v}''\mathbf{v}''}$ denotes the Reynolds stress tensor. The second term on the right-hand side of this equation constitutes the transfer of kinetic energy to microturbulent eddies. Since the kinetic energy of the turbulent flow cannot accumulate without limitation, there must also be an energy sink. This sink can only arise from dissipation by molecular viscosity. Therefore, it is reasonable to associate the transfer term with dissipative heating. Then, we obtain for the Reynolds-averaged equation for inner energy:

$$\frac{\partial}{\partial t}(\bar{\rho}c_v\hat{T}) + \nabla \cdot (\hat{\mathbf{v}}\bar{\rho}c_v\hat{T}) = -L_v\hat{S}_{l,v} - \bar{p}\nabla \cdot \hat{\mathbf{v}}$$

$$- \mathbf{J}_T : \nabla\hat{\mathbf{v}} + c_p\bar{\Pi}\nabla \cdot (\bar{\rho}K_\theta\nabla\hat{\theta}) - \nabla \cdot \overline{\mathbf{j}_R}$$

$$(1.251)$$

It remains to show that the dissipative heating is positive. We get

$$-\mathbf{J}_T : \nabla\hat{\mathbf{v}}$$

$$= K_m\left(\nabla\hat{\mathbf{v}} + \nabla\hat{\mathbf{v}}^T - \frac{2}{3}\mathbf{I}\nabla \cdot \hat{\mathbf{v}}\right) : \nabla\hat{\mathbf{v}}$$

$$= \frac{K_m}{2}\left(\nabla\hat{\mathbf{v}} + \nabla\hat{\mathbf{v}}^T - \frac{2}{3}\mathbf{I}\nabla \cdot \hat{\mathbf{v}}\right) : (\nabla\hat{\mathbf{v}} + \nabla\hat{\mathbf{v}}^T + \nabla\hat{\mathbf{v}} - \nabla\hat{\mathbf{v}}^T)$$

$$= \frac{K_m}{2}\left(\nabla\hat{\mathbf{v}} + \nabla\hat{\mathbf{v}}^T - \frac{2}{3}\mathbf{I}\nabla \cdot \hat{\mathbf{v}}\right) : (\nabla\hat{\mathbf{v}} + \nabla\hat{\mathbf{v}}^T)$$

$$= \frac{K_m}{2}\left(\nabla\hat{\mathbf{v}} + \nabla\hat{\mathbf{v}}^T - \frac{2}{3}\mathbf{I}\nabla \cdot \hat{\mathbf{v}}\right) : \left(\nabla\hat{\mathbf{v}} + \nabla\hat{\mathbf{v}}^T - \frac{2}{3}\mathbf{I}\nabla \cdot \hat{\mathbf{v}}\right) > 0$$

$$(1.252)$$

The third identity in (1.252) holds because the double dot product between a symmetric and an asymmetric tensor vanishes. The last identity in (1.252) results from the vanishing trace of the Reynolds stress tensor. Figure 1.17 shows the energy conversions and transfers when the microturbulent exchange is considered. We see that for an energetically closed system, total energy is conserved.[j]

One may postulate a second law of thermodynamics for Reynolds-averaged quantities. However, the microturbulent parameterization

---

[j]By slightly altering the scheme one can ensure exact energy conversion, e.g. by shifting Π behind the nabla operator in the turbulent flux term.

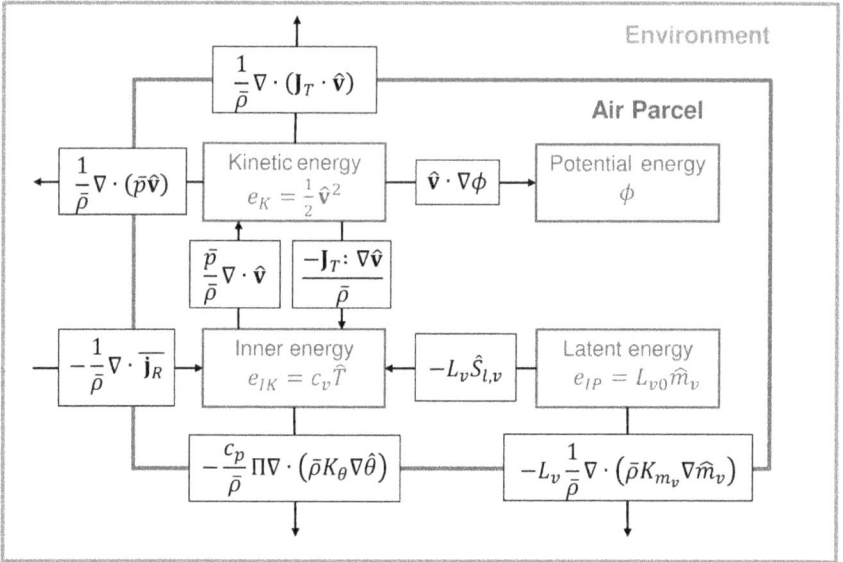

Fig. 1.17.   Energy conversions in a fluid parcel and energy transfers to the surroundings when microturbulent exchange is included.

does not necessarily comply with a second law of thermodynamics. In the simplified case without moisture and radiation, the production of Reynolds-averaged entropy becomes

$$\hat{s}_i = K_m \frac{\left(\nabla\hat{\mathbf{v}} + \nabla\hat{\mathbf{v}}^T - \frac{2}{3}\mathbf{I}\nabla\cdot\hat{\mathbf{v}}\right):\nabla\hat{\mathbf{v}}}{\bar{\rho}\hat{T}} + K_\theta\frac{\Pi\nabla\hat{\theta}\cdot\nabla\hat{T}}{\bar{\rho}\hat{T}^2} \qquad (1.253)$$

While we have shown that the first term on the right-hand side is always positive, we cannot assert this for the second one. For stable stratification $(\partial\hat{\theta}/\partial z > 0)$ and a negative vertical gradient of temperature $(\partial\hat{T}/\partial z < 0)$, the second term becomes negative. This is, however, the usual case in the troposphere. The first term must exceed the amplitude of the second term to ensure positive entropy production but it is not clear that this is always satisfied.

Finally, the eddy exchange coefficients K_m, K_θ and K_{m_v} must be determined for a complete parameterization scheme. A suitable approach was introduced by Lilly (1962) which is based on energetical grounds. In his scheme, the eddy exchange coefficient takes the form

$$K_m = l_m^2 \Sigma\sqrt{1 - \text{PrRi}} \qquad (1.254)$$

where $\Sigma = [1/2(\nabla\hat{\mathbf{v}} + \nabla\hat{\mathbf{v}}^T - 2/3\mathbf{I}\nabla\cdot\hat{\mathbf{v}}):(\nabla\hat{\mathbf{v}} + \nabla\hat{\mathbf{v}}^T - 2/3\mathbf{I}\nabla\cdot\hat{\mathbf{v}})]^{\frac{1}{2}}$ and means the square root of the double dot product of the deformation tensor and Ri the **Richardson number** which is defined by

$$\text{Ri} = \frac{g}{\bar{\hat{\theta}}}\frac{\partial\hat{\theta}}{\partial z}\Big/\Sigma^2 \tag{1.255}$$

With this parameterization K_m decreases with increasing stability of stratification and actually vanishes for $\partial\hat{\theta}/\partial z > \Sigma^2\hat{\theta}/(g\text{Pr})$. Then, microturbulence breaks down completely.

The parameterization of microturbulence by Lilly (1962) also provides reasonable results for mesoscale flows including clouds. However, inside clouds, the Richardson number should be modified by replacing potential temperature with equivalent potential temperature because the latter quantity is approximately conserved in this case and its gradient is a measure for the stability of the stratification. For large-scale flows, the parameterization can be simplified by neglecting the horizontal derivatives since the horizontal microturbulent exchange has nearly no impact on the dynamics. Then, the prognostic governing equations become

$$\frac{\partial\hat{\mathbf{v}}}{\partial t} + \hat{\mathbf{v}}\cdot\nabla\hat{\mathbf{v}} + 2\mathbf{\Omega}\times\hat{\mathbf{v}} = -\frac{1}{\bar{\rho}}\nabla\bar{p} - \nabla\phi + \frac{1}{\bar{\rho}}\frac{\partial}{\partial z}\left(\bar{\rho}K_m\frac{\partial\hat{\mathbf{v}}_h}{\partial z}\right) \tag{1.256}$$

$$\frac{\partial\hat{\theta}}{\partial t} + \hat{\mathbf{v}}\cdot\nabla\hat{\theta} = -\frac{L_v}{c_p\bar{\Pi}}\hat{S}_{l,v} - \frac{1}{c_p\bar{\Pi}\bar{\rho}}\nabla\cdot\overline{\mathbf{j}_R} + \frac{1}{\bar{\rho}}\frac{\partial}{\partial z}\left(\bar{\rho}K_\theta\frac{\partial\hat{\theta}}{\partial z}\right) \tag{1.257}$$

$$\frac{\partial\hat{m}_v}{\partial t} + \hat{\mathbf{v}}\cdot\nabla\hat{m}_v = \hat{S}_{l,v} + \frac{1}{\bar{\rho}}\frac{\partial}{\partial z}\left(\bar{\rho}K_{m_v}\frac{\partial\hat{m}_v}{\partial z}\right) \tag{1.258}$$

In (1.256), we have neglected the vertical component of the turbulence term since it contributes marginally to the momentum budget of large-scale flows. The parameterization for the eddy exchange coefficients reduces in the case of neglected horizontal variations to

$$K_m = l_m^2\left|\frac{\partial\hat{\mathbf{v}}_h}{\partial z}\right|\sqrt{1 - \text{PrRi}} \tag{1.259}$$

$$K_\theta = \text{Pr}^{-1}l_m^2\left|\frac{\partial\hat{\mathbf{v}}_h}{\partial z}\right|\sqrt{1 - \text{PrRi}} \tag{1.260}$$

where the Richardson number becomes

$$\mathrm{Ri} = \frac{g}{\hat{\theta}} \frac{\partial \hat{\theta}}{\partial z} \left| \frac{\partial \hat{\mathbf{v}}_h}{\partial z} \right|^{-2} \qquad (1.261)$$

It is reasonable to assume that the eddy exchange coefficients for potential temperature and water vapor are identical so that $K_{m_v} = K_\theta$. The boundary conditions (1.243)–(1.245) can be used to describe the turbulent eddy fluxes at the surface. In the subsequent chapters, we assume that all variables are Reynolds or Hesselberg averages. Therefore, the averaging symbols $^-$ and $\hat{\ }$ do not appear anymore.

1.14 Vortex Dynamics

Vortices dominate large-scale weather in the mid-latitudes. Therefore, it is reasonable to consider equations which describe the dynamics of vortices better than the governing equations derived in the previous sections. One well-known quantity to measure the intensity of vortices is the **circulation** which is defined by the following line integral:

$$C = \oint_{\partial a} \mathbf{v} \cdot \mathbf{t} \, dl \qquad (1.262)$$

where ∂a denotes the boundary of the area a, $\mathbf{t}$ the tangent unit vector and dl the differential line element (cf. Eq. (A.73)). For simplicity, we restrict the analysis to planar surfaces but the subsequent argumentation also holds for curved surfaces like the surface of a sphere. Figure 1.18 sketches that a closed streamline yields a nontrivial circulation when the integral is applied along this streamline. Anticlockwise and clockwise flow directions are referred to as **cyclonic circulation** $(C > 0)$ and **anticyclonic circulation** $(C < 0)$, respectively. However, other flow configurations like a simple shear flow also have nonzero circulation. By Stokes' law, we get another formula for the circulation:

$$C = \iint_a (\nabla \times \mathbf{v}) \cdot \mathbf{n} \, da \qquad (1.263)$$

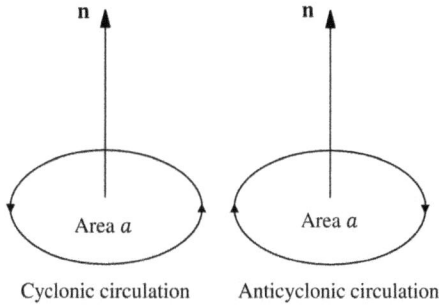

Cyclonic circulation   Anticyclonic circulation

Fig. 1.18.   Sketch of cyclonic and anticyclonic circulations. The circulation direction refers to the direction of the surface normal inside of the streamline.

The curl of the wind field constitutes the **vorticity vector** $\xi = \nabla \times \mathbf{v}$. Let us assume that the surface normal directs parallel to $\mathbf{e}_z$. Then, the circulation becomes

$$C = \iint_a \xi \cdot \mathbf{e}_z da = \iint_a \xi_z da = \iint_a \frac{\partial v}{\partial x} - \frac{\partial u}{\partial y} da \qquad (1.264)$$

The vertical component of the vorticity vector, ξ_z is referred to as **relative vorticity** in atmospheric sciences. It can be solely derived from the **stream function** ψ of the wind field by using the Helmholtz decomposition (A.54):

$$\xi_z = \frac{\partial v}{\partial x} - \frac{\partial u}{\partial y} = \frac{\partial}{\partial x}\left(\frac{\partial \psi}{\partial x} + \frac{\partial \chi}{\partial y}\right) - \frac{\partial}{\partial y}\left(-\frac{\partial \psi}{\partial y} + \frac{\partial \chi}{\partial x}\right) = \nabla_h^2 \psi$$
$$(1.265)$$

where $\nabla_h^2 = \partial^2/\partial x^2 + \partial^2/\partial y^2$ denotes the horizontal Laplace operator.

Applying Gauss's divergence theorem in two dimensions to (1.264) yields another expression for the circulation:

$$C = \iint_a \nabla_h \cdot (\nabla_h \psi) da = \oint_{\partial a} \nabla_h \psi \cdot (\mathbf{t} \times \mathbf{e}_z)\, dl \qquad (1.266)$$

Since $\mathbf{t} \times \mathbf{e}_z$ points outward we can conclude that the streamfunction decreases (increases) inward of a closed streamfunction contour if the circulation is cyclonic (anticyclonic).

Circulation is also related to **angular momentum**. In the case of a circular area having its center at the origin, it can be shown by using rule (A.17) that

$$C = \oint_{\partial a} \mathbf{v} \cdot \mathbf{t}\, dl = \oint_{\partial a} \mathbf{v} \cdot \mathbf{t}\, r d\alpha = \oint_{\partial a} \mathbf{v} \cdot (\mathbf{e}_z \times \mathbf{r}) d\alpha$$

$$= \oint_{\partial a} (\mathbf{r} \times \mathbf{v}) \cdot \mathbf{e}_z d\alpha \qquad (1.267)$$

where r denotes the radius of the circle and α the angle with respect to one horizontal axis so that the differential arc length becomes $dl = r d\alpha$. Consequently, the circulation coincides in this case with the azimuthally integrated vertical component of specific angular momentum $\mathbf{m} = \mathbf{r} \times \mathbf{v}$.

To diagnose vortex intensification, it is necessary to determine the time derivative of the circulation. We obtain for the time derivative in the case that the boundary of the area is a material curve

$$\frac{dC}{dt} = \oint_{\partial a} \frac{D\mathbf{v}}{Dt} \cdot \mathbf{t}\, dl + \oint_{\partial a} \mathbf{v} \cdot \frac{D(\mathbf{t}\, dl)}{Dt}$$

$$= \oint_{\partial a} \frac{D\mathbf{v}}{Dt} \cdot \mathbf{t}\, dl + \oint_{\partial a} \mathbf{v} \cdot d\mathbf{v} \qquad (1.268)$$

The last conversion is obtained because $\mathbf{t}\, dl = d\mathbf{r}$ describes the differential vector difference vector between two points along the material curve. As sketched in Fig. 1.19 it changes in time by the difference of the velocity vectors. The second closed line integral vanishes since

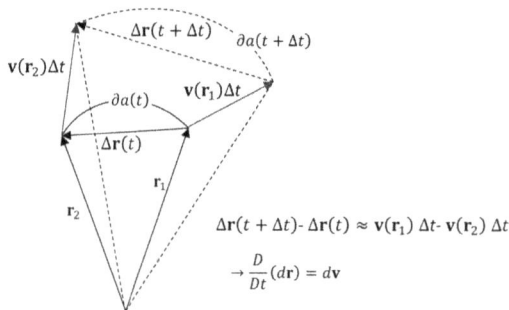

$$\Delta\mathbf{r}(t + \Delta t) - \Delta\mathbf{r}(t) \approx \mathbf{v}(\mathbf{r}_1)\, \Delta t - \mathbf{v}(\mathbf{r}_2)\, \Delta t$$

$$\rightarrow \frac{D}{Dt}(d\mathbf{r}) = d\mathbf{v}$$

Fig. 1.19.   Sketch of the proof that the material time derivative of $\mathbf{t}\, dl$ yields $d\mathbf{v}$ (see text).

$\mathbf{v} \cdot d\mathbf{v} = d\mathbf{v}^2/2$. Therefore, we get

$$\frac{dC}{dt} = \oint_{\partial a} \frac{D\mathbf{v}}{Dt} \cdot d\mathbf{r} \qquad (1.269)$$

and by using the momentum equation (1.256) we find

$$\frac{dC}{dt} = -\oint_{\partial a} 2\mathbf{\Omega} \times \mathbf{v} \cdot d\mathbf{r} - \oint_{\partial a} \frac{1}{\rho}\nabla p \cdot d\mathbf{r} - \oint_{\partial a} \nabla\phi \cdot d\mathbf{r} + \oint_{\partial a} \mathbf{f}_T \cdot d\mathbf{r} \qquad (1.270)$$

where $\mathbf{f}_T$ denotes the force that results from the microturbulent exchange of momentum. The third closed line integral vanishes because $\nabla\phi \cdot d\mathbf{r} = d\phi$.

The first line integral on the right-hand side of (1.270) describes the impact of the Coriolis force on the circulation. It can be converted into a more convenient form by applying the rule (A.17) for the triple-product

$$\oint_{\partial a} 2\mathbf{\Omega} \times \mathbf{v} \cdot d\mathbf{r} = \oint_{\partial a} 2\mathbf{\Omega} \cdot \mathbf{v} \times d\mathbf{r} = \oint_{\partial a} 2\mathbf{\Omega} \cdot \mathbf{n}_a v_\perp dl \qquad (1.271)$$

where $\mathbf{n}_a$ denotes the unit vector in the direction of $\mathbf{v} \times d\mathbf{r}$ and $v_\perp$ the velocity component perpendicular to $\mathbf{t}$. $v_\perp dl$ yields the area change per time unit so that $2\mathbf{\Omega} \cdot \mathbf{n}_a v_\perp dl = 2\mathbf{\Omega} \cdot \mathbf{n}_a d(da)/dt$. Let us assume the simplified case when the vertical wind component can be neglected and the area is flat with the surface normal directing parallel to the vertical unit vector. Then, $\mathbf{n}_a = \mathbf{e}_z$ and (1.271) becomes due to Gauss's divergence theorem in two dimensions

$$\oint_{\partial a} 2\mathbf{\Omega} \times \mathbf{v} \cdot d\mathbf{r} = 2\mathbf{\Omega} \cdot \mathbf{e}_z \iint_{a_F} \nabla_h \cdot \mathbf{v}_h da = 2\mathbf{\Omega} \cdot \mathbf{e}_z \frac{da_F}{dt} \qquad (1.272)$$

where a_F is the area content that results from the distortion of the boundary line by the horizontal flow. We see that the vortex can intensify when the planetary rotation axis has a component in the vertical direction and the area shrinks. This happens due to the Coriolis force which points to the tangential direction for a flow toward the center of the area. The spin-up by this force is the main mechanism for the formation of extratropical and tropical cyclones.

We can also write the second line integral on the right-hand side of Eq. (1.270) in another form. Using Stokes' theorem and rule (A.49) we get

$$\oint_{\partial a} \frac{1}{\rho} \nabla p \cdot d\mathbf{r} = \iint_a \nabla \times \left(\frac{1}{\rho} \nabla p \right) \cdot \mathbf{e}_z da$$

$$= - \iint_a \frac{1}{\rho^2} (\nabla \rho \times \nabla p) \cdot \mathbf{e}_z da \qquad (1.273)$$

Therefore, the pressure gradient force changes the circulation if isolines of density and pressure do not run parallel. In this case, the fluid is **baroclinic**. On the other hand, the fluid is **barotropic** if the density is only a function of pressure, i.e. $\rho = \rho(p)$. Then, the integral (1.273) vanishes since $\nabla \rho = d\rho/dp \nabla p$. The integrand of the integral is a measure for **baroclinicity**. The baroclinicity can also be measured by **solenoids**. The solenoids form little parallelograms bounded by lines with constant pressure and specific volume (see Fig. 1.20). If the pressure and specific volume differences between those lines are Δp and Δv, respectively, we obtain for the integral over the solenoid:

$$\oint_{\partial a} \frac{1}{\rho} \nabla p \cdot d\mathbf{r} = \oint_{\partial a} v \, dp = \Delta p \Delta v \qquad (1.274)$$

One must distinguish between positive solenoids causing a circulation increase and negative ones causing a circulation decrease. The contributions of positive and negative solenoids cancel each other. The more solenoids fit into the area of interest the higher is the time change of circulation due to baroclinicity.

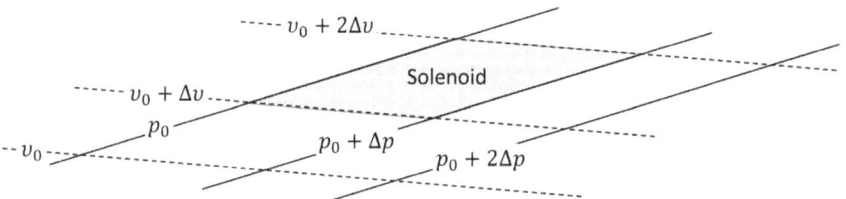

Fig. 1.20. Sketch of a solenoid. These take the form of a parallelogram for small enough differences Δp and Δv.

Rewriting the circulation tendency equation by using (1.272) and (1.274) yields **Bjerknes' circulation theorem** (Bjerknes, 1898):

$$\frac{dC}{dt} = -2\mathbf{\Omega} \cdot \mathbf{e}_z \frac{da_F}{dt} - \oint_{\partial a_F} \upsilon\, dp + \oint_{\partial a_F} \mathbf{f}_T \cdot d\mathbf{r} \qquad (1.275)$$

Friction due to microturbulent exchange (last term on the right-hand side of Eq. (1.275)) usually decreases the circulation.

Bjerknes' circulation theorem reduces to **Kelvin's theorem** (Thomson, 1867) in the case of a barotropic and frictionless flow ($\mathbf{f}_T = 0$). Then, the **absolute circulation** $C_a = C + 2\mathbf{\Omega} \cdot \mathbf{e}_z a_F$ is conserved, i.e.

$$\frac{dC_a}{dt} = 0 \qquad (1.276)$$

Kelvin's theorem corresponds to angular momentum conservation for circular areas and axisymmetric flows.

For a more detailed analysis of vortex dynamics, it is useful to consider the **vorticity equation** which was first derived by Helmholtz (1858) for a homogenous fluid. It results after the application of the curl operator to the equation of motion (1.256). We obtain

$$\frac{\partial \boldsymbol{\xi}}{\partial t} + \mathbf{v} \cdot \nabla \boldsymbol{\xi} + (2\mathbf{\Omega} + \boldsymbol{\xi})\nabla \cdot \mathbf{v} - (2\mathbf{\Omega} + \boldsymbol{\xi}) \cdot \nabla \mathbf{v} = \frac{\nabla \rho \times \nabla p}{\rho^2} + \nabla \times \mathbf{f}_T$$
$$(1.277)$$

For the derivation, we made use of the calculation rules (A.57) and (A.58). The identity

$$\mathbf{v} \cdot \nabla \mathbf{v} = \frac{1}{2}\nabla(\mathbf{v} \cdot \mathbf{v}) - \mathbf{v} \times \boldsymbol{\xi} \qquad (1.278)$$

results from (A.57). Since the curl of a gradient and the divergence of a curl vanish, we find by using (A.58) that

$$-\nabla \times (\mathbf{v} \cdot \nabla \mathbf{v}) = \nabla \times (\mathbf{v} \times \boldsymbol{\xi}) = -\boldsymbol{\xi}\nabla \cdot \mathbf{v} + \boldsymbol{\xi} \cdot \nabla \mathbf{v} - \mathbf{v} \cdot \nabla \boldsymbol{\xi}$$
$$(1.279)$$

We can write the vorticity equation in a more convenient way by recognizing that the vorticity of Earth's rotation becomes

$$\nabla \times \mathbf{v}_E = \nabla \times (\mathbf{\Omega} \times \mathbf{r}) = \mathbf{\Omega}\nabla \cdot \mathbf{r} - \mathbf{\Omega} \cdot \nabla \mathbf{r} = 2\mathbf{\Omega} \qquad (1.280)$$

Therefore, the vorticity of the flow in the nonrotating absolute frame takes the form

$$\boldsymbol{\zeta} = \boldsymbol{\xi} + 2\boldsymbol{\Omega} \tag{1.281}$$

This vorticity constitutes the **absolute vorticity vector** while $\boldsymbol{\xi}_E = 2\boldsymbol{\Omega}$ is referred to as the **planetary vorticity vector**. Using this definition, the vorticity equation can be recast in

$$\frac{D\boldsymbol{\zeta}}{Dt} = \boldsymbol{\zeta} \cdot \nabla \mathbf{v} - \boldsymbol{\zeta} \nabla \cdot \mathbf{v} + \frac{\nabla \rho \times \nabla p}{\rho^2} + \nabla \times \mathbf{f}_T \tag{1.282}$$

The last two terms on the right-hand side describe the change of vorticity by baroclinicity and microturbulent exchange, respectively. They have corresponding counterparts in Bjerknes' circulation theorem (see Eq. (1.278)). To understand the first two terms on the right-hand side of the vorticity equation it is useful to consider a local coordinate system where the vertical axis directs parallel to the absolute vorticity vector. Then, we get

$$\boldsymbol{\zeta} \cdot \nabla \mathbf{v} - \boldsymbol{\zeta} \nabla \cdot \mathbf{v} = \zeta_z \left(\frac{\partial u}{\partial z} \mathbf{e}_x + \frac{\partial v}{\partial z} \mathbf{e}_y \right) - \zeta_z \left(\frac{\partial u}{\partial x} + \frac{\partial v}{\partial y} \right) \mathbf{e}_z \tag{1.283}$$

Note that terms including vertical velocity cancel each other out in this case. The last term on the right-hand side is related to the circulation change by the shrinking of the area enclosed by a material line. Indeed, by assuming vertical independence we can recover Bjerknes' circulation theorem by integrating over an area enclosed with a material line and using Reynolds transport theorem in two dimensions.

The first term on the right hand describes the tilting of the vorticity vector in the horizontal direction by the vertical shear. Let us assume that only the x-component has vertical shear and that there is no horizontal divergence. Then, the vorticity change becomes approximately after the time step Δt:

$$\Delta\boldsymbol{\zeta} \approx \zeta_z \frac{\partial u}{\partial z} \mathbf{e}_x \Delta t \tag{1.284}$$

At the same time, the fluid parcel moved at $z = z + \Delta z$ over the distance $\Delta x \approx \partial u/\partial z \Delta z \Delta t$ in x-direction relative to the parcel at the height z. Hence, we obtain for the tilt angle α of the material line

$$\alpha = \tan^{-1}\left(\frac{\partial u}{\partial z}\Delta t\right) = \tan^{-1}\left(\frac{\Delta \zeta_x}{\zeta_z}\right) \tag{1.285}$$

Therefore, the tilting of the vorticity vector coincides with the tilting of the material line. In this context, it is appropriate to introduce the term **vortex filament**. A vortex filament is a field line of the vorticity vector field, i.e. a filament results from the integration[k]

$$\mathbf{l}(l) = \mathbf{l}(0) + \int_0^l \zeta dl' \tag{1.286}$$

Vortex filaments are material lines in a frictionless, barotropic and nondivergent fluid ($\mathbf{f}_T = 0$, $\nabla \rho \times \nabla p = 0$, $\nabla \cdot \mathbf{v} = 0$). This becomes evident by considering the time evolution of a small material line element $\Delta \mathbf{l}$. To first order we have

$$\Delta \mathbf{l}(t + \Delta t) \approx \Delta \mathbf{l}(t) + \Delta \mathbf{v}(t)\Delta t \tag{1.287}$$

The difference in the velocity vector in space results from the directional derivative

$$\Delta \mathbf{v} \approx \Delta \mathbf{l} \cdot \nabla \mathbf{v} \tag{1.288}$$

Therefore, we obtain

$$\frac{\Delta \mathbf{l}(t + \Delta t) - \Delta \mathbf{l}(t)}{\Delta t} - \Delta \mathbf{l} \cdot \nabla \mathbf{v} \approx 0 \tag{1.289}$$

In the limit of infinitesimal line elements and time increments, this relation turns into the identity

$$\frac{D(d\mathbf{l})}{Dt} = d\mathbf{l} \cdot \nabla \mathbf{v} \tag{1.290}$$

The vorticity vector ζ is parallel to $d\mathbf{l}$ in the case that the material line forms a vortex filament because $d\mathbf{l} = \zeta dl$. Therefore, the vector

---

[k]Note that the curve parameter l has the physical dimension ms.

product $\boldsymbol{\zeta} \times d\mathbf{l}$ vanishes and we get for a frictionless, barotropic and nondivergent fluid due to (1.282):

$$\frac{D}{Dt}(\boldsymbol{\zeta} \times d\mathbf{l}) = (\boldsymbol{\zeta} \cdot \nabla \mathbf{v}) \times d\mathbf{l} + \boldsymbol{\zeta} \times (d\mathbf{l} \cdot \nabla \mathbf{v})$$

$$= dl\,(\boldsymbol{\zeta} \cdot \nabla \mathbf{v}) \times \boldsymbol{\zeta} + dl\,\boldsymbol{\zeta} \times (\boldsymbol{\zeta} \cdot \nabla \mathbf{v}) = \mathbf{0} \quad (1.291)$$

Hence, the vorticity vector stays parallel to the material line element and the vortex filament represents a material line in this case.

A **vortex tube** is a volume for which vortex filaments are parallel to the lateral surfaces of the volume. Due to Gauss's divergence theorem, the circulation of the cutting areas at the beginning and end of the volume must be identical since the vorticity vector field has no divergence. Figure 1.21(a) sketches such a vortex tube while Fig. 1.21(b) shows a photograph of a tornado funnel cloud which demonstrates impressively the deformation of a vortex tube and the material line character of a vortex filament.

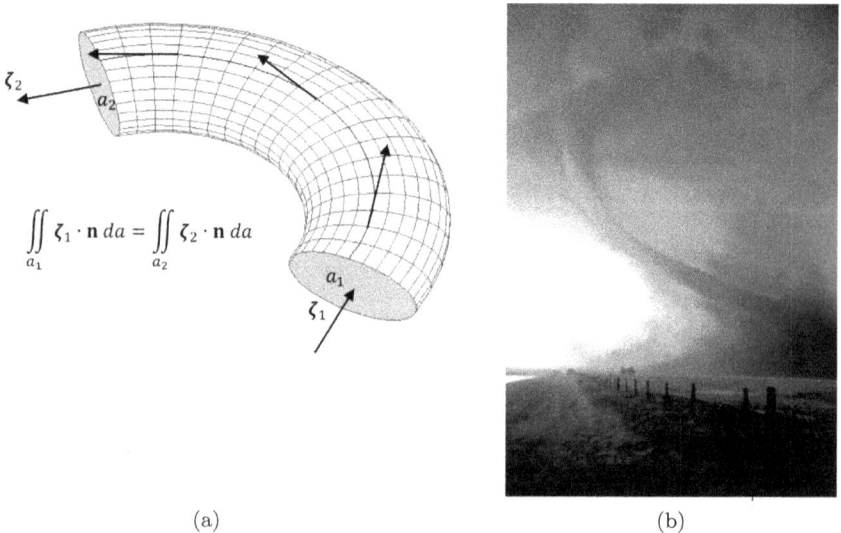

(a)   (b)

Fig. 1.21.   (a) Sketch of a vortex tube. The black line on the surface shows a vortex filament. (b) Photograph of a tornado funnel cloud. The rotating flow is parallel to the surface of the cloud.

Source: (b) https://www.noaa.gov/media/digital-library-photo/pl23nssl0052jpg.

The vorticity vector mainly points in horizontal direction in the atmosphere since the vertical shear is much larger than the horizontal shear. Therefore, one should use another equation to describe the dynamics of synoptic scale vortices having their rotation axes in vertical direction. A useful quantity to describe such vortices is the **potential vorticity (PV)**. First, we define this quantity and derive a PV theorem before its meaning is explained. PV combines the vorticity vector $\boldsymbol{\zeta}$ with a field function $b = b(\mathbf{r}, t)$. It is defined by

$$P_b = \frac{\boldsymbol{\zeta} \cdot \nabla b}{\rho} \qquad (1.292)$$

Therefore, PV is given by the scalar product between the vorticity vector $\boldsymbol{\zeta}$ and the gradient of b divided by the density ρ.

To derive the PV theorem, we assume that the field quantity b is governed by the budget equation.

$$\frac{Db}{Dt} = \dot{b} \qquad (1.293)$$

Applying the directional derivative $\boldsymbol{\zeta} \cdot \nabla$ to (1.293) yields[1]

$$\boldsymbol{\zeta} \cdot \frac{D\nabla b}{Dt} + (\boldsymbol{\zeta} \cdot \nabla \mathbf{v}) \cdot \nabla b = \boldsymbol{\zeta} \cdot \nabla \dot{b} \qquad (1.294)$$

On the other hand, eliminating the divergence in the vorticity equation (1.282) by using the continuity equation (1.23) leads to

$$\frac{D\boldsymbol{\zeta}}{Dt} - \frac{\boldsymbol{\zeta}}{\rho}\frac{D\rho}{Dt} = \rho\frac{D}{Dt}\left(\frac{\boldsymbol{\zeta}}{\rho}\right) = \boldsymbol{\zeta} \cdot \nabla \mathbf{v} + \frac{\nabla\rho \times \nabla p}{\rho^2} + \nabla \times \mathbf{f}_T \qquad (1.295)$$

We find the PV theorem by taking the scalar product between (1.295) and ∇b and combining the result with (1.294)

$$\frac{DP_b}{Dt} = \frac{\boldsymbol{\zeta} \cdot \nabla \dot{b}}{\rho} + \frac{(\nabla\rho \times \nabla p) \cdot \nabla b}{\rho^3} + \frac{(\nabla \times \mathbf{f}_T) \cdot \nabla b}{\rho} \qquad (1.296)$$

---

[1]Note that $(\boldsymbol{\zeta} \cdot \nabla \mathbf{v}) \cdot \nabla b$ does not coincide with $\boldsymbol{\zeta} \cdot (\nabla \mathbf{v} \cdot \nabla b)$.

Ertel derived this theorem in 1942 (Ertel, 1942) and for this reason P_b is sometimes called **Ertel PV**. From the theorem, we can deduce that PV is a conserved quantity if

(1) the field quantity b represents a conserved fluid property,
(2) the fluid is barotropic or b can be written as a function of density ρ and pressure p, i.e. $b = b(\rho, p)$,
(3) friction by microturbulent exchange can be neglected.

It is obvious that the first and third conditions cannot exactly be fulfilled in the real atmosphere. However, synoptic-scale flows above the boundary layer can be approximately considered as being frictionless and adiabatic over a short period.

Let us assume that all conditions are met. Then, PV is a conserved quantity. Moreover, we assume that the field property b has a stratified structure, i.e. isosurfaces of b draw small angles with the horizontal plane. For example, temperature has a stratified structure since the vertical component of the gradient largely exceeds the horizontal components in its absolute value. Figure 1.22 outlines the consequences of PV conservation. Shown is a vortex tube between two isosurfaces of b. The vertical component of vorticity mainly contributes to the scalar product $\zeta \cdot \nabla b$ since the field has a stratified structure. The isosurfaces move vertically in the subsequent development so that the distance between them increases. Since the

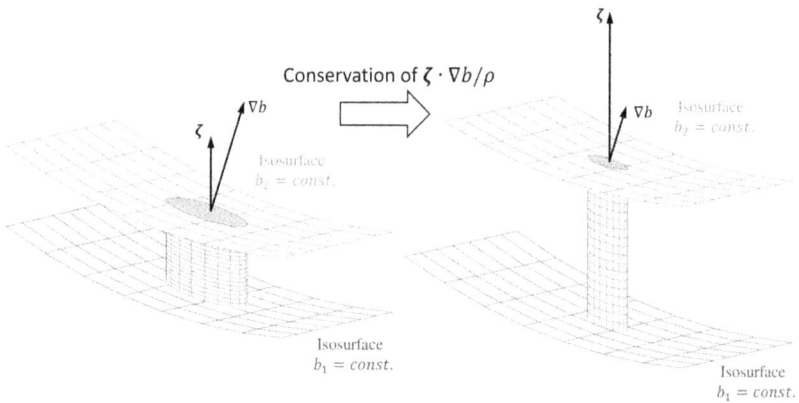

Fig. 1.22.   Sketch of a vortex tube embedded between two isosurfaces of a conserved fluid property and the consequence of PV conservation for the vorticity vector when the isosurfaces increase their distance from each other.

isosurfaces are material the vortex tube stretches and its diameter shrinks due to mass conservation. The vorticity must increase according to Kelvin's theorem because solenoids cannot appear on the isosurface when $b = b(\rho, p)$. Indeed, PV conservation leads to the same outcome. The gradient of b decreases by the vertical movement of the isosurfaces and the vorticity vector $\boldsymbol{\zeta}$ must elongate to keep PV constant.

Two possibilities are of interest for the choice of the field quantity b. The first one is to use it as a coordinate and the second one is to identify b with a thermodynamic quantity.

In the first case, we obtain a budget equation for the vorticity component of the selected coordinate. The vertical component has the largest relevance for studying synoptic-scale scale weather systems. By choosing $b = z^{\mathrm{m}}$ we get from the PV theorem (1.296)

$$\frac{D}{Dt}\left(\frac{\zeta_z}{\rho}\right) = \frac{\boldsymbol{\zeta} \cdot \nabla w}{\rho} + \frac{(\nabla \rho \times \nabla p) \cdot \mathbf{e}_z}{\rho^3} + \frac{(\nabla \times \mathbf{f}_T) \cdot \mathbf{e}_z}{\rho} \quad (1.297)$$

With the continuity equation (1.16), this equation can be converted to the vorticity equation for the vertical component:

$$\frac{\partial \zeta_z}{\partial t} + \nabla \cdot (\mathbf{v}\zeta_z) = \boldsymbol{\zeta} \cdot \nabla w + \frac{(\nabla \rho \times \nabla p) \cdot \mathbf{e}_z}{\rho^2} + (\nabla \times \mathbf{f}_T) \cdot \mathbf{e}_z \quad (1.298)$$

which is equivalent to

$$\frac{D\zeta_z}{Dt} = -\zeta_z \nabla \cdot \mathbf{v} + \boldsymbol{\zeta} \cdot \nabla w + \frac{(\nabla \rho \times \nabla p) \cdot \mathbf{e}_z}{\rho^2} + (\nabla \times \mathbf{f}_T) \cdot \mathbf{e}_z \quad (1.299)$$

The same equation would result if we had simply taken the vertical component of (1.282). In a Cartesian coordinate frame, this equation can be written in the form

$$\frac{D\zeta_z}{Dt} = \underbrace{-\zeta_z \nabla_h \cdot \mathbf{v}_h}_{\substack{\text{Stretching} \\ \text{term}}} + \underbrace{\frac{\partial u}{\partial z}\frac{\partial w}{\partial y} - \frac{\partial v}{\partial z}\frac{\partial w}{\partial x}}_{\text{Tilting term}}$$

$$+ \underbrace{\frac{\partial v}{\partial x}\frac{\partial p}{\partial y} - \frac{\partial v}{\partial y}\frac{\partial p}{\partial x}}_{\text{Solenoid term}} + \underbrace{\frac{\partial f_{Ty}}{\partial x} - \frac{\partial f_{Tx}}{\partial y}}_{\text{Friction term}} \quad (1.300)$$

---

[m]More appropriate for large scale atmospheric dynamics are spherical coordinates which will be treated in Chapter 2.

We have five different terms that change the vertical component of the absolute vorticity of a fluid parcel. The **stretching term** describes the increase of vorticity by stretching of a vortex tube. By this process, the cross-section area decreases and the vorticity increases due to Kelvin's circulation theorem as sketched in Fig. 1.22. The **tilting term** results from the tilting of a vortex filament horizontally or vertically by the wind shear (see the real case example in Fig. 1.21). The **solenoid term** yields the vorticity change by the baroclinicity in the horizontal plane. Crossings of isobars and density isolines lead to a vorticity change (see Fig. 1.20). The **friction term** describes the change of vorticity by microturbulent exchange of horizontal momentum. All terms are included in Bjerknes' circulation theorem except for the tilting term. This is because the divergence term in (1.275) only results with neglected vertical velocity which is, on the other hand, necessary for the tilting.

The field b is a thermodynamic state function in the second case. It has the advantage that it can often be expressed as a function of pressure and density only, at least approximately. In this case, the solenoid term vanishes in the PV theorem. Many applications of PV employ potential temperature θ since it is approximately a conserved quantity. With $b = \theta$ the PV theorem becomes

$$\frac{DP_\theta}{Dt} = \frac{\boldsymbol{\zeta} \cdot \nabla\dot{\theta}}{\rho} + \frac{(\nabla \times \mathbf{f}_T) \cdot \nabla\theta}{\rho} \qquad (1.301)$$

where $P_\theta = \boldsymbol{\zeta} \cdot \nabla\theta/\rho$. PV based on potential temperature has much significance for the dynamics of extratropical weather systems. Especially, the **invertibility principle** states that it is possible to deduce approximately the horizontal wind field by only knowing the PV distribution and the potential temperature at the surface.

Saturated equivalent potential temperature θ_e represents another useful field quantity to substitute b. θ_e is approximately conserved in saturated air and it depends only on pressure and temperature if one neglects the liquid water content as in the pseudoadiabatic case. P_{θ_e} is a suitable diagnostic in cloud bands, fronts and tropical cyclones. In these flow systems, air often rises slantwise along θ_e isosurfaces.

Chapter 2

Scale Analysis, Coordinate Systems and the Primitive Equations

2.1 Introduction

To better understand the dynamics of atmospheric weather systems, it is useful to simplify the governing equations by neglecting unimportant terms. Scale analysis serves to estimate the magnitude of the various terms and lays the foundation for simpler equations. Further simplification and a better manageability can be reached by introducing an alternative vertical coordinate. Before doing this, we must formulate the governing equations in a suitable coordinate system. The spherical coordinates are most appropriate because of the spherical shape of the atmosphere.

2.2 Formulation of the Governing Equations in Spherical Coordinates

The **spherical coordinates** form an orthogonal curvilinear coordinate system (see Section A.3.6) and are illustrated in Fig. 2.1. They comprise **longitude** λ, **latitude** φ and **radius** r. Longitude constitutes the angle between the location vector projected on the horizontal plane and the x-axis while latitude forms the angle between the location vector and the horizontal plane. The radius is just the length of the location vector. From geometrical considerations, we

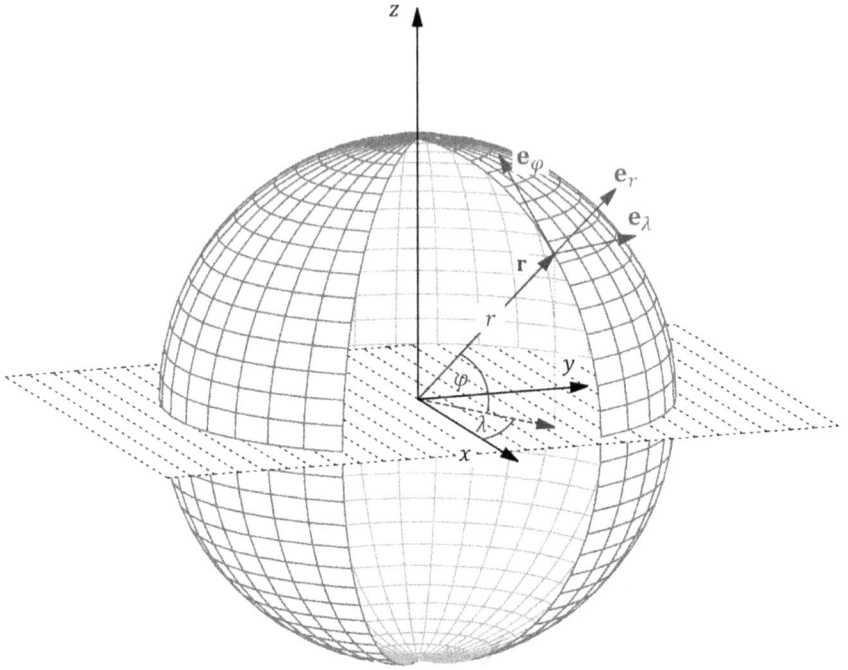

Fig. 2.1.   Spherical coordinates.

find the following transformation equations:

$$\lambda\left(x,y,z\right) = \tan^{-1}\left(\frac{y}{x}\right) \tag{2.1}$$

$$\varphi\left(x,y,z\right) = \tan^{-1}\left(\frac{z}{\sqrt{x^2+y^2}}\right) \tag{2.2}$$

$$r\left(x,y,z\right) = \sqrt{x^2+y^2+z^2} \tag{2.3}$$

It is also possible to recover the Cartesian coordinates from spherical coordinates by the relations

$$x\left(\lambda,\varphi,r\right) = r\cos\lambda\cos\varphi \tag{2.4}$$

$$y\left(\lambda,\varphi,r\right) = r\sin\lambda\cos\varphi \tag{2.5}$$

$$z\left(\lambda,\varphi,r\right) = r\sin\varphi \tag{2.6}$$

Using (2.1)–(2.6), we obtain for the gradient of the spherical coordinates

$$\nabla \lambda = \nabla \left[\tan^{-1} \left(\frac{y}{x} \right) \right] = \frac{\frac{1}{x}\mathbf{e}_y - \frac{y}{x^2}\mathbf{e}_x}{1 + \left(\frac{y}{x} \right)^2} = \frac{x\mathbf{e}_y - y\mathbf{e}_x}{x^2 + y^2}$$

$$= \frac{\cos \lambda \mathbf{e}_y - \sin \lambda \mathbf{e}_x}{r \cos \varphi} \tag{2.7}$$

$$\nabla \varphi = \nabla \left[\tan^{-1} \left(\frac{z}{\sqrt{x^2 + y^2}} \right) \right] = \frac{\frac{\mathbf{e}_z}{\sqrt{x^2+y^2}} - \frac{zx\mathbf{e}_x + zy\mathbf{e}_y}{\sqrt{x^2+y^2}^3}}{1 + \frac{z^2}{x^2+y^2}}$$

$$= \frac{\cos \varphi \mathbf{e}_z - \cos \lambda \sin \varphi \mathbf{e}_x - \sin \lambda \sin \varphi \mathbf{e}_y}{r} \tag{2.8}$$

$$\nabla r = \nabla \left(\sqrt{x^2 + y^2 + z^2} \right) = \frac{x\mathbf{e}_x + y\mathbf{e}_y + z\mathbf{e}_z}{r}$$

$$= \cos \lambda \cos \varphi \mathbf{e}_x + \sin \lambda \cos \varphi \mathbf{e}_y + \sin \varphi \mathbf{e}_z \tag{2.9}$$

We can use these results to calculate the Lamé coefficients [cf. Eq. (A.63)]

$$h_\lambda = \frac{1}{|\nabla \lambda|} = r \cos \varphi, \ h_\varphi = \frac{1}{|\nabla \varphi|} = r, \ h_r = \frac{1}{|\nabla r|} = 1 \tag{2.10}$$

and the unit vectors [cf. Eq. (A.62)]

$$\mathbf{e}_\lambda = \frac{\nabla \lambda}{|\nabla \lambda|} = -\sin \lambda \mathbf{e}_x + \cos \lambda \mathbf{e}_y \tag{2.11}$$

$$\mathbf{e}_\varphi = \frac{\nabla \varphi}{|\nabla \varphi|} = -\cos \lambda \sin \varphi \mathbf{e}_x - \sin \lambda \sin \varphi \mathbf{e}_y + \cos \varphi \mathbf{e}_z \tag{2.12}$$

$$\mathbf{e}_r = \frac{\nabla r}{|\nabla r|} = \cos \lambda \cos \varphi \mathbf{e}_x + \sin \lambda \cos \varphi \mathbf{e}_y + \sin \varphi \mathbf{e}_z \tag{2.13}$$

It is easy to verify that these unit vectors form an orthogonal basis. In the following, we make use of the notation $\mathbf{i} = \mathbf{e}_\lambda$, $\mathbf{j} = \mathbf{e}_\varphi$ and $\mathbf{k} = \mathbf{e}_r$. The partial derivatives of the unit vectors are sometimes needed. They become

$$\frac{\partial \mathbf{i}}{\partial \lambda} = \sin \varphi \mathbf{j} - \cos \varphi \mathbf{k}, \quad \frac{\partial \mathbf{i}}{\partial \varphi} = 0, \quad \frac{\partial \mathbf{i}}{\partial r} = 0 \qquad (2.14)$$

$$\frac{\partial \mathbf{j}}{\partial \lambda} = -\sin \varphi \mathbf{i}, \quad \frac{\partial \mathbf{j}}{\partial \varphi} = -\mathbf{k}, \quad \frac{\partial \mathbf{j}}{\partial r} = 0 \qquad (2.15)$$

$$\frac{\partial \mathbf{k}}{\partial \lambda} = \cos \varphi \mathbf{i}, \quad \frac{\partial \mathbf{k}}{\partial \varphi} = \mathbf{j}, \quad \frac{\partial \mathbf{k}}{\partial r} = 0 \qquad (2.16)$$

The components of the velocity vector can be determined in spherical coordinates by applying the material time derivative to the location vector

$$\mathbf{v} = \frac{D\mathbf{r}}{Dt} = \frac{D}{Dt}(r\mathbf{k}) = \frac{Dr}{Dt}\mathbf{k} + r\frac{D\mathbf{k}}{Dt} \qquad (2.17)$$

The second term in (2.17) does not vanish since the unit vectors of spherical coordinates depend on location. We obtain

$$\frac{D\mathbf{k}}{Dt} = \frac{D\lambda}{Dt}\frac{\partial \mathbf{k}}{\partial \lambda} + \frac{D\varphi}{Dt}\frac{\partial \mathbf{k}}{\partial \varphi} + \frac{Dr}{Dt}\frac{\partial \mathbf{k}}{\partial r} = \cos \varphi \frac{D\lambda}{Dt}\mathbf{i} + \frac{D\varphi}{Dt}\mathbf{j} \qquad (2.18)$$

The latter identity can be verified with (2.16). Finally, the velocity vector becomes

$$\mathbf{v} = r\cos\varphi\frac{D\lambda}{Dt}\mathbf{i} + r\frac{D\varphi}{Dt}\mathbf{j} + \frac{Dr}{Dt}\mathbf{k} \equiv u\mathbf{i} + v\mathbf{j} + w\mathbf{k} \qquad (2.19)$$

In this equation, we have introduced the velocity components u, v and w which, in meteorology, are referred to as the **zonal wind**, **meridional wind** and **vertical wind**, respectively.

The equation of motion contains the acceleration vector. We must apply the material time derivative to (2.19) to obtain its components in spherical coordinates

$$\frac{D\mathbf{v}}{Dt} = \frac{Du}{Dt}\mathbf{i} + \frac{Dv}{Dt}\mathbf{j} + \frac{Dw}{Dt}\mathbf{k} + u\frac{D\mathbf{i}}{Dt} + v\frac{D\mathbf{j}}{Dt} + w\frac{D\mathbf{k}}{Dt}$$

$$= \frac{Du}{Dt}\mathbf{i} + \frac{Dv}{Dt}\mathbf{j} + \frac{Dw}{Dt}\mathbf{k} + u\left(\frac{\partial \mathbf{i}}{\partial \lambda}\frac{D\lambda}{Dt} + \frac{\partial \mathbf{i}}{\partial \varphi}\frac{D\varphi}{Dt}\right)$$

$$+ v \left(\frac{\partial \mathbf{j}}{\partial \lambda} \frac{D\lambda}{Dt} + \frac{\partial \mathbf{j}}{\partial \varphi} \frac{D\varphi}{Dt} \right) + w \left(\frac{\partial \mathbf{k}}{\partial \lambda} \frac{D\lambda}{Dt} + \frac{\partial \mathbf{k}}{\partial \varphi} \frac{D\varphi}{Dt} \right)$$

$$= \frac{Du}{Dt}\mathbf{i} + \frac{Dv}{Dt}\mathbf{j} + \frac{Dw}{Dt}\mathbf{k} + u \left(\sin \varphi \mathbf{j} - \cos \varphi \mathbf{k} \right) \frac{D\lambda}{Dt}$$

$$- v \left(\sin \varphi \frac{D\lambda}{Dt}\mathbf{i} + \frac{D\varphi}{Dt}\mathbf{k} \right) + w \left(\cos \varphi \frac{D\lambda}{Dt}\mathbf{i} + \frac{D\varphi}{Dt}\mathbf{j} \right) \quad (2.20)$$

This expression can be solely formulated in terms of the components u, v and w by considering Eq. (2.19). We get

$$\frac{D\mathbf{v}}{Dt} = \left(\frac{Du}{Dt} - \tan \varphi \frac{uv}{r} + \frac{uw}{r} \right) \mathbf{i} + \left(\frac{Dv}{Dt} + \tan \varphi \frac{u^2}{r} + \frac{vw}{r} \right) \mathbf{j}$$

$$+ \left(\frac{Dw}{Dt} - \frac{u^2 + v^2}{r} \right) \mathbf{k} \quad (2.21)$$

We see that additional acceleration terms arise besides the material derivative of the velocity components. These result from the curvature of the coordinate lines. A parcel following such a coordinate line executes accelerations. If, for example, a parcel moves along a latitude circle with a constant speed, i.e. $u = \text{const.}, v = 0, w = 0$, then the parcel executes a **centripetal acceleration** with the magnitude $u^2/(r \cos \varphi)$. The meridional and vertical components of this acceleration are $u^2/r \tan \varphi$ and $-u^2/r$, respectively, which accords with (2.21).

The components of the angular velocity vector $\mathbf{\Omega}$ are needed to determine the Coriolis force components in spherical coordinates. In the Cartesian coordinate system, this vector is given by

$$\mathbf{\Omega} = \Omega \mathbf{e}_z \quad (2.22)$$

With (2.12) and (2.13), it is easy to show that $\mathbf{e}_z = \cos \varphi \mathbf{j} + \sin \varphi \mathbf{k}$. Therefore,

$$\mathbf{\Omega} = \Omega \cos \varphi \mathbf{j} + \Omega \sin \varphi \mathbf{k} \equiv \frac{f_\varphi}{2}\mathbf{j} + \frac{f_r}{2}\mathbf{k} \quad (2.23)$$

where $f_\varphi = 2\Omega \cos \varphi$ and $f_r = 2\Omega \sin \varphi$ denote the **Coriolis param-eters**. With this notation, we obtain for the Coriolis force

$$
\begin{aligned}
2\Omega \times \mathbf{v} &= (f_\varphi \mathbf{j} + f_r \mathbf{k}) \times (u\mathbf{i} + v\mathbf{j} + w\mathbf{k}) \\
&= f_\varphi u\mathbf{j} \times \mathbf{i} + f_\varphi w\mathbf{j} \times \mathbf{k} + f_r u\mathbf{k} \times \mathbf{i} + f_r v\mathbf{k} \times \mathbf{j} \\
&= (f_\varphi w - f_r v)\,\mathbf{i} + f_r u\mathbf{j} - f_\varphi u\mathbf{k}
\end{aligned}
\tag{2.24}
$$

The differential operators gradient, divergence and curl can be determined in spherical coordinates with the Lamé coefficients (2.10). They become [cf. (A.64), (A.66) and (A.69)]

$$
\begin{aligned}
\nabla b &= \frac{1}{h_\lambda}\frac{\partial b}{\partial \lambda}\mathbf{i} + \frac{1}{h_\varphi}\frac{\partial b}{\partial \varphi}\mathbf{j} + \frac{1}{h_r}\frac{\partial b}{\partial r}\mathbf{k} \\
&= \frac{1}{r\cos\varphi}\frac{\partial b}{\partial \lambda}\mathbf{i} + \frac{1}{r}\frac{\partial b}{\partial \varphi}\mathbf{j} + \frac{\partial b}{\partial r}\mathbf{k}
\end{aligned}
\tag{2.25}
$$

$$
\begin{aligned}
\nabla \cdot \mathbf{a} &= \frac{1}{h_\lambda h_\varphi h_r}\left[\frac{\partial}{\partial \lambda}(h_\varphi h_r a_\lambda) + \frac{\partial}{\partial \varphi}(h_\lambda h_r a_\varphi) + \frac{\partial}{\partial r}(h_\lambda h_\varphi a_r)\right] \\
&= \frac{1}{r\cos\varphi}\frac{\partial a_\lambda}{\partial \lambda} + \frac{1}{r\cos\varphi}\frac{\partial}{\partial \varphi}(\cos\varphi a_\varphi) + \frac{1}{r^2}\frac{\partial}{\partial r}(r^2 a_r) \quad (2.26)
\end{aligned}
$$

$$
\begin{aligned}
\nabla \times \mathbf{a} &= \frac{1}{h_\varphi h_r}\left[\frac{\partial}{\partial \varphi}(a_r h_r) - \frac{\partial}{\partial r}(a_\varphi h_\varphi)\right]\mathbf{i} \\
&\quad + \frac{1}{h_\lambda h_r}\left[\frac{\partial}{\partial r}(a_\lambda h_\lambda) - \frac{\partial}{\partial \lambda}(a_r h_r)\right]\mathbf{j} \\
&\quad + \frac{1}{h_\lambda h_\varphi}\left[\frac{\partial}{\partial \lambda}(a_\varphi h_\varphi) - \frac{\partial}{\partial \varphi}(a_\lambda h_\lambda)\right]\mathbf{k} \\
&= \frac{1}{r}\left[\frac{\partial a_r}{\partial \varphi} - \frac{\partial}{\partial r}(r a_\varphi)\right]\mathbf{i} + \left[\frac{1}{r}\frac{\partial}{\partial r}(r a_\lambda) - \frac{1}{r\cos\varphi}\frac{\partial a_r}{\partial \lambda}\right]\mathbf{j} \\
&\quad + \left[\frac{1}{r\cos\varphi}\frac{\partial a_\varphi}{\partial \lambda} - \frac{1}{r\cos\varphi}\frac{\partial}{\partial \varphi}(\cos\varphi a_\lambda)\right]\mathbf{k}
\end{aligned}
\tag{2.27}
$$

With the aforementioned relations, we have sufficient information to express the governing equations in spherical coordinates. We obtain the following for the momentum and continuity equations:

- Zonal momentum equation

$$\frac{Du}{Dt} + \left(f_\varphi + \frac{u}{r} \right) w - \left(f_r + \tan\varphi \frac{u}{r} \right) v$$

$$= -\frac{1}{r\cos\varphi} \left(\frac{1}{\rho} \frac{\partial p}{\partial \lambda} + \frac{\partial \phi}{\partial \lambda} \right) + f_{T,\lambda} \qquad (2.28)$$

- Meridional momentum equation

$$\frac{Dv}{Dt} + \frac{vw}{r} + \left(f_r + \tan\varphi \frac{u}{r} \right) u = -\frac{1}{r} \left(\frac{1}{\rho} \frac{\partial p}{\partial \varphi} + \frac{\partial \phi}{\partial \varphi} \right) + f_{T,\varphi} \quad (2.29)$$

- Vertical momentum equation

$$\frac{Dw}{Dt} - \frac{u^2 + v^2}{r} - f_\varphi u = -\frac{1}{\rho} \frac{\partial p}{\partial r} + \frac{\partial \phi}{\partial r} + f_{T,r} \qquad (2.30)$$

- Continuity equation

$$\frac{\partial \rho}{\partial t} + \frac{1}{r\cos\varphi} \frac{\partial}{\partial \lambda} (\rho u) + \frac{1}{r\cos\varphi} \frac{\partial}{\partial \varphi} (\cos\varphi \rho v) + \frac{1}{r^2} \frac{\partial}{\partial r} (r^2 \rho w) = 0$$
$$(2.31)$$

where the material time derivative is written as

$$\frac{D}{Dt} = \frac{\partial}{\partial t} + \frac{u}{r\cos\varphi} \frac{\partial}{\partial \lambda} + \frac{v}{r} \frac{\partial}{\partial \varphi} + w \frac{\partial}{\partial r} \qquad (2.32)$$

In (2.28)–(2.30), $f_{T,\lambda}$, $f_{T,\varphi}$ and $f_{T,r}$ denote the components of the force $\mathbf{f}_T$ that results from the microturbulent exchange of momentum. The temperature tendency equation, the budget equation for water vapor and the thermal equation of state keep their form except that the gradient and divergence operators must be evaluated according to (2.25) and (2.26).

2.3   Scale Analysis and the Shallow Atmosphere Approximation

Atmospheric motions take place within a thin spherical shell and **synoptic-scale** weather systems[a] have a horizontal extension that

---

[a]The physical meaning of the synoptic scale will be explained in a subsequent chapter.

is much larger than their vertical extension. These facts justify the **shallow atmosphere approximation** that will be explained in the following.

The relative change of radius r is small because the atmosphere is contained in a thin spherical shell. Therefore, it is justified to neglect the deviation of r when it appears in a factor of the various terms. We see in (2.28)–(2.32) that the radius mostly appears in the form of the factor $1/r$. Therefore, we can make the approximation

$$\frac{1}{r} = \frac{1}{r_E} + \frac{1}{r} - \frac{1}{r_E} = \frac{1}{r_E} - \frac{z}{r_E r} \approx \frac{1}{r_E} \qquad (2.33)$$

where r_E denotes the mean **Earth radius** and $z = r - r_E$ the **height**.[b] The Earth's radius increases from $r = 6357$ km at the pole to $r = 6378$ km at the equator. The troposphere has a vertical depth of less than 18 km. Therefore, the maximum variation of z is about 40 km. The mean Earth radius is $r_E = 6371$ km and, therefore, the relative change of radius r is less than 1% which justifies the approximation (2.33).

The radius r also appears in the form

$$\frac{1}{r^2} \frac{\partial}{\partial r} \left(r^2 b \right) = \frac{\partial b}{\partial z} + \frac{2b}{r} \approx \frac{\partial b}{\partial z} \qquad (2.34)$$

The second summand can be neglected if the vertical change of b within the atmosphere has the same order of magnitude as that of b itself. This is usually the case.

The shallowness of large-scale weather systems justifies the **quasi-hydrostatic approximation**, namely, that the pressure results from the solution of the vertical component of the hydrostatic balance equation (1.55). A **scale analysis** becomes necessary to derive this approximation. For this purpose, the typical fluctuation range of the various physical quantities must be estimated. Figure 2.2(a) shows a weather map on 26th of January 2008 for the North Atlantic region. The chart displays density isolines and the horizontal wind vector at a height of 1000 m. A prominent synoptic-scale **extra-tropical cyclone** appears at the east coast of North America.

---

[b]From now on, z denotes the vertical coordinate in the spherical coordinate system and not in the Cartesian where the vertical axis coincides with Earth's axis of rotation.

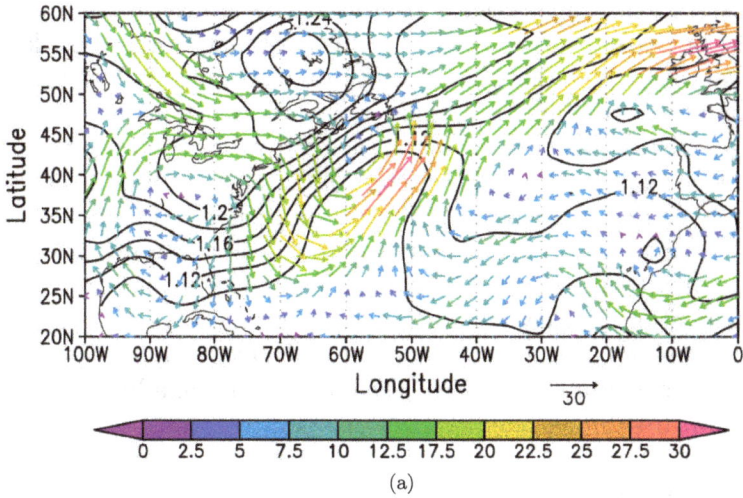

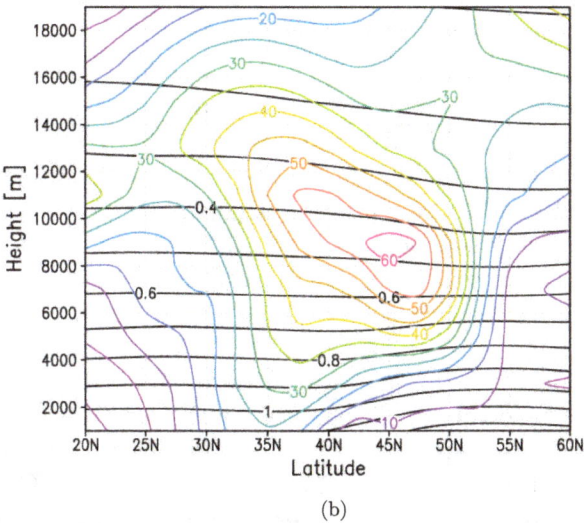

Fig. 2.2. Density and wind fields on 26th of January 2008, 00UTC: (a) Density (isolines, contour interval 0.02 kg/m^3) and horizontal wind vector (colored arrows) at a height of 1000 m. (b) Density (black isolines, contour interval 0.1 kg/m^3) and horizontal wind speed in a meridional cross-section at 60 W longitude. The data are derived from NCEP Reanalysis [see Kalnay *et al.* (1996) for a description of the Reanalysis data set].

We see that the typical wind velocities have a magnitude of 10 m/s while the density has a typical fluctuation range of about 0.1 kg/m^3. A zone with a high-density gradient separates cold air with high density in the north from warm air with low density in the south. The characteristic diameter of low- and high-pressure systems is about 1000 km. Note that 10 degrees shift of latitude corresponds to a distance of 1112 km. Figure 2.2(b) shows a meridional cross section at 60 W longitude. The density isolines run approximately in the horizontal direction demonstrating that the density variation in the vertical direction is larger than in the horizontal direction. A prominent wind speed maximum can be seen at a height of about 9000 m. This is the **jet stream** that evolves as a consequence of a high meridional temperature gradient.

All in all, we can deduce the following scales for synoptic-scale weather systems:

Length scale L $= 10^6$ m
Height scale H $= 10^4$ m
Horizontal wind velocity scale U $= 10$ m/s
Vertical velocity scale W $= 0.01$ m/s
Time scale (advective) T $= \frac{L}{U} = 10^5$ s
Horizontal density scale $\rho_h = 0.1$ kg/m^3
Vertical density scale $\rho_v = 1$ kg/m^3
Scale of vertical gravity acceleration G $= 10$ m/s^2
Scale of centrifugal force F$_C = \frac{\Omega^2 r_E}{2} = 0.01$ m/s^2

This list does not contain scales for pressure fluctuations. The reason is that we estimate these by taking the pressure gradient force as a residual in the equation of motion. The small vertical velocity scale can be justified in terms of the continuity equation. Only a small part of the horizontal flow has nonzero divergence but vertical motion arises in atmospheric dynamics primarily from the divergence of the horizontal flow because of the mass budget. For a rough estimation, we can write

$$\rho \nabla \cdot \mathbf{v}_\chi \approx \frac{\partial}{\partial z}(\rho w) \qquad (2.35)$$

where $\mathbf{v}_\chi$ denotes the divergent part of the horizontal wind vector. The left-hand side of this equation has the scale $\rho_v U_\chi / L$ while the right-hand takes the scale $\rho_v W/H$. Now, the scale $W = 0.01\,\text{m/s}$ appears reasonable since $H/L = 0.01$ and $U_\chi \ll U$.

We have seen that the horizontal scale of density is much smaller than its vertical scale. Therefore, it is useful to subtract a horizontally independent base state from each variable in order to estimate the magnitude of the various terms more accurately. A large part of the vertical pressure gradient compensates for the gravity force which explains the high pressure decay in the vertical direction. Therefore, the base state is suitably chosen to be a solution to the hydrostatic balance equation:

$$\nabla p_0 = -\rho_0 \nabla \phi \qquad (2.36)$$

where $p_0 = p_0(\mathbf{r})$ and $\rho_0 = \rho_0(z)$ denote the base state pressure and density, respectively. The base state density $\rho_0(z)$ results from global and time averaging. Therefore, the sum of pressure gradient and gravity forces can be written in the following way:

$$-\frac{1}{\rho}\nabla p - \nabla\phi = -\frac{1}{\rho}\nabla\left(p' + p_0\right) - \nabla\phi = -\frac{1}{\rho}\nabla p' - \frac{\rho'}{\rho}\nabla\phi \qquad (2.37)$$

where p' and ρ' denote the pressure and density anomalies of the base state, respectively. The primed variables fluctuate in the range given by the horizontal scales. With this information, we can estimate the magnitudes of the various terms in the equations of motion:

$$
\underbrace{\frac{\partial u}{\partial t}}_{\substack{U^2/L \\ =10^{-4}\frac{m}{s^2}}}
+ \underbrace{\frac{u}{r_E \cos\varphi}\frac{\partial u}{\partial\lambda}}_{\substack{U^2/L \\ =10^{-4}\frac{m}{s^2}}}
+ \underbrace{\frac{v}{r_E}\frac{\partial u}{\partial\varphi}}_{\substack{U^2/L \\ =10^{-4}\frac{m}{s^2}}}
+ \underbrace{w\frac{\partial u}{\partial z}}_{\substack{UW/H \\ =10^{-5}\frac{m}{s^2}}}
+ \underbrace{f_\varphi w}_{\substack{\Omega W \\ =10^{-6}\frac{m}{s^2}}}
+ \underbrace{\frac{uw}{r_E}}_{\substack{UW/r_E \\ =10^{-8}\frac{m}{s^2}}}
- \underbrace{f_r v}_{\substack{\Omega U \\ =10^{-3}\frac{m}{s^2}}}
$$

$$
- \underbrace{\tan\varphi\frac{uv}{r_E}}_{\substack{U^2/r_E \\ =10^{-5}\frac{m}{s^2}}}
= - \underbrace{\frac{1}{\rho r_E \cos\varphi}\frac{\partial p'}{\partial\lambda}}_{10^{-3}\frac{m}{s^2}}
- \underbrace{\frac{\rho'}{\rho r_E \cos\varphi}\frac{\partial\phi}{\partial\lambda}}_{\substack{\rho_h F_C/\rho_v \\ =10^{-3}\frac{m}{s^2}}}
+ \underbrace{f_{T,\lambda}}_{\text{vague}}
$$

$$(2.38)$$

$$\underbrace{\frac{\partial v}{\partial t}}_{\substack{U^2/L \\ =10^{-4}\frac{m}{s^2}}} + \underbrace{\frac{u}{r_E \cos\varphi}\frac{\partial v}{\partial \lambda}}_{\substack{U^2/L \\ =10^{-4}\frac{m}{s^2}}} + \underbrace{\frac{v}{r_E}\frac{\partial v}{\partial \varphi}}_{\substack{U^2/L \\ =10^{-4}\frac{m}{s^2}}} + \underbrace{w\frac{\partial v}{\partial z}}_{\substack{UW/H \\ =10^{-5}\frac{m}{s^2}}} + \underbrace{\frac{vw}{r_E}}_{\substack{UW/r_E \\ =10^{-8}\frac{m}{s^2}}} + \underbrace{f_r u}_{\substack{\Omega U \\ =10^{-3}\frac{m}{s^2}}}$$

$$+ \underbrace{\tan\varphi\frac{u^2}{r_E}}_{\substack{U^2/r_E \\ =10^{-5}\frac{m}{s^2}}} = -\underbrace{\frac{1}{\rho r_E}\frac{\partial p'}{\partial \varphi}}_{\substack{10^{-3}\frac{m}{s^2}}} - \underbrace{\frac{\rho'}{\rho r_E}\frac{\partial \phi}{\partial \varphi}}_{\substack{<\rho_h F_C/\rho_v \\ =10^{-3}\frac{m}{s^2}}} + \underbrace{f_{T,\varphi}}_{\text{vague}} \qquad (2.39)$$

$$\underbrace{\frac{\partial w}{\partial t}}_{\substack{UW/L \\ =10^{-7}\frac{m}{s^2}}} + \underbrace{\frac{u}{r_E \cos\varphi}\frac{\partial w}{\partial \lambda}}_{\substack{UW/L \\ =10^{-7}\frac{m}{s^2}}} + \underbrace{\frac{v}{r_E}\frac{\partial w}{\partial \varphi}}_{\substack{UW/L \\ =10^{-7}\frac{m}{s^2}}} + \underbrace{w\frac{\partial w}{\partial z}}_{\substack{W^2/H \\ =10^{-8}\frac{m}{s^2}}} - \underbrace{\frac{u^2+v^2}{r_E}}_{\substack{U^2/r_E \\ =10^{-5}\frac{m}{s^2}}} \underbrace{-f_\varphi u}_{\substack{\Omega U \\ =10^{-3}\frac{m}{s^2}}}$$

$$= -\underbrace{\frac{1}{\rho}\frac{\partial p'}{\partial z}}_{1\frac{m}{s^2}} - \underbrace{\frac{\rho'}{\rho r_E}\frac{\partial \phi}{\partial z}}_{\substack{\rho_h G/\rho_v \\ =1\frac{m}{s^2}}} + \underbrace{f_{T,r}}_{\ll 1\frac{m}{s^2}} \qquad (2.40)$$

To derive approximated equations, we neglect all terms in the equation which are three orders of magnitudes smaller than the terms having the largest magnitude. This yields the **quasi-hydrostatic approximation** by which the equations of motion become

$$\frac{Du}{Dt} - \left(f + \tan\varphi\frac{u}{r_E}\right)v = -\frac{1}{r_E\cos\varphi}\left(\frac{1}{\rho}\frac{\partial p}{\partial \lambda} + \frac{\partial \phi}{\partial \lambda}\right) + f_{T,\lambda} \quad (2.41)$$

$$\frac{Dv}{Dt} + \left(f + \tan\varphi\frac{u}{r_E}\right)u = -\frac{1}{r_E}\left(\frac{1}{\rho}\frac{\partial p}{\partial \varphi} + \frac{\partial \phi}{\partial \varphi}\right) + f_{T,\varphi} \qquad (2.42)$$

$$\frac{\partial p}{\partial z} = -g\rho \qquad (2.43)$$

where we use from now on the notation $f = f_r = 2\Omega\sin\varphi$ for the Coriolis parameter and $g = \partial\phi/\partial z$ for the gravity acceleration.

The other governing equations do not suffer simplifications due to the shallow atmosphere approximation except for using the simplifications (2.33) and (2.34). The simplified set of equations is called **primitive equations** and can be summarized as follows:

Equation of motion

$$\left(\frac{D\mathbf{v}_h}{Dt}\right)_h + f\mathbf{k} \times \mathbf{v}_h = -\frac{1}{\rho}\nabla_h p - \nabla_h \phi + \boldsymbol{f}_T \qquad (2.44)$$

Hydrostatic balance equation

$$\frac{\partial p}{\partial z} = -g\rho \qquad (2.45)$$

Continuity equation

$$\frac{\partial \rho}{\partial t} + \nabla \cdot (\rho \mathbf{v}) = 0 \qquad (2.46)$$

Temperature tendency equation

$$c_p \frac{DT}{Dt} = -L_v S_{l,v} + \frac{1}{\rho}\frac{Dp}{Dt} - \frac{1}{\rho}\nabla \cdot \mathbf{j}_R + D_T \qquad (2.47)$$

Thermal equation of state

$$\rho = \frac{p}{R_d T} \qquad (2.48)$$

Budget equation for water vapor

$$\frac{Dm_v}{Dt} = S_{l,v} + D_{m_v} \qquad (2.49)$$

where D_T and D_{m_v} denote the divergence of turbulent fluxes of enthalpy and water vapor, respectively. The index h means that the vertical component of the vector is omitted. The primitive equations proved to be suitable for global weather forecasting (e.g. Kalnay, 2003). Even nowadays, many global weather forecast models are based on primitive equations.[c] However, the equations in the present

---

[c]However, state-of-the-art weather forecast models do not include all simplifications we made in Section 1.12.

form are not suitable for a numerical solution because no equation for vertical velocity exists. Vertical velocity w only occurs in the advection and flux terms of the other prognostic equations but it is not possible to solve this equation for w when the time tendency is unknown. Introducing a new vertical coordinate represents a possible solution to circumvent this problem.

2.4 Transformation of the Vertical Coordinate

For diagnostic, analytic and numerical purposes, it is often convenient to replace the vertical coordinate z by another one. First, the transformation will be outlined without specifying the new vertical coordinate which we denote with q_3. The applicability of this new vertical coordinate depends on the strictly monotonic character of the function $q_3(z)$. Therefore, q_3 must either increase or decrease with height and we can express the height z uniquely as a function of the new vertical coordinate. Hence, the field functions

$$q_3 = q_3\left(\lambda, \varphi, z, t\right) \tag{2.50}$$

and

$$z_{[q_3]} = z\left(\lambda, \varphi, q_3, t\right) \tag{2.51}$$

both exist. The index $[q_3]$ symbolizes the field as a function of the new vertical coordinate. Now, every field function b can be formally written as a function of the new vertical coordinate by

$$b_{[q_3]}\left(\lambda, \varphi, q_3, t\right) = b\left[\lambda, \varphi, z\left(\lambda, \varphi, q_3, t\right), t\right] \tag{2.52}$$

With the chain rule of differentiation, we can find the partial derivative with respect to any coordinate except q_3

$$\frac{\partial b_{[q_3]}}{\partial s} = \frac{\partial b}{\partial s} + \frac{\partial b}{\partial z}\frac{\partial z_{[q_3]}}{\partial s} \quad \text{for} \quad s = \lambda, \varphi \text{ or } t \tag{2.53}$$

For the vertical derivative, we obtain the following transformation rule:

$$\frac{\partial b_{[q_3]}}{\partial q_3} = \frac{\partial b}{\partial z}\frac{\partial z_{[q_3]}}{\partial q_3} \quad \leftrightarrow \quad \frac{\partial b}{\partial z} = \left(\frac{\partial z_{[q_3]}}{\partial q_3}\right)^{-1}\frac{\partial b_{[q_3]}}{\partial q_3} \tag{2.54}$$

Substituting the second expression in (2.53) yields

$$\frac{\partial b}{\partial s} = \frac{\partial b_{[q3]}}{\partial s} - \left(\frac{\partial z_{[q3]}}{\partial q_3}\right)^{-1} \frac{\partial b_{[q3]}}{\partial q_3} \frac{\partial z_{[q3]}}{\partial s} \quad \text{for } s = \lambda, \varphi \text{ or } t \qquad (2.55)$$

All partial derivatives can be transformed in the primitive equations by using Eq. (2.54) or (2.55). The transformation of the advection terms requires special attention. The material derivative of a scalar field function is invariant to the change of the vertical coordinate. Therefore,

$$\frac{Db_{[q3]}}{Dt} = \frac{Db}{Dt} \qquad (2.56)$$

Executing the material time derivative in the system with the new vertical coordinate leads to

$$\frac{Db_{[q3]}}{Dt} = \frac{\partial b_{[q3]}}{\partial t} + \frac{D\lambda}{Dt} \frac{\partial b_{[q3]}}{\partial \lambda} + \frac{D\varphi}{Dt} \frac{\partial b_{[q3]}}{\partial \varphi} + \frac{Dq_3}{Dt} \frac{\partial b_{[q3]}}{\partial q_3}$$

$$= \frac{\partial b_{[q3]}}{\partial t} + \frac{u_{[q3]}}{r_E \cos\varphi} \frac{\partial b_{[q3]}}{\partial \lambda} + \frac{v_{[q3]}}{r_E} \frac{\partial b_{[q3]}}{\partial \varphi} + \dot{q}_3 \frac{\partial b_{[q3]}}{\partial q_3} \qquad (2.57)$$

where $\dot{q}_3$ constitutes the vertical velocity in the transformed system.

We conduct the transformation by using four prominent examples (i) $q_3 = p$ (pressure), (ii) $q_3 = p/p_s$ (sigma), (iii) $q_3 = [1 - \Pi(p)] H_a$ (pseudo height) and (iv) $q_3 = \theta$ (potential temperature), where p_s denotes the surface pressure and H_a the height of the isentropic atmosphere. The vertical coordinates (i)–(iii) change strictly monotonic with height because of the hydrostatic balance equation (2.43). The potential temperature increases with height above the boundary layer and can also be used as the vertical coordinate for the free atmosphere.

2.4.1 $q_3 = p$ *(pressure)*

The vertical coordinate pressure is widely used for analysis of the synoptic weather situation. Most weather maps except for the surface chart visualize fields on isobaric surfaces. Radiosondes measure the properties horizontal wind, pressure, temperature and humidity only and this represents one reason for choosing pressure as the analysis coordinate. A coordinate set with $q_3 = p$ is referred to as **isobaric coordinates**.

The transformation relations (2.54) and (2.55) become for $q_3 = p$:

$$\frac{\partial b}{\partial z} = -g\rho_{[p]}\frac{\partial b_{[p]}}{\partial p} \tag{2.58}$$

$$\frac{\partial b}{\partial s} = \frac{\partial b_{[p]}}{\partial s} + g\rho_{[p]}\frac{\partial b_{[p]}}{\partial p}\frac{\partial z_{[p]}}{\partial s} \text{ for } s = \lambda, \varphi \text{ or } t \tag{2.59}$$

In these equations, we made use of the hydrostatic balance equation $\partial p/\partial z = \left(\partial z_{[p]}/\partial p\right)^{-1} = -g\rho_{[p]}$. Applying these formulas to the right-hand side of the equation of motion gives

$$-\frac{1}{\rho}\nabla_h p - \nabla_h \phi + \boldsymbol{f}_T$$

$$= -g\nabla_h z_{[p]} - \nabla_h \phi_{[p]} - g\rho_{[p]}\frac{\partial \phi_{[p]}}{\partial p}\nabla_h z_{[p]} + \boldsymbol{f}_{T[p]}$$

$$= -g\nabla_h z_{[p]} - \nabla_h \phi_{[p]} + \frac{\partial \phi}{\partial z}\nabla_h z_{[p]} + \boldsymbol{f}_{T[p]}$$

$$= -\nabla_h \phi_{[p]} + \boldsymbol{f}_{T[p]} \tag{2.60}$$

because $\partial\phi/\partial z = g$. We see that the pressure gradient force and gravity force can be united to the horizontal gradient of the geopotential field in isobaric coordinates. This outcome is explained schematically in Fig. 2.3 and results because the isobaric surface has a slope in the presence of a horizontal pressure gradient. It is exactly the horizontal geopotential gradient in isobaric coordinates that replaces the pressure gradient force in isobaric coordinates if geopotential isosurfaces have no slope, which is approximately true. Using also the transformation rule for the material derivative (2.57), we get the equation of motion in isobaric coordinates:

$$\frac{\partial \boldsymbol{v}_{h[p]}}{\partial t} + \left(\boldsymbol{v}_{h[p]} \cdot \nabla \boldsymbol{v}_{h[p]}\right)_h + \omega\frac{\partial \boldsymbol{v}_{h[p]}}{\partial p} + f\boldsymbol{k}\times\boldsymbol{v}_{h[p]}$$

$$= -\nabla_h \phi_{[p]} + \boldsymbol{f}_{T[p]} \tag{2.61}$$

where ω denotes the vertical velocity in isobaric coordinates. The hydrostatic balance equation must be expressed in another way since

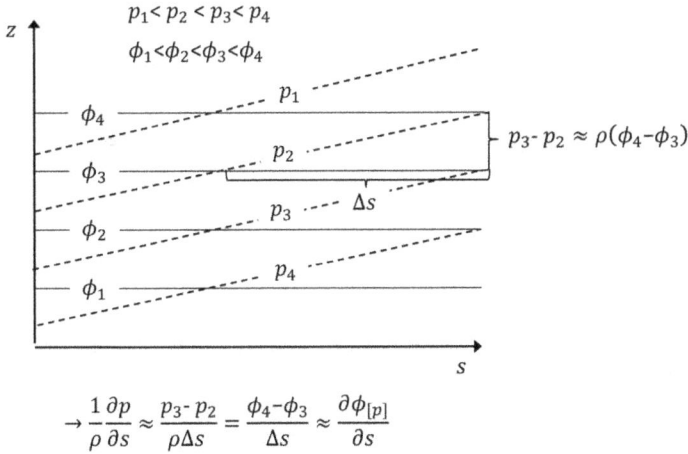

Fig. 2.3. Sketch of the transformation of the pressure gradient from height to pressure vertical coordinate. The approximative relation turns into an identity when the finite differences become infinitesimally small.

pressure is not an unknown variable in isobaric coordinates anymore. Substituting the geopotential in (2.58) yields

$$\frac{\partial \phi}{\partial z} = -g\rho_{[p]} \frac{\partial \phi_{[p]}}{\partial p} \quad \leftrightarrow \quad \frac{\partial \phi_{[p]}}{\partial p} = -\frac{1}{\rho_{[p]}} \tag{2.62}$$

Combining this with the thermal equation of state leads to

$$\frac{\partial \phi_{[p]}}{\partial p} = -\frac{R_d T_{[p]}}{p} \tag{2.63}$$

The transformation of the continuity equation requires more tedious calculations. First, we transform the divergence of the velocity vector to isobaric coordinates

$$\nabla_h \cdot \mathbf{v}_h + \frac{\partial w}{\partial z} = \nabla_h \cdot \mathbf{v}_{h[p]} + g\rho_{[p]} \left(\frac{\partial \mathbf{v}_{h[p]}}{\partial p} \cdot \nabla_h z_{[p]} - \frac{\partial w_{[p]}}{\partial p} \right)$$

$$= \nabla_h \cdot \mathbf{v}_{h[p]} - g\rho_{[p]} \left[\frac{D}{Dt} \left(\frac{\partial z_{[p]}}{\partial p} \right) + \frac{\partial \omega}{\partial p} \frac{\partial z_{[p]}}{\partial p} \right]$$

$$= \nabla_h \cdot \mathbf{v}_{h[p]} + \frac{\partial \omega}{\partial p} - \frac{1}{\rho_{[p]}} \frac{D\rho_{[p]}}{Dt} \tag{2.64}$$

The conversions in (2.64) require that variations of gravity acceleration g can be neglected. The material time derivative is invariant with respect to a coordinate transform and, therefore, the transformed continuity equation becomes

$$\frac{\partial \rho}{\partial t} + \nabla \cdot (\rho \mathbf{v}) = \frac{D\rho}{Dt} + \rho \nabla \cdot \mathbf{v} = \rho_{[p]} \left(\nabla_h \cdot \mathbf{v}_{h[p]} + \frac{\partial \omega}{\partial p} \right) = 0 \quad (2.65)$$

Since the density is nonzero, we find that the continuity equation in isobaric coordinates reduces to

$$\nabla_h \cdot \mathbf{v}_{h[p]} + \frac{\partial \omega}{\partial p} = 0 \qquad (2.66)$$

Hence, if the isobaric coordinates were interpreted to constitute an Euclidean space, the velocity field would be nondivergent. This is a great simplification that makes isobaric coordinates better manageable than height coordinates. The vertical velocity ω can be found by vertical integration of (2.66):

$$\omega = \int_0^p \nabla_h \cdot \mathbf{v}_{h[p]} dp' \qquad (2.67)$$

where the boundary condition of a vanishing vertical velocity at the "top" of the atmosphere ($p = 0$) has been used.

The temperature tendency equation changes only slightly after transformation to isobaric coordinates. It becomes

$$\frac{\partial T_{[p]}}{\partial t} + \mathbf{v}_{h[p]} \cdot \nabla T_{[p]} + \left(\frac{\partial T_{[p]}}{\partial p} - \frac{R_d T_{[p]}}{c_p p} \right) \omega = \frac{Q_{[p]}}{c_p} \qquad (2.68)$$

where $Q_{[p]}$ denotes the heating term that contains all irreversible and diabatic processes. The factor in front of the vertical velocity is related to a vertical change of potential temperature since

$$\frac{\partial \theta_{[p]}}{\partial p} = \frac{1}{\Pi} \frac{\partial T_{[p]}}{\partial p} + T_{[p]} \frac{\partial}{\partial p} \left(\frac{1}{\Pi} \right) = \frac{1}{\Pi} \left(\frac{\partial T_{[p]}}{\partial p} - \frac{R_d T_{[p]}}{c_p p} \right) \qquad (2.69)$$

Therefore, rising air ($\omega < 0$) cools in the typical case when the potential temperature increases with height ($\partial \theta_{[p]}/\partial p < 0$) and vice versa. This leads to a static stability effect which is very important for the dynamics of synoptic-scale weather systems. The budget equation for water vapor (2.49) keeps its original form except for a different evaluation of the material derivative. In summary, we find the following set of primitive equations in isobaric coordinates.

Equation of motion

$$\frac{\partial \mathbf{v}_h}{\partial t} + (\mathbf{v}_h \cdot \nabla \mathbf{v}_h)_h + \omega \frac{\partial \mathbf{v}_h}{\partial p} + f\mathbf{k} \times \mathbf{v}_h = -\nabla_h \phi + \boldsymbol{f}_T \qquad (2.70)$$

Hydrostatic balance equation

$$\frac{\partial \phi}{\partial p} = -\frac{R_d T}{p} \qquad (2.71)$$

Continuity equation

$$\nabla_h \cdot \mathbf{v}_h + \frac{\partial \omega}{\partial p} = 0 \qquad (2.72)$$

Temperature tendency equation

$$\frac{\partial T}{\partial t} + \mathbf{v}_h \cdot \nabla T + \Pi \frac{\partial \theta}{\partial p} \omega = \frac{Q}{c_p} \qquad (2.73)$$

Budget equation for water vapor

$$\frac{\partial m_v}{\partial t} + \mathbf{v}_h \cdot \nabla m_v + \omega \frac{\partial m_v}{\partial p} = S_{l,v} + D_{m_v} \qquad (2.74)$$

The index $[p]$ has been omitted in these equations for a better lucidity. Note that the thermal equation of state is not listed since the density has been eliminated and, therefore, this equation becomes superfluous. The unknown variables are horizontal wind $\mathbf{v}_h$, geopotential ϕ, vertical velocity ω, temperature T and specific humidity m_v. Isobaric coordinates proved to be advantageous for diagnostic purposes. Weather observations typically comprise temperature, pressure, horizontal wind and humidity measurements but height is not measured directly. Therefore, it is easier to compile measurements on pressure than on height levels. Furthermore, the vertical velocity ω can be deduced diagnostically from Eq. (2.72). In height coordinates, such a simple method does not exist and direct measurements of vertical velocity w are not practicable due to its small magnitude. A disadvantage of isobaric coordinates is the complicated boundary condition at the Earth's surface. The lower boundary is neither an isobaric surface nor is the pressure at the boundary time-independent. Isobaric surfaces intersect the ground even in the case of a flat lower boundary. Therefore, isobaric coordinates are inappropriate for a weather forecast based on a numerical solution of the primitive equations. Instead, the sigma coordinate explained in the following represents a suitable coordinate for a numerical solution technique because of the simple boundary conditions.

2.4.2 $q_3 = p/p_s$ *(sigma)*

The **sigma coordinate** was introduced by Phillips (1957) and is defined by

$$q_3 = \sigma = \frac{p}{p_s} \tag{2.75}$$

where p_s denotes the surface pressure. Figure 2.4 illustrates the shape of the sigma coordinate. The advantage of the sigma coordinate is obviously the simple boundary condition $\sigma = 1$ at the surface and the adaptation of the sigma surface to the surface orography. At the upper boundary, sigma takes the value $\sigma = 0$ but it should be noted that the upper boundary of the atmosphere is not clearly defined. Therefore, the upper boundary rather refers to a very small pressure so that the approximation $\sigma \approx 0$ is justified. Since the upper and lower boundary can be considered as material surfaces, we also obtain simple boundary conditions for vertical velocity $\dot{\sigma}$, namely

$$\dot{\sigma} = 0 \quad \text{at} \quad \sigma = 0, 1 \tag{2.76}$$

To simplify the derivation, it is appropriate to make the transformation from isobaric coordinates to sigma coordinates. Then, we have the transformation rules:

$$\frac{\partial b_{[p]}}{\partial p} = \left(\frac{\partial p_{[\sigma]}}{\partial \sigma}\right)^{-1} \frac{\partial b_{[\sigma]}}{\partial \sigma} = \frac{1}{p_s}\frac{\partial b_{[\sigma]}}{\partial \sigma} \tag{2.77}$$

$$\frac{\partial b_{[p]}}{\partial s} = \frac{\partial b_{[\sigma]}}{\partial s} - \left(\frac{\partial p_{[p]}}{\partial \sigma}\right)^{-1} \frac{\partial b_{[\sigma]}}{\partial \sigma}\frac{\partial p_{[\sigma]}}{\partial s}$$

$$= \frac{\partial b_{[\sigma]}}{\partial s} - \frac{\sigma}{p_s}\frac{\partial b_{[\sigma]}}{\partial \sigma}\frac{\partial p_s}{\partial s} \quad \text{for} \quad s = \lambda, \varphi \,\text{or}\, t \tag{2.78}$$

The relation $p_{[p]} = \sigma p_s$ has been used for the conversions in these equations. Note that the surface pressure does not depend on the

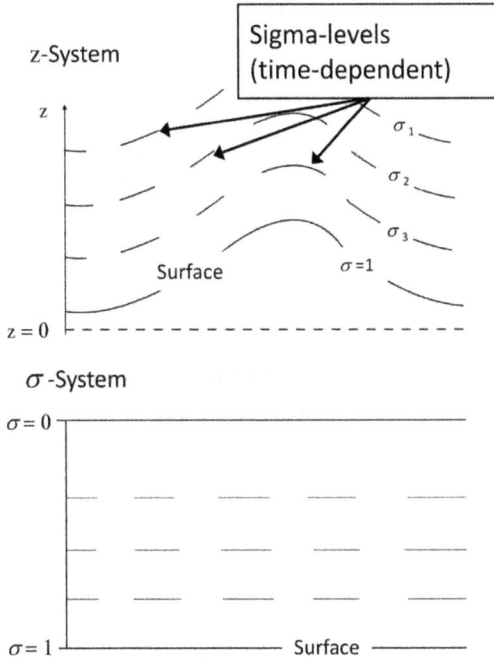

Fig. 2.4.   Arrangement of sigma surfaces in the atmosphere.

vertical coordinate and, therefore, the index $[\sigma]$ is obsolete. Applying
the transformation rules to the right-hand side of the equation of
motion yields

$$-\nabla_h \phi_{[p]} + \boldsymbol{f}_{T[p]} = -\nabla_h \phi_{[\sigma]} + \frac{\sigma}{p_s}\frac{\partial \phi_{[\sigma]}}{\partial \sigma}\nabla_h p_s + \boldsymbol{f}_{T[\sigma]}$$

$$= -\nabla_h \phi_{[\sigma]} + \sigma\frac{\partial \phi_{[p]}}{\partial p}\nabla_h p_s + \boldsymbol{f}_{T[\sigma]}$$

$$= -\nabla_h \phi_{[\sigma]} - \frac{R_d T}{p_s}\nabla_h p_s + \boldsymbol{f}_{T[\sigma]} \qquad (2.79)$$

where the hydrostatic balance equation (2.71) has been substituted.
For transformation of the continuity equation (2.72), it is necessary
to express the vertical velocity ω in the new coordinate system

$$\omega_{[\sigma]} = \frac{Dp_{[\sigma]}}{Dt} = \sigma\frac{Dp_s}{Dt} + p_s\dot{\sigma} \qquad (2.80)$$

Inserting this in the continuity equation gives

$$\nabla_h \cdot \mathbf{v}_{h[p]} + \frac{\partial \omega_{[p]}}{\partial p}$$

$$= \nabla_h \cdot \mathbf{v}_{h[\sigma]} - \frac{\sigma}{p_s}\frac{\partial \mathbf{v}_{h[\sigma]}}{\partial \sigma} \cdot \nabla_h p_s + \frac{1}{p_s}\frac{Dp_s}{Dt} + \frac{\sigma}{p_s}\frac{\partial \mathbf{v}_{h[\sigma]}}{\partial \sigma} \cdot \nabla_h p_s$$

$$+ \frac{\partial \dot{\sigma}}{\partial \sigma} = \frac{1}{p_s}\left(\frac{\partial p_s}{\partial t} + \mathbf{v}_{h[\sigma]} \cdot \nabla_h p_s\right) + \nabla_h \cdot \mathbf{v}_{h[\sigma]} + \frac{\partial \dot{\sigma}}{\partial \sigma} = 0$$

$$(2.81)$$

We see that the continuity equation in sigma coordinates becomes a prognostic equation for surface pressure. It can be written as

$$\frac{\partial p_s}{\partial t} + \nabla_h \cdot (p_s \mathbf{v}_{h[\sigma]}) + p_s \frac{\partial \dot{\sigma}}{\partial \sigma} = 0 \qquad (2.82)$$

By using the boundary condition (2.76) and vertical integration, we find that

$$\frac{\partial p_s}{\partial t} = - \int_0^1 \nabla_h \cdot (p_s \mathbf{v}_{h[\sigma]}) d\sigma \qquad (2.83)$$

Therefore, the local surface pressure tendency coincides with the vertical integral of the convergence of horizontal wind multiplied with surface pressure. This is understandable because the mass of an atmospheric column is proportional to surface pressure as becomes evident by the calculation

$$\int_{z_s}^{\infty} \rho dz = \int_{p_s}^0 \rho\left(\frac{\partial p}{\partial z}\right)^{-1} dp = -\int_{p_s}^0 \frac{1}{g} dp = \frac{p_s}{g} \qquad (2.84)$$

We can also find a diagnostic equation for vertical velocity by integrating (2.82) from 0 to σ:

$$\dot{\sigma} = -\frac{1}{p_s}\int_0^{\sigma}\frac{\partial p_s}{\partial t} + \nabla_h \cdot (p_s \mathbf{v}_{h[\sigma]}) d\sigma'$$

$$= \frac{\sigma}{p_s}\int_0^1 \nabla_h \cdot (p_s \mathbf{v}_{h[\sigma]}) d\sigma' - \frac{1}{p_s}\int_0^{\sigma} \nabla_h \cdot (p_s \mathbf{v}_{h[\sigma]}) d\sigma' \qquad (2.85)$$

It is obvious that this equation fulfills the boundary condition (2.76).

The temperature tendency equation (2.73) can be written in sigma coordinates as

$$\frac{\partial T_{[\sigma]}}{\partial t} + \mathbf{v}_{h[\sigma]} \cdot \nabla T_{[\sigma]} + \dot{\sigma}\frac{\partial T_{[\sigma]}}{\partial \sigma} = \frac{R_d T_{[\sigma]}}{c_p p_s}\left(\frac{Dp_s}{Dt} + p_s\frac{\dot{\sigma}}{\sigma}\right) + \frac{Q_{[\sigma]}}{c_p}$$

$$= \frac{R_d T_{[\sigma]}}{c_p p_s}\left(\mathbf{v}_{h[\sigma]} \cdot \nabla p_s - \frac{1}{\sigma}\int_0^\sigma \nabla_h \cdot \left(p_s\mathbf{v}_{h[\sigma]}\right)d\sigma'\right) + \frac{Q_{[\sigma]}}{c_p} \quad (2.86)$$

where the identity (2.85) has been substituted. The transformation of the water vapor budget equation turns out to be straightforward. So, we can summarize the primitive equations in sigma coordinates as follows:

Equation of motion

$$\frac{\partial \mathbf{v}_h}{\partial t} + (\mathbf{v}_h \cdot \nabla\mathbf{v}_h)_h + \dot{\sigma}\frac{\partial \mathbf{v}_h}{\partial \sigma} + f\mathbf{k}\times\mathbf{v}_h = -\nabla_h\phi - \frac{R_d T}{p_s}\nabla_h p_s + \mathbf{f}_T \quad (2.87)$$

Hydrostatic balance equation

$$\frac{\partial \phi}{\partial \sigma} = -\frac{R_d T}{\sigma} \quad (2.88)$$

Surface pressure tendency equation

$$\frac{\partial p_s}{\partial t} = -\int_0^1 \nabla_h \cdot (p_s\mathbf{v}_h)\, d\sigma \quad (2.89)$$

Diagnostic equation for vertical velocity

$$\dot{\sigma} = \frac{\sigma}{p_s}\int_0^1 \nabla_h \cdot (p_s\mathbf{v}_h)\, d\sigma - \frac{1}{p_s}\int_0^\sigma \nabla_h \cdot (p_s\mathbf{v}_h)\, d\sigma' \quad (2.90)$$

Temperature tendency equation

$$\frac{\partial T}{\partial t} + \mathbf{v}_h \cdot \nabla T + \dot{\sigma}\frac{\partial T}{\partial \sigma} = \frac{R_d T}{c_p p_s}$$

$$\times \left(\mathbf{v}_h \cdot \nabla p_s - \frac{1}{\sigma}\int_0^\sigma \nabla_h \cdot (p_s\mathbf{v}_h)\, d\sigma'\right) + \frac{Q}{c_p} \quad (2.91)$$

Budget equation for water vapor

$$\frac{\partial m_v}{\partial t} + \mathbf{v}_h \cdot \nabla m_v + \dot{\sigma}\frac{\partial m_v}{\partial \sigma} = S_{l,v} + D_{m_v} \quad (2.92)$$

Again, the coordinate index $[\sigma]$ has been omitted to simplify the notation. The primitive equations in sigma coordinates form the basis of many weather forecast models (e.g. Shuman and Hovermale, 1968) and general circulation models (e.g. Hoskins and Simmons, 1975). However, the **hybrid sigma-pressure coordinate** introduced by Simmons and Burridge (1981) represents a more advantageous vertical coordinate. It is close to sigma in the lower part of the atmosphere but turns into pressure at larger heights. This has the advantage that the coordinate surface still adapts to the surface but the hybrid levels are not perturbed by surface orography where it is unnecessary. The hybrid sigma-pressure coordinate comes into operation in the Global Forecast System (GFS) — Global Spectral Model (GSM) (Sela, 2010), the ECMWF model (Simmons *et al.*, 1989) and in the climate model ECHAM (Roeckner *et al.*, 2003).

2.4.3 $q_3 = [1 - \Pi(p)] H_a$ *(pseudo height)*

The **pseudo height** was introduced by Hoskins and Bretherton (1972) in a frontogenesis study. It is defined by

$$q_3 = z_p = [1 - \Pi(p)] H_a = \left[1 - \left(\frac{p}{p_0}\right)^{\frac{R_d}{c_p}}\right] \frac{c_p \theta_0}{g} \qquad (2.93)$$

where θ_0 is a reference potential temperature. Pseudo height coincides with true height if p_0 is the pressure at zero height and θ_0 the vertically independent potential temperature. We can prove this with the hydrostatic balance equation. Using the definition of potential temperature (1.141), we get

$$\frac{\partial p}{\partial z} = -g \frac{p}{R_d \theta_0 \Pi} = -g \frac{p}{R_d \theta_0} \left(\frac{p}{p_0}\right)^{-\frac{R_d}{c_p}} \quad \leftrightarrow \quad \frac{c_p}{R_d} \frac{\partial \Pi}{\partial z} = -\frac{g}{R_d \theta_0} \qquad (2.94)$$

Vertical integration of this equation yields

$$\frac{c_p}{R_d}(\Pi - 1) = -\frac{gz}{R_d \theta_0} \quad \leftrightarrow \quad z = (1 - \Pi) \frac{c_p \theta_0}{g} \qquad (2.95)$$

Obviously, $H_a = c_p \theta_0 / g$ is the height of the isentropic atmosphere having the potential temperature θ_0 and surface pressure p_0.

The transformation of the primitive equations does not require many manipulations since the pseudo height depends on pressure

p only. Therefore, the transformation rules from isobaric coordinates become

$$\frac{\partial b_{[p]}}{\partial p} = \left(\frac{\partial p_{[z_p]}}{\partial z_p}\right)^{-1} \frac{\partial b_{[z_p]}}{\partial z_p} = -\frac{R_d H_a \Pi_{[z_p]}}{c_p p_{[z_p]}} \frac{\partial b_{[z_p]}}{\partial z_p} \tag{2.96}$$

$$\frac{\partial b_{[p]}}{\partial s} = \frac{\partial b_{[z_p]}}{\partial s} \quad \text{for} \quad s = \lambda, \varphi \, \text{or} \, t \tag{2.97}$$

The partial derivatives with respect to all coordinates except the vertical remain unchanged because coordinate surfaces in isobaric and pseudo-height coordinates coincide. Therefore, the equation of motion has the same form as in isobaric coordinates except for the vertical advection term. The hydrostatic balance equation can be transformed by using (2.96)

$$\frac{\partial \phi_{[p]}}{\partial p} = -\frac{R_d H_a \Pi_{[z_p]}}{c_p p_{[z_p]}} \frac{\partial \phi_{[z_p]}}{\partial z_p} = -\frac{R_d T_{[z_p]}}{p_{[z_p]}} \tag{2.98}$$

This equation is equivalent to

$$\frac{\partial \phi_{[z_p]}}{\partial z_p} = \frac{c_p \theta_{[z_p]}}{H_a} = g \frac{\theta_{[z_p]}}{\theta_0} \tag{2.99}$$

For transformation of the continuity equation (2.72), one must determine the relation between the vertical velocities of the two coordinate systems. We get

$$w_p \equiv \frac{Dz_p}{Dt} = \left(\frac{\partial p_{[z_p]}}{\partial z_p}\right)^{-1} \omega_{[z_p]} = -\frac{R_d H_a \Pi_{[z_p]}}{c_p p_{[z_p]}} \omega_{[z_p]} = -\frac{\omega_{[z_p]}}{g \rho_p (z_p)} \tag{2.100}$$

where $\rho_p (z_p)$ denotes the **pseudo density** which is given by

$$\rho_p (z_p) \equiv \frac{p_0}{R_d \theta_0} \left(1 - \frac{z_p}{H_a}\right)^{\frac{c_p}{R_d} - 1} \tag{2.101}$$

The pseudo density corresponds to the true density in the case of an isentropic atmosphere having the potential temperature θ_0 and surface pressure p_0. Using the result (2.100) in the continuity equation

(2.72) yields

$$\nabla_h \cdot \mathbf{v}_{h[z_p]} + \left(\frac{\partial p_{[z_p]}}{\partial z_p}\right)^{-1} \frac{\partial}{\partial z_p}(-g\rho_p w_p)$$

$$= \nabla_h \cdot \mathbf{v}_{h[z_p]} + \frac{1}{\rho_p}\frac{\partial}{\partial z_p}(\rho_p w_p) = 0 \qquad (2.102)$$

Consequently, we can write the primitive equation with the pseudo height vertical coordinate as follows:

Equation of motion

$$\frac{\partial \mathbf{v}_h}{\partial t} + (\mathbf{v}_h \cdot \nabla \mathbf{v}_h)_h + w_p \frac{\partial \mathbf{v}_h}{\partial z_p} + f\mathbf{k}\times\mathbf{v}_h = -\nabla_h\phi + \boldsymbol{f}_T \qquad (2.103)$$

Hydrostatic balance equation

$$\frac{\partial \phi}{\partial z_p} = g\frac{\theta}{\theta_0} \qquad (2.104)$$

Continuity equation

$$\nabla_h \cdot \mathbf{v}_h + \frac{1}{\rho_p}\frac{\partial}{\partial z_p}\left(\rho_p w_p\right) = 0 \qquad (2.105)$$

Temperature tendency equation

$$\frac{\partial \theta}{\partial t} + \mathbf{v}_h \cdot \nabla\theta + w_p\frac{\partial \theta}{\partial z_p} = \frac{Q}{c_p\Pi} \qquad (2.106)$$

Budget equation for water vapor

$$\frac{\partial m_v}{\partial t} + \mathbf{v}_h \cdot \nabla m_v + w_p\frac{\partial m_v}{\partial z_p} = S_{l,v} + D_{m_v} \qquad (2.107)$$

These equations have the advantage of a linear relationship between the vertical gradient of geopotential height and potential temperature. Therefore, pseudo height has been used for theoretical studies which favor easily manageable equations. It is notable that the primitive equations based on pseudo height have the same form as the **anelastic equations** derived by Ogura and Phillips (1962) except that these authors do not make a quasihydrostatic approximation and use true height as the vertical coordinate.[d] Consequently, the anelastic approximation made by Ogura and Phillips (1962) can be circumvented by using pseudo height. However, this advantage is compensated by the drawback of complicated boundary conditions since constant pseudo height corresponds to constant pressure which varies at the surface.

2.4.4 $q_3 = \theta$ (potential temperature)

The replacement of height by potential temperature yields **isentropic coordinates**. Potential temperature increases with height above the boundary layer since the atmosphere has always a stable stratification there. Isentropic coordinates have the advantage that the vertical velocity can be approximately neglected since θ is nearly conserved above the planetary boundary. Furthermore, regions with sharp horizontal temperature contrasts do not pose a big problem for numerical modeling because they vanish on isentropic surfaces.

Applying the transformation rule (2.53) to the sum of horizontal pressure gradient force and body force yields

$$-\frac{1}{\rho}\nabla_h p - \nabla_h\phi = -\frac{\nabla_h p_{[\theta]}}{\rho_{[\theta]}} + \frac{\partial p}{\partial z}\frac{\nabla_h z_{[\theta]}}{\rho} - \nabla_h\phi_{[\theta]} + \frac{\partial\phi}{\partial z}\nabla_h z_{[\theta]}$$

$$= -\frac{1}{\rho_{[\theta]}}\nabla_h p_{[\theta]} - \nabla_h\phi_{[\theta]} \qquad (2.108)$$

Therefore, the right-hand side takes the same form as in the original equation set.[e] However, we can manipulate the pressure gradient force further by using the definition of potential temperature (1.141) and the thermal equation of state (2.48).

$$-\frac{1}{\rho_{[\theta]}}\nabla_h p_{[\theta]} = -\frac{R_d T_{[\theta]}}{p_{[\theta]}}\nabla_h\left[p_0\left(\frac{T_{[\theta]}}{\theta}\right)^{\frac{c_p}{R_d}}\right] = -c_p\nabla_h T_{[\theta]} \qquad (2.109)$$

---

[d]We will treat the anelastic equations in Section 6.2.
[e]Indeed, this is true for every new vertical coordinate.

With this relation, we can write the sum of pressure gradient and body force as a gradient of the **Montgomery potential** ϕ_M (Montgomery, 1937):

$$-\frac{1}{\rho_{[\theta]}}\nabla_h p_{[\theta]} - \nabla_h \phi_{[\theta]} = -\nabla_h \phi_M \quad \text{with } \phi_M = c_p T_{[\theta]} + \phi_{[\theta]} \quad (2.110)$$

Transforming the hydrostatic balance equation leads to

$$\frac{\partial p}{\partial z} = \left(\frac{\partial z_{[\theta]}}{\partial \theta}\right)^{-1}\frac{\partial p_{[\theta]}}{\partial \theta} = -g\rho_{[\theta]}$$

$$\leftrightarrow \quad \frac{\partial p_{[\theta]}}{\partial \theta} = -\frac{\partial \phi}{\partial z}\rho_{[\theta]}\frac{\partial z_{[\theta]}}{\partial \theta} = -\rho_{[\theta]}\frac{\partial \phi_{[\theta]}}{\partial \theta} \quad (2.111)$$

On the other hand, the derivative of pressure with respect to potential temperature gives

$$\frac{\partial p_{[\theta]}}{\partial \theta} = \frac{\partial}{\partial \theta}\left[p_0\left(\frac{T_{[\theta]}}{\theta}\right)^{\frac{c_p}{R_d}}\right] = -\frac{c_p}{R_d}\left(\frac{p_{[\theta]}}{\theta} - \frac{p_{[\theta]}}{T_{[\theta]}}\frac{\partial T_{[\theta]}}{\partial \theta}\right)$$

$$= -c_p\rho_{[\theta]}\left(\Pi_{[\theta]} - \frac{\partial T_{[\theta]}}{\partial \theta}\right) \quad (2.112)$$

Therefore, the hydrostatic balance equation can also be written as

$$\frac{\partial \phi_M}{\partial \theta} = c_p\Pi_{[\theta]} \quad (2.113)$$

The vertical velocity in isentropic coordinates is simply

$$\dot{\theta} = \frac{Q_{[\theta]}}{c_p\Pi_{[\theta]}} \quad (2.114)$$

Consequently, under adiabatic conditions, the vertical velocity vanishes identically and the parcel moves along isentropic surfaces.

We obtain for the divergence of the velocity field in isentropic coordinates.

$$\nabla_h \cdot \mathbf{v}_h + \frac{\partial w}{\partial z} = \nabla_h \cdot \mathbf{v}_{h[\theta]} - \left(\frac{\partial z_{[\theta]}}{\partial \theta}\right)^{-1} \left(\frac{\partial \mathbf{v}_{h[\theta]}}{\partial \theta} \cdot \nabla_h z_{[\theta]} - \frac{\partial w_{[\theta]}}{\partial \theta}\right)$$

$$= \nabla_h \cdot \mathbf{v}_{h[\theta]} + \frac{\partial \dot{\theta}}{\partial \theta} + \left(\frac{\partial z_{[\theta]}}{\partial \theta}\right)^{-1} \frac{D}{Dt}\left(\frac{\partial z_{[\theta]}}{\partial \theta}\right)$$

$$(2.115)$$

The last term can be summarized with the material derivative of density

$$\left(\frac{\partial z_{[\theta]}}{\partial \theta}\right)^{-1} \frac{D}{Dt}\left(\frac{\partial z_{[\theta]}}{\partial \theta}\right) + \frac{1}{\rho_{[\theta]}} \frac{D\rho_{[\theta]}}{Dt} = \frac{D}{Dt}\left[\ln\left(\rho_{[\theta]}\frac{\partial z_{[\theta]}}{\partial \theta}\right)\right]$$

$$= \frac{D}{Dt}\left\{\ln\left[\rho_{[\theta]}\left(\frac{\partial p}{\partial z}\right)^{-1}\frac{\partial p_{[\theta]}}{\partial \theta}\right]\right\} = \frac{D}{Dt}\left[\ln\left(-\frac{1}{g}\frac{\partial p_{[\theta]}}{\partial \theta}\right)\right] \quad (2.116)$$

Consequently, the continuity equation in isentropic coordinates becomes

$$\frac{D}{Dt}\left[\ln\left(-\frac{1}{g}\frac{\partial p_{[\theta]}}{\partial \theta}\right)\right] + \nabla_h \cdot \mathbf{v}_{h[\theta]} + \frac{\partial \dot{\theta}}{\partial \theta} = 0 \quad\quad (2.117)$$

or alternatively

$$\frac{\partial}{\partial t}\left(\frac{\partial p_{[\theta]}}{\partial \theta}\right) + \nabla_h \cdot \left(\mathbf{v}_{h[\theta]}\frac{\partial p_{[\theta]}}{\partial \theta}\right) + \frac{\partial}{\partial \theta}\left(\dot{\theta}\frac{\partial p_{[\theta]}}{\partial \theta}\right) = 0 \quad (2.118)$$

To obtain a prognostic equation for pressure, one must integrate this equation vertically downward from the top of the atmosphere

$$\frac{\partial p_{[\theta]}}{\partial t} - \int_{\theta}^{\theta_t} \nabla_h \cdot \left(\mathbf{v}_{h[\theta]}\frac{\partial p_{[\theta]}}{\partial \theta}\right) d\theta' + \dot{\theta}\frac{\partial p_{[\theta]}}{\partial \theta} = 0 \quad (2.119)$$

where θ_t denotes the potential temperature at the top of the atmosphere which should theoretically be infinite but is in practical modeling a finite number. At the top of the atmosphere, we can assume that the heating $Q_{[\theta]}$ vanishes and, therefore, also the vertical velocity $\dot{\theta}$.

Finally, we can write the primitive equations in isentropic coordinates as follows:
Equation of motion

$$\frac{\partial \mathbf{v}_h}{\partial t} + (\mathbf{v}_h \cdot \nabla \mathbf{v}_h)_h + \dot{\theta}\frac{\partial \mathbf{v}_h}{\partial \theta} + f\mathbf{k}\times\mathbf{v}_h = -\nabla_h \phi_M + \mathbf{f}_T \quad (2.120)$$

Hydrostatic balance equation

$$\frac{\partial \phi_M}{\partial \theta} = c_p \Pi \quad (2.121)$$

Definition of Montgomery potential (see Eq. 2.110)

$$\phi_M = c_p \theta \Pi + \phi \quad (2.122)$$

Pressure tendency equation

$$\frac{\partial p}{\partial t} - \int_\theta^{\theta_t} \nabla_h \cdot \left(\mathbf{v}_h \frac{\partial p}{\partial \theta}\right) d\theta' + \dot{\theta}\frac{\partial p_{[\theta]}}{\partial \theta} = 0 \quad (2.123)$$

Diagnostic equation for vertical velocity

$$\dot{\theta} = \frac{Q}{c_p \Pi} \quad (2.124)$$

Budget equation for water vapor

$$\frac{\partial m_v}{\partial t} + \mathbf{v}_h \cdot \nabla m_v + \dot{\theta}\frac{\partial m_v}{\partial \theta} = S_{l,v} + D_{m_v} \quad (2.125)$$

An important feature of isentropic coordinates is the convenient derivation of a PV equation in the adiabatic case. Then, the vertical velocity vanishes and the PV advection can be analyzed in two dimensions. To obtain an equation for PV, we must first derive the vorticity equation. For this purpose, we recast the momentum advection terms as follows:

$$\mathbf{v}_h \cdot \nabla_h u - \tan\varphi \frac{uv}{r_E} = \frac{1}{r_E \cos\varphi}\frac{\partial}{\partial\lambda}\left(\frac{1}{2}\mathbf{v}_h^2\right) - \mathbf{k}\cdot\nabla_h \times \mathbf{v}_h v \quad (2.126)$$

$$\mathbf{v}_h \cdot \nabla_h v + \tan\varphi \frac{u^2}{r_E} = \frac{1}{r_E}\frac{\partial}{\partial\varphi}\left(\frac{1}{2}\mathbf{v}_h^2\right) + \mathbf{k}\cdot\nabla_h \times \mathbf{v}_h u \quad (2.127)$$

Therefore, we can write the equation of motion (2.120) also in the form

$$\frac{\partial \mathbf{v}_h}{\partial t} + \dot{\theta}\frac{\partial \mathbf{v}_h}{\partial \theta} + (f + \xi_\theta)\mathbf{k}\times\mathbf{v}_h = -\nabla_h\left(\phi_M + \frac{1}{2}\mathbf{v}_h^2\right) + \mathbf{f}_T \quad (2.128)$$

where $\xi_\theta = \mathbf{k}\cdot\nabla_h\times\mathbf{v}_h$ denotes the vertical component of relative vorticity in isentropic coordinates. Applying the curl operator $\mathbf{k}\cdot\nabla_h\times$

to this equation yields the vorticity equation in isentropic coordinates

$$\frac{\partial \xi_\theta}{\partial t} + \mathbf{v}_h \cdot \nabla_h \left(f + \xi_\theta \right) + \dot{\theta} \frac{\partial \xi_\theta}{\partial \theta} + \mathbf{k} \cdot \nabla_h \dot{\theta} \times \frac{\partial \mathbf{v}_h}{\partial \theta}$$

$$= -(f + \xi_\theta) \nabla_h \cdot \mathbf{v}_h + \mathbf{k} \cdot \nabla_h \times \boldsymbol{f}_T \qquad (2.129)$$

In this vorticity equation, no solenoid term occurs because on isentropic surfaces the density becomes a function of pressure only. We can also write the vorticity equation in the Lagrangian frame of reference:

$$\frac{D}{Dt} (f + \xi_\theta) = -(f + \xi_\theta) \nabla_h \cdot \mathbf{v}_h - \mathbf{k} \cdot \nabla_h \dot{\theta} \times \frac{\partial \mathbf{v}_h}{\partial \theta} + \mathbf{k} \cdot \nabla_h \times \boldsymbol{f}_T$$

$$(2.130)$$

The stretching term on the right-hand side can be eliminated by using the continuity equation (2.117). This leads to

$$\frac{D}{Dt} (f + \xi_\theta) = \frac{f + \xi_\theta}{\frac{\partial p}{\partial \theta}} \frac{D}{Dt} \left(\frac{\partial p}{\partial \theta} \right) + (f + \xi_\theta) \frac{\partial \dot{\theta}}{\partial \theta} - \mathbf{k} \cdot \nabla_h \dot{\theta} \times \frac{\partial \mathbf{v}_h}{\partial \theta}$$

$$+ \mathbf{k} \cdot \nabla_h \times \boldsymbol{f}_T \qquad (2.131)$$

Merging both material time derivatives yields the following PV equation:

$$\frac{D P_\theta}{Dt} = -g \left(\frac{\partial p}{\partial \theta} \right)^{-1} \left[(f + \xi_\theta) \frac{\partial \dot{\theta}}{\partial \theta} - \mathbf{k} \cdot \nabla_h \dot{\theta} \times \frac{\partial \mathbf{v}_h}{\partial \theta} + \mathbf{k} \cdot \nabla_h \times \boldsymbol{f}_T \right]$$

$$(2.132)$$

where the PV is defined by

$$P_\theta = -g \left(f + \xi_\theta \right) \left(\frac{\partial p}{\partial \theta} \right)^{-1} \qquad (2.133)$$

This expression for PV is referred to as **isentropic PV** (IPV) and is individually conserved under adiabatic and frictionless conditions. IPV can be calculated in a much simpler way as Ertel's PV since only the vertical component of vorticity and the vertical derivative of potential temperature are needed. Therefore, IPV became a popular diagnostic for extratropical weather analysis (see Hoskins *et al.*, 1985).

Figure 2.5(a) shows the IPV field on the 330 K isentropic surface for the same date and time as chosen for Fig. 2.2. The unit for

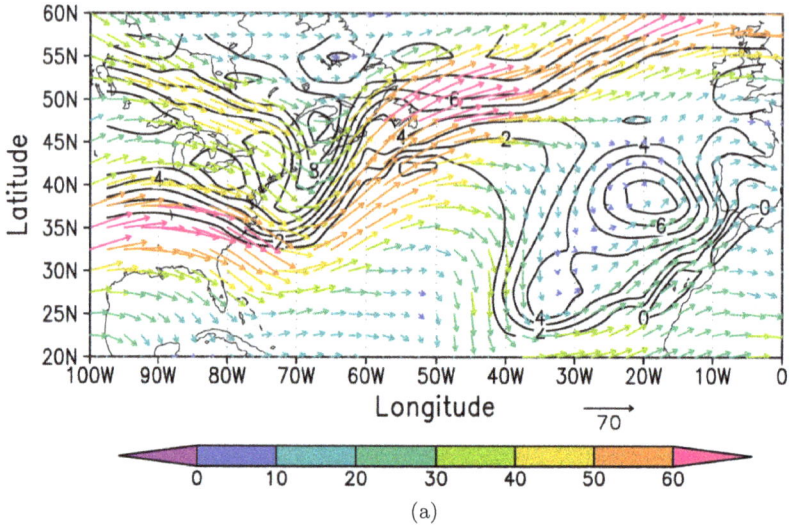

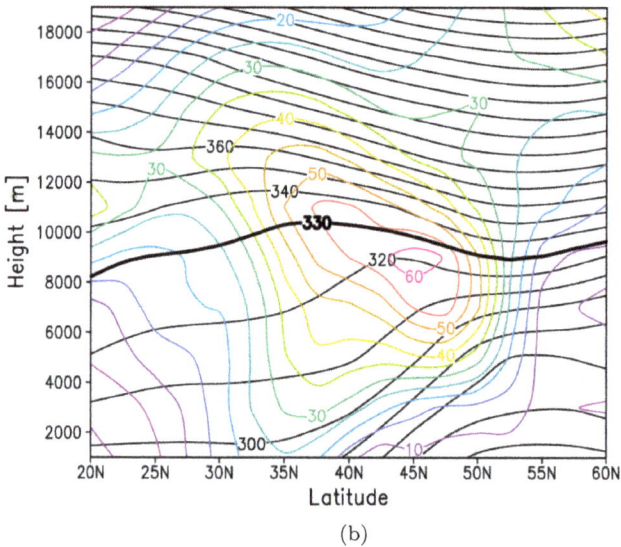

Fig. 2.5. (a) IPV (isolines, contour interval 1 PVU) and wind field on 26th of January 2008, 00UTC on the 330 K isentropic surface, (b) potential temperature (black isolines, contour interval 10 K) and wind speed in a meridional cross-section at 60 W longitude. The bold black isoline highlights the cutline of the 330 K surface.

IPV is 1 PVU $= 10^{-6}\,\mathrm{km}^2\,\mathrm{kg}^{-1}\,\mathrm{s}^{-1}$ (one **potential vorticity unit**). An enhanced gradient of IPV appears in regions with high horizontal wind speeds. This region marks the **polar front** that separates subtropical from polar air. The latter has higher IPV since air on the 330 K isentropic surface is already in the stratosphere where the vertical gradient of potential temperature attains larger values. This can be clearly seen in the cross-section displayed in Fig. 2.5(b) showing isolines of potential temperature and wind speed. The highlighted 330 K contour crosses the jet stream at a height of about 9 km. There, the vertical distance between the isolines shrinks dramatically in the northward direction because north of the jet stream the tropopause height is low, and consequently, the 330 K contour runs through the stratosphere. A strong positive IPV anomaly appears at 70 W and 45 N just upstream of the low-level extratropical cyclone seen in Fig. 2.2(a). This anomaly is connected with an upper-level trough that leads to further intensification of the cyclone by baroclinic instability which will be explained in Section 4.6. Another significant positive IPV anomaly occurs west of Spain which represents a cut-off low. Cut-off lows develop from wave breaking and isolate polar air equatorward of the polar front. The cutting-off process cannot take place without irreversible processes like mixing. This shows that the assumption of IPV conservation is not in agreement with cut-off low development. However, IPV anomalies persist longer than anomalies of other quantities like vorticity or pressure and turn out to be favorable for diagnosing air masses and predicting synoptic-scale weather systems.

2.5 Energetics Based on the Primitive Equations

The energetics presented in Section 1.12 requires some modifications to be consistent with the approximations based on the scale analysis. First of all, there is no equation of motion for the vertical velocity. Therefore, kinetic energy only involves the horizontal components of the velocity vector so that

$$e_K\left(\mathbf{v}_h^2\left(\mathbf{r},t\right)t\right) = \frac{1}{2}\mathbf{v}_h^2\left(\mathbf{r},t\right) \qquad (2.134)$$

However, the prognostic energy equations (1.190)–(1.193) keep its form because the kinetic energy conversion terms involving vertical velocity, i.e. $-w\partial\phi/\partial z$ and $-w/\rho\partial p/\partial z$, are offset against each other because of the hydrostatic balance equation (2.43).

There is another peculiarity of the primitive equations regarding energetics. Inner energy and potential energy are independent quantities in the system without the shallow atmosphere approximation. However, in the primitive equation system, these energies are related to each other. To show this, we evaluate the potential energy of an atmospheric column having the base area a_F:

$$E_P = \iiint_{V_F} \rho\phi dV = \iint_{a_F} \int_{z_s}^{\infty} \rho\phi dz \, da = \frac{1}{g} \iint_{a_F} \int_0^{p_s} \phi dp \, da$$

$$(2.135)$$

The last conversion includes a substitution of the integration variable height by pressure. Now, we find by partial integration and the hydrostatic balance equation in isobaric coordinates

$$\int_0^{p_s} \phi dp = [p\phi]\big|_0^{p_s} - \int_0^{p_s} p\frac{\partial\phi}{\partial p} dp = [p\phi]\big|_0^{p_s} + \int_0^{p_s} R_d T dp \quad (2.136)$$

The term in brackets is not determined since the upper boundary of the atmosphere is not properly defined. However, it is reasonable to assume that the pressure vanishes there while the geopotential remains finite. Then, the potential energy becomes

$$E_P = \frac{1}{g} \iint_{a_F} \int_0^{p_s} R_d T dp da + \frac{1}{g} \iint_{a_F} p_s \phi_s da \qquad (2.137)$$

Consequently, the potential energy is related to inner kinetic energy as follows:

$$E_P = \frac{R_d}{c_v} E_{IK} + \frac{1}{g} \iint_{a_F} p_s \phi_s da \qquad (2.138)$$

Hence, an increase of inner energy results in an increase of potential energy and vice versa if the surface geopotential ϕ_s is zero. This condition can be assumed in a small area with a flat bottom. The relation has the consequence that the primitive equations cannot describe **sound waves** since these stand out by a periodic conversion of kinetic energy into inner energy without the incorporation of potential energy. This shortcoming turns out to be a big advantage for solving the primitive equations numerically with a computer. Sound waves have large phase speeds ($\sim$300 m/s) and the numerical simulation of a vertically propagating sound wave would require a tiny time step so that the numerical solution would include a large number of calculation steps.

Margules (1905) united inner and potential energies into a single energy form which he called **total potential energy** and is given by

$$E_{TP} = \frac{1}{g} \iint_{a_F} \int_0^{p_s} c_p T dp da + \frac{1}{g} \iint_{a_F} p_s \phi_s da \qquad (2.139)$$

With this expression, we can formulate global energy equations by integrating over the entire Earth's surface, a_E. They become

$$\frac{dE_{TP}}{dt} = -\frac{1}{g} \iint_{a_E} \int_0^{p_s} L_v S_{l,v} dp da$$

$$+ \frac{1}{g} \iint_{a_E} \int_0^{p_s} \mathbf{v}_h \cdot \left(\frac{1}{\rho} \nabla_h p + \nabla_h \phi \right) dp da$$

$$+ \iint_{a_E} j_{R,s} - j_{R,t} da + \iint_{a_E} j_{e_I,s} da \qquad (2.140)$$

$$\frac{dE_{IP}}{dt} = \frac{1}{g} \iint_{a_E} \int_0^{p_s} L_v S_{l,v} dp da + \iint_{a_E} j_{m_v,s} da \qquad (2.141)$$

$$\frac{dE_K}{dt} = -\frac{1}{g} \iint_{a_E} \int_0^{p_s} \mathbf{v}_h \cdot \left(\frac{1}{\rho} \nabla_h p + \nabla_h \phi \right) dp da$$

$$+ \frac{1}{g} \iint_{a_E} \int_0^{p_s} \mathbf{v}_h \cdot \boldsymbol{f}_T dp da \qquad (2.142)$$

where $j_{R,s}$, $j_{e_I,s}$ and $j_{m_v,s}$ denote the fluxes of **net radiation** (upward minus downward), **sensible heat** (inner energy flux) and **latent heat** (water vapor) at the surface, respectively, while $j_{R,t}$ is the flux of net radiation at the top of the atmosphere. Note that the contribution by the vertical velocity in the second integral of (2.140) vanishes because of the hydrostatic balance equation. Obviously, the energy equations for the complete atmosphere only involve three

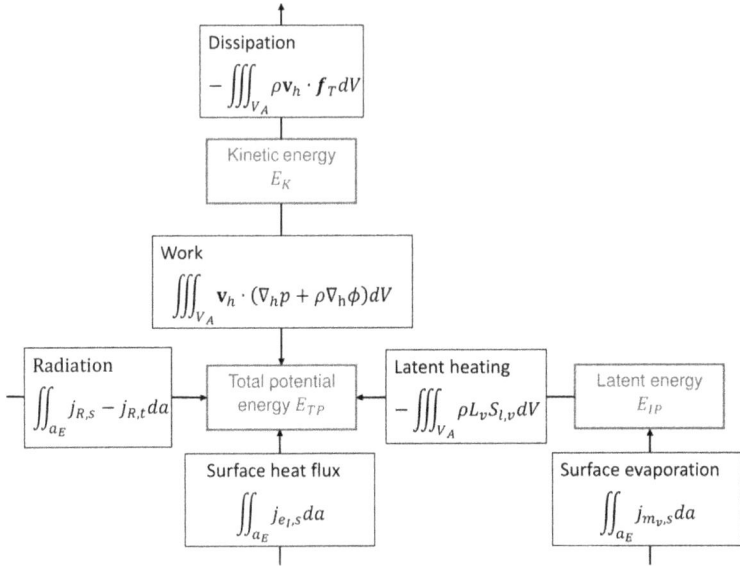

Fig. 2.6. Energetical conversions in the global atmosphere as described by the primitive equations.

forms of energy. The conversions between these energies are sketched in Fig. 2.6. Total potential energy builds up by radiation, sensible surface heat flux and latent heating (condensation or sublimation). Latent heating reduces latent energy that is recovered by evaporation at the surface. Kinetic energy is converted into total potential energy by work. Finally, dissipation diminishes kinetic energy. However, within the primitive equations, the conversion of dissipated kinetic energy into inner energy is neglected, but can be introduced. Table 2.1 shows average values for the various energy forms. We see that inner and potential energies have the largest magnitudes. The ratio between inner and potential energy yields 2.6 which is close to the ratio $c_v/R_d = 2.5$ (see Eq. 2.138).

2.6 Available Potential Energy

Lorenz (1955) introduced **available potential energy** (APE) which refers to the part of total potential energy that is available for conversion into kinetic energy. He followed an idea by Margules (1905) that

Table 2.1. Typical energy values.

Energy form	Value (J/m^2)
Inner energy, E_{IK}	180.3
Potential energy, E_P	69.3
Latent energy, E_{IP}	6.38
Kinetic energy, E_K	0.123
Total energy, $E_{IK} + E_P + E_{IP} + E_K$	256.1

Source: After Peixoto and Oort (1992).

there is a minimum amount of total potential energy in the atmosphere if only adiabatic processes take place. At this minimum, the isentropic surfaces coincide with geopotential isosurfaces as sketched in Fig. 2.7. Then, the sum of horizontal pressure and geopotential gradient forces vanishes so that no kinetic energy change takes place anymore. An adiabatic transformation of the atmosphere to the state of minimum potential energy could happen by mass transport along isentropic surfaces which induces a vertical displacement of these surfaces. Because of this conception, it is suitable to evaluate the available potential energy in isentropic coordinates. We get for the enthalpy of the atmosphere

$$\frac{1}{g} \iint_{a_E} \int_0^{p_s} c_p T \, dp \, da = -\frac{1}{g} \iint_{a_E} \int_{\theta_s}^{\infty} c_p \theta \left(\frac{p}{p_0}\right)^{\kappa} \frac{\partial p}{\partial \theta} \, d\theta \, da$$

$$= -\frac{1}{g} \iint_{a_E} \int_{\theta_s}^{\infty} \frac{c_p}{(\kappa + 1) p_0^{\kappa}} \theta \frac{\partial p^{\kappa+1}}{\partial \theta} \, d\theta \, da$$

(2.143)

where $\kappa = R_d/c_p$. To obtain a more convenient expression, Lorenz (1955) introduced formally a massless layer at the surface in which the potential temperature decreases from θ_s to 0. Then, integration by parts yields

$$\frac{1}{g} \iint_{a_E} \int_0^{p_s} c_p T \, dp \, da = -\frac{1}{g} \iint_{a_E} \int_0^{\infty} \frac{c_p}{(\kappa + 1) p_0^{\kappa}} \theta \frac{\partial p^{\kappa+1}}{\partial \theta} \, d\theta \, da$$

$$= \frac{c_p}{g (\kappa + 1) p_0^{\kappa}} \left\{ \iint_{a_E} \int_0^{\infty} p^{\kappa+1} d\theta - \left[\theta p^{\kappa+1}\right]_0^{\infty} da \right\}$$

(2.144)

The term in squared brackets vanishes because the temperature remains finite at the upper boundary of the atmosphere. With this

Typical atmospheric state   Reference state
                                          (after adiabatic transformation)

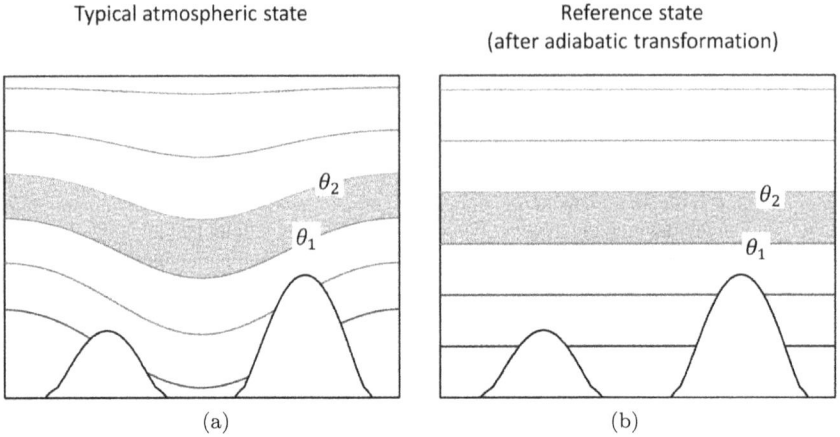

(a)   (b)

Fig. 2.7.   Sketch of the adiabatic transformation: (a) isentropes (isolines) of a
typical atmospheric state; (b) isentropes of the reference state having horizontally
uniform fields.

result, we obtain for the available potential energy

$$E_A = E_{TP} - E_{TP,r}$$

$$= \frac{1}{g} \iint_{a_F} (p_s - p_{s,r})\, \phi_s da + \frac{c_p}{g\,(\kappa + 1)\, p_0^\kappa}$$

$$\times \iint_{a_E} \int_0^\infty p^{\kappa+1} - p_r^{\kappa+1} d\theta da \qquad (2.145)$$

where the index r denotes the state after adiabatic transformation.
The globally averaged pressure does not change due to mass conservation. Therefore, the pressure p_r after adiabatic transformation is
identical to the globally averaged pressure when the isentropic surface does not intersect mountains. Lorenz (1955) ignored mountains
in his proof that E_A has a positive sign.[f] Then, we can set $\phi_s = 0$
and get the relations

$$p_r = \frac{1}{a_E} \iint_{a_E} p\, da \quad \text{and} \quad \iint_{a_E} p^{\kappa+1} - p_r^{\kappa+1} da \geq 0. \qquad (2.146)$$

---

[f]Taylor (1979) proved that E_A has a positive sign even over uneven topography.

This inequality relation proves that available potential energy is positive and can be explained by the higher amplitudes of positive anomalies of $p^{\kappa+1} - p_r^{\kappa+1}$ compared to negative anomalies of the same quantity when $\kappa + 1 > 1$. The available potential energy takes a simpler form after applying a Taylor approximation to the second order:

$$E_A = \frac{c_p}{g\,(\kappa+1)\,p_0^\kappa} \iint_{a_E} \int_0^\infty p^{\kappa+1} - p_r^{\kappa+1} d\theta da$$

$$\approx \frac{\kappa c_p}{2gp_0^\kappa} \iint_{a_E} \int_0^\infty p_r^{\kappa+1} \left(\frac{p'}{p_r}\right)^2 d\theta da \qquad (2.147)$$

where p' denotes the deviation from the reference pressure p_r on an isentropic surface. Now, the positive sign of E_A is obvious. Lorenz (1955) found that the error introduced by this approximation is low even in extreme cases. Usually, available potential energy is considered in isobaric coordinates. To transform the integrand, we first note that the time derivative of pressure in isentropic coordinates is related to the time derivative of potential temperature in isobaric coordinates as follows (cf. Eq. 2.55):

$$\frac{\partial p_{[\theta]}}{\partial t} = -\frac{\partial p_{[\theta]}}{\partial \theta}\frac{\partial \theta_{[p]}}{\partial t} \qquad (2.148)$$

Time integration over the period of the adiabatic transformation yields approximately a relation between the anomalies

$$p'_{[\theta]} \approx -\frac{\partial p_{r[\theta]}}{\partial \theta}\theta'_{[p]} \qquad (2.149)$$

With this approximation, we find for the available potential energy in isobaric coordinates

$$E_A \approx -\frac{\kappa c_p}{2gp_0^\kappa} \iint_{a_E} \int_0^{p_{s,r}} \theta_r^2 p^{\kappa-1} \left(\frac{\theta'}{\theta_r}\right)^2 \left(\frac{\partial \theta_r}{\partial p}\right)^{-2} \frac{\partial \theta}{\partial p} dpda$$

$$\approx \frac{\kappa c_p}{2gp_0^\kappa} \iint_{a_E} \int_0^{p_{s,r}} \theta_r^2 p^{\kappa-1} \left(\frac{\theta'}{\theta_r}\right)^2 \left(-\frac{\partial \theta_r}{\partial p}\right)^{-1} dpda \qquad (2.150)$$

The vertical gradient of potential temperature can be recast using Eq. (2.69):

$$\frac{\partial \theta_r}{\partial p} = \left(\frac{p_0}{p}\right)^{\kappa} \left(\frac{\partial T_r}{\partial p} - \frac{R_d T_r}{c_p p}\right) = -\kappa \frac{c_p \theta_r}{gp} \left(\frac{g}{c_p} + \frac{\partial T_r}{\partial z}\right)$$

$$= -\kappa \frac{c_p \theta_r}{gp} (\Gamma_d - \Gamma) \qquad (2.151)$$

where $\Gamma_r = -\partial T_r/\partial z$ and $\Gamma_d = g/c_p$ denote the temperature lapse rate of the reference state and the isentropic atmosphere, respectively. Since $\theta'/\theta_r = T'/T_r$ on isobaric surfaces, we get for the available potential energy

$$E_A \approx \frac{1}{2} \iint_{a_E} \int_0^{p_{s,r}} \frac{T_r}{\Gamma_d - \Gamma} \left(\frac{T'}{T_r}\right)^2 dp \, da \qquad (2.152)$$

Lorenz (1955) found that the typical magnitude of E_A is only about 0.5% to that of the total potential energy E_T. This is still larger by the factor of seven than the kinetic energy E_K. Therefore, a large amount of E_A remains. The formula (2.152) is not suitable for consistent energy budgets since it involves several approximations and the associated error can have a magnitude comparable to that of kinetic energy.

Chapter 3

Waves in the Barotropic Atmosphere

3.1 Introduction

The first successful numerical weather forecast by Charney *et al.* (1950) was conducted with a barotropic model. It seems that barotropic dynamics suffice to mimic the movement of extratropical weather systems although baroclinic processes play a relevant role in their development. Therefore, we devote this chapter exclusively to the dynamics based on barotropic primitive equations. In this way, we obtain an easier understanding of atmospheric waves which are important for weather dynamics. First, we describe how the primitive equations simplify in the barotropic case and show that these equations are nearly identical to the shallow water equations. These form the basis for further analysis where we consider various wave types, adjustment mechanisms and instabilities. Finally, the quasi-geostrophic approximation forms an adequate method for filtering atmospheric noise so that weather systems evolve on the basis of Rossby wave dynamics. Finally, we consider instabilities in barotropic flows.

3.2 Barotropic Primitive Equations

The primitive equations have barotropic solutions if heating vanishes. Then, the potential temperature is conserved and the isentropic atmosphere ($\theta = \theta_0 =$ const.) represents the initial state of a possible barotropic solution. The isentropic atmosphere is barotropic since the density becomes a function of pressure only. This can be seen with the thermal equation of state by substituting the Exner function $\Pi = (p/p_0)^{R_d/c_p}$

$$\rho = \frac{p}{R_d T} = \frac{p}{R_d \theta \Pi} = \frac{p}{R_d \theta_0 \Pi} = \rho(p) \qquad (3.1)$$

The sole dependency on pressure leads to a simplification of the pressure gradient force

$$\frac{1}{\rho}\nabla p = \frac{1}{\rho_0}\nabla p_\rho$$

$$\text{with} \quad p_\rho = \int_0^p \frac{\rho_0}{\rho(p')}\,dp' = \int_0^p \frac{\rho_0 R_d \theta_0 \Pi(p')}{p'}\,dp' = c_p \rho_0 \theta_0 \Pi \quad (3.2)$$

where ρ_0 is the mean density. Therefore, the hydrostatic balance equation becomes

$$\frac{1}{\rho}\frac{\partial p}{\partial z} = \frac{1}{\rho_0}\frac{\partial p_\rho}{\partial z} = c_p \theta_0 \frac{\partial \Pi}{\partial z} = -g \qquad (3.3)$$

This equation can be easily integrated so that we obtain the vertical temperature profile

$$T = \theta_0 \Pi = \theta_0 \Pi_s - \frac{g}{c_p}(z - z_s) = T_s - \Gamma_d(z - z_s) \qquad (3.4)$$

where the index s indicates the surface. Consequently, the temperature decreases vertically with the **dry adiabatic lapse rate** $\Gamma_d = g/c_p$ in the isentropic atmosphere. The height where pressure vanishes defines the upper boundary of the isentropic atmosphere. Hence the depth of the isentropic atmosphere is given by

$$H_a = \frac{\theta_0 \Pi_s}{\Gamma_d} \qquad (3.5)$$

With the previous calculations, we can now write the equation of motion for the isentropic atmosphere as follows:

$$\left(\frac{D\mathbf{v}_h}{Dt}\right)_h + f\mathbf{k} \times \mathbf{v}_h = -g\nabla_h(H_a + z_s) - \nabla_h\phi + \boldsymbol{f}_T \qquad (3.6)$$

The horizontal geopotential gradient mainly results from the centrifugal acceleration by the Earth's rotation. We can assume vertical independence of this force within the shallow atmosphere approximation (see Section 2.2). Furthermore, the ocean adapts to the centrifugal force so that the ocean surface coincides approximately with a geopotential surface. Then, $\phi = g(z - z_s) + \text{const.}$ holds. Therefore, it is appropriate to define a mean sea level z_{MSL} such that

$$g\nabla_h z_{\text{MSL}} = -\nabla_h \phi \qquad (3.7)$$

Then, the equation of motion becomes

$$\left(\frac{D\mathbf{v}_h}{Dt}\right)_h + f\mathbf{k} \times \mathbf{v}_h = -g\nabla_h H_a - g\nabla_h H_s + \boldsymbol{f}_T \qquad (3.6)$$

where H_s denotes the height of the Earth's surface above sea level. The right-hand side of this equation is independent of height except for the contribution from the turbulent momentum exchange. However, this term and those of the left-hand side become vertically independent if the horizontal wind is vertically independent too and the surface drag is neglected. This is usually done in applications of the barotropic primitive equations. Consequently, the horizontal wind remains vertically independent if this is the case once.

The depth of the isentropic atmosphere, H_a, is an unknown that still needs an equation for its determination. Using (3.1) and the vertical independence of horizontal wind we can write the surface pressure tendency equation (2.83) as

$$\frac{\partial \Pi_s^{\frac{1}{\kappa}}}{Dt} = -\nabla_h \cdot \left(\Pi_s^{\frac{1}{\kappa}} \mathbf{v}_h\right) = -\mathbf{v}_h \cdot \nabla_h \Pi_s^{\frac{1}{\kappa}} - \Pi_s^{\frac{1}{\kappa}} \nabla_h \cdot \mathbf{v}_h \qquad (3.7)$$

which is identical to

$$\frac{D\Pi_s}{Dt} = -\kappa \Pi_s \nabla_h \cdot \mathbf{v}_h \qquad (3.8)$$

or

$$\frac{DH_a}{Dt} = -\kappa H_a \nabla_h \cdot \mathbf{v}_h \qquad (3.9)$$

Equations (3.6) and (3.9) form the governing equations of the barotropic atmospheric model. The widely used **shallow water**

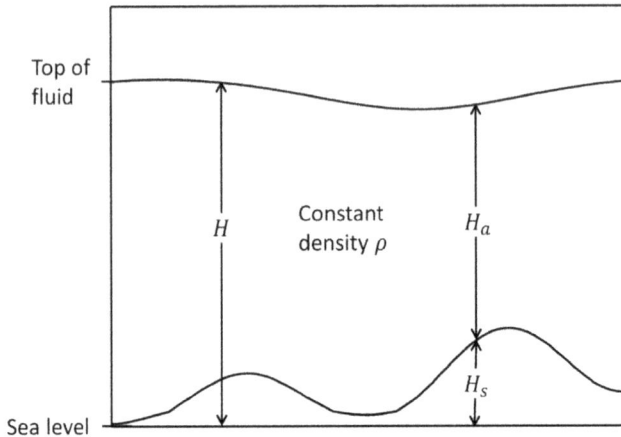

Fig. 3.1. Sketch of the shallow water model.

equations result by setting $\kappa = 1$:

$$\left(\frac{D\mathbf{v}_h}{Dt}\right)_h + f\mathbf{k} \times \mathbf{v}_h = -g\nabla_h H_a - g\nabla_h H_s + \mathbf{f}_T \quad (3.10)$$

$$\frac{DH_a}{Dt} = -H_a \nabla \cdot \mathbf{v}_h \quad (3.11)$$

It can be easily verified in (3.1) that the density is constant in this case. Therefore, we should interpret $\kappa = 1$ as the limit where the fluid is incompressible.[a] Figure 3.1 sketches the **shallow water model** based on Eqs. (3.10) and (3.11). Obviously, it is better suitable for the ocean than for the atmosphere since the constant density assumption is acceptable for many applications in ocean dynamics. However, the shallow water equations provide some insight into the behavior of atmospheric waves. The difference to the primitive equations of an isentropic atmosphere appears only in the factor κ in Eq. (3.9) which essentially leads to a different phase speed of inertia gravity waves. In most cases, derivations in the shallow water model can be easily repeated with $\kappa \neq 1$. Therefore, the achieved conclusions also apply for the isentropic atmosphere and offer a basic understanding of synoptic scale atmospheric waves.

---

[a]Note that $\kappa = 1$ turns out to be nonsensical for an ideal gas.

An alternative form of the shallow water equations results by using the height of the fluid surface above sea level, $H = H_a + H_s$:

$$\left(\frac{D\mathbf{v}_h}{Dt}\right)_h + f\mathbf{k} \times \mathbf{v}_h = -g\nabla_h H + \mathbf{f}_T \tag{3.12}$$

$$\frac{\partial H}{\partial t} = -\nabla \cdot [(H - H_s)\mathbf{v}_h] \tag{3.13}$$

In this system the surface topography H_s occurs in the continuity equation instead of the momentum equation. With a flat bottom, both forms are identical.

3.3 Conserved Quantities in the Shallow Water Model

The shallow water equations conserve potential vorticity, total energy and potential enstrophy in the inviscid case ($\mathbf{f}_T = 0$). These quantities are also conserved quantities of the primitive equations but they take a simpler form in the shallow water model.

3.3.1 *Potential vorticity*

We start with the equation for the vertical vorticity component to derive the potential vorticity conservation:

$$\frac{D}{Dt}(f + \xi) = -(f + \xi)\nabla_h \cdot \mathbf{v}_h \tag{3.14}$$

The derivation of this equation is likewise as for Eq. (2.130) but here we obtain a simpler result because the horizontal wind is vertically independent and the fluid is barotropic. Therefore, the solenoid and tilting terms vanish identically. The divergence term on the right-hand side can be eliminated by the continuity equation (3.11). Then, we obtain

$$\frac{D}{Dt}(f + \xi) = \frac{f + \xi}{H_a}\frac{DH_a}{Dt} \tag{3.15}$$

Therefore, the potential vorticity

$$P_H = \frac{f + \xi}{H_a} \tag{3.16}$$

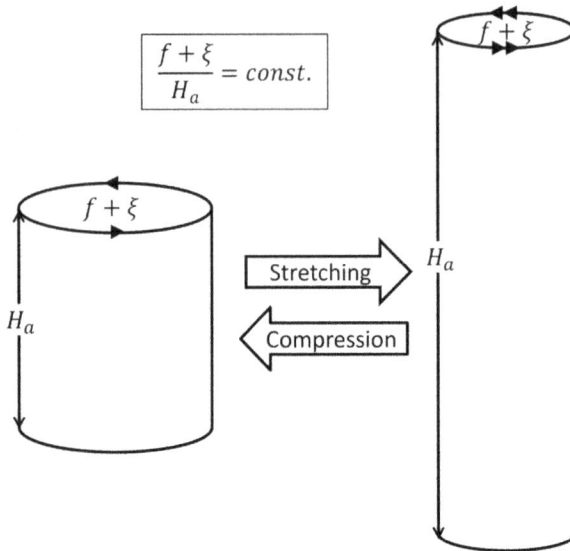

$$\frac{f + \xi}{H_a} = const.$$

Fig. 3.2. Vorticity change by stretching and compression of a vortex tube explains the conservation of Rossby potential vorticity.

is conserved along trajectories in the inviscid shallow water model.[b] The conservation of this quantity was derived by Rossby (1936) and, therefore, it is referred to as **Rossby potential vorticity**. The conservation mechanism is sketched in Fig. 3.2 and resembles that of Ertel PV. The absolute vorticity increases (decreases) when a fluid column is stretched (compressed). Stretching and compression lead to a change in the fluid depth H_a which explains the conservation of P_H. Potential vorticity constitutes an individually conserved quantity, i.e., the potential vorticity value remains constant along a fluid trajectory but can take other values along other trajectories.

3.3.2 *Energy*

Only kinetic and potential energy exist as energy forms in the shallow water model. The energy conversions can be deduced from the

---

[b]In the isentropic atmosphere the conserved potential vorticity becomes $P_H = (f + \xi)/H_a^{1/\kappa}$. Obviously, the expression in (3.16) represents the special case $\kappa = 1$.

governing equations. Multiplying (3.10) with the velocity vector yields the prognostic equation for specific kinetic energy

$$\frac{De_K}{Dt} = -g\mathbf{v}_h \cdot \nabla_h H + \mathbf{v}_h \cdot \boldsymbol{f}_T \tag{3.17}$$

With (3.11) one can write the kinetic energy equation in flux form

$$\frac{\partial}{\partial t}(H_a e_K) + \nabla \cdot (\mathbf{v}_h H_a e_K) = -g H_a \mathbf{v}_h \cdot \nabla_h H + H_a \mathbf{v}_h \cdot \boldsymbol{f}_T \tag{3.18}$$

The work by the pressure gradient force can be recast as

$$
\begin{aligned}
-g H_a \mathbf{v}_h \cdot \nabla_h H &= -\nabla_h \cdot (\mathbf{v}_h g H_a H) + g H \nabla_h \cdot (\mathbf{v}_h H_a) \\
&= -\nabla_h \cdot (\mathbf{v}_h g H_a H) - \frac{\partial}{\partial t}\left(\frac{g H^2}{2}\right)
\end{aligned}
\tag{3.19}
$$

Consequently, the energy equation becomes

$$\frac{\partial}{\partial t}\left(H_a e_K + \frac{g H^2}{2}\right) + \nabla_h \cdot [\mathbf{v}_h H_a(e_K + g H)] = H_a \mathbf{v}_h \cdot \boldsymbol{f}_T \tag{3.20}$$

Area integration over the complete surface of the Earth yields

$$\frac{d}{dt}\iint_{a_E} H_a e_K da + \frac{d}{dt}\iint_{a_E} \frac{g H^2}{2} da = \iint_{a_E} H_a \mathbf{v}_h \cdot \boldsymbol{f}_T da \tag{3.21}$$

We can relate the second term on the left-hand side to the potential energy tendency because potential energy can be calculated as follows:

$$E_P = \iint_{a_E}\int_{H_s}^{H} \rho g z\, dz\, da = \iint_{a_E} \frac{g\rho}{2}(H^2 - H_s{}^2)\, da \tag{3.22}$$

Therefore, we get for the volume integrated energies

$$\frac{d}{dt}(E_K + E_P) = \iint_{a_E} \rho H_a \mathbf{v}_h \cdot \boldsymbol{f}_T da \tag{3.23}$$

This equation states that the sum of kinetic and potential energy is conserved in the absence of turbulent momentum exchange.[c]

One can also determine the available potential energy in the shallow water model. The reference state with minimum potential energy is as expected associated with an even upper fluid surface. Then, the available potential energy becomes

$$E_A = \iint_{a_E} \frac{g\rho}{2}(H^2 - H_r{}^2)da \qquad (3.24)$$

where $H_r = $ const. denotes the height of the upper surface in the reference state. To show that E_A has a positive sign we insert $H = H_r + H'$:

$$E_A = \iint_{a_E} \frac{g\rho}{2}\left[(H_r + H')^2 - H_r{}^2\right]da$$

$$= \iint_{a_E} \frac{g\rho}{2}\left(H'^2 + 2H'H_r\right)da$$

$$= \iint_{a_E} \frac{g\rho}{2}H'^2 da + gH_r\iint_{a_E} \rho(H - H_r)da \qquad (3.25)$$

The second integral vanishes because of mass conservation

$$M = \iint_{a_E} \frac{g\rho}{2}\underbrace{(H - H_S)}_{H_a}da = M_r = \iint_{a_E} \frac{g\rho}{2}(H_r - H_s)da \qquad (3.26)$$

Therefore,

$$E_A = \iint_{a_E} \frac{g\rho}{2}H'^2 da \geq 0 \qquad (3.27)$$

Hence, in the shallow water model only deflections of the upper fluid surface can provide an energy source for kinetic energy (see Fig. 3.3). This is distinct from the baroclinic atmosphere where also horizontal density gradients give rise to nonzero available potential energy.

---

[c]The energetics for the primitive equations still applies for the isentropic case ($\kappa < 1$) in which the kinetic energy and the total potential energy become $E_K = \iint_{a_E} \frac{p_s}{2g}v_h^2 da$ and $E_{TP} = \iint_{a_E} p_s\left(\frac{\kappa}{\kappa+1}H_a + H_s\right)da$. Again, one can show that the limit $\kappa \to 1$ yields the formulation of the shallow water model.

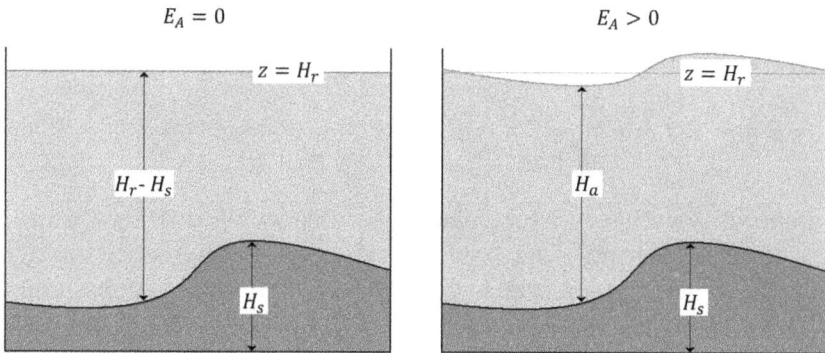

Fig. 3.3. States with zero and nonzero available potential energy in the shallow water model.

3.3.3 *Potential enstrophy*

Potential enstrophy measures the eddy activity in the fluid. It results from the integral

$$Z = \iint_{a_E} \frac{\rho H_a}{2} P_H{}^2 da \qquad (3.28)$$

The individual conservation of potential vorticity guarantees the conservation of potential enstrophy in the inviscid case. Note that the planetary vorticity also contributes to potential enstrophy. Therefore, a fluid at rest is not associated with the minimum potential enstrophy. Instead, a fluid with many anticyclones could have less potential enstrophy.

3.4 Linearization of the Shallow Water Equations

The shallow water equations have **wave solutions**. Essentially, an arbitrary flow can be understood as the superposition of many waves having different wavelengths and phase speeds. The waves can be amplified by **nonlinear interactions**. However, the basic dynamics of the waves already result from **linearized equations** which can be tackled more easily. The linearized equations are valid for waves with small amplitudes since the nonlinear terms have a small magnitude and can be neglected. The linearization procedure refers to a **base state** and is straightforward in shallow wave models since it only includes nonlinearities in the form of a product between two variables. Then, the linearization becomes

$$ab = (\overline{a} + a')(\overline{b} + b')$$
$$= \overline{a}\overline{b} + \overline{a}b' + \overline{b}a' + a'b' \approx \overline{a}\overline{b} + \overline{a}b' + \overline{b}a' \qquad (3.29)$$

where the overbar $^-$ symbolizes the base state and the prime $'$ the **perturbation**.[d] The base state has the property that it fulfills the governing equations. In addition, it is useful to assume a time-independent base state. Then, the linearized equations have time-independent coefficients so that the solution technique simplifies drastically. The most convenient base state is a fluid at rest. In this case, the base state variable must fulfill the following equation:

$$0 = -g\nabla_h \overline{H} \qquad (3.30)$$

Therefore, $\overline{\mathbf{v}}_h = 0$ and $\overline{H} = \text{const.}$ constitutes a time-independent base state. In further analysis, we consider an inviscid fluid ($\mathbf{f}'_T = 0$) and neglect surface orography ($H_s = 0$). Then, linearizing the governing equations (3.12) and (3.13) according to (3.29) results in

$$\frac{\partial \mathbf{v}'_h}{\partial t} + f\mathbf{k} \times \mathbf{v}'_h = -g\nabla_h H' \qquad (3.31)$$

$$\frac{\partial H'}{\partial t} = -\overline{H}\nabla_h \cdot \mathbf{v}'_h \qquad (3.32)$$

These are **Laplace's tidal equations** and they can be used to determine the ocean tides for a realistic tidal forcing.[e] One can easily show that these linearized equations still conserve the sum of kinetic energy E_K and available potential energy E_A if H_a is replaced by $\overline{H}$. Consequently, perturbations remain small and the linearization continues to be valid.

We will first find solutions for a simple Cartesian geometry since solving the equations in spherical geometry requires tedious calculations. Simplifying the geometry is an accepted practice to gain a

---

[d]These symbols should not be confused with the Reynolds average and its perturbation introduced in Section 1.13.

[e]Laplace's tidal equations result also in the case of an isentropic atmosphere but the mean **scale height** $\overline{H}_S = \kappa \overline{H}_a = R_d \overline{T}_s/g$ must be inserted in (3.32) instead of the mean fluid height $\overline{H}$.

better understanding of the phenomena. However, this simplification is not suitable for the prediction of observed large-scale flows.

3.5   Wave Solutions for an f-Plane Geometry

The effects of spherical geometry can be neglected if the considered domain has limited horizontal dimensions. Then, the **f-plane approximation** sketched in Fig. 3.4 is valid. With this approximation, the Earth surface is projected on a tangential surface with its origin located at the longitude λ_0 and latitude φ_0. The Cartesian coordinates of the f-plane becomes

$$x = r_E \cos \varphi_0 (\lambda - \lambda_0), \quad y = r_E (\varphi - \varphi_0) \tag{3.33}$$

The Coriolis parameter f takes the constant value $f_0 = 2\Omega \sin \varphi_0$ on the f-plane and this is the reason why it has this designation. In contrast, the **β-plane approximation** includes in addition a meridional linear change of the Coriolis parameter. This extended approximation has the advantage that it mimics more characteristics of the full spherical geometry. We will discuss this concept in a subsequent section. On the f-plane Laplace's tidal equations are written as

$$\frac{\partial u'}{\partial t} - f_0 v' = -g \frac{\partial H'}{\partial x} \tag{3.34}$$

$$\frac{\partial v'}{\partial t} + f_0 u' = -g \frac{\partial H'}{\partial y} \tag{3.35}$$

$$\frac{\partial H'}{\partial t} = -\overline{H} \left(\frac{\partial u'}{\partial x} + \frac{\partial v'}{\partial y} \right) \tag{3.36}$$

Applying curl and divergence to the linearized momentum equation yields the vorticity and divergence equations, respectively.

$$\frac{\partial \xi'}{\partial t} = -f_0 D' \tag{3.37}$$

$$\frac{\partial D'}{\partial t} = f_0 \xi' - g \nabla_h^2 H' \tag{3.38}$$

where D' denotes the divergence of the horizontal flow.

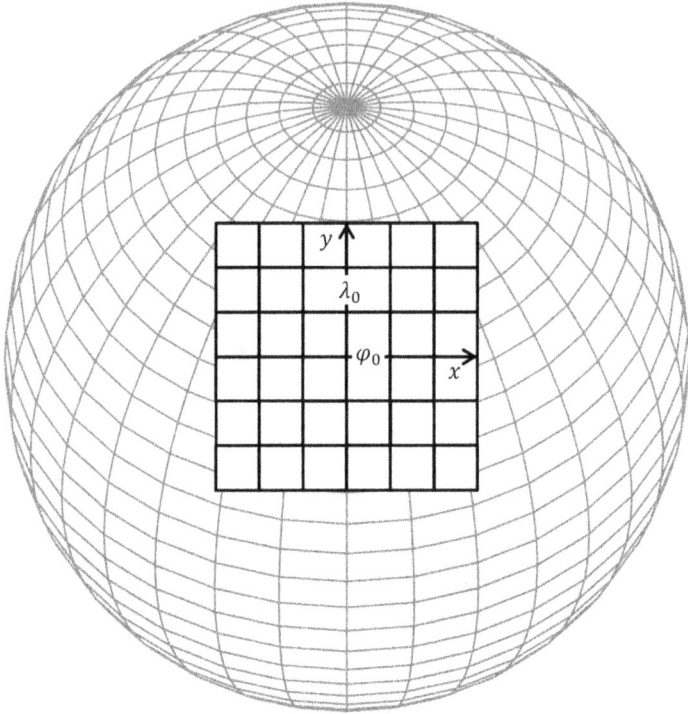

Fig. 3.4.   Sketch of the f-plane coordinate system which is a projection on a tangential surface at longitude λ_0 and latitude φ_0.

The variables can be represented by Fourier series in x and y if the f-plane domain has a rectangular shape. Each Fourier term yields a trigonometric wave function. Therefore, it is reasonable to try a **wave ansatz** of the form

$$\xi' = \Re(\hat{\xi}e^{ikx}e^{ily}e^{-i\nu t}), \quad D' = \Re(\hat{D}e^{ikx}e^{ily}e^{-i\nu t}),$$
$$H' = \Re(\hat{H}e^{ikx}e^{ily}e^{-i\nu t}) \tag{3.39}$$

where $\hat{\xi}$, $\hat{D}$ and $\hat{H}$ are the complex-valued amplitudes of the wave while k denotes the **zonal wavenumber**, l the **meridional wavenumber** and ν the **frequency**. The solution appears as the real part of a complex-valued wave function. It is a convenient approach to consider an imaginary part of the variables which also fulfills the equations. However, only the real part forms the desired eventual solution. The advantage of this approach is that we can replace the differential operators by multiplication with a factor since the derivative of an exponential function yields the exponential function itself.[f] Hence, we have

$$\frac{\partial}{\partial t} = -i\nu, \quad \frac{\partial}{\partial x} = ik, \quad \frac{\partial}{\partial y} = il \tag{3.40}$$

Extending Eqs. (3.36)–(3.38) to complex numbers and inserting the wave ansatz (3.39) yields an algebraic equation set:

$$-i\nu\hat{H} + \overline{H}\hat{D} = 0 \tag{3.41}$$

$$-i\nu\hat{\xi} + f_0\hat{D} = 0 \tag{3.42}$$

$$-i\nu\hat{D} - f_0\hat{\xi} - g(k^2 + l^2)\hat{H} = 0 \tag{3.43}$$

This set of equations can also be written in matrix form

$$\begin{pmatrix} -i\nu & 0 & \overline{H} \\ 0 & -i\nu & f_0 \\ -g(k^2 + l^2) & -f_0 & -i\nu \end{pmatrix} \cdot \begin{pmatrix} \hat{H} \\ \hat{\xi} \\ \hat{D} \end{pmatrix} = 0 \tag{3.44}$$

Nontrivial solutions require a vanishing determinant of the matrix.[g] This leads to the **characteristic equation**:

$$\begin{vmatrix} -i\nu & 0 & \overline{H} \\ 0 & -i\nu & f_0 \\ -g(k^2 + l^2) & -f_0 & -i\nu \end{vmatrix}$$

$$= -i\nu \begin{vmatrix} -i\nu & f_0 \\ -f_0 & -i\nu \end{vmatrix} + \overline{H} \begin{vmatrix} 0 & -i\nu \\ -g(k^2 + l^2) & -f_0 \end{vmatrix}$$

$$= (i\nu)^3 - [f_0^2 + g\overline{H}(k^2 + l^2)](i\nu) = 0 \tag{3.45}$$

The characteristic equation has solutions for frequency as a function of the wavenumbers k and l:

$$\nu_{1,2} = \pm\sqrt{f_0^2 + g\overline{H}(k^2 + l^2)} \tag{3.46}$$

$$\nu_3 = 0 \tag{3.47}$$

---

[f] Note Euler's formula $e^{i\alpha} = \cos\alpha + i\sin\alpha$.
[g] It is assumed that the reader is familiar with the foundation of linear algebra and calculation of determinants.

These solutions constitute **dispersion relations** since they describe the dependency of wave frequency on wavenumber. The first two solutions constitute the frequencies of an **inertia-gravity wave** while the third solution is associated with a steady-state **planetary wave**. The full solution results from the algebraic equations (3.41)–(3.43). It can be assumed without loss of generality that $\hat{H} = A$ where A is a real amplitude. Then, we obtain for the other amplitudes

$$\hat{D} = \frac{i\nu}{\overline{H}}A, \quad \hat{\xi} = -\left[\frac{g}{f_0}(k^2 + l^2) - \frac{\nu^2}{f_0\overline{H}}\right]A \qquad (3.48)$$

and the fields of the perturbation become

$$H' = A\cos(kx + ly - \nu t) \qquad (3.49)$$

$$\xi' = -\left[\frac{g}{f_0}(k^2 + l^2) - \frac{\nu^2}{f_0\overline{H}}\right]A\cos(kx + ly - \nu t) \qquad (3.50)$$

$$D' = -\frac{\nu}{\overline{H}}A\sin(kx + ly - \nu t) \qquad (3.51)$$

We can apply the Helmholtz theorem (A.54) to determine the velocity components. The theorem relates vorticity and divergence with the velocity vector:

$$\mathbf{v}'_h = \nabla_h\chi' + \mathbf{k} \times \nabla_h\psi', \quad \xi' = \nabla_h^2\psi', \quad D' = \nabla_h^2\chi' \qquad (3.52)$$

It is easy to invert the Laplace operator since $\nabla_h^2 = -(k^2 + l^2)$ holds in the considered case. Therefore, the velocity components become

$$u' = \frac{k}{k^2 + l^2}\frac{\nu}{\overline{H}}H' - \left[\frac{g}{f_0} - \frac{\nu^2}{f_0\overline{H}(k^2 + l^2)}\right]\frac{\partial H'}{\partial y} \qquad (3.53)$$

$$v' = \frac{l}{k^2 + l^2}\frac{\nu}{\overline{H}}H' + \left[\frac{g}{f_0} - \frac{\nu^2}{f_0\overline{H}(k^2 + l^2)}\right]\frac{\partial H'}{\partial x} \qquad (3.54)$$

We can set l to zero without loss of generality since the solution is a **plane wave** with straight isolines of H' and $l = 0$ can be reached by a simple rotation of the coordinate system. Then, the inertia-gravity

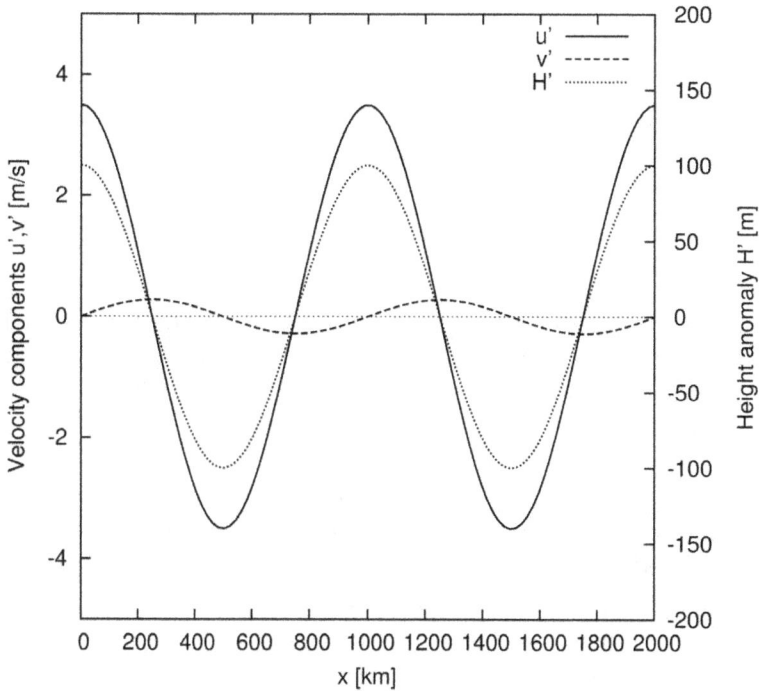

Fig. 3.5. Zonal profiles of an eastward propagating ($\nu > 0$) inertia-gravity wave. The model parameters $f_0 = 10^{-4}\mathrm{s}^{-1}, \overline{H} = 8\,\mathrm{km}, k = 6.283 \times 10^{-6}\mathrm{m}^{-1}$ and $g = 9.81\,\mathrm{m/s}^2$ have been used for the calculation.

wave has the velocity components

$$u'_{1,2} = \pm\frac{\sqrt{f_0^2 + g\overline{H}k^2}}{k\overline{H}}A\cos(kx - \nu t) \qquad (3.55)$$

$$v'_{1,2} = \frac{f_0}{k\overline{H}}A\sin(kx - \nu t) \qquad (3.56)$$

We see in Fig. 3.5 that the zonal velocity of the wave with positive frequency is in phase with the height field while the meridional velocity is phase shifted by 90°. Positive frequency coincides with a positive **phase speed** $c = \nu/k$ which means that the wave propagates in x-direction with the speed c. We can easily separate the inertia and gravity mechanisms of the wave by setting the gravity acceleration and the Coriolis parameter to zero, respectively.

In the first case, we get

$$u'_{1,2} = \pm \frac{f_0}{k\overline{H}} A \cos(kx \mp f_0 t) \tag{3.57}$$

$$v'_{1,2} = \frac{f_0}{k\overline{H}} A \sin(kx \mp f_0 t) \tag{3.58}$$

This wave describes an **inertial oscillation** where the velocity vector rotates with the frequency of f_0 in a clockwise direction at a fixed location. The oscillation results from the acceleration by the Coriolis forces. For example, a fluid parcel moving eastward experiences a southward Coriolis force. The resulting acceleration yields a southward velocity component which by itself give rise to a westward acceleration. The parcel moves along a circular trajectory as a consequence of these accelerations. The trajectory sketched in Fig. 3.6 is called **inertia circle** and has a radius of V_0/f_0 where V_0 is the velocity amplitude of the inertial oscillation. Inertial oscillation can appear in the nocturnal boundary layer where the turbulence breaks down due to high static stability (see Blackadar, 1957).

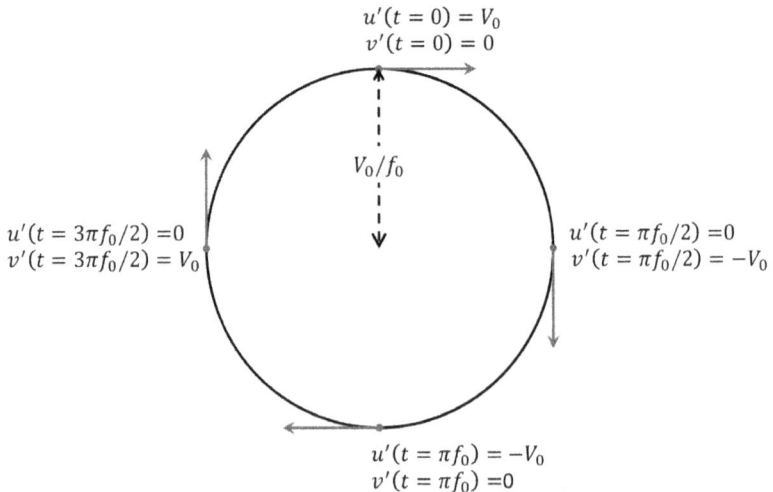

$$u'(t = 0) = V_0$$
$$v'(t = 0) = 0$$

$$V_0/f_0$$

$$u'(t = 3\pi f_0/2) = 0$$
$$v'(t = 3\pi f_0/2) = V_0$$

$$u'(t = \pi f_0/2) = 0$$
$$v'(t = \pi f_0/2) = -V_0$$

$$u'(t = \pi f_0) = -V_0$$
$$v'(t = \pi f_0) = 0$$

Fig. 3.6.   Sketch of the inertia circle. The circle is a trajectory of a parcel subjected to an inertial oscillation. The circle has a radius of V_0/f_0.

Setting the Coriolis parameter f_0 to zero produces a **gravity wave**. Then, the solution becomes

$$u'_{1,2} = \pm \frac{c_G}{\overline{H}} A \cos[k(x \mp c_G t)] \tag{3.59}$$

$$v'_{1,2} = 0 \tag{3.60}$$

where $c_G = \sqrt{g\overline{H}}$ denote the phase speed of a gravity wave in the shallow water system. The velocity field of the gravity wave is curl-free and has, therefore, only a component in the direction of the propagation. The gravity wave shows no **dispersion**, i.e. waves with different wavelengths propagate with the same phase speed. Consequently, the shape of the gravity wave does not change in time even if it has no sinusoidal form. The phase speed of a gravity wave has a very large value compared to the flow speed. A reasonable height $\overline{H} = 8\,\mathrm{km}$ yields a gravity wave speed of $280\,\mathrm{m/s}$.

The inertia-gravity wave includes both the mechanism of an inertial oscillation and a gravity wave. However, it depends sensitively on the spatial scale of the wave whose mechanism dominates. This can be seen by rewriting the frequency in the form

$$\nu_{1,2} = \pm c_G \sqrt{\frac{1}{L_R^2} + k^2} \tag{3.61}$$

where L_R denotes the **Rossby radius of deformation** which is given by

$$L_R = \frac{\sqrt{g\overline{H}}}{f_0} = \frac{c_G}{f_0} \tag{3.62}$$

Obviously, a long wave with a zonal wavelength of $L_x = 2\pi/k \gg L_R$ has the character of an inertial oscillation while a short wave with a wavelength of $L_x = 2\pi/k \ll L_R$ behaves like a gravity wave.

The third frequency solution (3.47) is associated with a planetary wave that has the velocity components

$$u'_3 = -\frac{g}{f_0} \frac{\partial H'}{\partial y} \tag{3.63}$$

$$v'_3 = \frac{g}{f_0} \frac{\partial H'}{\partial x} \tag{3.64}$$

The solution just describes the **geostrophic balance**, i.e. the Coriolis force is balanced by the pressure gradient force. In vector form the geostrophic balance becomes

$$\mathbf{v}'_g = \frac{g}{f_0} \mathbf{k} \times \nabla_h H' \tag{3.65}$$

where $\mathbf{v}_g$ denotes the **geostrophic wind**. Figure 3.7 sketches the geostrophic flow of the planetary wave and the associated force balance. Obviously, the wind vector is parallel to the isolines of the upper surface height. The flow remains steady since no other forces appear in the linearized equations.[h] Every arbitrary geostrophic flow can be represented by a superposition of planetary plane waves in a Fourier expansion. This has the consequence that any geostrophic flow remains in a steady state if it is a solution of the Laplace tidal equations on the f-plane. Therefore, these linear equations do not suffice for the prediction of synoptic weather systems which are approximately in geostrophic balance.

The plane wave has an unrealistic shape and does not appear in reality. However, the superposition of several plane wave solutions also yields a solution of the linear equations and can describe a more familiar pattern. For example, by the addition theorem for trigonometric functions, we find that

$$H' = \frac{A}{2}[\cos{(kx + ly - \nu t)} + \cos(kx - ly - \nu t)]$$
$$= A \cos(ly) \cos(kx - \nu t) \tag{3.66}$$

This combination of the trigonometric function leads to a checkerboard pattern with alternating highs and lows. The resulting wave propagates only in zonal direction with the phase speed $c = \nu/k$ while the wave is stationary in meridional direction. The **midlatitude channel** is a widely used concept in theoretical studies. The channel has rigid walls at the northern and southern boundary of the channel while in the zonal direction periodicity is assumed. At the rigid boundary the meridional flow must vanish. Indeed, a wave solution of the form (3.66) has a vanishing meridional flow at several

---

[h]Even the inclusion of nonlinear terms does not disturb the balance in the plane wave case.

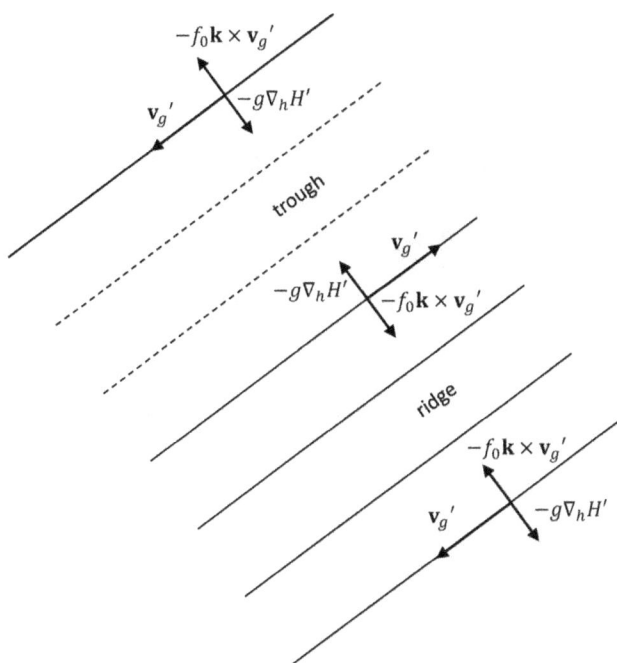

Fig. 3.7. Sketch of the geostrophic flow and force balance in the plane planetary wave.

latitudes. Therefore, one can fit this wave into a channel by adjusting the meridional wavenumber and the latitude of the coordinate system origin. Figure 3.8 shows the height and velocity fields of the previously discussed wave solutions in a channel with a width of 2000 km. It can be seen that the wind vector field of the inertia-gravity wave is neither curl-free nor source-free while a pure gravity wave has a vanishing curl. The eastward propagating wave has a positive zonal velocity component at the maximum of surface height while the opposite is true for the westward propagating wave. The planetary wave reveals high- and low-pressure systems which have anticyclonic and cyclonic circulations, respectively. The vector field has zero divergence and the vectors always direct parallel to the isolines of the height anomaly. Consequently, these isolines form streamlines of the geostrophic flow and by comparison of (3.65) with (A.54) we can identify the **geostrophic streamfunction**

$$\psi'_g = \frac{g}{f_0} H' \qquad (3.67)$$

Eastward propagating inertia–gravity wave

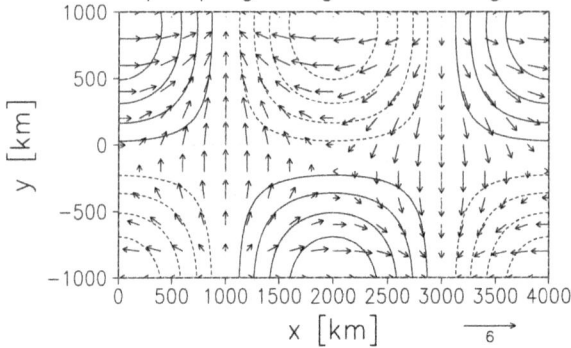

Westward propagating inertia–gravity wave

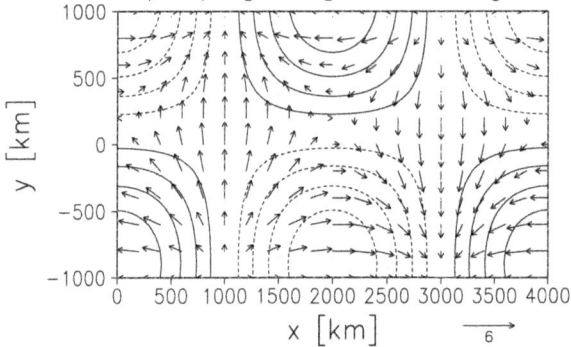

Steady–state planetary wave

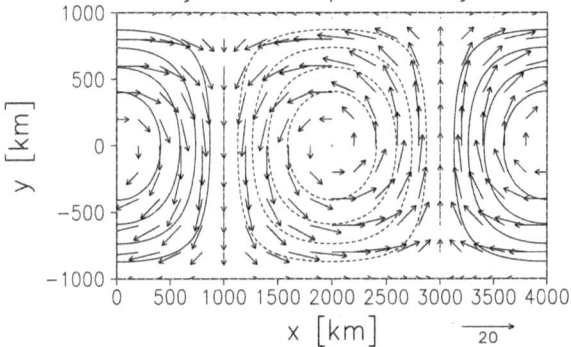

Fig. 3.8.   Height anomaly (isolines, contour interval 20 m) and horizontal velocity (arrows, scale in m/s) of the eastward propagating inertia-gravity wave (upper panel), westward propagating inertia-gravity wave (middle panel) and planetary wave (bottom panel) in a channel with 2000 km width. Negative isolines are dashed.

The channel system allows for a further wave solution type, namely the **Kelvin wave**. The Kelvin wave includes only a zonal velocity perturbation while in the meridional direction geostrophic balance prevails. This has the consequence that this wave behaves like a gravity wave in a zonal direction and propagates with the phase speed c_G. A lateral boundary is required for the existence of the Kelvin wave and, therefore, it is predominantly a phenomenon that occurs in the ocean. One exception is the **equatorial Kelvin wave** which does not require lateral boundaries. This wave will be discussed in more detail in Section 3.7.

3.6 Geostrophic Adjustment

Atmospheric flows having horizontal scales larger or equal to the Rossby radius are approximately in geostrophic balance in the absence of significant microturbulent momentum exchange. Therefore, there should be a mechanism by which the flow turns into geostrophic balance. The **geostrophic adjustment** represents such a mechanism and it can already be explained on the basis of the linearized shallow water equations. The geostrophic adjustment takes place in a localized region where the geostrophic balance is disturbed by any process. The resulting flow can be represented by the superposition of inertia-gravity and planetary waves. The inertia-gravity waves propagate with a large phase speed rapidly away from the region while the planetary wave remains stationary. Therefore, the flow in the considered region becomes geostrophic eventually.

The geostrophic flow after the geostrophic adjustment process can be determined with the linearized potential vorticity (PV). With Eqs. (3.36) and (3.37) we find the following perturbation PV equation:

$$\frac{\partial}{\partial t} P'_H = 0 \quad \text{where} \quad P'_H = \frac{\xi'}{\overline{H}} - \frac{f_0}{\overline{H}^2} H' = \frac{\nabla_h^2 \psi' - \psi'_g / L_R^2}{\overline{H}} \quad (3.68)$$

Therefore, the perturbation PV P'_H remains constant at every location. It is the appropriate linear approximation of the PV anomaly $P_H - \overline{P_H} = (f_0 + \xi)/H_a - f_0/\overline{H}$. With the conservation of P'_H one can determine the final geostrophic flow. Let us assume that the initial flow has the PV field P'_{H0}. Then, the geostrophic streamfunction of the final state results as a solution of the following differential equation:

$$\left(\nabla_h^2 - \frac{1}{L_R^2}\right)\psi'_g = \overline{H}P'_{H0} \tag{3.69}$$

This differential equation is the inhomogeneous Helmholtz equation which can be solved with a Fourier expansion of the fields. For a trigonometric function with wavenumbers k and l the solution becomes

$$\psi'_g = -\frac{\overline{H}}{K^2 + 1/L_R^2}\Re(\hat{P}'_{H0}e^{ikx}e^{ily}) \tag{3.70}$$

where $K = \sqrt{k^2 + l^2}$ denotes the **total wavenumber**. On the other hand, the initial perturbation PV is a function of the initial stream function ψ'_0 and initial geostrophic stream function ψ'_{g0}. Therefore,

$$\psi'_g = \frac{K^2\psi'_0 + 1/L_R^2\psi'_{g0}}{K^2 + 1/L_R^2} = \frac{(KL_R)^2\psi'_0 + \psi'_{g0}}{(KL_R)^2 + 1} \tag{3.71}$$

We see that both the initial wind field (ψ'_0) and the initial mass field (ψ'_{g0}) contribute to the final geostrophic flow. Which of these dominate depends on the ratio between the horizontal scale of the perturbation and the Rossby Radius of deformation, i.e. KL_R. For initial perturbations which have a small scale compared to L_R we get $KL_R \gg 1$. Then, the initial wind field dominates in the final state and we get

$$\psi'_g \approx \psi'_0 \tag{3.72}$$

This kind of adjustment prevails in the tropics since the Rossby Radius of deformation is very large due to the small Coriolis parameter f_0. Therefore, the wind field gives more information about the nondivergent flow than the mass field. This is the reason why a streamlined analysis is usually applied for tropical weather regions while geopotential height fields are displayed for mid-latitude regions.

On the other hand, perturbations having a large scale compared to L_R ($KL_R \ll 1$) cause an adjustment where the initial mass field dominates the final state. Then, the approximation

$$\psi'_g \approx \psi'_{g0} = \frac{g}{f_0} H'_0 \tag{3.73}$$

holds. This type of adjustment takes place in the extratropics where planetary waves can be reasonably deduced from geopotential height fields.

It is not clear so far how the geostrophic adjustment takes place. Indeed, an initially unbalanced single wave of the form (3.39) does not reach the geostrophic balance. For example, let us assume an initial wave of the form

$$H'|_{t=0} = \hat{H}\cos(kx), \quad \xi'|_{t=0} = \hat{\xi}\cos(kx), \quad D'|_{t=0} = \hat{D}\sin(kx) \tag{3.74}$$

Then, the time-dependent solution results by projection onto the wave solutions (3.49)–(3.51). This just means that the height anomaly H' can be expressed as

$$H' = A_1\cos(kx - \nu_1 t) + A_2\cos(kx - \nu_2 t) + A_3\cos(kx - \nu_3 t) \tag{3.75}$$

subject to the initial condition (3.74). Applying this also to the vorticity anomaly (3.50) and divergence anomaly (3.51) yields the following linear equation set

$$\begin{pmatrix} 1 & 1 & 1 \\ f_0/\overline{H} & f_0/\overline{H} & -gk^2/f_0 \\ -\nu_1/\overline{H} & -\nu_2/\overline{H} & 0 \end{pmatrix} \cdot \begin{pmatrix} A_1 \\ A_2 \\ A_3 \end{pmatrix} = \begin{pmatrix} \hat{H} \\ \hat{\xi} \\ \hat{D} \end{pmatrix} \tag{3.76}$$

Let us assume that the initial perturbation has only a nontrivial height anomaly, i.e. $\hat{\xi} = \hat{D} = 0$. Then, we find the solution

$$A_1 = A_2 = \frac{(kL_R)^2}{2[(kL_R)^2 + 1]}\hat{H}, \quad A_3 = \frac{1}{[(kL_R)^2 + 1]}\hat{H} \tag{3.77}$$

If, on the other hand, the initial perturbation has only a nontrivial vorticity anomaly, i.e. $\hat{H} = \hat{D} = 0$ we obtain

$$A_1 = A_2 = \frac{\overline{H}/f_0}{2[(kL_R)^2 + 1]}\hat{\xi}, \quad A_3 = -\frac{\overline{H}/f_0}{[(kL_R)^2 + 1]}\hat{\xi} \tag{3.78}$$

Therefore, in both cases, inertia-gravity waves are also triggered by the initialization. These sum up to a **standing wave** since

$$A_1 \cos(kx - \nu_1 t) + A_1 \cos(kx + \nu_1 t) = 2A_1 \cos(kx) \cos(\nu_1 t) \quad (3.79)$$

A standing wave does not propagate but it executes an amplitude oscillation with the frequency ν_1. Hence, geostrophic adjustment does not take place when the initial state is a plane wave. Consequently, this process can only happen with localized initial perturbations if at all.

To investigate the geostrophic adjustment of localized perturbations analytically we consider the following initial height perturbation:

$$H'|_{t=0} = A \cos^n(kx) = \frac{A}{2^{n-1}} \sum_{j=0}^{\frac{n-1}{2}} \binom{n}{j} \cos[(n-2j)kx] \quad (3.80)$$

where n is a positive odd-numbered integer.[i] Obviously, the localized perturbation at $x = 0$ is a superposition of $(n-1)/2$ plane waves. Since the superposition of several solutions represents another solution of the linear equation (**superposition principle**) we obtain from (3.75) and (3.77) for the time-dependent height field

$$H' = \frac{A}{2^{n-1}} \sum_{j=0}^{\frac{n-1}{2}} \frac{(k_j L_R)^2}{[(k_j L_R)^2 + 1]} \binom{n}{j} \cos(k_j x) \cos(\nu_j t)$$

$$+ \frac{A}{2^{n-1}} \sum_{j=0}^{\frac{n-1}{2}} \frac{1}{[(k_j L_R)^2 + 1]} \binom{n}{j} \cos(k_j x) \quad (3.81)$$

where $k_j = (n-2j)k$ and $\nu_j = (f_0^2 + g\overline{H}k_j^2)^{1/2}$. The solution reveals the simplest behavior in the case $f_0 = 0$. Then, the Rossby radius of deformation becomes infinitely large and the planetary wave part

---

[i]The odd number has the consequence that a mirror perturbation with opposite sign appears at $x = \pi/k$. However, this additional perturbation does not affect the geostrophic adjustment if n is sufficiently large.

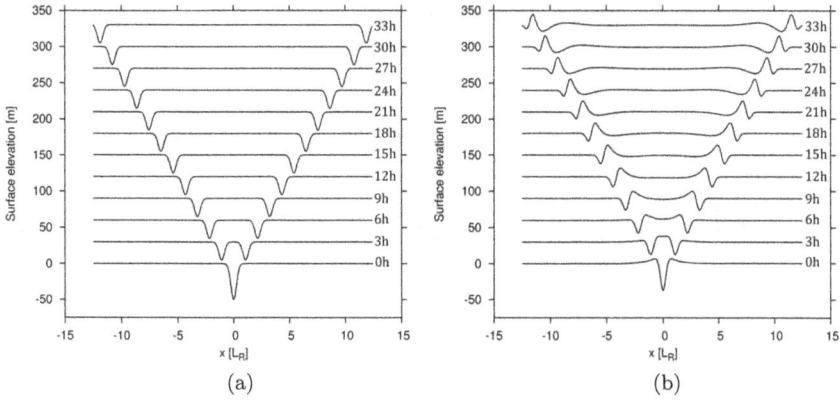

Fig. 3.9. Zonal surface height profiles of the inertia-gravity wave during the geostrophic adjustment of the perturbation (3.80) for (a) without planet rotation ($f_0 = 0\,\mathrm{s}^{-1}$) and (b) with planet rotation ($f_0 = 10^{-4}\,\mathrm{s}^{-1}$). The chosen parameters are $A = -50\,\mathrm{m}, n = 1001, k = 2\pi/L_R, L_R = 2801$ km and $\overline{H} = 8$ km. Numbers at the right end of the profile indicate the time and the profiles have been shifted vertically after each time step to visualize the propagation of the wave.

vanishes. As seen in Fig. 3.9(a) the initial perturbation divides into east- and westward propagating gravity waves. These waves conserve its form since the phase speed is independent of wavenumber. In this case, the solution simply becomes

$$H' = \frac{A}{2^n} \sum_{j=0}^{\frac{n-1}{2}} \binom{n}{j} \cos[k_j(x - c_G t)]$$

$$+ \frac{A}{2^n} \sum_{j=0}^{\frac{n-1}{2}} \binom{n}{j} \cos[k_j(x + c_G t)] \qquad (3.82)$$

The situation is more complicated in the general case since the phase velocity depends upon the wavenumber. We also see in Fig. 3.9(b) that east and westward traveling inertia-gravity waves occur. However, the initial minima decay while new maxima and minima form to the left and right of the eastward and westward propagating wave, respectively. This behavior is referred to as **dispersion**.

The dispersion of a **wave packet** can be described as the Fourier transform

$$H'(x,t) = \frac{1}{\sqrt{2\pi}}\Re\left\{\int_{-\infty}^{\infty} \hat{H}(k)\exp[ikx - i\nu(k)t]dk\right\} \qquad (3.83)$$

which applies for an infinitely large domain. The function $\hat{H}(k)$ constitutes the **spectrum** of the wave packet. A typical wave packet can be generated by a spectrum having the form of a Gaussian bell:

$$\hat{H}(k) = \exp\left[-\frac{(k-k_0)^2}{\Delta k^2}\right] \qquad (3.84)$$

For a narrow maximum $(\Delta k^2 \ll k_0^2)$ the frequency $\nu(k)$ can be approximated by a Taylor series up to the second order

$$\nu(k) \approx \nu|_{k=k_0} + \left.\frac{d\nu}{dk}\right|_{k=k_0}(k-k_0) + \frac{1}{2}\left.\frac{d^2\nu}{dk^2}\right|_{k=k_0}(k-k_0)^2$$

$$= \nu_0 + c_g k' + \delta k'^2 \qquad (3.85)$$

where $k' = k - k_0$; c_g is the **group velocity** and δ a constant that measures the dispersion of the wave packet. Inserting this approximation yields

$$\frac{1}{\sqrt{2\pi}}\int_{-\infty}^{\infty} \hat{H}(k)\exp[ikx - i\nu(k)t]dk$$

$$= \frac{e^{i(k_0 x - \nu_0 t)}}{\sqrt{2\pi}}\int_{-\infty}^{\infty}\exp\left(-\frac{k'^2}{\Delta k^2} + ik'(x - c_g t) - i\delta k'^2 t\right)dk'$$

$$= \frac{e^{i(k_0 x - \nu_0 t)}}{\sqrt{2\pi}}\int_{-\infty}^{\infty}\exp(-ak'^2 + ibk')dk'$$

$$= \frac{e^{i(k_0 x - \nu_0 t)}}{\sqrt{2\pi}}\sqrt{\frac{\pi}{a}}\exp\left(\frac{-b^2}{4a}\right)$$

$$= \frac{1}{\sqrt{2\left(\frac{1}{\Delta k^2} + i\delta t\right)}}\exp\left[-\frac{(x - c_g t)^2}{4\left(\frac{1}{\Delta k^2} + i\delta t\right)}\right]e^{i(k_0 x - \nu_0 t)} \qquad (3.86)$$

where the indefinite integral has been evaluated by consulting a mathematical handbook (Bronshtein *et al.*, 2015). The first factor can be recast as

$$\frac{1}{\sqrt{2\left(\frac{1}{\Delta k^2} + i\delta t\right)}} \exp\left[-\frac{(x - c_g t)^2}{4\left(\frac{1}{\Delta k^2} + i\delta t\right)}\right]$$

$$= \frac{e^{-i\tan^{-1}(\delta t \Delta k^2)/2}}{\sqrt[4]{\left(\frac{4}{\Delta k^4} + 4\delta^2 t^2\right)}} \exp\left[-\frac{\Delta k^{-2}(x - c_g t)^2}{\frac{4}{\Delta k^4} + 4\delta^2 t^2}\right]$$

$$\times \exp\left[\frac{i\delta t (x - c_g t)^2}{\frac{4}{\Delta k^4} + 4\delta^2 t^2}\right] \tag{3.87}$$

Therefore, the height anomaly of this wave packet becomes

$$H'(x, t)$$

$$= \underbrace{\frac{e^{-\frac{\Delta k^{-2}(x-c_g t)^2}{\frac{4}{\Delta k^4} + 4\delta^2 t^2}}}{\sqrt[4]{\frac{4}{\Delta k^4} + 4\delta^2 t^2}}}_{\text{Envelope}} \underbrace{\cos\left(-\frac{\tan^{-1}(\delta t \Delta k^2)}{2} + \frac{\delta t(x - c_g t)^2}{\frac{4}{\Delta k^4} + 4\delta^2 t^2} + k_0 x - v_0 t\right)}_{\text{Carrier wave}}$$

$$\tag{3.88}$$

We see that the evolving wave packet splits into the **carrier wave** and the **envelope** both of which depend on space and time. The special case $\nu(k) = c_g k(k_0 = \delta = \nu_0 = 0)$ just leads to a Gaussian bell moving with the group velocity c_g. This conforms to the pure gravity wave case shown in Fig. 3.9(a). In general, the envelope connects the maxima of the wave packet and is a function of time t and $x - c_g t$. Therefore, the envelope migrates with the group velocity c_g and changes its shape by the time dependence. The envelope has the form of a Gaussian bell that broadens and flattens in the course of time. The larger the parameter δ is, the faster is this process. For the inertia-gravity waves we find

$$c_{g1,2} = \pm c_G \frac{L_R k}{(1 + L_R^2 k^2)^{1/2}}, \quad \delta_{1,2} = \pm c_G \frac{L_R}{(1 + L_R^2 k^2)^{3/2}} \tag{3.89}$$

These formulas show that with decreasing wavelength the group velocity approaches the phase speed of gravity waves and dispersion decreases. Furthermore, the group velocity is always smaller in its absolute value than the phase speed since

$$|c_{g1,2}| - |c_{1,2}| = c_G \frac{L_R k}{(1 + L_R^2 k^2)^{\frac{1}{2}}} - c_G \frac{(1 + L_R^2 k^2)^{\frac{1}{2}}}{L_R k}$$

$$= -c_G \frac{1}{L_R k (1 + L_R^2 k^2)^{\frac{1}{2}}} < 0 \qquad (3.90)$$

Consequently, the wave packet envelope migrates slower than the phases of the carrier wave. This is clearly seen in Fig. 3.9(b) because the new highs and lows of the wave packet form always on the side which is opposite to the propagation direction. The flattening of the envelope is less obvious since the parameter δ is too small for the chosen case. Figure 3.10 displays an example for a wave packet of the form (3.88) where an appreciable dispersion is seen. Note that the propagation of this packet covers a distance that corresponds to several orbits around the earth.

We can deduce that an unbalanced local perturbation adjusts geostrophically since inertia-gravity wave packets move away and leave behind the planetary wave that is in geostrophic balance. Figure 3.11(a) shows the planetary wave part of the perturbation (3.80) for various values of the Coriolis parameter f_0. Obviously, the planetary wave attains a larger size and decreases in amplitude with decreasing f_0. This happens because the Rossby radius L_R increases relative to the size of the initial perturbation. Indeed, displaying the height as a function of x scaled with Rossby radius yields similar height profiles that only differ in their amplitude (see Fig. 3.11(b)).

In this section, we found that geostrophic adjustment takes place when the fluid is perturbed locally. However, the emitted gravity wave packets persist for a long time and may return to the location of the initial perturbation due to the sphericity of the Earth. Therefore, the recurrent appearance of unbalanced perturbations should lead to a flow that is far away from geostrophic balance. Consequently, other explanations must be considered to understand why the observed synoptic-scale flow is close to geostrophic balance.

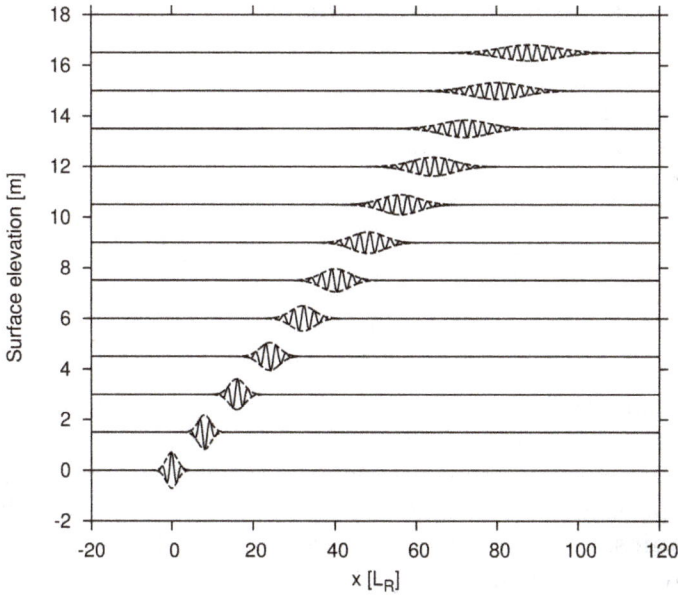

Fig. 3.10. Zonal surface height profiles of an inertia-gravity wave packet as described by Eq. (3.88) for the parameters $k_0 = 2.5/L_R$, $\Delta k_0 = 1/L_R$, $L_R = 2801$ km and $\overline{H} = 8$ km. The profiles have been shifted vertically after each time step to visualize the propagation of the wave.

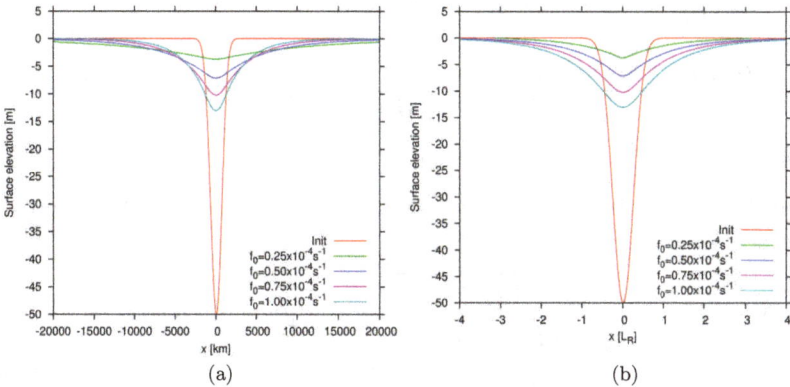

Fig. 3.11. Zonal surface height profiles of the planetary wave part of the perturbation as described by Eq. (3.80) for various values of the Coriolis parameter f_0. The chosen parameters are $A = -50$m, $n = 1001$, $k = 2\pi/L_R$ and $\overline{H} = 8$ km. (a) Profiles as a function of x and (b) profiles as a function of x/L_R. The red curve shows the initial height profile for $f_0 = 10^{-4}\,\mathrm{s}^{-1}$.

3.7   Wave Solutions on the Sphere

It is far more complicated to obtain wave solutions of Laplace tidal equations on the sphere than for the f-plane geometry. The reason becomes obvious when we derive the vorticity and divergence equations in spherical geometry from (3.31) using $f = 2\Omega \sin \varphi$:

$$\frac{\partial \xi'}{\partial t} + \frac{2\Omega}{r_E} \cos \varphi v' = -2\Omega \sin \varphi D' \tag{3.91}$$

$$\frac{\partial D'}{\partial t} + \frac{2\Omega}{r_E} \cos \varphi u' = 2\Omega \sin \varphi \xi' - g\nabla_h^2 H' \tag{3.92}$$

We see that the factors in front of the variables related to the Coriolis parameter depend on latitude. Consequently, the wave fields cannot be described by simple trigonometric functions. Longuet–Higgins (1968) presented a numerical solution method based on **spherical harmonics**. He also derived an approximate analytical solution for a large Rossby radius L_R. The procedure of this approach is described in the following.

Equations (3.91) and (3.92) can be recast with stream function and velocity potential as unknown variables using the Helmholtz theorem (3.52):

$$\frac{\partial}{\partial t} \nabla_h^2 \psi' + \frac{2\Omega}{r_E^2} \left(\frac{\partial \psi'}{\partial \lambda} + \cos \varphi \frac{\partial \chi'}{\partial \varphi} \right) = -2\Omega \sin \varphi \nabla_h^2 \chi' \quad (3.93)$$

$$\frac{\partial}{\partial t} \nabla_h^2 \chi' + \frac{2\Omega}{r_E^2} \left(\frac{\partial \chi'}{\partial \lambda} - \cos \varphi \frac{\partial \psi'}{\partial \varphi} \right) = 2\Omega \sin \varphi \nabla_h^2 \psi' - g\nabla_h^2 H'$$

$$\tag{3.94}$$

The fields can be represented by spherical harmonics Y_n^m that are defined as follows:

$$Y_n^m(\lambda, \varphi) = \sqrt{\frac{2n+1}{4\pi} \frac{(n-m)!}{(n+m)!}} P_n^m(\sin \varphi) e^{im\lambda} \quad \text{for } n \geq m \geq 0$$

$$\tag{3.95}$$

where m denotes the **zonal wavenumber**, n the **total wavenumber** and $P_n^m(\sin \varphi)$ the **associated Legendre polynomial**.[j]

---

[j]Note that m and n are integers.

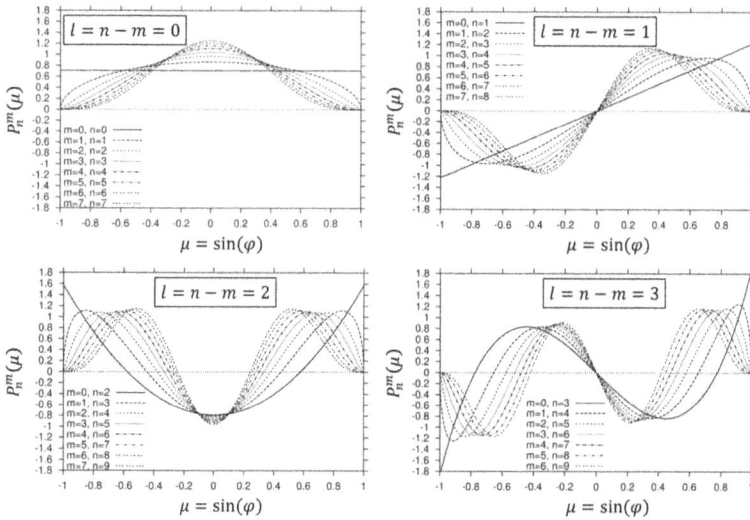

Fig. 3.12. Meridional profiles of various associated Legendre polynomials for $n - m = 0, 1, 2$ and 3.

These are given by

$$P_n^m(\mu) = \frac{1}{2^n n!} (1 - \mu^2)^{\frac{m}{2}} \frac{\partial^{m+n}}{\partial \mu^{m+n}} (\mu^2 - 1)^n \qquad (3.96)$$

where $\mu = \sin \varphi$. Obviously, the spherical harmonics have sinusoidal shape in zonal direction. However, the latitudinal dependence is described by the more complicated associated Legendre polynomials. The difference between the total and zonal wavenumber, $n - m$, yields the number of zero crossings between the north and south poles. Figure 3.12 displays the meridional profiles of some associated Legendre polynomials. Evidently, functions with even $n - m$ are symmetric about the equator while those with odd $n - m$ have an antisymmetric profile. The higher the zonal wavenumber, the closer the distance of the extrema to the equator.

Spherical harmonics have the advantage of being an **eigenfunction** of the Laplace operator such that

$$\nabla_h^2 Y_n^m = -\frac{n(n+1)}{r_E^2} Y_n^m \qquad (3.97)$$

Another useful relationship is

$$\frac{\partial Y_n^m}{\partial \lambda} = imY_n^m \tag{3.98}$$

Here, we consider a representation for a fixed zonal wavenumber m since waves with different zonal wavenumbers do not interact in the linearized system. We assume that

$$\psi' = \sum_{n=m}^{\infty} \psi_n'^m P_n^m(\mu) \cos(m\lambda - \nu t) \tag{3.99}$$

$$\chi' = \sum_{n=m}^{\infty} \chi_n'^m P_n^m(\mu) \sin(m\lambda - \nu t) \tag{3.100}$$

$$H' = \sum_{n=m}^{\infty} H_n'^m P_n^m(\mu) \cos(m\lambda - \nu t) \tag{3.101}$$

Obviously, the perturbations in the zonal direction have the form of sinusoidal waves propagating with the angular phase speed ν/m. The following recursion relations are beneficial for finding a solution:

$$(1 - \mu^2)\frac{\partial P_n^m}{\partial \mu} = \frac{(n+1)(n+m)}{2n+1} P_{n-1}^m$$
$$- \frac{n(n-m+1)}{2n+1} P_{n+1}^m \tag{3.102}$$

$$\mu P_n^m = \frac{n+m}{2n+1} P_{n-1}^m + \frac{n-m+1}{2n+1} P_{n+1}^m \tag{3.103}$$

It is possible to find similar recursion relations for spherical harmonics but the resulting formulas differ because of the normalization factor in Eq. (3.95). These relations can be used to evaluate the linear operator $(1-\mu^2)\partial P_n^m/\partial\mu + \mu \nabla_h^2$ that appears in (3.93) and (3.94) due to the existence of the Coriolis force.

Inserting the representation (3.99)–(3.101) in (3.93)–(3.94) and equating coefficients of the same associated Legendre polynomial to zero yields by consideration of (3.97), (3.98), (3.102) and (3.103)

$$\left[\tilde{\nu} + \frac{m}{n(n+1)}\right] \psi_n'^m = -\frac{(n+2)(n+m+1)}{(2n+3)(n+1)} \chi_{n+1}'^m$$
$$- \frac{(n-1)(n-m)}{n(2n-1)} \chi_{n-1}'^m \tag{3.104}$$

$$\left[\tilde{\nu} + \frac{m}{n(n+1)}\right]\chi_n'^m = -\frac{(n+2)(n+m+1)}{(2n+3)(n+1)}\psi_{n+1}'^m$$

$$-\frac{(n-1)(n-m)}{n(2n-1)}\psi_{n-1}'^m + \tilde{H}_n'^m \quad (3.105)$$

where $\tilde{\nu} = \nu/(2\Omega)$ denotes a dimensionless frequency and $\tilde{H}_n'^m = gH_n'^m/(2\Omega)$. The height anomaly $\tilde{H}_n'^m$ results from the continuity equation (3.32).

$$\tilde{\nu}\tilde{H}_n'^m = \frac{L_R^2}{r_E^2}n(n+1)\chi_n'^m \quad (3.106)$$

where $L_R = \sqrt{g\overline{H}}/(2\Omega)$ is the Rossby radius at the pole. We see that the coefficients $\psi_n'^m$ and $\tilde{H}_n'^m$ with even total wavenumbers coupled with coefficients $\chi_n'^m$ having odd total wavenumbers and vice versa. Consequently, the velocity potential is symmetric (antisymmetric) about the equator if stream function and height anomalies are anti-symmetric (symmetric) about the equator (see Fig. 3.12). Finally, we can write the linear equations in matrix form for each symmetry as

$$(\overset{\leftrightarrow}{D} - \tilde{\nu}\overset{\leftrightarrow}{I})\cdot\vec{X} = 0 \quad (3.107)$$

where $\vec{X}$ denotes the **state vector** containing all unknown variables as components, $\overset{\leftrightarrow}{D}$ the **Jacobi matrix**, and $\overset{\leftrightarrow}{I}$ the unit matrix.[k] This constitutes an **eigenvalue problem** where the **eigenvalues** $\tilde{\nu}$ result from the characteristic equation

$$|\overset{\leftrightarrow}{D} - \tilde{\nu}\overset{\leftrightarrow}{I}| = 0 \quad (3.108)$$

The solution belonging to a certain eigenvalue can be determined from the matrix equation (3.107) yielding the **eigenvector** or **normal mode** $\vec{X}_{\tilde{\nu}}$ of the eigenvalue $\tilde{\nu}$. Hence, the solution procedure is similar to that described in Section 3.5 but now the matrix has an infinite number of elements and no analytical solution can be found. To find an approximated solution, the matrix must be truncated

---

[k]We use the notation $\vec{X}$ and $\overset{\leftrightarrow}{I}$ for coefficient vectors and matrices, respectively to distinguish them from vectors and tensors in physical space.

at a certain maximum total wavenumber and the eigenvalues and eigenvectors must be determined by a numerical technique. We will not explain this technique but rather consider it as a black box in the form of a computer program. The truncation wavenumber must be chosen carefully. It will be appropriate if the result does not change significantly when this number is further increased.

Figure 3.13 shows the resulting frequencies $\tilde{\nu}$ as a function of L_R/r_E for zonal wavenumber $m = 5$. In general, the frequencies increase with increasing Rossby radius. We can detect three wave types: (i) westward traveling inertia-gravity waves, (ii) eastward traveling inertia-gravity waves and (iii) westward traveling planetary waves. The planetary waves have in contrast to the f-plane geometry nonzero frequencies. They move in comparison with inertia-gravity waves very slowly in westward direction (note the logarithmic axis scaling in Fig. 3.13). These waves are called **Rossby waves** or **Rossby–Haurwitz waves**. However, we note that these slowly propagating planetary waves on the sphere were originally derived by Margules (1893). Rossby (1939) introduced the so-called **β-plane geometry** with which some features of spherical planetary waves are preserved. Haurwitz (1940) presented the solution for full spherical geometry. The numbers marking the curves refer to the mode of the respective wave. The higher the mode the larger the number of zero crossings in the meridional direction. It can be asserted that the mode number coincides with $n - m$ of the most dominant Legendre polynomial in the expansion of stream function or velocity potential. The frequencies of the inertia-gravity waves show an approximately linear increase for large Rossby radii $L_R \gg r_E$ while the frequencies of the Rossby waves become independent of this radius. At such large radii, the interaction between modes having different total wavenumbers can be neglected and the frequencies become approximately

$$\nu_{1,2} = \pm\sqrt{g\overline{H}\frac{n(n+1)}{r_E^2}}, \quad \nu_3 = -\frac{2\Omega m}{n(n+1)} \tag{3.109}$$

It turns out that $\nu_{1,2}$ coincide with the frequencies of a gravity wave on a nonrotating sphere while ν_3 is the frequency of the Rossby wave in the shallow water model with an upper rigid lid in which no vertical displacements of the fluid are possible.

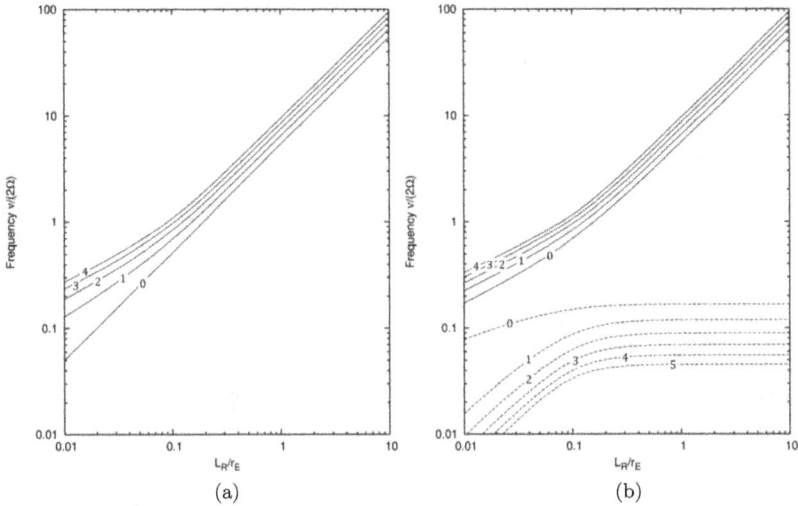

Fig. 3.13. Frequencies resulting from the eigenvalue problem (3.108) as a function of L_R/r_E for zonal wavenumber $m = 5$. (a) Eastward traveling waves and (b) westward traveling waves. The black lines display frequencies of inertia-gravity waves while the dashed lines show frequencies of planetary waves. The number indicates the mode.

The fields belonging to the frequencies of the gravity wave take the form

$$\psi'_{1,2} = 0, \quad \chi'_{1,2} = AP_n^m(\mu)\sin(m\lambda - \nu_{1,2}t),$$

$$H'_{1,2} = \frac{\overline{H}n(n+1)}{\nu_{1,2}r_E^2} AP_n^m(\mu)\cos(m\lambda - \nu_{1,2}t) \qquad (3.110)$$

while the fields

$$\psi'_3 = AP_n^m(\mu)\cos(m\lambda - \nu_3 t), \quad \chi'_3 = 0, \quad H'_3 = 0 \qquad (3.111)$$

describe the Rossby wave in the model with a rigid lid.

Figure 3.14 shows various wave patterns resulting from the eigenvalue analysis for $m = 5$ and mode 1. The height anomaly patterns of the two inertia-gravity waves (Figs. 3.14(a) and (c)) are antisymmetric about the equator and coincide due to the very large Rossby radius ($L_R = 10r_E$) nearly with the spherical harmonic Y_6^5. The solution (3.110) yields a good approximation for these waves. In contrast, the Rossby wave has for this L_R (Fig. 3.14(e)) a nonzero

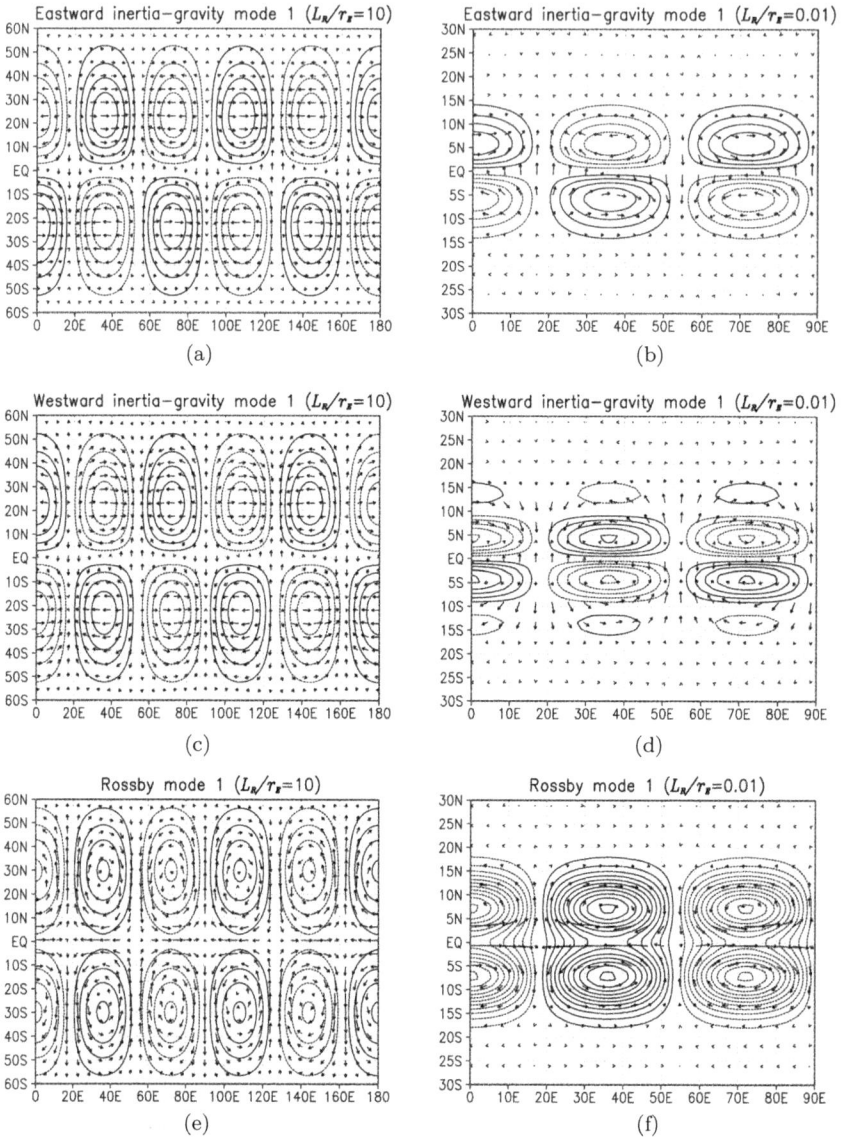

Fig. 3.14.   Height anomaly (isolines) and horizontal velocity (vectors) of various wave patterns resulting from the eigenvalue analysis for zonal wavenumber $m = 5$ and mode 1. (a) and (b) Eastward propagating inertia-gravity wave, (c) and (d) westward propagating inertia-gravity wave, (e) and (f) Rossby wave. The left (right) column shows results for $L_R/r_E = 10$ ($L_R/r_E = 0.01$).

height anomaly although it is absent in the rigid-lid solution (3.111). The height anomaly is responsible to keep the flow approximately in geostrophic balance as is the case for the planetary wave on the f-plane. For small L_R the structure of the wave pattern changes significantly. The meridional extension of the wave is limited to the equatorial band between $-20°$S and $20°$N (note the different latitudinal ranges in the left and right columns of Fig. 3.14). Therefore, the global analysis reveals **equatorial waves** when the Rossby radius becomes very small compared to the earth radius. Indeed, an analysis based on the equatorial β-plane approximation as done by Matsuno (1966) yields approximately the same results. The east- and westward traveling inertia-gravity modes have different structures in the meridional direction in contrast to very large L_R where the fields of these modes have the shape of the same spherical harmonic. Indeed, the westward propagating mode 1 is rather similar to the eastward propagating mode 3 for very small L_R and Matsuno's analysis confirms this. The equatorial waves have some peculiarities that can be deduced from the dispersion diagram in Fig. 3.15(a). The frequency of the eastward propagating mode 0 increases linearly with zonal wavenumber m so that the phase speed coincides with that of a pure gravity wave. This type of wave is referred to as the **equatorial Kelvin wave** and it has nonzero meridional flow. We will discuss this wave in more detail below. On the other hand, the Rossby mode 0 becomes a **mixed Rossby-gravity wave** since the character of this mode is rather like that of an inertia-gravity wave at low zonal wavenumbers m This wave type has also the name **Yanai wave** because of one of its discoverers (Yanai and Maruyama, 1966).

The dispersion diagrams at large L_R (Figs. 3.15(b) and 3.15(c)) show that the frequencies of all inertia gravity waves increase almost linearly with zonal wavenumber m while the frequencies of the Rossby waves are comparatively very small. These curves are well reproduced by the approximation (3.109). Equatorial waves play no role in barotropic weather forecast models since the Rossby radius $L_R = \sqrt{g\overline{\overline{H}}}/(2\Omega)$ has a magnitude similar to the earth's radius r_E and, therefore, $L_R/r_E = o(1)$. However, these waves become important in the baroclinic atmosphere that allows for **internal inertia-gravity waves** (treated in Section 4.1). Internal gravity waves experience smaller restoring forces in the vertical direction

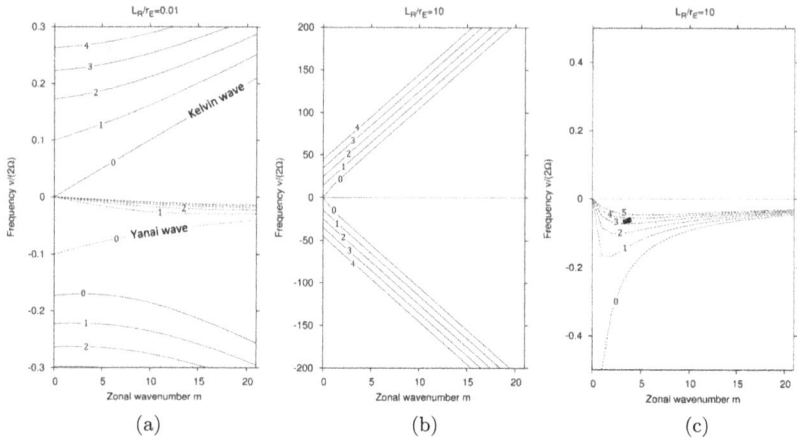

Fig. 3.15. Frequencies resulting from the eigenvalue problem (3.108) as a function of zonal wavenumber $m = 5$. (a) $L_R/r_E = 0.01$, (b) $L_R/r_E = 10$, (c) as (b) but a smaller frequency range is shown. The black lines display frequencies of inertia-gravity waves while the dashed lines show frequencies of planetary waves. The number indicates the mode.

than gravity waves with surface deflections. Therefore, the **inertia-gravity waves** are related to a much smaller Rossby radius. Such motions can be triggered by condensational heating in the equatorial region which gives rise to the excitation of equatorial waves. These waves have intraseasonal time scales because of their small frequencies.

The equatorial Kelvin wave can easily be derived from Laplace's tidal equations for the equatorial β-lane geometry. The geometry is nearly Cartesian in the vicinity of the equator. Then, we can use the coordinates

$$x = r_E \lambda, \quad y = r_E \varphi \tag{3.112}$$

and the Coriolis parameter becomes approximately

$$f = 2\Omega\varphi \equiv \beta y \tag{3.113}$$

where $\beta = 2\Omega/r_E$. Equatorial Kelvin waves have no flow component in the meridional direction. Therefore, the linearized equations take the form

$$\frac{\partial u'}{\partial t} = -g\frac{\partial H'}{\partial x} \tag{3.114}$$

$$\beta y u' = -g \frac{\partial H'}{\partial y} \tag{3.115}$$

$$\frac{\partial H'}{\partial t} = -\overline{H} \frac{\partial u'}{\partial x} \tag{3.116}$$

Obviously, the zonal flow must be in geostrophic balance to keep the meridional velocity zero. With the solution ansatz

$$u' = \hat{u}(y)\cos(kx - \nu t), \quad H' = \hat{H}(y)\cos(kx - \nu t) \tag{3.117}$$

we obtain the eigenfrequencies

$$\nu_{1,2} = \pm\sqrt{g\overline{H}}\, k = \pm c_G k \tag{3.118}$$

and the wave amplitudes are related to each other by

$$\hat{u}(y) = \mp\sqrt{\frac{g}{\overline{H}}}\,\hat{H}(y) \tag{3.119}$$

Therefore, the geostrophic balance equation (3.115) leads to the linear ordinary differential equation

$$\frac{d\hat{H}}{dy} \mp \frac{\beta y}{c_G}\hat{H} = 0 \tag{3.120}$$

The solution is

$$\hat{H} = A\exp\left(\mp\frac{\beta y^2}{2c_G}\right) \tag{3.121}$$

Only the wave solution with the minus sign describes an equatorial wave while the other sign yields a questionable increase of amplitude with increasing latitude and must, therefore, be rejected. Consequently, the amplitude function is represented by a Gaussian bell with the width $\sqrt{c_G/\beta} = \sqrt{L_R r_E}$. Now, it is understandable that $L_R \ll r_E$ is a requirement for the existence of the equatorial Kelvin wave. This statement also holds for all other equatorial waves (see Matsuno, 1966).

Figure 3.16 shows the height and wind fields of an equatorial Kelvin wave. Zonal wind and height anomalies maximize at the equator. In the zonal direction, the profiles of the quantities are identical to those of a plane eastward traveling gravity wave. The meridional decay of zonal wind and height anomaly is in such a way that geostrophic balance holds exactly.

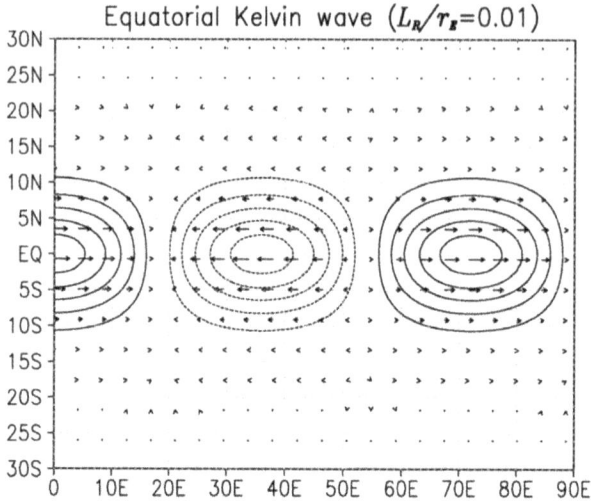

Fig. 3.16. Height anomaly (isolines) and horizontal velocity (vectors) of the equatorial Kelvin wave mode resulting from the eigenvalue analysis for zonal wavenumber $m = 5$ and $L_R/r_E = 0.01$.

3.8 Midlatitude Rossby Waves and the Quasigeostrophic Approximation

Synoptic-scale weather systems in the midlatitude can be well represented by the superposition of Rossby wave modes. However, the linear Laplace tidal equations are not suited for a prediction of these systems. This becomes clear by the fact that Rossby wave modes travel westward while real weather systems move usually eastward. A better description of these systems can be gained with the quasigeostrophic equations for a midlatitude β-plane geometry although they are still not sufficient for weather prediction. We will first study Rossby waves on the midlatitude β-plane and derive the quasigeostrophic equations afterward.

3.8.1 *Rossby waves in a nondivergent barotropic fluid*

The basic mechanism of Rossby wave propagation can already be understood in a nondivergent barotropic fluid. The shallow water model becomes nondivergent if no surface orography exists and the upper boundary forms a **rigid lid**. Then, the depth of the fluid

H_a is a constant and the divergence vanishes due to the continuity equation (3.13). Consequently, the horizontal velocity can be exclusively represented by the stream function ψ and the vorticity equation (3.14) becomes

$$\frac{\partial}{\partial t}\nabla_h^2\psi + \mathbf{k}\times\nabla_h\psi\cdot\nabla(f+\nabla_h^2\psi) = 0 \tag{3.122}$$

This equation has analytical Rossby wave solutions even for spherical geometry. Here, we consider the midlatitude β-plane geometry instead. This means that we employ the coordinates (3.33) of a tangential surface as for the f-plane case but we allow for a meridional gradient of the Coriolis parameter such that

$$f = f_0 + \left.\frac{\partial f}{\partial y}\right|_{y=0} y \equiv f_0 + \beta y \tag{3.123}$$

Then, we obtain for the vorticity equation

$$\frac{\partial}{\partial t}\nabla_h^2\psi + J_{xy}(\psi,\nabla_h^2\psi) + \beta\frac{\partial\psi}{\partial x} = 0 \tag{3.124}$$

where $J_{xy}(a,b)$ denotes the **Jacobi-operator** which is defined by

$$J_{xy}(a,b) \equiv \begin{vmatrix} \dfrac{\partial a}{\partial x} & \dfrac{\partial a}{\partial y} \\[2mm] \dfrac{\partial b}{\partial x} & \dfrac{\partial b}{\partial y} \end{vmatrix} = \frac{\partial a}{\partial x}\frac{\partial b}{\partial y} - \frac{\partial b}{\partial x}\frac{\partial a}{\partial y} \tag{3.125}$$

Note that the Jacobi-operator has the following properties:

$$J_{xy}(a,a) = 0 \tag{3.126}$$

$$J_{xy}(a,b) = -J_{xy}(b,a) \tag{3.127}$$

$$J_{xy}(a,b+c) = J_{xy}(a,b) + J_{xy}(a,c) \tag{3.128}$$

$$J_{xy}(a,bc) = bJ_{xy}(a,c) + cJ_{xy}(a,b) \tag{3.129}$$

Usually, an eastward wind prevails in the mid-latitudes. Therefore, the perturbation analysis should be made with respect to a base state including such a flow. For simplicity, we assume a constant base state zonal velocity U and get

$$\overline{\psi} = -Uy \tag{3.130}$$

Then, the linearized vorticity equation becomes

$$\frac{\partial}{\partial t}\nabla_h^2 \psi' + U\frac{\partial}{\partial x}\nabla_h^2 \psi' + \beta\frac{\partial \psi'}{\partial x} = 0 \tag{3.131}$$

Inserting a wave solution of the form

$$\psi' = A\cos ly \cos(kx - \nu t) \tag{3.132}$$

yields the dispersion relation[1]

$$\nu = Uk - \frac{\beta k}{k^2 + l^2} \tag{3.133}$$

Obviously, the rigid lid acts as a filter that removes the inertia-gravity waves. For $U = 0$ and $\beta = 0$, the planetary wave on the f-plane as shown in the bottom panel of Fig. 3.8 is recovered. Figure 3.17 shows dispersion diagrams for $\beta = 2\Omega\cos(45°)/r_E = 1.6182 \times 10^{-11}(\text{ms})^{-1}$ and different meridional wavenumbers l. The wave propagates westward for nonzero β and no base state flow (Fig. 3.17(a)). The frequency decreases with increasing meridional wavenumber and it maximizes at a certain zonal wavenumber. An exception is the plane wave solution for $l = 0$ where the frequency approaches infinity with increasing wavelength. However, such a solution is unrealistic since a β-plane cannot have a meridionally infinite extension which is required for $l = 0$. The inclusion of a nonzero base state flow U allows for eastward propagating waves. Very short waves feel mainly the zonal flow and the β-effect can be neglected. Long waves may propagate westward and a zonal wavenumber k_s could exist where the wave becomes stationary. We obtain this wavenumber by setting ν to zero in (3.133):

$$k_s = \sqrt{\frac{\beta}{U} - l^2} \tag{3.134}$$

We see, however, that the Rossby wave travels eastward for every zonal wavenumber at meridional wavenumbers larger than $\sqrt{\beta/U}$.

The mechanism of Rossby wave propagation is illustrated in Fig. 3.18. It shows zonal profiles of meridional velocity and vorticity at $y = 0$. There, the magnitude of the meridional flow and the

---

[1]The wave (3.132) also fulfills the nonlinear vorticity equation (3.124) with the same dispersion relation since $J_{xy}(\psi', \nabla_h^2\psi') = -(k^2 + l^2)J_{xy}(\psi', \psi') = 0$.

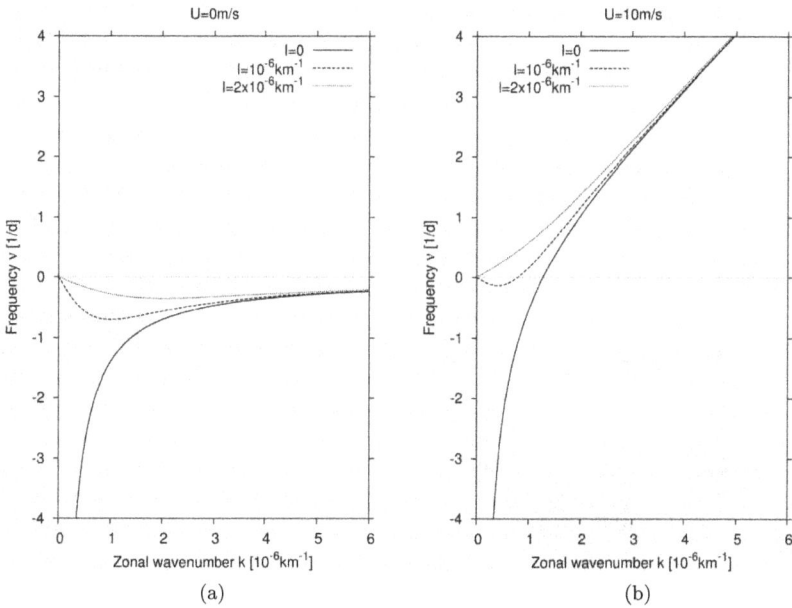

Fig. 3.17. Dispersion diagrams for Rossby wave solutions of the nondivergent vorticity equation for different meridional wavenumbers. (a) Basic state flow $U=0$ and (b) basic state flow $U=10\,\mathrm{m/s}$.

planetary vorticity advection $-\beta v$ maximize. The vorticity wave is phase-shifted by $90°$ in westward direction. At the points marked with 1 low planetary vorticity from the south is advected north- wards which leads to a decrease of relative vorticity while at the points marked with 2, the opposite is the case. The vorticity decrease (increase) causes a decrease (increase) of zonal shear $\partial v/\partial x$ at points 1 (3.2). Consequently, the maxima and minima of the merid- ional wind shift westward due to this change. The same is true for the relative vorticity maxima and minima.

3.8.2   Rossby waves in a barotropic fluid with a free surface

Removing the rigid upper lid complicates the derivation of Rossby wave solutions due to additional processes. For simplicity, we neglect the basic flow U. In this case, the linearized shallow water equations

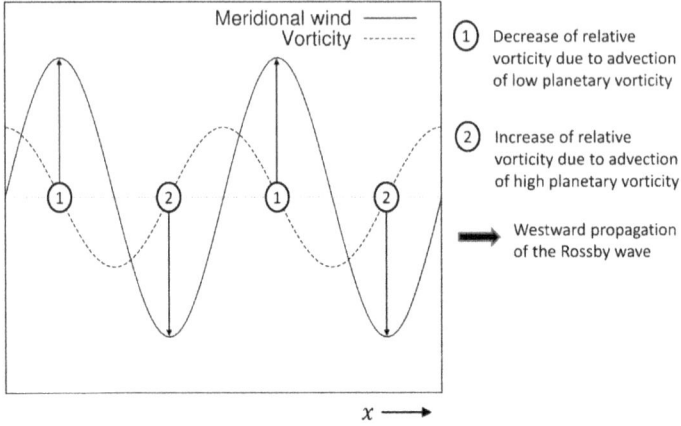

Fig. 3.18. Sketch of the mechanism for Rossby wave propagation. The solid and dashed curves show zonal profiles of meridional wind and relative vorticity respectively. Symbols 1 and 2 mark positions of extremal planetary vorticity advection causing the westward propagation.

on the β-plane become

$$\frac{\partial u'}{\partial t} - (f_0 + \beta y)v' = -g\frac{\partial H'}{\partial x} \tag{3.135}$$

$$\frac{\partial v'}{\partial t} + (f_0 + \beta y)u' = -g\frac{\partial H'}{\partial y} \tag{3.136}$$

$$\frac{\partial H'}{\partial t} = -\overline{H}\left(\frac{\partial u'}{\partial x} + \frac{\partial v'}{\partial y}\right) \tag{3.137}$$

The factors in these equations depend partly on y and, therefore, it is not possible to apply a standard solution method as in the f-plane case. However, we can make a separation ansatz of the form

$$(u', v', H') = \Re\{[\hat{u}(y), \hat{v}(y), \hat{H}(y)]e^{i(kx-\nu t)}\} \tag{3.138}$$

Inserting this ansatz leads to

$$-i\nu\hat{u} - (f_0 + \beta y)\hat{v} = -igk\hat{H} \tag{3.139}$$

$$-i\nu\hat{v} + (f_0 + \beta y)\hat{u} = -g\frac{d\hat{H}}{dy} \tag{3.140}$$

$$-i\nu\hat{H} = -\overline{H}\left(ik\hat{u} + \frac{d\hat{v}}{dy}\right) \tag{3.141}$$

Eliminating $\hat{u}$ in (3.140) and (3.141) yields

$$[\nu^2 - (f_0 + \beta y)^2]\hat{v} = -i(f_0 + \beta y)gk\hat{H} - ig\nu\frac{d\hat{H}}{dy} \qquad (3.142)$$

$$(\nu^2 - c_G^2 k^2)\hat{H} = ik\overline{H}(f_0 + \beta y)\hat{v} - i\nu\overline{H}\frac{d\hat{v}}{dy} \qquad (3.143)$$

The second equation can easily be solved for $\hat{H}$ if $\nu^2 - c_G^2 k^2 \neq 0$ holds, which excludes Kelvin waves. Then, we obtain by inserting the result in (3.142)

$$\frac{d^2\hat{v}}{dy^2} - \left[\frac{(f_0 + \beta y)^2}{c_G^2} - \frac{\nu^2}{c_G^2} + k^2 + \frac{\beta k}{\nu}\right]\hat{v} = 0 \qquad (3.144)$$

This equation yields for $\beta = 0$ the same solutions as in Section 3.5. For $f_0 = 0$ the equation coincides with that found by Matsuno (1966) to derive the equatorial waves on the equatorial β-plane. We can even recover Matsuno's equation in the case $f_0 \neq 0$ by the simple coordinate transform $y \to y - f_0/\beta$. Therefore, a superposition of solutions of Matsuno's analysis can be used to construct the midlatitude waves. Here, we restrict to a narrow latitudinal range in the mid-latitudes so that $\beta y \ll f_0$. This could be realized in a zonal channel having a width $L \ll f_0/\beta$. Then, the approximation $f_0 + \beta y \approx f_0$ can be made in Eq. (3.144). In this case, the equation has a constant coefficient in front of $\hat{v}$ and the solution takes the form

$$\hat{v} = A_n \sin\left[\frac{n\pi}{L}\left(y + \frac{L}{2}\right)\right] \qquad (3.145)$$

The solution must fulfill the channel boundary condition $\hat{v}(-L/2) = \hat{v}(L/2) = 0$. The boundary conditions allow only for discrete modes $n = 1, 2, 3\ldots$. Inserting this solution in (3.144) yields a cubic equation for the frequency:

$$\nu^3 - [c_G^2(k^2 + l_n^2) + f_0^2]\nu - c_G^2\beta k = 0 \qquad (3.146)$$

where $l_n = n\pi/L$. Approximate solutions result for high and low frequencies. We can neglect the last term on the left-hand side for high frequencies $\nu^3 \gg c_G^2\beta k$. Then, we find the frequencies of the inertia-gravity waves:

$$\nu_{1,2} = \pm\sqrt{c_G^2(k^2 + l_n^2) + f_0^2} \qquad (3.147)$$

On the other hand, for low frequencies $\nu^3 \ll c_G^2\beta k$ we can neglect the first term on the left-hand side of Eq. (3.146). In this case, the

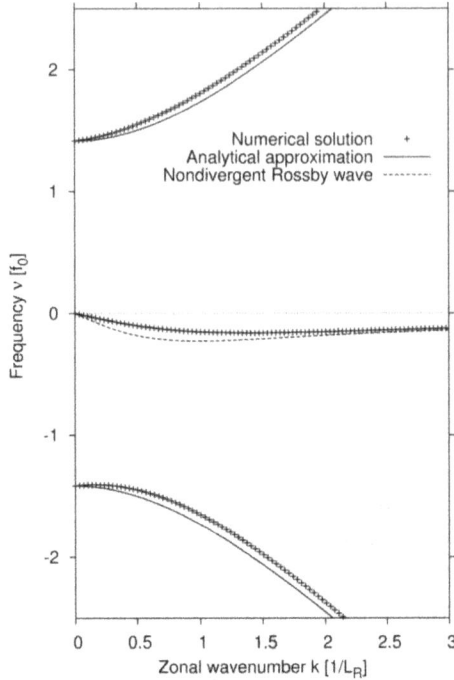

Fig. 3.19.   Dispersion diagram for solutions of the linearized shallow water equations on the β-plane for $\beta = 1.6182 \times 10^{-11}(\mathrm{ms})^{-1}$, $f_0 = 10^{-4}\,\mathrm{s}^{-1}$ and $\overline{H} = 8000\,\mathrm{m}$. The crosses show the numerical solutions of Eq. (3.146) while the solid line displays the analytical approximation. The dashed line displays the result for the nondivergent Rossby wave.

frequency of the Rossby wave results

$$\nu_3 = -\frac{\beta k}{\frac{1}{L_R^{\,2}} + k^2 + l_n^{\,2}} \tag{3.148}$$

By comparing this result with (3.133) we see that the rigid lid assumption leads to faster westward Rossby wave propagation and filtering of inertia-gravity waves. Figure 3.19 shows the dispersion diagram for the approximated solutions and the accurate solution of the frequency equation (3.146) for mode $n = 1$. Obviously, the approximation introduces little error for the chosen parameters, especially for the Rossby wave. The nondivergent Rossby wave has a more negative frequency so that the wave moves faster westward.

3.8.3 Quasigeostrophic approximation

For understanding synoptic-scale weather systems it is convenient to employ filtered model equations that do not allow for inertia-gravity waves since they play no role for these systems. The inclusion of a rigid lid already filters the inertia-gravity wave but the phase speeds of Rossby waves are overestimated. The wind in these weather systems is approximately geostrophic. Therefore, one may think that replacing the horizontal wind with the geostrophic wind in the equations could lead to an appropriate model. However, the model would be overdetermined in this case because the geostrophic wind is a unique function of the height anomaly. Instead, the **quasi-geostrophic approximation** that was first formulated by Charney (1948) represents a suitable way to filter inertia-gravity waves. It results from scale analysis and neglection of small terms. We assume as in Section 2.3 scales for the various variables. Besides the velocity scale U, length scale L and advective time scale T = L/U we need a scale H for the height anomaly. For estimating this we make use of the approximately valid geostrophic wind balance $f\mathbf{k} \times \mathbf{v}_h = -g\nabla_h H'$ where H' denotes the deviation from the reference state H_r. Then, we obtain H = $f_0 UL/g$. It is convenient to **nondimensionalize** all variables by the scales. Then, we find the following nondimensional variables:

$$\tilde{\mathbf{v}}_h \equiv \frac{\mathbf{v}_h}{U}, \quad (\tilde{x}, \tilde{y}) \equiv \frac{(x,y)}{L}, \quad \tilde{t} \equiv \frac{Ut}{L}, \quad \tilde{H}' \equiv \frac{H'}{H} \qquad (3.149)$$

Beside these definitions we introduce three nondimensional parameters:

- **Rossby number** Ro $= \frac{U}{f_0 L}$
- **Burger number** Bu $= L_R^2/L^2$,
- β-parameter $\tilde{\beta} = \beta L^2/U$.

By substituting the nondimensional variables in the shallow water equations (3.12) and (3.13) for a β-plane geometry and $\boldsymbol{f}_T = 0$ we get

$$\frac{\partial \tilde{\mathbf{v}}_h}{\partial \tilde{t}} + \tilde{\mathbf{v}}_h \cdot \tilde{\nabla}_h \tilde{\mathbf{v}}_h + \frac{1}{\text{Ro}}(1 + \text{Ro}\tilde{\beta}\tilde{y})\mathbf{k} \times \tilde{\mathbf{v}}_h = -\frac{1}{\text{Ro}}\tilde{\nabla}_h \tilde{H}' \quad (3.150)$$

$$\frac{\partial \tilde{H}'}{\partial \tilde{t}} = -\tilde{\nabla}_h \cdot [(\tilde{H}' - \tilde{H}_s)\tilde{\mathbf{v}}_h] - \frac{\text{Bu}}{\text{Ro}}\tilde{\nabla}_h \cdot \tilde{\mathbf{v}}_h \qquad (3.151)$$

The quasigeostrophic approximation is valid if

$$\text{Ro} \ll 1, \quad \text{Bu} = \text{o}(1), \quad \tilde{\beta} = \text{o}(1) \qquad (3.152)$$

The first condition states that the ratio of the inertia force to the Coriolis force is very small while the second condition has the consequence of a length scale in the order of the Rossby radius. The last condition is in agreement with the previously made requirement of a narrow zonal channel so that $\text{L} \ll f_0/\beta = \tan\varphi_0 r_E \approx r_E$.

To derive the quasigeostrophic approximation, we make first the following expansion:

$$\tilde{\mathbf{v}}_h = \tilde{\mathbf{v}}_h^{(0)} + \tilde{\mathbf{v}}_h^{(1)}\text{Ro} + \tilde{\mathbf{v}}_h^{(2)}\text{Ro}^2 + \text{o}(\text{Ro}^3), \qquad (3.153)$$

$$\tilde{H}' = \tilde{H}'^{(0)} + \tilde{H}'^{(1)}\text{Ro} + \tilde{H}'^{(2)}\text{Ro}^2 + \text{o}(\text{Ro}^3) \qquad (3.154)$$

It is assumed that the coefficients of these expansions are of order unity and that terms of the same order can be equated after insertion in the model equations. Then, a truncation of the expansion at a finite order yields an adequate approximation of the fields in the case $\text{Ro} \ll 1$. Substitution of (3.153) and (3.154) into (3.150) and (3.151) leads to

$$\left[\frac{\partial}{\partial \tilde{t}} + (\tilde{\mathbf{v}}_h^{(0)} + \tilde{\mathbf{v}}_h^{(1)}\text{Ro} + \cdots) \cdot \tilde{\nabla}_h \right] \left(\tilde{\mathbf{v}}_h^{(0)} + \tilde{\mathbf{v}}_h^{(1)}\text{Ro} + \cdots \right)$$

$$+ \frac{1}{\text{Ro}} \left(1 + \text{Ro}\tilde{\beta}\tilde{y} \right) \mathbf{k} \times \left(\tilde{\mathbf{v}}_h^{(0)} + \tilde{\mathbf{v}}_h^{(1)}\text{Ro} + \cdots \right)$$

$$= -\frac{1}{\text{Ro}} \tilde{\nabla}_h \left(\tilde{H}'^{(0)} + \tilde{H}'^{(1)}\text{Ro} + \cdots \right) \qquad (3.155)$$

$$\frac{\partial}{\partial \tilde{t}} \left(\tilde{H}'^{(0)} + \tilde{H}'^{(1)}\text{Ro} + \cdots \right)$$

$$= -\tilde{\nabla}_h \cdot \left[\left(\tilde{H}'^{(0)} + \tilde{H}'^{(1)}\text{Ro} + \cdots - \tilde{H}_s \right) \left(\tilde{\mathbf{v}}_h^{(0)} + \tilde{\mathbf{v}}_h^{(1)}\text{Ro} + \cdots \right) \right]$$

$$- \frac{\text{Bu}}{\text{Ro}} \tilde{\nabla}_h \cdot \left(\tilde{\mathbf{v}}_h^{(0)} + \tilde{\mathbf{v}}_h^{(1)}\text{Ro} + \cdots \right) \qquad (3.156)$$

Equating terms having the lowest order yield

$$\mathbf{k} \times \tilde{\mathbf{v}}_h^{(0)} = -\tilde{\nabla}_h \tilde{H}'^{(0)} \qquad (3.157)$$

$$\tilde{\nabla}_h \cdot \tilde{v}_h^{(0)} = 0 \qquad (3.158)$$

These equations state that the flow of zero order is geostrophic and nondivergent. For the next order, we find

$$\frac{\partial \tilde{\mathbf{v}}_h^{(0)}}{\partial \tilde{t}} + \tilde{\mathbf{v}}_h^{(0)} \cdot \tilde{\nabla}_h \tilde{\mathbf{v}}_h^{(0)} + \mathbf{k} \times \tilde{\mathbf{v}}_h^{(1)} + \tilde{\beta} \tilde{y} \mathbf{k} \times \tilde{\mathbf{v}}_h^{(0)} = -\tilde{\nabla}_h \tilde{H}'^{(1)}$$

$$(3.159)$$

$$\frac{\partial}{\partial \tilde{t}} \tilde{H}'^{(0)} + \tilde{\mathbf{v}}_h^{(0)} \cdot \tilde{\nabla}_h (\tilde{H}'^{(0)} - \tilde{H}_s) = -\mathrm{Bu} \tilde{\nabla}_h \cdot \tilde{\mathbf{v}}_h^{(1)} \quad (3.160)$$

We identify the zero-order wind $\tilde{\mathbf{v}}_h^{(0)}$ with geostrophic wind $\tilde{\mathbf{v}}_g$ and zero-order height anomaly $\tilde{H}'^{(0)}$ with geostrophic stream function $\tilde{\psi}_g$ while first-order wind $\tilde{\mathbf{v}}_h^{(1)}$ is interpreted as the **ageostrophic wind** $\tilde{\mathbf{v}}_a$. With these notations the quasigeostrophic equations become

$$\tilde{\mathbf{v}}_g = \mathbf{k} \times \tilde{\nabla}_h \tilde{\psi}_g \quad (3.161)$$

$$\frac{\partial \tilde{\mathbf{v}}_g}{\partial \tilde{t}} + \tilde{\mathbf{v}}_g \cdot \tilde{\nabla}_h \tilde{\mathbf{v}}_g + \mathbf{k} \times \tilde{\mathbf{v}}_a + \tilde{\beta} \tilde{y} \mathbf{k} \times \tilde{\mathbf{v}}_g = -\tilde{\nabla}_h \tilde{H}'^{(1)} \quad (3.162)$$

$$\frac{\partial \tilde{\psi}_g}{\partial \tilde{t}} + \tilde{\mathbf{v}}_g \cdot \tilde{\nabla}_h (\tilde{\psi}_g - \tilde{H}_s) = -\mathrm{Bu} \tilde{\nabla}_h \cdot \tilde{\mathbf{v}}_a \quad (3.163)$$

We see that the quasigeostrophic equations do not depend on the Rossby number anymore. Therefore, the solutions for different small Rossby numbers are identical although the dimensioned fields differ because the scales differ. This system is not closed since the first-order height anomaly is unknown. However, we can eliminate this unknown by deriving the **quasigeostrophic potential vorticity equation**. The quasigeostrophic vorticity equation results after applying the curl-operator to (3.162)

$$\frac{\partial}{\partial \tilde{t}} \nabla_h^2 \tilde{\psi}_g + \tilde{\mathbf{v}}_g \cdot \tilde{\nabla}_h (\nabla_h^2 \tilde{\psi}_g) + \tilde{\beta} \tilde{\mathbf{v}}_g = -\tilde{\nabla}_h \cdot \tilde{\mathbf{v}}_a \quad (3.164)$$

The divergence of the ageostrophic wind can be eliminated using the quasigeostrophic continuity equation (3.163) so that the quasigeostrophic potential vorticity equation results

$$\left(\frac{\partial}{\partial \tilde{t}} + \tilde{\mathbf{v}}_g \cdot \tilde{\nabla}_h \right) \left[\nabla_h^2 \tilde{\psi}_g + \frac{1}{\mathrm{Ro}} + \tilde{\beta} \tilde{y} - \frac{1}{\mathrm{Bu}} (\tilde{\psi}_g - \tilde{H}_s) \right] = 0 \quad (3.165)$$

In dimensioned form this equation becomes

$$\left(\frac{\partial}{\partial t} + \mathbf{v}_g \cdot \nabla_h\right) P_g = 0 \qquad (3.166)$$

where

$$P_g = \nabla_h^2 \psi_g + f_0 + \beta y - \frac{1}{L_R^2}\left(\psi_g - \frac{gH_s}{f_0}\right)$$

defines the **quasigeostrophic potential vorticity**. It is the linearized approximation of Rossby's potential vorticity (3.16) for a geostrophic flow. It is notable that the derivation of (3.166) does not involve any approximation. However, the geostrophic wind and height anomaly predicted by this equation does not coincide with the geostrophic wind and height anomaly deduced from the original shallow water equations because the latter also includes contributions from higher-order terms. The solution of (3.166) at high Rossby numbers has little in common with the true solution since high-order terms dominate in this case. The scale analysis in Section 2.3 suggests a Rossby number Ro = 0.1. The length scale L is close to the Rossby radius L_R and L ≈ $0.16 f_0/\beta$. Therefore, the conditions for the quasi-geostrophic approximation are only roughly fulfilled. The β-plane approximation is also a rather conceptual approach so the quasi-geostrophic vorticity equation (3.166) hardly suffices for the forecast of real weather systems. However, the quasigeostrophic model on the β-plane provides a conceptual understanding of synoptic-scale weather dynamics and it can indeed be applied for diagnostic purposes.

We can deduce an interesting property of the quasigeostrophic shallow water model, namely, that the same mathematical equation results if we set (i) $f = f_0 + \beta y$ and $H_s = 0$ or we set, (ii) $f = f_0, H_s = L_R^2 f_0 \beta y / g$. Therefore, a surface orography with a linear meridional increase has the same effect as the meridional increase of the Coriolis parameter. The reason is that the potential vorticity shows in both cases a positive meridional gradient due to these factors. Case (ii) facilitates **topographic Rossby waves** which share all properties with conventional Rossby waves.

Linearizing the quasigeostrophic potential vorticity equation with respect to the base state

$$\overline{\psi}_g = -Uy, \quad H_s = 0 \qquad (3.167)$$

describing a zonal basic flow U yields

$$\left(\frac{\partial}{\partial t} + U\frac{\partial}{\partial x}\right)\left(\nabla_h^2\psi_g' - \frac{1}{L_R^2}\psi_g'\right) + \left(\beta + \frac{U}{L_R^2}\right)\frac{\partial\psi_g'}{\partial x} = 0 \qquad (3.168)$$

with the wave ansatz $\psi_g' = A\cos ly\cos(kx - \nu t)$, we find the dispersion relation

$$\nu = Uk - \frac{(\beta + U/L_R^2)k}{1/L_R^2 + k^2 + l^2} \qquad (3.169)$$

This relation coincides in the case $U = 0$ with the Rossby wave frequency as given by (3.148). Therefore, the quasigeostrophic approximation does not falsify the phase velocity of Rossby waves in this case in contrast to the rigid lid assumption that reveals faster Rossby waves. The basic flow U enhances the westward phase speed relative to the velocity of the fluid parcel. This is explained by the meridional downward slope of the fluid surface due to the geostrophic balance of the zonal flow U. The same arguments hold for the higher relative westward phase speed as for the topographic Rossby waves.

The quasigeostrophic Rossby waves have the following phase speed and group velocity:

$$c = \frac{\nu}{k} = U - \frac{\hat{\beta}}{\frac{1}{L_R^2} + k^2 + l^2},$$

$$c_g = \frac{\partial\nu}{\partial k} = U - \frac{\hat{\beta}\left(\frac{1}{L_R^2} + l^2 - k^2\right)}{\left(\frac{1}{L_R^2} + k^2 + l^2\right)^2} \qquad (3.170)$$

where $\hat{\beta} = \beta + U/L_R^2$. We see that the wave phases move faster westward relative to the basic flow than the wave groups. The dispersion diagram in Fig. 3.20 shows frequency as a function of zonal and meridional wavenumbers for $U = 0$. The absolute frequency values are below the maximum $\beta L_R/2$ at $k = 1/L_R$ and $l = 0$. The dashed line crossing the maximum separates long waves with westward group velocity from short waves having eastward group velocity. Adding a basic flow would move the dashed line to smaller zonal wavenumbers. Therefore, most Rossby waves have eastward group velocities.

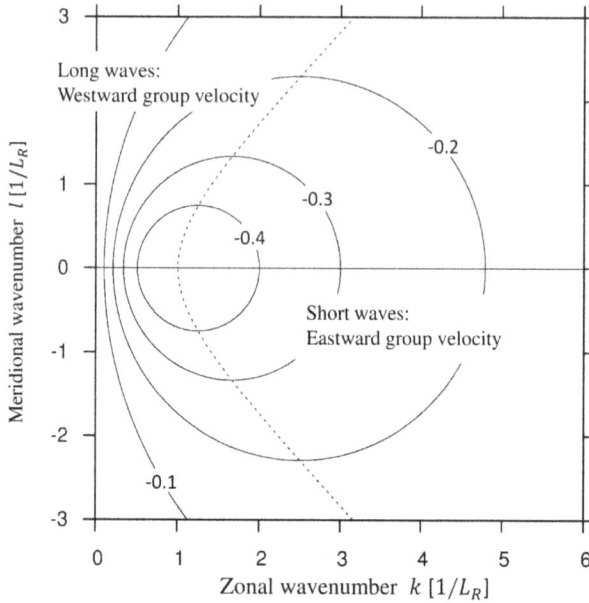

Fig. 3.20. Rossby wave frequency as a function of zonal and meridional wavenumber for $U = 0$. The frequency has been nondimensionalized with βL_R. The dashed line separates Rossby waves with westward group velocity from those having eastward group velocity.

Eastward propagation of wave groups together with the result $c_g > c$ give rise to the phenomenon **downstream development**. It defines the consecutive formation of lows and highs east of the Rossby wave packet while lows and highs consecutively decay west of the Rossby wave packet. Downstream development can be observed on weather maps. Figure 3.21 shows 500 hPa geopotential height for selected dates in April 1984. On April 5, 12UTC we see a strong trough at 80 W and a weaker one at 40 W while ridges occur east and west of the troughs. This is obviously a Rossby wave packet having its center over the eastern part of North America. After 36 h the wave troughs and ridges have traveled slightly eastward. Furthermore, the trough over the North Atlantic has intensified. Further intensification takes place after another 36 h while the trough over North America has weakened. The downstream development can be better visualized in a **Hovmöller diagram** shown in Fig. 3.21(b). This space–time diagram displays the field as a function of longitude

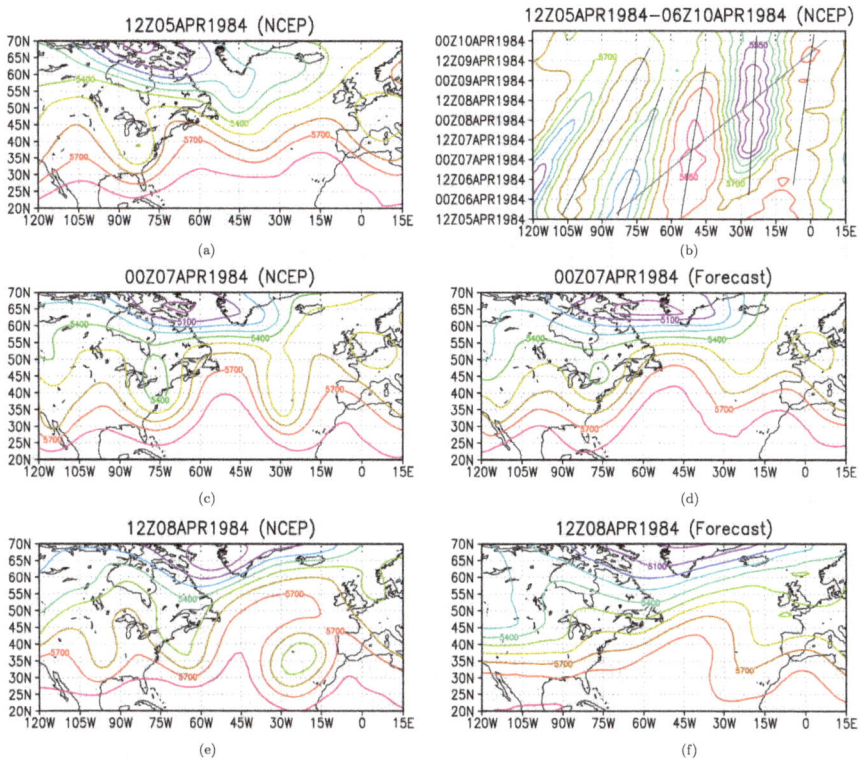

Fig. 3.21. Maps showing 500 hPa geopotential height (contour interval 100 m). (a) NCEP reanalysis, April 5, 1984, 12 UTC, (c) NCEP reanalysis, April 7, 00 UTC, (d) numerical forecast, April 7, 00 UTC, (e) NCEP reanalysis, April 8, 12 UTC, (f) numerical forecast, April 8, 12 UTC. The forecast model was initialized with NCEP reanalysis data on April 5, 1984, 12 UTC. (b) Longitude-time diagram of 500 hPa geopotential height at 30 N.

and time at a fixed latitude. In this diagram, the individual troughs and ridges exhibit a life cycle with growth and decay. Connecting the local maxima and minima of these systems by a line yields the propagation of the wave group. On the other hand, the propagation of the phases can be traced by connecting the positions of the troughs and ridges in the diagram. Evidently, the wave group moves much faster eastward as the wave phases. This is in agreement with the result (3.170).

Figures 3.21(d) and 3.21(f) show forecasts of a numerical model based on the shallow water equations. The forecast model was initialized with NCEP reanalysis data from April 5, 12UTC. We see that the model has some skill to predict the positions of the troughs and ridges. However, the amplitude of the Rossby wave is significantly underestimated. This suggests that Rossby wave propagation and dispersion can be reproduced realistically with a barotropic model but that the growth of waves cannot be predicted with sufficient skill. The latter can indeed result in the presence baroclinic processes as we will understand in the next chapter.

3.9 The Ekman Layer and its Impact on the Barotropic Flow

The effect of surface drag cannot be neglected in weather forecast models since the interaction of the free atmosphere with the boundary layer weakens the flow. The Ekman layer theory describes this process in a simple way and it is possible to consider it in quasi-geostrophic models. The **Ekman spiral**[m] represents an analytical solution for horizontal wind in the Ekman layer assuming a constant eddy exchange coefficient K_m and a constant density ρ. In this case, the momentum and continuity equations become

$$\left(\frac{D\mathbf{v}_h}{Dt}\right)_h + (f_0 + \beta y)\mathbf{k} \times \mathbf{v}_h = f_0 \mathbf{k} \times \mathbf{v}_g + K_m \frac{\partial^2 \mathbf{v}_h}{\partial z^2} \quad (3.171)$$

$$\nabla_h \cdot \mathbf{v}_h + \frac{\partial w}{\partial z} = 0 \quad (3.172)$$

We can use the same scales as in the previous section except for the height scale. The height scale is now the vertical depth of the Ekman layer, H_E. The continuity equation suggests the vertical wind scale $W = UH_E/L$. Nondimensionalizing (3.171) and (3.172) yields with these scales

$$\left(\frac{D\tilde{\mathbf{v}}_h}{D\tilde{t}}\right)_h + \frac{1}{\mathrm{Ro}}(1 + \mathrm{Ro}\tilde{\beta}\tilde{y})\mathbf{k} \times \tilde{\mathbf{v}}_h = \frac{1}{\mathrm{Ro}}\mathbf{k} \times \tilde{\mathbf{v}}_g + \frac{\mathrm{Ek}}{\mathrm{Ro}}\frac{\partial^2 \tilde{\mathbf{v}}_h}{\partial \tilde{z}^2}$$

$$(3.173)$$

$$\tilde{\nabla}_h \cdot \tilde{\mathbf{v}}_h + \frac{\partial \tilde{w}}{\partial \tilde{z}} = 0 \quad (3.174)$$

---

[m]Ekman (1905) derived this solution for ocean currents. It is, however, also applicable for atmospherics flows.

where Ek $= K_m/(f_0 H_E^2)$ denotes the **vertical Ekman number**. Turbulent exchange becomes relevant if Ek $= o(1)$. Therefore, the depth of the Ekman layer is of the order of $H_E = \sqrt{K_m/f_0}$ given the fact that the horizontal wind at the surface nearly vanishes and becomes geostrophic at the upper boundary. Inserting expansions of the form (3.153) leads to the following zero-order equations:

$$\mathbf{k} \times \tilde{\mathbf{v}}_h^{(0)} = \mathbf{k} \times \tilde{\mathbf{v}}_g + \text{Ek} \frac{\partial^2 \tilde{\mathbf{v}}_h^{(0)}}{\partial \tilde{z}^2} \qquad (3.175)$$

$$\tilde{\nabla}_h \cdot \tilde{\mathbf{v}}_h^{(0)} + \frac{\partial \tilde{w}^{(0)}}{\partial \tilde{z}} = 0 \qquad (3.176)$$

The geostrophic wind can be considered as an exogenous variable since it is vertically independent due to the constant density and results from the dynamics of the overlying atmosphere. Consequently, the zero-order solution already suffices to determine the boundary layer flow. The solution of (3.175) yields the Ekman spiral. The dimensioned component form of (3.175) can be written as

$$-f_0 v_a = K_m \frac{\partial^2 u_a}{\partial z^2} \qquad (3.177)$$

$$f_0 u_a = K_m \frac{\partial^2 v_a}{\partial z^2} \qquad (3.178)$$

where u_a and v_a denote the ageostrophic wind components. Eliminating v_a gives

$$\frac{\partial^4 u_a}{\partial z^4} + \left(\frac{f_0}{K_m}\right)^2 u_a = 0 \qquad (3.179)$$

with the solution ansatz $u_a = \Re[\hat{u}_a \exp(imz)]$ we obtain for the vertical wavenumber

$$m^4 = -\left(\frac{f_0}{K_m}\right)^2 \rightarrow m_{1,2,3,4} = \pm(1 \pm i)\sqrt{\frac{f_0}{2K_m}} \qquad (3.180)$$

The solutions with a minus sign in front of the i must be rejected since they describe wind profiles approaching infinity for $z \rightarrow \infty$. Therefore, the solution becomes

$$u_a = \left[\hat{u}_{a1} \cos\left(\frac{\pi z}{H_E}\right) + \hat{u}_{a2} \sin\left(\frac{\pi z}{H_E}\right)\right] \exp\left(-\frac{\pi z}{H_E}\right) \qquad (3.181)$$

where $H_E = \pi\sqrt{2K_m/f_0}$ is the depth of the Ekman layer. Inserting this solution in (3.177) yields the meridional component

$$v_a = \left[-\hat{u}_{a1} \sin\left(\frac{\pi z}{H_E}\right) + \hat{u}_{a2} \cos\left(\frac{\pi z}{H_E}\right) \right] \exp\left(-\frac{\pi z}{H_E}\right) \quad (3.182)$$

To determine the unknown coefficients, we can use the boundary condition that the horizontal flow vanishes at the surface (assumed to be at $z = 0$). Therefore, the equations

$$u_a(0) = \hat{u}_{a1} = -u_g \quad (3.183)$$

$$v_a(0) = \hat{u}_{a2} = -v_g \quad (3.184)$$

must be fulfilled. Consequently, the wind components of the Ekman spiral take the form

$$u = u_g - \left[u_g \cos\left(\frac{\pi z}{H_E}\right) + v_g \sin\left(\frac{\pi z}{H_E}\right) \right] \exp\left(-\frac{\pi z}{H_E}\right) \quad (3.185)$$

$$v = v_g - \left[v_g \cos\left(\frac{\pi z}{H_E}\right) - u_g \sin\left(\frac{\pi z}{H_E}\right) \right] \exp\left(-\frac{\pi z}{H_E}\right) \quad (3.186)$$

In vector form, the Ekman spiral reads

$$\mathbf{v}_h = \left[1 - \cos\left(\frac{\pi z}{H_E}\right) \exp\left(-\frac{\pi z}{H_E}\right) \right] \mathbf{v}_g$$

$$+ \sin\left(\frac{\pi z}{H_E}\right) \exp\left(-\frac{\pi z}{H_E}\right) \mathbf{k} \times \mathbf{v}_g \quad (3.187)$$

Figure 3.22 shows wind vectors of the Ekman spiral for the case of a purely zonal geostrophic wind. Near the surface, the wind vector draws an angle of 45° with respect to the geostrophic wind. The flow has a component which is directed towards lower pressure. Therefore, there is a mass flux in the Ekman layer that reduces the horizontal pressure gradient. The wind vector veers to the right with increasing height and at $z = H_E$ the flow is nearly geostrophic. Only a slight variation of the wind vector occurs above this level. Therefore, one can identify H_E with the depth of the Ekman layer.

The vertical wind can be deduced by inserting (3.187) in the continuity equation (3.172). Only the second part of the wind vector (3.187) contributes to vertical motion since the divergence of the

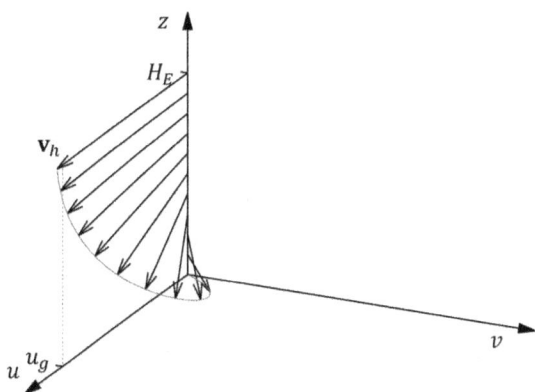

Fig. 3.22. Horizontal wind vectors of the Ekman spiral as a function of height.

geostrophic wind vanishes. Therefore, we obtain by vertical integration of (3.172)

$$w = -\int_0^z \nabla \cdot \mathbf{v}_h dz' = \frac{H_E}{2\pi} \left\{ 1 - \left[\sin\left(\frac{\pi z}{H_E}\right) + \cos\left(\frac{\pi z}{H_E}\right) \right] \right.$$
$$\left. \times \exp\left(-\frac{\pi z}{H_E}\right) \right\} \xi_g \qquad (3.188)$$

where we have used the boundary condition $w(0) = 0$. Obviously, the vertical motion in the Ekman layer is proportional to the geostrophic relative vorticity. This result is known as **Ekman pumping** and can explain the damping of high- and low-pressure systems since the associated mass fluxes of cross-isobaric flow weaken the pressure anomalies.

The Ekman layer can be coupled to a quasigeostrophic model by evaluating the vertical velocity as given by (3.188) at the height $z = H_E$ and using it for the lower boundary condition of the free atmosphere layer. The associated mass flux of the Ekman pumping modifies the vortex stretching term which leads to damping of vorticity. The quasigeostrophic shallow water model coupled to a subjacent Ekman layer is sketched in Fig. 3.23. The sloping bottom is problematic since the boundary layer flow parallel to the surface includes a vertical component while the pressure gradient not balancing gravity is still horizontal. On the other hand, steep mountain slopes cannot be part of a quasigeostrophic model since the perturbations triggered are clearly subsynoptic. However, the synoptic-scale

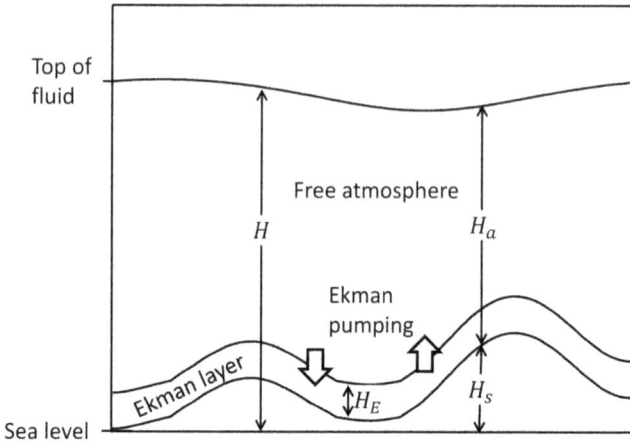

Fig. 3.23.   Sketch of the shallow water model coupled with a subjacent Ekman layer.

part of the orography slopes at a low angle of less than 0.5°. Then, the Ekman solution can be approximately derived as in the flat case. However, the vertical velocity of the up- and downhill motion must be considered in the vortex stretching term.

We have to integrate the continuity equation (3.172) vertically across the free atmosphere layer to take the Ekman layer coupling into account. This yields the following prognostic equation:

$$\frac{DH}{Dt} - \frac{DH_s}{Dt} - w_E + (H - H_s - H_E)\nabla_h \cdot \mathbf{v}_h = 0 \qquad (3.189)$$

where w_E is the vertical velocity at the top of the Ekman layer due to Ekman pumping (cf. Eq. (3.188)). The quasigeostrophic approximation becomes (cf. Eq. (3.163))

$$\frac{\partial H'}{\partial t} + \mathbf{v}_g \cdot \nabla_h (H' - H_s) - w_E = -(\overline{H} - H_E)\nabla_h \cdot \mathbf{v}_a \qquad (3.190)$$

with the geostrophic balance and quasigeostrophic vorticity equation (3.164) we can derive the potential vorticity equation

$$\left(\frac{\partial}{\partial t} + \mathbf{v}_g \cdot \nabla_h\right)\left[\nabla_h^2 \psi_g + f_0 + \beta y - \frac{1}{L_R^2}\left(\psi_g - \frac{gH_s}{f_0}\right)\right]$$

$$= -\frac{f_0}{\overline{H} - H_E} w_E \qquad (3.191)$$

Here, the Rossby radius L_R relates to the mean depth of the free atmospheric layer, $\overline{H}_f = \overline{H} - H_E$. By evaluating the Ekman vertical velocity at $z = H_E + H_s$ we find

$$\left(\frac{\partial}{\partial t} + \mathbf{v}_g \cdot \nabla_h\right)\left[\nabla_h^2 \psi_g + f_0 + \beta y - \frac{1}{L_R^2}\left(\psi_g - \frac{gH_s}{f_0}\right)\right]$$

$$= -\frac{\nabla_h^2 \psi_g}{\tau_E} \qquad (3.192)$$

where $\tau_E \approx 2\pi(\overline{H} - H_E)/(f_0 H_E)$ denotes the **Ekman spin-down time**. The additional term on the right-hand side of the potential vorticity equation (3.192) leads to a damping of vorticity anomalies with the time scale τ_E. In order to understand this, we examine Rossby wave solutions for a flat bottom surface ($H_s = 0$). With the wave ansatz $\psi_g = A\Re\{\cos ly \exp[i(kx - \nu t)]\}$ we find the dispersion relation

$$\nu = -\frac{\beta k + i(k^2 + l^2)/\tau_E}{1/L_R^2 + k^2 + l^2} \qquad (3.193)$$

Note that the solution already fulfills the nonlinear equation (3.192) since streamlines form also isolines of relative vorticity in the case of a monochromatic wave. Then, relative vorticity advection vanishes identically. We see that the frequency has an imaginary part which leads to a damping of the wave. We can write the solution as follows:

$$\psi_g = A \exp\left(-\frac{k^2 + l^2}{1/L_R^2 + k^2 + l^2}\frac{t}{\tau_E}\right)\cos ly \cos(kx - \nu_r t) \qquad (3.194)$$

where ν_r denotes the real part of the frequency. The wave amplitude decays with the time scale τ_E. However, the **e-folding time** (measuring the time of decay by the factor e^{-1}) is larger than τ_E due to the presence of a variable top surface. One part of the Ekman mass flux fills the low or drains the high while the other part causes flow divergence or convergence immediately above the boundary layer. Only the latter process causes the decay of the vorticity anomaly.

3.10 Orographically Forced Rossby Waves

The extratropical flow includes stationary Rossby waves of significant amplitude besides the eastward propagating weather systems.

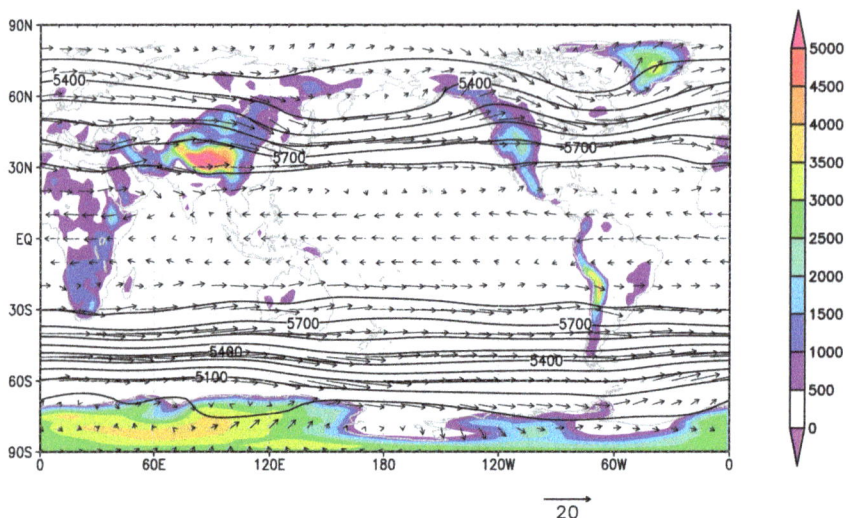

Fig. 3.24.   Long-term averaged (1981–2010) geopotential height (isolines, contour interval 100 m) and horizontal wind vector on the 500 hPa isobaric surface deduced from NCEP reanalysis data. The colored shadings display the height of the surface orography.

Figure 3.24 shows the long-term averaged 500 hPa geopotential height map and the height of the surface orography. We see that the 500 hPa flow has prominent wave perturbations in the Northern Hemisphere while the flow is approximately zonal in the Southern Hemisphere. At the same time, we remark that the position of the ridges roughly coincides with the maxima of the mountain ranges and that a trough occurs further downstream.

To understand the excitation of stationary Rossby waves by orography we employ the quasigeostrophic shallow water model as has been originally done by Charney and Eliassen (1949). The quasigeostrophic potential vorticity equation (3.192) linearized with respect to a uniform zonal flow U becomes

$$\left(\frac{\partial}{\partial t} + U\frac{\partial}{\partial x}\right)\left(\nabla_h^2 \psi_g' - \frac{1}{L_R^2}\psi_g'\right) + \left(\beta + \frac{U}{L_R^2}\right)\frac{\partial \psi_g'}{\partial x}$$

$$= -\frac{f_0 U}{\overline{H}_f}\frac{\partial H_s}{\partial x} - \frac{\nabla_h^2 \psi_g'}{\tau_E} \tag{3.195}$$

In this equation, we have treated the orography H_s as a small perturbation although in reality, this is not the case. However, the model still yields reasonable results.

First, we determine stationary solutions without Ekman layer friction ($\tau_E \to \infty$). The orography on the β-plane and geostrophic stream function can be represented by a Fourier expansion:

$$H_s = \Re \left[\sum_{n=0}^{\infty} \sum_{m=0}^{\infty} \hat{H}_{m,n} \exp\left(ik_m x\right) \exp(il_n y) \right] \tag{3.196}$$

$$\psi_g' = \Re \left[\sum_{n=0}^{\infty} \sum_{m=0}^{\infty} \hat{\psi}_{m,n} \exp\left(ik_m x\right) \exp(il_n y) \right] \tag{3.197}$$

where k_m and l_n are wavenumbers that conform with the boundary conditions of the domain considered. Inserting these expansions in (3.195) gives for each wavenumber pair the following relation:

$$\hat{\psi}_{m,n} = \frac{f_0 \hat{H}_{m,n} / \overline{H}_f}{k_m^2 + l_n^2 - \beta/U} \tag{3.198}$$

We see that the orography forces a stationary Rossby wave which has for $k_m^2 + l_n^2 > \beta/U \equiv K_S^2$ its ridge above the mountain and the trough above the valley. This is partly consistent with Fig. 3.24 where ridges appear above the mountain ranges. However, the model without the Ekman layer exhibits a **resonance catastrophe** since the denominator in (3.198) becomes zero for $k_m^2 + l_n^2 = K_S^2$. Therefore, the amplitude rises to infinity if the total wavenumber of the orography is identical with that of a stationary Rossby wave (cf. 3.134). For $k_m^2 + l_n^2 < K_S^2$ the Rossby wave has a trough above the mountain which is obviously not observed in the real atmosphere.

The model becomes more realistic when it includes Ekman friction. Then, we obtain for the Fourier coefficient of the stationary Rossby wave

$$\hat{\psi}_{m,n} = \frac{f_0 \hat{H}_{m,n} / \overline{H}_f}{K^2 - K_S^2 - i\frac{K^2}{U k_m \tau_E}} = \frac{f_0}{\overline{H}_f} \frac{K^2 - K_S^2 + i\frac{K^2}{U k_m \tau_E}}{(K^2 - K_S^2)^2 + \left(\frac{K^2}{U k_m \tau_E}\right)^2} \hat{H}_{m,n} \tag{3.199}$$

where $K^2 = k_m^2 + l_n^2$. Now, the denominator cannot become zero so that a resonance catastrophe cannot occur. Furthermore, the stream

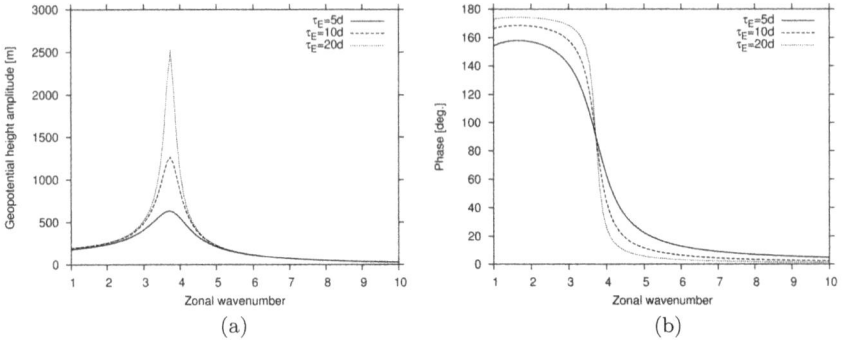

Fig. 3.25.   (a) Amplitude and (b) phase of the steady state Rossby waves forced by a sinusoidal shaped surface orography as a function of zonal wavenumber for different Ekman spin-down time scales, τ_E. The model parameters are $\beta = 1.6182 \times 10^{-11}\,(\mathrm{ms})^{-1}, f_0 = 10^{-4}\,\mathrm{s}^{-1}, U = 15\,\mathrm{m/s}$ and $\overline{H}_f = 8000\,\mathrm{m}$. The β channel has a width of 5000 km and the center is located at 45° N.

function wave is phase shifted upstream with respect to the orography in the usual case $K^2 > K_S^2$. Consequently, the ridge appears somewhat upstream of the mountaintop. Figure 3.25 shows the wave amplitude and phase as a function of zonal wavenumber m for various Ekman spin-down time scales at 45° N. The domain is a β-channel with a width of 5000 km and the meridonal wavenumber is of the mode which fits with half a wavelength into the channel. The zonal wavenumber m indicates the zonal modes fulfilling the periodic boundary conditions of the latitude circle at 45°N. We see in Fig. 3.25(a) that the amplitude increases with decreasing Ekman friction. The tendency towards a resonance catastrophe is obvious for $K^2 \rightarrow K_S^2$ which appears between $m = 3$ and $m = 4$. At this wavenumber, the phase changes rapidly from an inversely phased response at low wavenumbers to a response in phase with the orography. The smaller the Ekman friction is, the sharper is the phase change around $K^2 = K_S^2$.

Figure 3.26 presents an application of the Charney–Eliassen model using real data. The orography as given in the NCEP reanalysis data set has been meridionally averaged over the latitude range from 30 N to 60 N and from the resulting profile (Fig. 3.26(b)). Fourier coefficients are determined which yield responses in the wave amplitude as described by (3.199) assuming 45° N as the reference latitude. However, the calculation involves a modification of

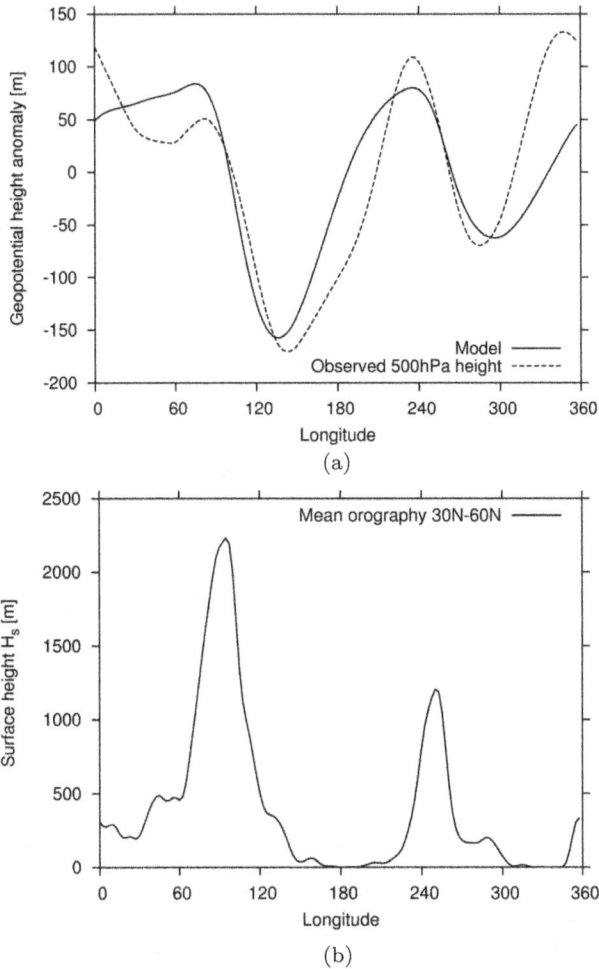

Fig. 3.26. (a) Zonal height profile of the Charney–Eliassen model forced by a realistic surface orography (solid line). The figure also displays the long-term mean of meridionally averaged 500 hPa geopotential height anomaly in January deduced from NCEP reanalysis data (dashed line). (b) Smoothed and meridionally averaged surface height profile based on NCEP reanalysis data used for the model calculation. The model parameters are $\beta = 1.6182 \times 10^{-11} (\text{ms})^{-1}, f_0 = 10^{-4} \text{s}^{-1}, U = 18 \text{ m/s}, \tau_E = 5$ days and $\overline{H}_f = 8000$ m. The β-channel has a width of 3558 km and the center is located at 45°N.

the orography term according to Charney and Eliassen (1949). This is because the real zonal mean geostrophic flow is usually weaker at the surface than aloft. Therefore, the zonal basic flow in front of the orography term in (3.195) has been reduced by a factor 0.4. The other model parameters are $U = 18\,\mathrm{m/s}$, $l_n = 8.829 \times 10^{-7}\mathrm{m}^{-1}$, $\tau_E = 5$ days, 1 and $\overline{H}_f = 8000\,\mathrm{m}$. Figure 3.26(a) shows the zonal height anomaly profile of the superposed forced Rossby waves together with the long-term January mean of 500 hPa geopotential height which has also been meridionally averaged from 30 N to 60 N. Model and observations exhibit an astonishing resemblance. Positions and amplitude of the ridges and troughs are quite similar except for the observed ridge between 300° and 360° longitude. On the other hand, we note that the results depend sensitively on the various model parameters and one may ask if the chosen parameters are a result of model tuning. However, the model parameters do not appear unrealistic and we can state at least that the model explains consistently the wavenumber of two stationary wave with the correct order of magnitude. Another reason which may cause such a wave pattern are zonally asymmetric heat sources. These stem from the arrangement with two continents and two interjacent oceans.

3.11   Barotropic Instabilities

A barotropic flow can become unstable under certain circumstances. We have already learned that perturbations cannot grow in the case of a uniform basic state at rest. However, perturbations can possibly gain energy if the base state has nontrivial kinetic energy. This can happen in the case of a shear flow. In this section, we discuss the **inertial instability** and the **shear instability** which can become relevant in atmospheric flows.

3.11.1   *Inertial instability*

Inertial instability occurs on subsynoptic scales and cannot be described by quasigeostrophic equations or shallow water equations. The reason is that a divergent horizontal flow contributes to the instability. The restoring forces of the resulting surface elevations inhibit the instability. However, the fluid can release the instability in the form of an **overturning circulation** which requires relaxing the assumption of a vertically independent horizontal flow. Therefore, we consider the Euler equations for a constant-density fluid on an f-plane

$$\frac{\partial \mathbf{v}}{\partial t} + \mathbf{v} \cdot \nabla \mathbf{v} + f_0 \mathbf{k} \times \mathbf{v}_h = -\frac{1}{\rho} \nabla p - g\mathbf{k} \qquad (3.200)$$

$$\nabla \cdot \mathbf{v} = 0 \qquad (3.201)$$

Furthermore, a rigid lid is assumed at the top of the fluid. A rigid lid is not a necessary ingredient of shear instability but it simplifies the mathematical problem drastically. The basic state includes a zonal geostrophic shear flow that has the form

$$\bar{u} = u_0 + \Lambda y \qquad (3.202)$$

where Λ denotes the meridional shear of zonal flow. This basic state constitutes a **Couette flow**. It is in geostrophic and hydrostatic balance since the basic state must fulfill the governing equations. Inertial instability is characterized by perturbations which do not vary in the direction of the basic flow. Therefore, we assume

$$\frac{\partial u'}{\partial x} = \frac{\partial v'}{\partial x} = \frac{\partial w'}{\partial x} = \frac{\partial p'}{\partial x} = 0 \qquad (3.203)$$

With these assumptions, we obtain for the linearized equations

$$\frac{\partial u'}{\partial t} + \Lambda v' - f_0 v' = 0 \qquad (3.204)$$

$$\frac{\partial v'}{\partial t} + f_0 u' = -\frac{1}{\rho} \frac{\partial p'}{\partial y} \qquad (3.205)$$

$$\frac{\partial w'}{\partial t} = -\frac{1}{\rho} \frac{\partial p'}{\partial z} \qquad (3.206)$$

$$\frac{\partial v'}{\partial y} + \frac{\partial w'}{\partial z} = 0 \qquad (3.207)$$

The pressure anomaly can be eliminated by deriving the prognostic equation for the zonal vorticity vector component. This yields

$$\frac{\partial \xi'_x}{\partial t} = f_0 \frac{\partial u'}{\partial z} \qquad (3.208)$$

On the other hand, the vanishing divergence allows us to formulate the meridional and vertical velocity components in terms of a stream function, i.e.

$$v' = \frac{\partial \psi'}{\partial z}, \quad w' = -\frac{\partial \psi'}{\partial y} \tag{3.209}$$

Therefore, the equations for zonal momentum and vorticity become

$$\frac{\partial u'}{\partial t} = (f_0 - \Lambda)\frac{\partial \psi'}{\partial z} \tag{3.210}$$

$$\frac{\partial}{\partial t}\left(\frac{\partial^2 \psi'}{\partial y^2} + \frac{\partial^2 \psi'}{\partial z^2}\right) = -f_0 \frac{\partial u'}{\partial z} \tag{3.211}$$

These two equations can be combined into one equation for stream function ψ'

$$\frac{\partial^2}{\partial t^2}\left(\frac{\partial^2 \psi'}{\partial y^2} + \frac{\partial^2 \psi'}{\partial z^2}\right) + f_0(f_0 - \Lambda)\frac{\partial^2 \psi'}{\partial z^2} = 0 \tag{3.212}$$

The flow of the perturbation must fulfill the boundary conditions

$$w = -\frac{\partial \psi'}{\partial y} = 0 \quad \text{at } z = 0, H \tag{3.213}$$

These can be achieved if the stream function is zero at these heights. Therefore, we make the following ansatz for the perturbation:

$$\psi' = \hat{\psi} R\left\{\sin\left(\frac{n\pi}{H}z\right)\exp[i(ly - \nu t)]\right\} \tag{3.214}$$

where n denotes the vertical mode and $n\pi/H$ represents the vertical wavenumber of this mode. We see in Fig. 3.27 that the perturbation describes a roll circulation in which the center axis is oriented parallel to the basic flow direction. Consequently, the horizontal shear of the basic flow cannot deform the perturbation by advection. Inserting this ansatz yields the following dispersion relation

$$\nu_{1,2} = \pm\sqrt{f_0(f_0 - \Lambda)\frac{\left(\frac{n\pi}{H}\right)^2}{\left(\frac{n\pi}{H}\right)^2 + l^2}} \tag{3.215}$$

We see that the frequency becomes imaginary when the argument under the square root is negative. In this case, the basic flow is **inertially unstable**. In the other case, the perturbation describes an

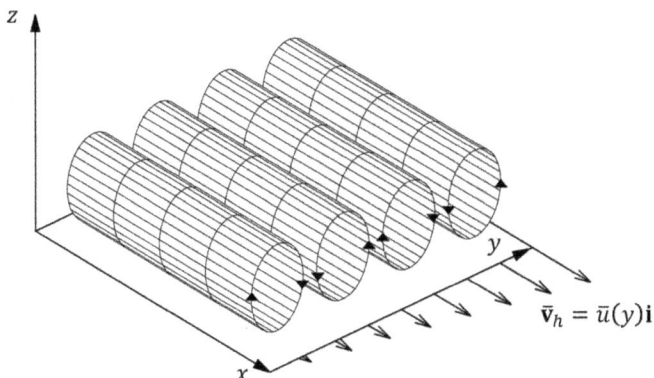

Fig. 3.27. Sketch of the horizontally oriented roll circulations that evolve due to the inertial instability of a horizontal shear flow.

inertial oscillation. Then, the flow is often referred to as **inertially stable** although it describes a neutral oscillation without decay. The sign of the absolute vorticity of the basic flow determines whether the flow is unstable or not. We get for the Northern Hemisphere ($f_0 > 0$)

$$\overline{\zeta} = f_0 - \Lambda < 0 \rightarrow \text{Inertial instability}$$

$$\overline{\zeta} = f_0 - \Lambda > 0 \rightarrow \text{Inertial stability}$$

In the unstable case, the frequency turns into the **growth rate** $\varsigma = -i\nu$. Then, the perturbation takes the form

$$\psi' = \hat{\psi} \sin\left(\frac{n\pi}{H}z\right) \cos(ly)e^{\varsigma t} \qquad (3.216)$$

Therefore, the flow pattern does not vary in time but the amplitude grows exponentially with the **e-folding time** $\tau = \varsigma^{-1}$. Figure 3.28 shows the growth rate as a function of meridional wavelength $2\pi/l$ for the first vertical mode $n = 1$. Obviously, the growth rate increases with wavelength and approaches the value $\sqrt{f_0(\Lambda - f_0)}$. This limiting value would result exactly if we had done the quasihydrostatic approximation. Higher modes have smaller growth rates. Consequently, the longest waves with vertical mode $n = 1$ grow fastest. It can be assumed that the fastest mode dominates in a real evolution. Therefore, the evolving overturning circulation extends vertically over the entire atmosphere and the width of the zone where $\overline{\zeta} < 0$.

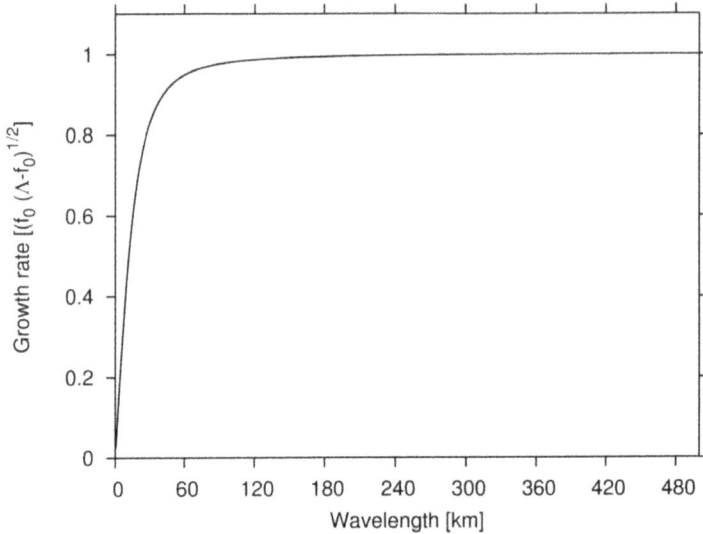

Fig. 3.28.   Growth rate of the inertially unstable flow (3.202) as a function of meridional wavelength. The height of the fluid is assumed to be $H = 10\,\text{km}$.

Figure 3.29 explains the mechanism of inertial instability schematically. Let us consider a parcel at the latitude where the basic flow vanishes. The zonal flow of this parcel increases by the Coriolis force if the parcel is displaced northward (grey arrow) over the distance Δy. The resulting velocity usually differs from the geostrophic wind at this location (black arrow). Consequently, a force occurs that displaces the parcel farther north in the case of inertial instability or back to the original latitude when no instability is present. We can calculate the criteria by considering the zonal and meridional momentum equations

$$\frac{Du}{Dt} - f_0 v = 0 \tag{3.217}$$

$$\frac{Dv}{Dt} + f_0(u - u_g) = 0 \tag{3.218}$$

The first equation states that $u - f_0 \Delta y$ is an individually conserved quantity. The second equation forms a differential equation for the displacement Δy, namely

$$\frac{D^2}{Dt^2}(\Delta y) + f_0(u - u_g) = 0 \tag{3.219}$$

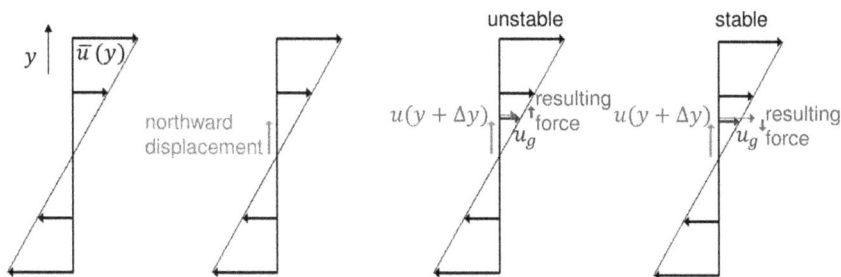

Fig. 3.29.   Mechanism of inertial instability explained by a parcel consideration.

By the linear approximation $u_g \approx u_{g0} + \Lambda\Delta y$ and $u - f_0\Delta y = u_{g0} = $ const. we find the ordinary differential equation

$$\frac{D^2}{Dt^2}(\Delta y) + f_0(f_0 - \Lambda)\Delta y = 0 \qquad (3.220)$$

Solving this equation yields the frequency $\nu = \pm\sqrt{f_0(f_0 - \Lambda)}$ which is the quasihydrostatic limit of Eq. (3.35). Consequently, we obtain a similar result with the simple parcel approach sketched in Fig. 3.29.

   Inertial instability occurs only in subsynoptic flows since absolute vorticity seldom falls below zero on synoptic scales ($L \approx 1000\,\mathrm{km}$). Blanchard *et al.* (1998) found that inertial instability can indeed trigger mesoscale rain bands.

3.11.2   *Shear instability*

Unidirectional shear flows may induce another type of instability, namely the shear instability triggering the growth of wave perturbations along the streamlines of the basic flow. The shear instability was mathematically analyzed by Rayleigh (1880) in a simple case of a **vortex strip** as displayed in Fig. 3.30. It is a band with uniform relative vorticity sandwiched between two regions having zero relative vorticity. A vortex strip oriented along the x-axis is associated with a zonal flow having uniform shear. Therefore, the basic flow of such a vortex strip has the form

$$\bar{u} = \begin{cases} -\Lambda L & y > L \\ -\Lambda y & -L \leq y \leq L \\ \Lambda L & y < -L \end{cases} \qquad (3.221)$$

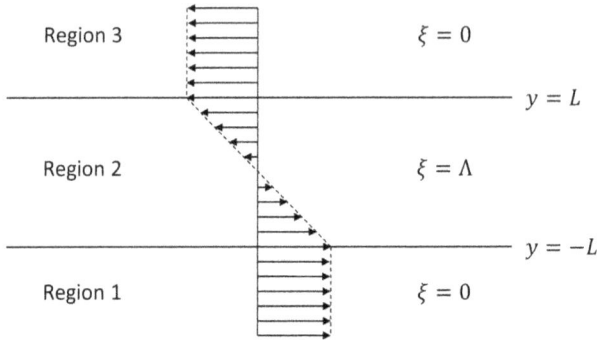

Fig. 3.30.   Sketch of the basic flow for the shear instability problem.

where Λ is the relative vorticity in the region of the vortex strip and at the same time it is the cyclonic shear of the basic flow.

We assume for simplicity a constant-density fluid with a rigid lid on the f-plane. Consequently, the linearized vorticity equation becomes for the three fluid regions

$$\left(\frac{\partial}{\partial t} + \bar{u}\frac{\partial}{\partial x}\right)\left(\frac{\partial^2 \psi_j'}{\partial x^2} + \frac{\partial^2 \psi_j'}{\partial y^2}\right) = 0 \quad \text{for } j = 1,2,3 \qquad (3.222)$$

where ψ' denotes the stream function of the horizontal nondivergent perturbation flow and j indicates the respective region (see Fig. 3.30). The intention of this subsection is to analyze perturbation growth due to instability. On the other hand, the relative vorticity of the perturbation cannot grow exponentially since Eq. (3.222) only describes zonal advection by the basic flow. Therefore, the growing disturbance must have zero relative vorticity, i.e.

$$\frac{\partial^2 \psi_j'}{\partial x^2} + \frac{\partial^2 \psi_j'}{\partial y^2} = 0 \qquad (3.223)$$

We insert the wave ansatz of the form $\psi'_j = \Re[\hat{\psi}_j(y)\exp(ikx - i\nu t)]$ and obtain the ordinary differential equation

$$\frac{d^2 \hat{\psi}_j}{dy^2} - k^2 \hat{\psi}_j = 0 \qquad (3.224)$$

The solution of this equation becomes

$$\hat{\psi}_j = A_j e^{ky} + B_j e^{-ky} \qquad (3.225)$$

Since the f-plane has no boundaries, we must set $B_1 = A_3 = 0$ in order to obtain finite perturbations for $y \to \pm\infty$. To ensure the continuity of meridional velocity the **kinematic boundary conditions**

$$\hat{\psi}_1(-L) = \hat{\psi}_2(-L), \quad \hat{\psi}_2(L) = \hat{\psi}_3(L) \tag{3.226}$$

must be fulfilled. Therefore, we get

$$A_1 = A_2 + B_2 e^{2kL} \tag{3.227}$$

$$B_3 = A_2 e^{2kL} + B_2 \tag{3.228}$$

Another equation is needed to solve the problem. At the boundary between the regions the pressure must be continuous which yields the **dynamic boundary condition**. Then, the right hand of the linearized zonal momentum equation

$$\left(\frac{\partial}{\partial t} + \bar{u}\frac{\partial}{\partial x} \right) u' + v'\frac{\partial \bar{u}}{\partial y} - f_0 v' = -\frac{1}{\rho}\frac{\partial p'}{\partial x} \tag{3.229}$$

must be continuous at $y = -L$ and $y = L$. Inserting the wave ansatz leads to the dynamic boundary conditions:

$$(\nu - \Lambda L k)\frac{d\hat{\psi}_1}{dy} - f_0 k\hat{\psi}_1$$

$$= (\nu - \Lambda L k)\frac{d\hat{\psi}_2}{dy} - (f_0 + \Lambda)k\hat{\psi}_2 \quad \text{at } y = -L \tag{3.230}$$

$$(\nu + \Lambda L k)\frac{d\hat{\psi}_3}{dy} - f_0 k\hat{\psi}_3$$

$$= (\nu + \Lambda L k)\frac{d\hat{\psi}_2}{dy} - (f_0 + \Lambda)k\hat{\psi}_2 \quad \text{at } y = L \tag{3.231}$$

With the solution (3.225) and the elimination of A_1 and B_3 using (3.227) and (3.228) we find the following matrix equation:

$$\begin{pmatrix} -[2(\Lambda L k + \nu) - \Lambda]e^{kL} & \Lambda e^{-kL} \\ \Lambda e^{-kL} & -[2(\Lambda L k - \nu) - \Lambda]e^{kL} \end{pmatrix}$$

$$\cdot \begin{pmatrix} A_2 \\ B_2 \end{pmatrix} = 0 \tag{3.232}$$

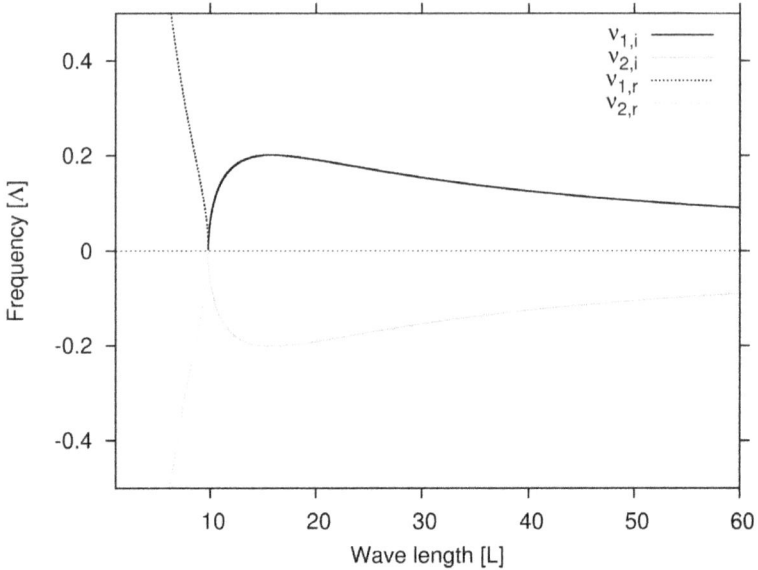

Fig. 3.31.   Real and imaginary parts of the frequency solutions (3.233) as a function of zonal wavelength.

Nontrivial solutions require a vanishing determinant and this condition yields the frequency solutions:

$$\nu_{1,2} = \pm\frac{\Lambda}{2}\sqrt{(2kL-1)^2 - e^{-4kL}} \qquad (3.233)$$

Obviously, the frequency becomes imaginary if

$$k < \frac{1 + e^{-2kL}}{2L} \qquad (3.234)$$

Therefore, longer wave perturbations trigger shear instability while shorter waves propagate in the east- or westward direction. Figure 3.31 shows the real and imaginary parts of the frequencies as a function of zonal wavelength. Instability sets in at a wavelength of about $10\,L$ while the maximum growth rate occurs approximately at $16\,L$.

   We can deduce the stream function field of the growing wave perturbation by solving the matrix equation (3.232) after insertion of $(3.233)^n$:

$$\psi_2 = Ae^{st}(e^{-2kL} + 2kL - 1)\cosh(ky)\cos(kx)$$

$$- Ae^{st}\frac{2\varsigma}{\Lambda}\sinh(ky)\sin(kx) \tag{3.235}$$

The fields ψ_1 and ψ_3 can be easily derived by the boundary conditions (3.226). Figure 3.32(a) shows the streamfunction of the wave perturbation having the largest growth rate. The wave has cyclones and anticyclones at the boundaries $y = -L$ and $y = L$. These are embedded in larger-scale cyclonic and anticyclonic disturbances in which the streamlines tilt against the shear of the basic flow. This is a characteristic property of the shear instability since it can explain the kinetic energy transfer from the basic flow to the wave as we will see later. Note that the vorticity of the perturbation vanishes although we see cyclones and anticyclones. This seems to contradict Stokes' law (1.264). However, the apparent contradiction can be explained by the existence of discontinuities in zonal velocity at $y = -L$ and $y = L$. There, the vorticity is indeed undetermined. Figure 3.32(b) displays a superposition of the basic flow with the wave perturbation which yields an elliptically shaped cyclone. The principal axis orients from the southwest to the northeast. This tilt can explain the mechanism of the instability. For this purpose, let us consider the meridional component of the momentum equation (2.44) in flux form

$$\frac{\partial v'}{\partial t} + \frac{\partial}{\partial x}(uv') + \frac{\partial}{\partial y}(v'^2) + f_0 u' = -\frac{1}{\rho}\frac{\partial p'}{\partial y} \tag{3.236}$$

Note that the two-dimensional flow is nondivergent so that the divergence of advective fluxes coincides with the usual advection term. Meridional integration of this equation over the latitude range where the perturbation has notable amplitude yields approximately

$$\int \frac{\partial v'}{\partial t}dy = -\int \frac{\partial}{\partial x}(uv')dy \tag{3.237}$$

The integrated Coriolis force vanishes because the zonal perturbation velocity is antisymmetric about $y = 0$. Therefore, net meridional acceleration takes place in the regions where the zonal flux

---

[n]Note that $\cosh(ky) = 1/2(e^{ky} + e^{-ky})$ and $\sinh(ky) = 1/2(e^{ky} - e^{-ky})$.

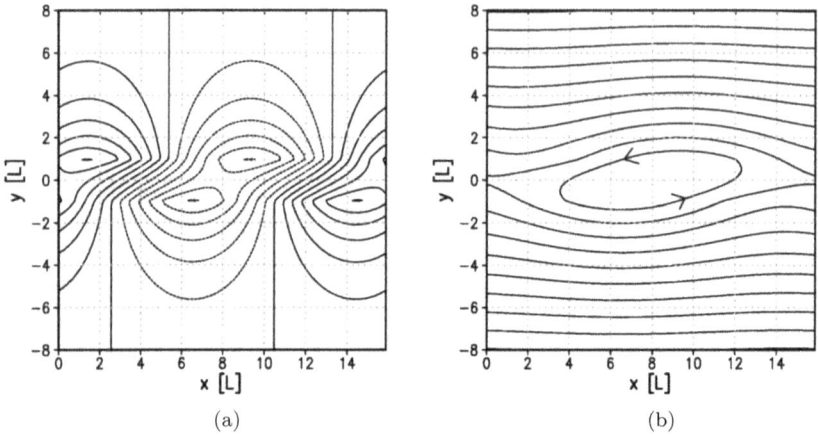

Fig. 3.32. (a) Streamfunction of the growing wave perturbation having the zonal wavelength $16\,L$, (b) wave perturbation in superposition with the basic flow.

of meridional momentum converges. From Fig. 3.32 we can deduce that u correlates with v' in the region of the elliptically shaped cyclone. Consequently, convergence (divergence) appears east (west) of the cyclone leading to northward (southward) acceleration of the flow. The acceleration enhances the perturbation and intensifies the cyclone which in turn causes larger momentum fluxes and even more intensification.

It is notable that the shear instability mode has some similarities with **frontal cyclones** frequently observed on synoptic weather charts. Figure 3.33(a) shows a surface weather chart where a temperature front runs from south to north near $20\,\mathrm{W}$. It exhibits undulations and coincides with the position of the pressure trough. Therefore, the geostrophic wind has cyclonic shear as in the previously treated case. That the front is a region with high cyclonic shear is a frequently observed property and can be explained with frontogenesis models introduced in Section 6.5. The two lows at the front could possibly intensify by barotropic shear instability. The northern cyclone actually intensifies during the next 24 hours as seen in Fig. 3.33(b). However, the intensification must not result from barotropic shear instability since the front also includes density gradients and the flow is therefore not barotropic. Orlanski (1968) beside others investigated the instability of a frontal discontinuity and found several instability modes with only one, which can be interpreted as

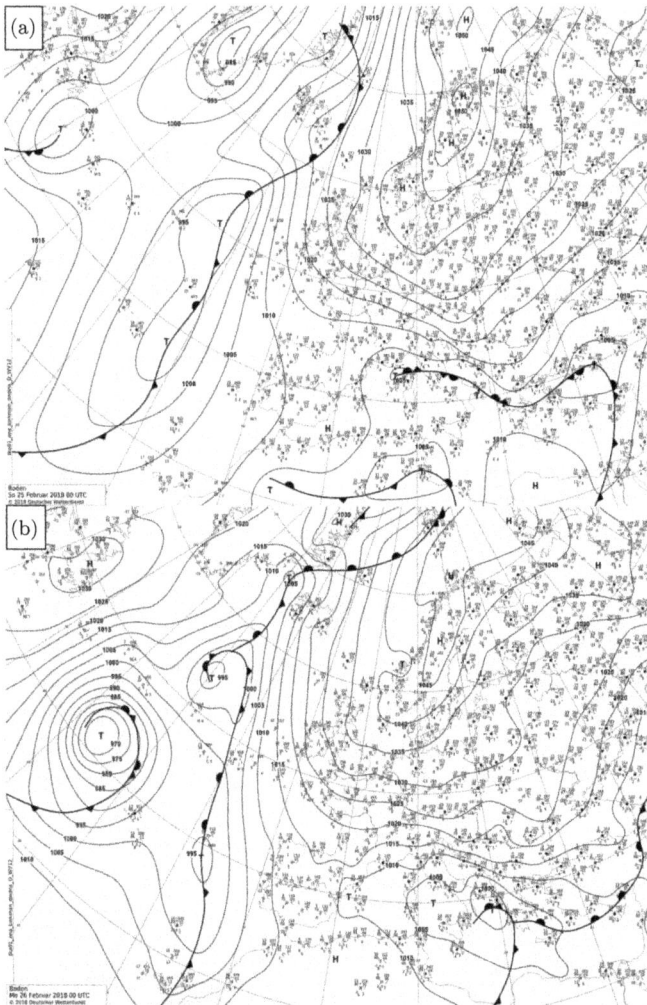

Fig. 3.33. Surface analyses by the German Weather Service for (a) February 25, 2018, 00UTC and (b) February 26, 2018, 00 UTC.
Source: Deutscher Wetterdienst. https://www.wetter3.de/archiv_dwd_dt.html.

Rayleigh's shear instability. Another type is the baroclinic instability which will be explained in Chapter 4.

The linearized equations have limited validity in the instability case since the perturbation grows exponentially and reaches eventually an amplitude that does not conform with the assumptions used to justify the linearization. Growth will decline due to the nonlinear terms which have been neglected. However, with the linear perturbation analysis, we cannot make a prediction when this takes place. Simulation with a numerical model can provide an answer to this question. Figure 3.34 shows the results of a model that integrates the barotropic vorticity equation numerically. The initial flow consists of a vortex strip as given by Eq. (3.221) plus perturbation. The initial vorticity field has been smoothed in order to avoid numerical errors in the model. At the beginning of the development, we see that the vortex strip is meridionally deflected by the perturbation. This results in an undulation of the zone with enhanced shear as seen on the weather map in Fig. 3.33. Later, the strip widens at the location of the cyclone and thins on the eastern and western sides. Finally, the thin vorticity strips wrap around the cyclone. The vorticity values in the thin strips decrease in the cause of time due to the diffusion in the numerical model which is necessary to dampen numerical noise. The strip is analogous to warm and cold fronts that separate warm air in the south from cold air in the north. Then, the vortex evolution has some similarities with the evolution of frontal cyclones. However, the **occlusion process** describing the observed merging of cold and warm fronts cannot be understood with the barotropic model.

We have learned that barotropic shear instability can lead to the formation of cyclones. However, so far it is not clear how the cyclones attain their kinetic energy. To tackle this question, we derive the energy equation for the zonal average. The zonally averaged zonal momentum equation becomes

$$\frac{\partial [u]_x}{\partial t} + \frac{\partial}{\partial y}[vu]_x = 0 \qquad (3.238)$$

where $[\cdot\cdot]_x = L_x^{-1}\int_0^{L_x} \cdot\cdot\, dx$ denotes the zonal averaging operator and L_x the length of the periodic domain in a zonal direction. Note that $[v]_x = [\partial\psi/\partial x]_x = 0$ in the nondivergent case considered here. By splitting the variable in zonal average and perturbation, i.e.

$$b = [b]_x + b^* \qquad (3.239)$$

we can simplify the zonally averaged momentum flux term

$$[vu]_x = [[v]_x[u]_x + [v]_x u^* + v^*[u]_x + v^* u^*]_x$$
$$= [v^* u^*]_x \qquad (3.240)$$

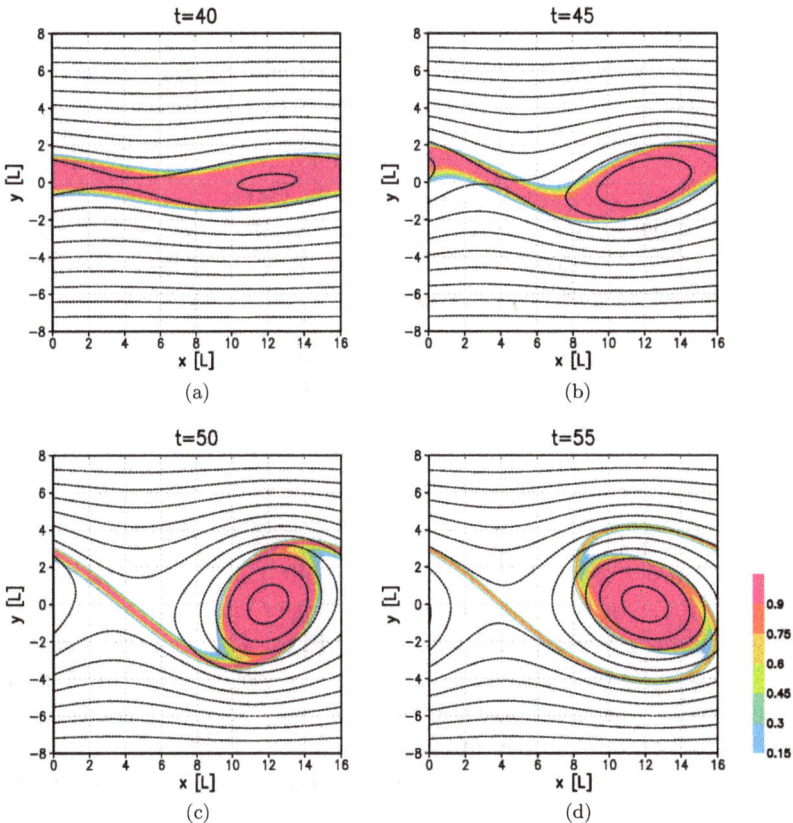

Fig. 3.34. Vorticity (colored shadings) and stream function (isolines) of the non-linear barotropic shear instability evolution simulated with a spectral barotropic model. Shown are the fields at (a) $t = 40\,\Lambda^{-1}$, (b) $t = 45\,\Lambda^{-1}$, (c) $t = 50\,\Lambda^{-1}$ and (d) $t = 55\,\Lambda^{-1}$.

Inserting this expression in (3.238) yields

$$\frac{\partial [u]_x}{\partial t} = -\frac{\partial}{\partial y}[v^* u^*]_x \qquad (3.241)$$

We see that only convergence of the meridional eddy momentum flux alters the zonally averaged flow. The zonal and meridional perturbation velocity components are correlated in the unstable case due to the tilt of the eddy axes (see Fig. 3.32(a)). Therefore, the eddy momentum flux decelerates the zonal flow and reduces the shear leaving eventually a stable flow. Multiplying Eq. (3.241) with $[u]_x$ gives the prognostic equation for zonal kinetic energy:

$$\frac{\partial e_Z}{\partial t} = -[u]_x \frac{\partial}{\partial y}[v^*u^*]_x \tag{3.242}$$

where $e_Z \equiv 1/2[u]_x^2$ denotes the zonal kinetic energy density. This equation states that kinetic energy is transferred from the zonal mean flow to the perturbation if the zonally averaged flow and the meridional momentum flux divergence have the same sign. The volume integrated energy equation becomes

$$\frac{dE_Z}{dt} = -\iint_{a_E} \rho H_a[u]_x \frac{\partial}{\partial y}[v^*u^*]_x da$$

$$= \iint_{a_E} \rho H_a[v^*u^*]_x \frac{\partial[u]_x}{\partial y} da \tag{3.243}$$

The second identity results after partial integration in the meridional direction and it has the consequence that the meridional momentum flux must be opposite to the gradient of the zonal mean flow in order to transfer kinetic energy to the eddies. These **down-gradient fluxes** are an essential element of turbulence theories since turbulence arises from shear instabilities (cf. Section 1.13). On the other hand, it is possible that momentum fluxes are in the direction of the zonal mean flow gradient. This circumstance can appear if a Rossby wave becomes by itself unstable as shown by Hoskins (1973). Then, the wave decays and forms a zonal mean shear flow. Indeed, upgradient momentum fluxes dominate on average in the synoptic scale.

3.11.3   *A criterion for barotropic shear stability*

The barotropic flow has usually some amount of shear. However, barotropic shear instability does not always occur and it would be helpful to have a criterion for excluding shear instability. Such a criterion has been derived by Rayleigh (1880) for a nonrotating fluid while Kuo (1949) extended this criterion by including the β-effect of planetary rotation. It applies to parallel basic flows in a zonal direction. Therefore, the linearized barotropic vorticity equation becomes

$$\left(\frac{\partial}{\partial t} + \bar{u}\frac{\partial}{\partial x}\right)\left(\frac{\partial^2 \psi'}{\partial x^2} + \frac{\partial^2 \psi'}{\partial y^2}\right) + \left(\beta - \frac{d^2\bar{u}}{dy^2}\right)\frac{\partial \psi'}{\partial x} = 0 \tag{3.244}$$

Inserting the usual wave ansatz $\psi' = \Re[\hat{\psi}(y)\exp(ikx - ikct)]$ yields

$$(\overline{u} - c)\left(\frac{d^2\hat{\psi}}{dy^2} - k^2\hat{\psi}\right) + \left(\beta - \frac{d^2\overline{u}}{dy^2}\right)\hat{\psi} = 0 \qquad (3.245)$$

By multiplying this result with the complex conjugate $\hat{\psi}^C \equiv \hat{\psi}_r - i\hat{\psi}_i$ we get

$$(\overline{u} - c)\left(\hat{\psi}^C\frac{d^2\hat{\psi}}{dy^2} - k^2|\hat{\psi}|^2\right) + \left(\beta - \frac{d^2\overline{u}}{dy^2}\right)|\hat{\psi}|^2 = 0 \qquad (3.246)$$

since $\hat{\psi}^C\hat{\psi} = |\hat{\psi}|^2$. This equation can be further manipulated by dividing it by $\overline{u} - c$ and partial integration over the meridional extension L_y of the perturbation. This leads to

$$\int_0^{L_y} k^2|\hat{\psi}|^2 + \left|\frac{d^2\hat{\psi}}{dy^2}\right|^2 dy = \int_0^{L_y}\left(\beta - \frac{d^2\overline{u}}{dy^2}\right)\frac{|\hat{\psi}|^2}{\overline{u} - c}dy \qquad (3.247)$$

Both real and imaginary parts must fulfill this equation and for the latter we obtain

$$c_i\int_0^{L_y}\left(\beta - \frac{d^2\overline{u}}{dy^2}\right)\frac{|\hat{\psi}|^2}{|\overline{u} - c|^2}dy = c_i\int_0^{L_y}\frac{d\overline{\zeta}}{dy}\frac{|\hat{\psi}|^2}{|\overline{u} - c|^2}dy = 0 \qquad (3.248)$$

Two possibilities exist to fulfill this equation. In the first case the imaginary part of the phase velocity, c_i vanishes. Then, instability is absent. In the second case the integral vanishes and $c_i \neq 0$. Then, the integrand must exhibit positive and negative values in the selected latitude range. Since real flows have continuous vorticity profiles,[°] we can state that shear instability can only occur if the meridional gradient of absolute vorticity changes its sign between $y = 0$ and $y = L_y$. In other words, the absolute vorticity must have a local extremum between these latitudes. Note, however, that this is a necessary but not sufficient criterion for barotropic shear instability. Obviously, the β-effect has a stabilizing effect since the curvature of the zonal wind profile, $d^2\overline{u}/dy^2$ must exceed the planetary vorticity gradient β somewhere. Without planetary rotation any curved wind profile having an inflection point can potentially be unstable.

[°]In contrast to the considered shear instability case in the previous subsection.

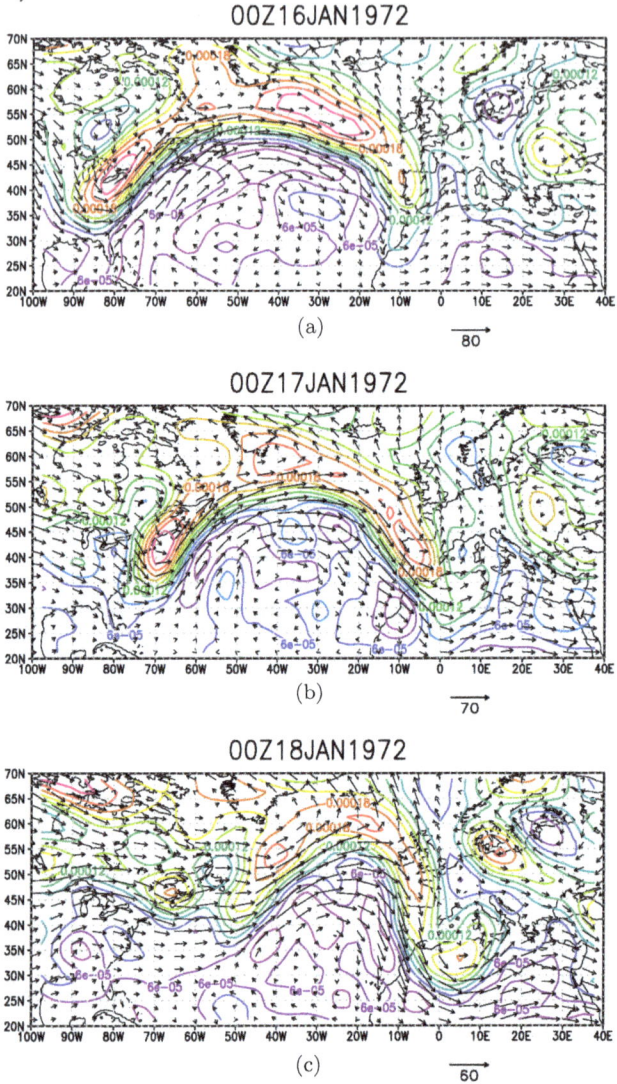

Fig. 3.35.   Absolute vorticity (colored isolines) and wind vectors on the 500 hPa pressure surface on (a) January 16, 1972, 00 UTC, (b) January 17, 1972, 00 UTC and (c) January 18, 1972, 00 UTC. The data are derived from NCEP Reanalysis.

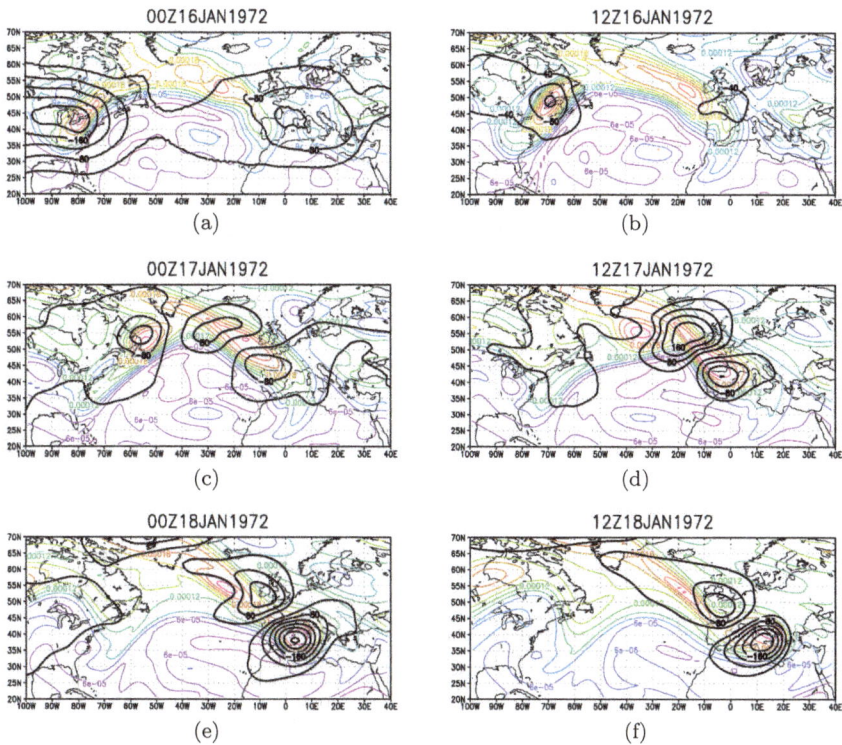

Fig. 3.36. Forecasts with a barotropic model initialized with NCEP reanalysis on January 16, 1972. Shown are absolute vorticity on the 500 hPa pressure surface (colored isolines) of the reference forecast and the 500 hPa geopotential height anomaly of the perturbed forecast on (a) January 16, 1972, 00 UTC, (b) January 17, 1972, 12 UTC, (c) January 17, 1972, 00 UTC, (d) January 17, 1972, 12 UTC, (e) January 18, 1972, 00 UTC and (f) January 18, 1972, 12 UTC.

Figure 3.35 shows a development of the 500 hPa flow that is likely influenced by barotropic instability. An anticyclonically curved shear flow extended from the east coast of North America over the Atlantic to Europe on January 16, 1972. The shear is so strong that the absolute vorticity has a maximum north of the jet. Therefore, the flow is potentially unstable with respect to a barotropic shear instability. The curvature of the positive vorticity increases on the subsequent days and eventually a trough forms over North Africa. Barotropic shear instability could play a role in this development. This possibility can be studied with a barotropic weather forecast model.

Figure 3.36 shows the results of two weather forecasts with a numerical model based on the shallow water equations. The first forecast has been initialized with NCEP reanalysis data on January 16, 1972. The other forecast uses the same initial data plus a random perturbation in geopotential height. The black isolines in Fig. 3.36 display the differences in geopotential height between the two forecasts. We see that the initial random perturbation disperses rapidly. However, 24 h later a wave pattern evolves along the curved band with enhanced planetary vorticity. This wave grows in amplitude and eventually, a low-pressure anomaly evolves over the Mediterranean Sea close to the position of the trough seen in Fig. 3.35(c). Therefore, it is likely that barotropic shear instability plays a role in this weather development and that this instability contributes to the limitation of predictability.

Chapter 4

Dynamics of Baroclinic Weather Systems

4.1 Introduction

Barotropic models can partially explain the behavior of synoptic-scale weather systems but such models cannot predict properly the development of these systems. We already learned that barotropic shear instability induces the growth of low-pressure systems that have some similarity with real frontal cyclones. However, there is no barotropic mechanism by which the horizontal shear of the parallel flow is maintained and cyclones develop in barotropically stable flows. In this chapter, we consider baroclinic quasigeostrophic equations to better understand the formation of extratropical weather systems. First, an overview of possible internal atmospheric oscillations is presented.

4.2 Internal Waves

Internal waves are generated in a **stably stratified atmosphere**, i.e. in an atmosphere where the potential temperature increases with height. The wave perturbation is baroclinic although the basic state can still be barotropic. For the analysis, we use the Euler equations on an f-plane with a tracer for the quasihydrostatic approximation.

$$\frac{D\mathbf{v}_h}{Dt} + f_0 \mathbf{k} \times \mathbf{v}_h = -\frac{1}{\rho} \nabla_h p \tag{4.1}$$

$$\delta \frac{Dw}{Dt} = -\frac{1}{\rho} \frac{\partial p}{\partial z} - g \qquad (4.2)$$

$$\frac{D\rho}{Dt} = -\rho \left(\nabla_h \cdot \mathbf{v}_h + \frac{\partial w}{\partial z} \right) \qquad (4.3)$$

$$\frac{DT}{Dt} = \frac{1}{c_p \rho} \frac{Dp}{Dt} \qquad (4.4)$$

$$\rho = \frac{p}{R_d T} \qquad (4.5)$$

where the trace-parameter $\delta = 0$ and $\delta = 1$ indicates the quasihydroststatic and the nonhydrostatic equation set, respectively. It is convenient to eliminate the temperature by inserting the equation of state (4.5) into the temperature tendency equation (4.4) which leads to

$$\frac{Dp}{Dt} = \frac{c_p p}{c_v \rho} \frac{D\rho}{Dt} = \frac{c_p}{c_v} R_d T \frac{D\rho}{Dt} \qquad (4.6)$$

For simplicity, we assume a horizontally uniform basic state at rest in hydrostatic balance. Then, the linearization of (4.1), (4.2), (4.3) and (4.6) yields

$$\frac{\partial \mathbf{v}'_h}{\partial t} + f_0 \mathbf{k} \times \mathbf{v}'_h = -\frac{1}{\bar{\rho}} \nabla_h p' \qquad (4.7)$$

$$\delta \frac{\partial w'}{\partial t} = -\frac{1}{\bar{\rho}} \frac{\partial p'}{\partial z} - g \frac{\rho'}{\bar{\rho}} \qquad (4.8)$$

$$\frac{1}{\bar{\rho}} \frac{\partial \rho'}{\partial t} + \frac{1}{\bar{\rho}} \frac{\partial \bar{\rho}}{\partial z} w' = -\nabla_h \cdot \mathbf{v}'_h - \frac{\partial w'}{\partial z} \qquad (4.9)$$

$$\frac{1}{\bar{\rho}} \frac{\partial p'}{\partial t} - g w' = -\frac{c_p}{c_v} R_d \bar{T} \left(\nabla_h \cdot \mathbf{v}'_h + \frac{\partial w'}{\partial z} \right) \qquad (4.10)$$

where we have substituted (2.37) in (4.2) prior to linearization. It is convenient to eliminate the density perturbation by combining (4.8),

(4.9) and (4.10) which gives

$$\left(\delta\frac{\partial^2}{\partial t^2} + N^2\right)w' = -\frac{1}{\bar{\rho}}\frac{\partial}{\partial t}\left(\frac{\partial p'}{\partial z} + \frac{g}{c_s^2}p'\right) \tag{4.11}$$

where $c_s = \sqrt{c_p/c_v R_d \bar{T}}$ denotes the phase speed of **sound waves** and N the **Brunt–Väisälä frequency** (Brunt, 1927) which is defined by

$$N = \sqrt{-\frac{g}{\bar{\rho}}\frac{\partial\bar{\rho}}{\partial z} - \frac{g^2}{c_s^2}} = \sqrt{\frac{g}{\bar{\theta}}\frac{\partial\bar{\theta}}{\partial z}} \tag{4.12}$$

In general, the linear system does not have constant coefficients but this can be achieved in the special case of an **isothermal atmosphere**. For this subsection we make this assumption and find that

$$\bar{p} = p_s e^{-\frac{z}{H_S}}, \quad \bar{\rho} = \rho_s e^{-\frac{z}{H_S}}, \tag{4.13}$$

fulfills the hydrostatic balance of an atmosphere with the constant temperature $\bar{T}$ and **scale height** $H_S = R_d\bar{T}/g$. Using this assumption and replacing the horizontal momentum equations by vorticity and divergence equations lead to the linear equations

$$\frac{\partial\xi'}{\partial t} = -f_0 D' \tag{4.14}$$

$$\frac{\partial D'}{\partial t} = f_0\xi' - \frac{1}{\bar{\rho}}\nabla_h^2 p' \tag{4.15}$$

$$\frac{1}{\bar{\rho}}\frac{\partial p'}{\partial t} - gw' = -c_s^2\left(D' + \frac{\partial w'}{\partial z}\right) \tag{4.16}$$

$$\left(\delta\frac{\partial^2}{\partial t^2} + N^2\right)w' = -\frac{1}{\bar{\rho}}\frac{\partial}{\partial t}\left(\frac{\partial p'}{\partial z} + \frac{g}{c_s^2}p'\right) \tag{4.17}$$

With the new perturbation variables $\tilde{\xi}' = \xi'\sqrt{\bar{\rho}/\rho_s}$, $\tilde{D}' = D'\sqrt{\bar{\rho}/\rho_s}$, $\tilde{w}' = w'\sqrt{\bar{\rho}/\rho_s}$ and $\tilde{p}' = p'/\sqrt{\bar{\rho}/\rho_s}$ we obtain the following system of linear differential equations with constant coefficients:

$$\frac{\partial \tilde{\xi}'}{\partial t} = -f_0 \tilde{D}' \tag{4.18}$$

$$\frac{\partial \tilde{D}'}{\partial t} = f_0 \tilde{\xi}' - \frac{1}{\rho_s} \nabla_h^2 \tilde{p}' \tag{4.19}$$

$$\frac{1}{\rho_s} \frac{\partial \tilde{p}'}{\partial t} = -c_s^2 \tilde{D}' - c_s^2 \left[\frac{\partial \tilde{w}'}{\partial z} - \frac{1}{\tilde{H}_S} \tilde{w}' \right] \tag{4.20}$$

$$\left(\delta \frac{\partial^2}{\partial t^2} + N^2 \right) \tilde{w}' = -\frac{1}{\rho_s} \frac{\partial}{\partial t} \left[\frac{\partial \tilde{p}'}{\partial z} + \frac{1}{\tilde{H}_S} \tilde{p}' \right] \tag{4.21}$$

where $\tilde{H}_S = 2H_S/(1 - 2N^2 H_S/g) = 2c_p/(c_v - R_d)H_S$ is a characteristic height scale of the problem.[a] As usual, we make a wave ansatz of the form

$$(\tilde{\xi}', \tilde{D}', \tilde{p}', \tilde{w}') = \Re[(\hat{\xi}, \hat{D}, \hat{p}, \hat{w})e^{i(kx+ly+mz-\nu t)}] \tag{4.22}$$

With this ansatz, we get the following matrix equation:

$$\begin{pmatrix} -i\nu & f_0 & 0 & 0 \\ -f_0 & -i\nu & -\frac{K^2}{\rho_s} & 0 \\ 0 & c_s^2 \rho_s & -i\nu & \rho_s c_s^2 \left(im - \frac{1}{\tilde{H}_S} \right) \\ 0 & 0 & -\frac{i\nu}{\rho_s} \left(im + \frac{1}{\tilde{H}_S} \right) & -\delta\nu^2 + N^2 \end{pmatrix} \cdot \begin{pmatrix} \hat{\xi} \\ \hat{D} \\ \hat{p} \\ \hat{w} \end{pmatrix} = 0 \tag{4.23}$$

where $K^2 = k^2 + l^2$. The determinant vanishes if the eigenvalue is zero (discussed later) or fulfills the polynomial equation

$$\delta\nu^4 - \left[\delta f_0^2 + N^2 + \left(\delta K^2 + m^2 + \frac{1}{\tilde{H}_S^2} \right) c_s^2 \right] \nu^2$$

$$+ f_0^2 \left[N^2 + \left(m^2 + \frac{1}{\tilde{H}_S^2} \right) c_s^2 \right] + N^2 K^2 c_s^2 = 0 \tag{4.24}$$

We consider special cases before discussing the general solution.

---

[a]Note that $N^2 = g^2/(c_p \bar{T})$ in the isothermal case.

4.2.1 *Pure sound waves* $(N = f_0 = 1/\tilde{H}_S = 0)$

In this case gravity and planetary rotation are neglected. Then, with $\delta = 1$ the frequency solutions become

$$\nu_{1,2} = \pm c_s \sqrt{k^2 + l^2 + m^2} \tag{4.25}$$

We see that the plane wave propagates exactly with the phase speed c_s in every direction. It is a solution of the wave equation

$$\frac{\partial^2 b}{\partial t^2} = c_s^2 \nabla^2 b \tag{4.26}$$

We can show with (4.19), (4.20) and (4.21) that $b = \nabla \cdot \mathbf{v}' = \nabla^2 \chi$ fulfills this equation under the given conditions. Therefore, wave phase isolines coincide with isolines of velocity potential χ. The vorticity vanishes exactly since there are no processes for generating vorticity. Hence, the velocity vector is parallel to the wave number vector which defines a **longitudinal wave**. Sound waves play no relevant role for weather dynamics. Therefore, weather modelers seek to filter or suppress these waves. Furthermore, they cause trouble in numerical solution schemes due to their high propagation speed in vertical direction.

4.2.2 *Pure internal gravity waves* $(f_0 = 1/c_s^2 = 0)$

The fluid becomes incompressible in the limit $c_s^2 \to \infty$. This is impossible for an ideal gas but a liquid fulfills the incompressibility condition quite well. Then, the divergence $\nabla \cdot \mathbf{v}'$ vanishes and density becomes an individually conserved quantity. However, density is not necessarily uniform and can have a vertical stratification as assumed in (4.13). The frequency solutions become in this case

$$\nu = \pm \sqrt{\frac{k^2 + l^2}{\delta(k^2 + l^2) + m^2 + \frac{1}{\tilde{H}_S^2}}} \, N \tag{4.27}$$

The frequencies approach $\pm N$ with increasing horizontal wavenumbers. The mechanism of internal gravity waves can be understood with the parcel method. In the vicinity of the reference level the vertical density profile can be approximated by

$$\bar{\rho} \approx \bar{\rho}(z_r) + \frac{d\bar{\rho}}{dz}\bigg|_{z=z_r} (z - z_r) = \bar{\rho}(z_r)\left[1 - \frac{N^2}{g}(z - z_r)\right] \tag{4.28}$$

Then, the vertical momentum equation of a fluid parcel deflected from the level $z = z_r$ can be written as

$$\frac{D^2 z}{Dt^2} = -\frac{1}{\rho}\frac{\partial p'}{\partial z} - g\frac{\rho - \bar{\rho}}{\rho} = -\frac{1}{\rho}\frac{\partial p'}{\partial z} - N^2(z - z_r) \qquad (4.29)$$

The second identity results because the parcel conserves its density in the incompressible case so that $\rho = \bar{\rho}(z_r)$. We assume that the pressure in the small parcel adapts immediately to the environment. Hence, the pressure anomaly p' vanishes and we obtain the following ordinary differential equation

$$\frac{D^2 z'}{Dt^2} + N^2 z' = 0 \qquad (4.30)$$

The solution is a harmonic oscillation with the frequency N. The mechanism is quite simple: The rising parcel conserves the density and attains negative **buoyancy**, $-g\rho'/\rho$ since $\rho > \bar{\rho}$. Therefore, the parcel accelerates back to the reference level and by inertia the deflection becomes negative. There, the buoyancy is positive since $\rho < \bar{\rho}$ and the parcel moves upward again. The result is a periodic oscillation that results from **static stability** defined by a positive N^2. One can also imagine the case $N^2 < 0$ that leads to **convective instability**. Then, density increases with height. We will discuss convective instability in more detail in Chapter 6.

4.2.3   *General case*

In the general case with $\delta = 1$ we obtain the four frequencies

$$\nu_{1,2,3,4} = \pm \left\{ \frac{1}{2}\left[f_0^2 + N^2 + \left(K^2 + m^2 + \frac{1}{\tilde{H}_S^2}\right)c_s^2 \right] \right.$$

$$\left. \pm\sqrt{\frac{1}{4}\left[f_0^2 - N^2 + \left(K^2 - m^2 - \frac{1}{\tilde{H}_S^2}\right)c_s^2 \right]^2 + K^2\left(m^2 + \frac{1}{\tilde{H}_S^2}\right)c_s^4} \right\}^{\frac{1}{2}}$$

$$\hspace{10cm} (4.31)$$

Obviously, the argument of the inner root is positive. Furthermore, it can be shown that the argument of the outer root is also positive. Therefore, all four frequencies are real and no instability can occur. Figure 4.1 shows the two positive frequencies as a function of

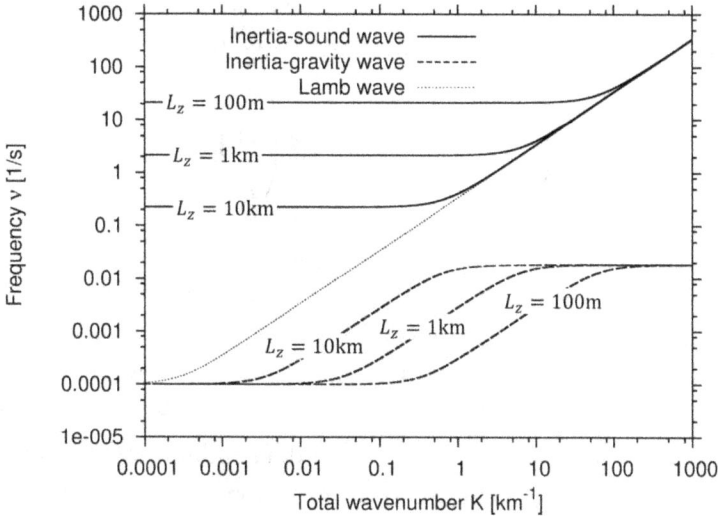

Fig. 4.1. Frequencies of the waves in the isothermal atmosphere as a function of total wavenumber K for different vertical wavelengths L_z (see label at the curves). The solid curves display the inertia-sound waves while the dashed curves show internal inertia-gravity waves. The dotted curve is the frequency of the Lamb wave. The chosen parameters are $f_0 = 10^{-4}\,\mathrm{s}^{-1}$ and $\bar{T} = 288.15\,\mathrm{K}$.

total wavenumber K for different vertical wavelengths in a log–log plot. The higher frequencies represent **inertia-sound waves** while the lower frequencies are associated with **internal inertia-gravity waves**. These two wave types behave like pure sound and internal gravity waves at very high total wavenumbers where planetary rotation has virtually no effect. At very low wavenumbers the inertial effect of planetary rotation dominates so that the frequency becomes wavenumber independent. There, the internal inertia-gravity wave becomes an inertial oscillation as described in Section 3.5. The frequency of the inertia-sound waves increases with decreasing vertical wavelength while that of the internal inertia-gravity wave becomes smaller. Note that the wave cannot propagate in vertical direction due to the boundary condition for vertical velocity. However, these waves can interfere with each other to yield a wave packet that can indeed propagate vertically.

The dotted curve displays the frequency of a special wave, namely the **Lamb wave** (Lamb, 1932). The Lamb wave results in our analysis if we set vertical velocity identical to zero. Then, Eq. (4.21) leads to $m = i/\tilde{H}_S$ so that the wave amplitude describes an exponential function with height (increasing for ξ', D', w' and decreasing for p'). In this case, the linearized equations reduce to

$$\begin{pmatrix} -i\nu & f_0 & 0 \\ -f_0 & -i\nu & -\frac{K^2}{\rho_s} \\ 0 & c_s^2\rho_s & -i\nu \end{pmatrix} \cdot \begin{pmatrix} \hat{\xi} \\ \hat{D} \\ \hat{p} \end{pmatrix} = 0 \qquad (4.32)$$

and the nontrivial frequency solutions become

$$\nu_{1,2} = \pm\sqrt{f_0^2 + c_s^2 K^2} = \pm\sqrt{f_0^2 + c_p/c_v R_d \bar{T} K^2} \qquad (4.33)$$

The Lamb wave is an inertia-sound wave that has no vertical velocity and is in hydrostatic balance. It has similarity with the inertia-gravity wave in the isentropic atmosphere (see Section 3.5) with the frequencies $\nu_{1,2} = \pm\sqrt{f_0^2 + R_d \bar{T}_s K^2}$ and a similarly fast phase speed.

With the quasihydrostatic approximation ($\delta = 0$) we get for the frequency solutions of (4.24):

$$\nu_{1,2} = \pm\sqrt{f_0^2 + \frac{N^2 K^2 c_s^2}{N^2 + \left(m^2 + \frac{1}{\bar{H}_S^2}\right)c_s^2}} \qquad (4.34)$$

These two eigenvalues are associated with internal inertia-gravity wave and they are very similar to those for $\delta = 1$ when the wavelength is small. However, long waves do not approach the frequency N as in the nonhydrostatic case. We see that the quasihydrostatic approximation filters the sound waves. This is a great advantage in numerical modelling because the high frequencies of sound waves pose a problem in such a way that extremely short time steps are necessary to keep the time integration scheme stable.

The linearized system (4.23) has also a planetary wave solution. Setting the frequency to zero leads to $\hat{D} = \hat{w} = 0$ and the geostrophic balance

$$f_0 \mathbf{k} \times \mathbf{v}_h' = -\frac{1}{\rho}\nabla_h p' \qquad (4.35)$$

The vertical structure of this wave is not determined and can have any shape. However, with nonzero β–effect the planetary wave becomes an internal Rossby wave with a defined vertical structure and westward propagation. This can be understood more clearly with a baroclinic quasigeostrophic model introduced in the next subsection.

4.3 Quasigeostrophic Approximation of the Primitive Equations

The quasigeostrophic equations are beneficial for diagnosing and understanding extratropical weather systems. In this section, we derive these equations by making the quasigeostrophic approximation of the baroclinic primitive equations formulated in isobaric coordinates (cf. Eqs. (2.70)–(2.73)). For β–plane geometry these equations become

$$\frac{\partial \mathbf{v}_h}{\partial t} + \mathbf{v}_h \cdot \nabla \mathbf{v}_h + \omega \frac{\partial \mathbf{v}_h}{\partial p} + (f_0 + \beta y)\mathbf{k} \times \mathbf{v}_h = -\nabla_h \phi + \mathbf{f}_T \quad (4.36)$$

$$\frac{\partial \phi}{\partial p} = -\frac{R_d T}{p} \quad (4.37)$$

$$\nabla_h \cdot \mathbf{v}_h + \frac{\partial \omega}{\partial p} = 0 \quad (4.38)$$

$$\frac{\partial T}{\partial t} + \mathbf{v}_h \cdot \nabla_h T + \Pi \frac{\partial \theta}{\partial p} \omega = \frac{Q}{c_p} \quad (4.39)$$

We have omitted the budget equation for water vapor since it is not needed for the subsequent analysis. It is important for the analysis to split temperature and geopotential into reference state and perturbation

$$T = T_r(p) + T'(x, y, p, t), \quad \phi = \phi_r(p) + \phi'(x, y, p, t) \quad (4.40)$$

for which the relation $\partial \phi_r/\partial p = -R_d T_r/p$ holds. Inserting these relations and eliminating temperature yield the following equation set:

$$\frac{\partial \mathbf{v}_h}{\partial t} + \mathbf{v}_h \cdot \nabla_h \mathbf{v}_h + \omega \frac{\partial \mathbf{v}_h}{\partial p} + (f_0 + \beta y)\mathbf{k} \times \mathbf{v}_h = -\nabla_h \phi' + \mathbf{f}_T \quad (4.41)$$

$$\nabla_h \cdot \mathbf{v}_h + \frac{\partial \omega}{\partial p} = 0 \quad (4.42)$$

$$\frac{\partial}{\partial t}\left(\frac{\partial \phi'}{\partial p}\right) + \mathbf{v}_h \cdot \nabla_h \left(\frac{\partial \phi'}{\partial p}\right) - \Pi \frac{R_d}{p}\frac{\partial \theta'}{\partial p}\omega = -\Xi\omega - \frac{R_d Q}{p c_p} \quad (4.43)$$

where $\Xi = -R_d\Pi/p\partial\theta_r/\partial p$ represents a stability parameter.[b] To nondimensionalize the variables we choose the following scales:

$$\tilde{\mathbf{v}}_h \equiv \frac{\mathbf{v}_h}{U}, \quad \tilde{\omega} \equiv \frac{\omega}{g\rho_v W}, \quad (\tilde{x}, \tilde{y}) \equiv \frac{(x, y)}{L}, \quad \tilde{t} \equiv \frac{Ut}{L}, \quad \tilde{p} \equiv \frac{p}{g\rho_v H},$$

$$\tilde{\phi}' \equiv \frac{\phi'}{f_0 LU}, \quad \tilde{\theta}' \equiv \frac{R_d\theta'}{f_0 LU}, \quad \tilde{\mathbf{f}}_T = \frac{L}{U^2}\mathbf{f}_T, \quad \tilde{Q} = \frac{R_d}{f_0 U^2}\frac{Q}{c_p\tilde{p}} \qquad (4.44)$$

With these scales the nondimensional primitive equations become

$$\mathrm{Ro}\left(\frac{\partial\tilde{\mathbf{v}}_h}{\partial\tilde{t}} + \tilde{\mathbf{v}}_h \cdot \tilde{\nabla}_h\tilde{\mathbf{v}}_h + \mathrm{r}_A\mathrm{Ro}\tilde{\omega}\frac{\partial\tilde{\mathbf{v}}_h}{\partial p}\right) + (1 + \mathrm{Ro}\tilde{\beta}\tilde{y})\mathbf{k} \times \tilde{\mathbf{v}}_h$$

$$= -\tilde{\nabla}_h\tilde{\phi}' + \mathrm{Ro}\tilde{\mathbf{f}}_T \qquad (4.45)$$

$$\tilde{\nabla}_h \cdot \tilde{\mathbf{v}}_h + \mathrm{r}_A\mathrm{Ro}\frac{\partial\tilde{\omega}}{\partial\tilde{p}} = 0 \qquad (4.46)$$

$$\frac{\partial}{\partial\tilde{t}}\left(\frac{\partial\tilde{\phi}'}{\partial\tilde{p}}\right) + \tilde{\mathbf{v}}_h \cdot \tilde{\nabla}_h\left(\frac{\partial\tilde{\phi}'}{\partial\tilde{p}}\right) - \Pi\frac{\mathrm{r}_A\mathrm{Ro}}{\tilde{p}}\frac{\partial\tilde{\theta}'}{\partial\tilde{p}}\tilde{\omega} = -\mathrm{Bu}\delta\tilde{\omega} - \tilde{Q} \quad (4.47)$$

In this system, the Rossby number Ro and β-parameter $\tilde{\beta}$ are defined as in Section 3.8.3. However, the Burger number Bu becomes

$$\mathrm{Bu} = \Xi\frac{g^2\rho_v^2 H^2}{f_0^2 L^2} = \Xi\frac{P^2}{f_0^2 L^2} = \frac{L_R^2}{L^2} \qquad (4.48)$$

where P denotes the pressure scale and $L_R = \sqrt{\Xi}P/f_0$ is now the **baroclinic Rossby radius** which depends on the temperature stratification. r_A measures the ratio of horizontal to vertical advection timescale divided by the Rossby number, i.e.

$$\mathrm{r}_A = \frac{L/U}{H/W}\frac{1}{\mathrm{Ro}} \qquad (4.49)$$

The quasigeostrophic approximation is valid if

$$\mathrm{Ro} \ll 1, \quad \mathrm{Bu} = \mathrm{o}(1), \quad \tilde{\beta} = \mathrm{o}(1) \quad \text{and} \quad \mathrm{r}_A = \mathrm{o}(1) \qquad (4.50)$$

The assumption $\mathrm{r}_A = \mathrm{o}(1)$ can be explained by the fact that only the divergent flow contributes to vertical velocity. The divergent flow

---

[b]Although $-\partial\phi'/\partial p$ is the specific volume anomaly, we will interpret it as the temperature variable of the quasigeostrophic model since it is proportional to the temperature anomaly T'.

is of order Ro and, therefore, the vertical advection timescale is $1/$Ro longer than the horizontal advection timescale since the former results from the nondivergent geostrophic flow.

As in Section 3.8, we expand in powers of the Rossby number:

$$\tilde{\mathbf{v}}_h = \tilde{\mathbf{v}}_h^{(0)} + \tilde{\mathbf{v}}_h^{(1)}\text{Ro} + \tilde{\mathbf{v}}_h^{(2)}\text{Ro}^2 + o(\text{Ro}^3) \tag{4.51}$$

$$\tilde{\phi}' = \tilde{\phi}'^{(0)} + \tilde{\phi}'^{(1)}\text{Ro} + \tilde{\phi}'^{(2)}\text{Ro}^2 + o(\text{Ro}^3) \tag{4.52}$$

$$\tilde{\omega} = \tilde{\omega}^{(0)} + \tilde{\omega}^{(1)}\text{Ro} + \tilde{\omega}^{(2)}\text{Ro}^2 + o(\text{Ro}^3) \tag{4.53}$$

Furthermore, we assume that $\tilde{\mathbf{f}}_T$ and $\tilde{Q}$ have order unity.[c] After insertion we obtain in the zeroth order

$$\mathbf{k} \times \tilde{\mathbf{v}}_h^{(0)} = -\tilde{\nabla}_h \tilde{\phi}'^{(0)} \tag{4.54}$$

$$\tilde{\nabla}_h \cdot \tilde{\mathbf{v}}_h^{(0)} = 0 \tag{4.55}$$

$$\frac{\partial}{\partial \tilde{t}}\left(\frac{\partial \tilde{\phi}'^{(0)}}{\partial \tilde{p}}\right) + \tilde{\mathbf{v}}_h^{(0)} \cdot \tilde{\nabla}_h\left(\frac{\partial \tilde{\phi}'^{(0)}}{\partial \tilde{p}}\right) = -\text{r}_A\text{Bu}\tilde{\omega}^{(0)} - \tilde{Q} \tag{4.56}$$

As in the quasigeostrophic barotropic model, we see that the zero-order flow is geostrophic and nondivergent. In addition, we obtain the temperature tendency equation in which horizontal advection only takes place by the zero-order geostrophic wind and temperature changes by vertical motion only incorporate a time-independent reference stratification. The first-order equations for momentum and mass continuity become

$$\frac{\partial \tilde{\mathbf{v}}_h^{(0)}}{\partial \tilde{t}} + \tilde{\mathbf{v}}_h^{(0)} \cdot \tilde{\nabla}_h \tilde{\mathbf{v}}_h^{(0)} + \mathbf{k} \times \tilde{\mathbf{v}}_h^{(1)} + \tilde{\beta}\tilde{y}\mathbf{k} \times \tilde{\mathbf{v}}_h^{(0)} = -\tilde{\nabla}_h \tilde{\phi}'^{(1)} + \tilde{\mathbf{f}}_T \tag{4.57}$$

$$\tilde{\nabla}_h \cdot \tilde{\mathbf{v}}_h^{(1)} + \text{r}_A\frac{\partial \tilde{\omega}^{(0)}}{\partial \tilde{p}} = 0 \tag{4.58}$$

---

[c]Note that this assumption is not appropriate in the boundary layer.

These zero and first-order equations form the quasigeostrophic equation set. With the notations

$$\mathbf{v}_g = U\tilde{\mathbf{v}}_h^{(0)}, \quad \mathbf{v}_a = \mathrm{RoU}\tilde{\mathbf{v}}_h^{(1)}, \quad \omega = g\rho_v W\tilde{\omega}^{(0)}, \quad \phi' = f_0 LU\tilde{\phi}'^{(0)}$$

$$(4.59)$$

we obtain the dimensional quasigeostrophic equations

$$\mathbf{v}_g = \frac{1}{f_0}\mathbf{k} \times \nabla_h \phi' \tag{4.60}$$

$$\frac{\partial \mathbf{v}_g}{\partial t} + \mathbf{v}_g \cdot \nabla_h \mathbf{v}_g + f_0 \mathbf{k} \times \mathbf{v}_a + \beta y \mathbf{k} \times \mathbf{v}_g = -\nabla_h \phi'^{(1)} + \mathbf{f}_T \tag{4.61}$$

$$\nabla_h \cdot \mathbf{v}_a + \frac{\partial \omega}{\partial p} = 0 \tag{4.62}$$

$$\frac{\partial}{\partial t}\left(\frac{\partial \phi'}{\partial p}\right) + \mathbf{v}_g \cdot \nabla_h \left(\frac{\partial \phi'}{\partial p}\right) = -\Xi\omega - \frac{R_d Q}{pc_p} \tag{4.63}$$

The indeterminate first-order geopotential can be eliminated by converting the momentum tendency equation into the vorticity equation

$$\frac{\partial \nabla_h^2 \psi_g}{\partial t} + \mathbf{v}_g \cdot \nabla_h (\nabla_h^2 \psi_g) + \beta v_g = -f_0 \nabla_h \cdot \mathbf{v}_a + \mathbf{k} \cdot (\nabla_h \times \mathbf{f}_T) \tag{4.64}$$

where $\psi_g = \phi'/f_0$ denotes the geostrophic streamfunction.

The quasigeostrophic approximation of the primitive equations leads to the following simplifications:

- The momentum equation does not include time tendencies of the ageostrophic wind.
- Horizontal advection only takes place by the geostrophic wind.
- The Coriolis force by the ageostrophic wind is calculated without β effect.
- Vertical advection is only considered in the temperature tendency equation and with a prescribed time–independent vertical stratification.

The quasigeostrophic equations can be combined to one equation, namely, the **quasigeostrophic potential vorticity equation**.

For this purpose, we solve the temperature tendency equation for ω and then differentiate with respect to pressure. This leads to

$$\frac{\partial \omega}{\partial p} = -\left(\frac{\partial}{\partial t} + \mathbf{v}_g \cdot \nabla_h\right)\frac{\partial}{\partial p}\left(\frac{1}{\Xi}\frac{\partial \phi'}{\partial p}\right) - \frac{1}{\Xi}\frac{\partial \mathbf{v}_g}{\partial p}\cdot\left(\frac{\partial \nabla_h \phi'}{\partial p}\right)$$
$$- \frac{\partial}{\partial p}\left(\frac{R_d Q}{\Xi p c_p}\right) \tag{4.65}$$

With the geostrophic balance equation (4.60) we see that the second term on the right-hand side vanishes since $\nabla_h \phi' = -f_0 \mathbf{k} \times \mathbf{v}_g$. By using this result and the continuity equation (4.62) we can substitute the divergence term into the vorticity equation and obtain the quasigeostrophic potential vorticity equation

$$\left(\frac{\partial}{\partial t} + \mathbf{v}_g \cdot \nabla_h\right) P_g = -f_0 \frac{\partial}{\partial p}\left(\frac{R_d Q}{\Xi p c_p}\right) + \mathbf{k} \cdot (\nabla_h \times \boldsymbol{f}_T) \tag{4.66}$$

where

$$P_g = \nabla_h^2 \psi_g + f_0 + \beta y + \frac{\partial}{\partial p}\left(\frac{f_0^2}{\Xi}\frac{\partial \psi_g}{\partial p}\right) \tag{4.67}$$

denotes the **quasigeostrophic potential vorticity** of the baroclinic atmosphere. Quasigeostrophic potential vorticity is conserved in the absence of turbulent momentum exchange ($\boldsymbol{f}_T = 0$) and heating ($Q = 0$). One needs boundary conditions to solve the quasigeostrophic potential vorticity equation. These can be deduced from the temperature tendency equation at the lower and upper boundaries. At the lower boundary we have

$$\left(\frac{\partial}{\partial t} + \mathbf{v}_{gs} \cdot \nabla_h\right)\left(\frac{\partial \psi_g}{\partial p}\bigg|_s\right) = -f_0 \Xi \omega_s + f_0 \frac{R_d Q_s}{p_s c_p} \tag{4.68}$$

where the index s indicates the surface. This equation is quite problematic to solve since ω_s is unknown and the surface does not coincide with an isobaric surface. However, the quasigeostrophic equations are rather used for conceptual understanding and diagnostics and not for weather forecasts. For these purposes, it is acceptable to simplify the boundary condition by assuming vanishing vertical velocity ω at a constant pressure level p_b close to the surface. Then, the boundary condition becomes

$$\left(\frac{\partial}{\partial t} + \mathbf{v}_g \cdot \nabla_h\right)\left(\frac{\partial \psi_g}{\partial p}\right) = f_0 \frac{R_d Q}{p c_p} \text{ at } p = p_b \qquad (4.69)$$

At the upper boundary, one can assume besides constant pressure a vanishing temperature anomaly or alternatively a vanishing vertical velocity (e.g. to describe the tropopause).

To solve the equations, we must first determine the geostrophic streamfunction from the potential vorticity field P_g and temperatures at the boundaries. Then, we also obtain the geostrophic wind and temperature fields. This characteristic is known as the **invertibility principle** (cf. section 1.14). With the obtained fields we can make a prediction of P_g and $\partial \psi_g/\partial p$ at the boundaries to proceed with a new inversion at a future time step.

Sometimes it is more convenient to use a different pressure-based vertical coordinate for the quasigeostrophic model. With the **log-pressure height** $z_l = -H_s \ln(p/p_0)$ the vertical derivative is substituted by

$$\frac{\partial}{\partial p} = -\frac{1}{g\rho_I}\frac{\partial}{\partial z_l} \qquad (4.70)$$

where $\rho_I = p/(R_d T_0) = p_0 \exp(-z_l/H_s)/(R_d T_0)$ denotes the density of the isothermal atmosphere having the temperature T_0. With this coordinate the quasigeostrophic potential vorticity equation is written as

$$\left(\frac{\partial}{\partial t} + \mathbf{v}_g \cdot \nabla_h\right) P_g = \frac{f_0}{\rho_I}\frac{\partial}{\partial z_l}\left(\rho_I \frac{gQ}{c_p T_0 N_l^2}\right) + \mathbf{k} \cdot (\nabla_h \times \mathbf{f}_T) \qquad (4.71)$$

where

$$P_g = \nabla_h^2 \psi_g + f_0 + \beta y + \frac{1}{\rho_I}\frac{\partial}{\partial z_l}\left(\rho_I \frac{f_0^2}{N_l^2}\frac{\partial \psi_g}{\partial z_l}\right) \qquad (4.72)$$

In these equations $N_l = \sqrt{g/\theta_0 \partial \theta_r/\partial z_l}$ denotes a frequency that is approximately identical with the Brunt–Väisälä frequency of the reference state. The boundary condition for temperature becomes at a pre-defined level z_b

$$\frac{\partial}{\partial t}\left(\frac{\partial \psi_g}{\partial z_l}\right) + \mathbf{v}_g \cdot \nabla_h\left(\frac{\partial \psi_g}{\partial z_l}\right) = \frac{g f_0}{T_0}\frac{Q_s}{c_p} \text{ at } z_l = z_b \qquad (4.73)$$

4.4 Rossby Waves in the Stratified Atmosphere

The quasigeostrophic equations are suitable for analyzing Rossby waves in the baroclinic atmosphere. For this purpose, we linearize these equations with respect to a uniform zonal basic flow U so that

$$\bar{\psi}_g = -Uy \qquad (4.74)$$

The boundaries with no vertical motion are at $z_l = z_s$ (surface) and $z_l = z_t$ (tropopause). Without heat sources and turbulent exchange, the linearized quasigeostrophic potential vorticity equation in the log-pressure coordinate becomes

$$\left(\frac{\partial}{\partial t} + U\frac{\partial}{\partial x}\right)\left[\nabla_h^2 \psi_g' + \frac{1}{\rho_I}\frac{\partial}{\partial z_l}\left(\rho_I \frac{f_0^2}{N_l^2}\frac{\partial \psi_g'}{\partial z_l}\right)\right] + \beta\frac{\partial \psi_g'}{\partial x} = 0 \quad (4.75)$$

and the boundary conditions are prescribed by

$$\frac{\partial \psi_g'}{\partial z_l} = 0 \text{ at } z_l = z_s, z_t \qquad (4.76)$$

Therefore, initial temperature anomalies do not appear at the boundaries. By assuming constant N_l and a perturbation of the form

$$\psi_g' = \mathfrak{R}\left\{\hat{\psi}\exp\left(\frac{z_l}{2H_s}\right)\exp[i(kx + ly + mz_l - \nu t)]\right\} \qquad (4.77)$$

we obtain the following frequency solution:

$$\nu = Uk - \frac{\beta k}{K^2 + \frac{f_0^2}{N_l^2}\left(m^2 + \frac{1}{4H_s^2}\right)} \qquad (4.78)$$

where $K^2 = k^2 + l^2$. There are discrete values of vertical wavenumber m for which the boundary condition is fulfilled. These indicate the **baroclinic modes** of the Rossby wave. There is also a vertically independent solution with $m = 0$ for which the frequency is

$$\nu = Uk - \frac{\beta k}{k^2 + l^2} \qquad (4.79)$$

This solution constitutes the barotropic Rossby wave which coincides with that of the nondivergent barotropic fluid (cf. Section 3.8.1).

Obviously, the upper boundary condition in (4.76) yields an effect similar to a rigid lid in the barotropic model. On the other hand, a flexible tropopause should lead to a dispersion relation similar to (3.170).

A relevant issue in atmospheric dynamics is the vertical propagation of Rossby waves from the troposphere into the stratosphere. Assuming a two-layer fluid with different Brunt–Väisälä frequencies N_l leads to a reasonable model for this purpose. Typically, Rossby waves are triggered in the troposphere by hydrodynamic instabilities, orographic forcing and heating. Then, a Rossby wave forcing appears at the lower boundary of the stratosphere where N_l increases significantly. Let ν_f be the frequency of the Rossby wave triggered in the troposphere. Then, the excited stratospheric Rossby must fulfill the dispersion relation

$$\nu_f = U_S k - \frac{\beta k}{K^2 + \frac{f_0^2}{N_S^2}\left(m_S^2 + \frac{1}{4H_s^2}\right)} \tag{4.80}$$

where U_S, N_S and m_S denote the zonal basic flow, Brunt–Väisälä frequency and vertical wavenumber in the stratosphere, respectively. Solving this equation for m_S yields:

$$m_S = \frac{N_S}{f_0}\sqrt{\frac{\beta}{U_S - \nu_f/k} - K^2 - \frac{f_0^2}{4N_S^2 H_s^2}} \tag{4.81}$$

The argument of the square root must be positive to obtain a real vertical wavenumber and, therefore, vertical propagation of wave energy into the stratosphere. Then, the zonal basic flow U_S must on the one hand be larger than ν_f/k but on the other hand it must be below an upper value depending on the total wavenumber K and Brunt–Väisälä frequency N_S. By assuming that the forcing wave is due to a barotropic mode we find that

$$U - \frac{\beta}{K^2} < U_S < U - \frac{\beta}{K^2}\frac{1}{1 + 4N_S^2 H_s^2 K^2/f_0^2} \tag{4.82}$$

Therefore, vertical propagation is only possibly in a narrow range for the zonal flow U_S. Figure 4.2(a) shows the region of vertical propagation in a diagram which has zonal wavenumber m (number of waves along a latitude circle at $45°N$) as abscissa and zonal basic

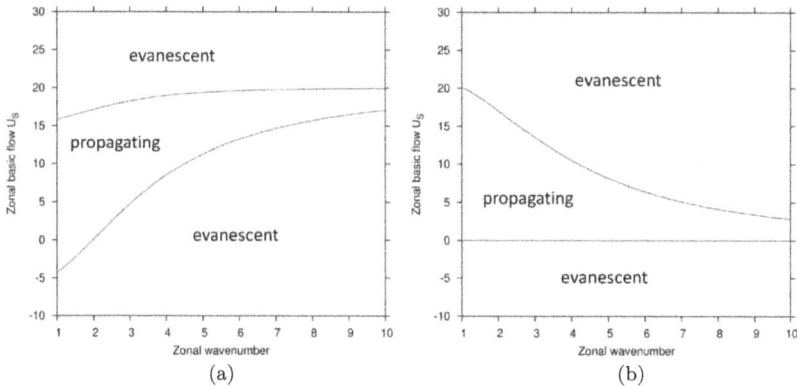

Fig. 4.2. Zonal mean flow range which allows vertical Rossby wave propagation into the stratosphere as a function of zonal wavenumber: (a) Forcing results from a barotropic Rossby wave in the troposphere, (b) forcing results from a stationary Rossby wave. For the calculation we assume $N_S = 0.021\,\mathrm{s}^{-1}, U = 20\,\mathrm{m/s}$, $l = 7.85 \times 10^{-7}\mathrm{m}^{-1}$ and the reference latitude 45°N.

flow U_S as ordinate. We see that the zonal flow range with vertical propagation is larger for small than for high zonal wavenumbers. At zonal wavenumber 1 vertical propagation is even possible for a weak easterly flow. The wave amplitude decays exponentially with height at the rate $-im_S$ when the condition (4.82) is not fulfilled. Then, the wave is referred to as **evanescent**.

In the case of stationary tropospheric waves ($\nu_f = 0$) we get

$$0 < U_S < \frac{\beta}{K^2 + f_0^2/(4N_S^2 H_s^2)} \qquad (4.83)$$

Figure 4.2(b) shows the same diagram as Fig. 4.2(a) but the region of vertical propagation for a stationary wave forcing is displayed. Now, the zonal flow must be a westerly flow for observing a real vertical wavenumber but again, the zonal flow range with vertical propagation shrinks with increasing zonal wavenumber. At wavenumber 10 the zonal flow cannot be larger than 3 m/s while wavenumber 1 allows a zonal flow up to 20 m/s. Vertical propagation of Rossby wave plays a role in stratospheric **sudden warming** events that appear in Northern Hemisphere spring. The flow in the **stratospheric polar vortex** falls into the range where stationary long Rossby waves can propagate vertically. These waves in turn cause a breakdown of the vortex and strong warming (for more details see James, 1994).

4.5 Quasigeostrophic Diagnostics

The quasigeostrophic equations form a useful tool for the diagnosis and short-term prediction of synoptic-scale weather systems although they are not suitable for numerical weather prediction. In this section, the most important methods are presented.

4.5.1 *Thermal wind*

The **thermal wind** results after combining the geostrophic and hydrostatic balance equations. The vertical derivative of the geostrophic balance equation (4.60) becomes in the log-pressure coordinate system:

$$\frac{\partial \mathbf{v}_g}{\partial z_l} = \frac{1}{f_0} \mathbf{k} \times \nabla_h \frac{\partial \phi}{\partial z_l} \tag{4.84}$$

On the other hand, the hydrostatic balance equation takes in the log-pressure coordinate system the form

$$\frac{\partial \phi}{\partial z_l} = -g\rho_l \frac{\partial \phi}{\partial p} = g\frac{\rho_l}{\rho} = g\frac{T}{T_0} \tag{4.85}$$

Insertion yields the **thermal wind equation**

$$\frac{\partial \mathbf{v}_g}{\partial z_l} = \frac{g}{f_0 T_0} \mathbf{k} \times \nabla_h T \tag{4.86}$$

Integrating this equation over the layer from z_1 to z_2 leads to

$$\Delta \mathbf{v}_g \equiv \mathbf{v}_g|_{z_2} - \mathbf{v}_g|_{z_1} = \frac{g\Delta z_l}{f_0 T_0} \mathbf{k} \times \nabla_h \bar{T}^z \tag{4.87}$$

where $\Delta z_l = z_2 - z_1$ is the depth of the considered layer, $\bar{T}^z = (\int_{z_1}^{z_2} T dz_l)/\Delta z_l$ the depth-averaged temperature and $\Delta \mathbf{v}_g$ the thermal wind. Obviously, the thermal wind is directed parallel to the isotherms of $\bar{T}^z$ and the colder air appears on the left (right)-hand side of this vector on the northern (southern) hemisphere (see Fig. 4.3). The thermal wind occurs because the pressure decreases faster with height in cold air than in warm air. Therefore, the **thickness** of the air layer between two pressure levels depends on temperature. The thickness can be determined by vertical integration of

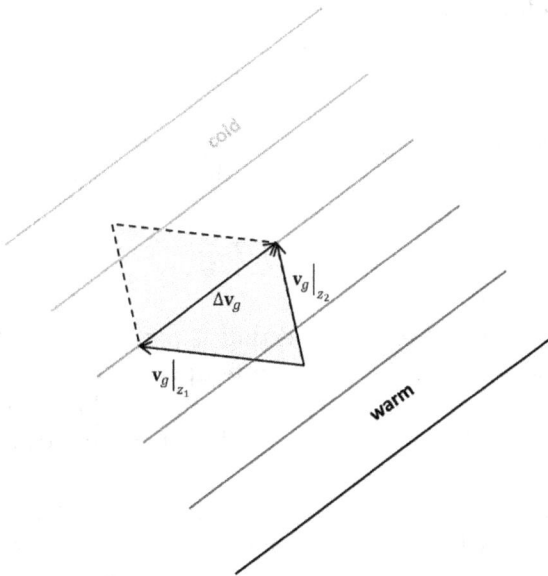

Fig. 4.3.   Sketch of the thermal wind $\Delta \mathbf{v}_g$ of the layer between the log-pressure height levels z_1 and z_2. The area of the parallelogram is proportional to the geostrophic temperature advection in this layer. This is also obvious by comparing the directions of the geostrophic wind vectors with the directions of the isotherms.

the hydrostatic balance equation

$$\Delta z \equiv \frac{\phi|_{z_2} - \phi|_{z_1}}{g} = \frac{\bar{T}^z}{T_0} \Delta z_l \qquad (4.88)$$

This relation shows that thickness is proportional to the depth-averaged temperature $\bar{T}^z$.

The thermal wind equation can also be used for short-term temperature prediction. The quasigeostrophic temperature tendency equation becomes in the absence of heat sources and vertical motion

$$\frac{\partial T}{\partial t} = -\mathbf{v}_g \cdot \nabla_h T = \frac{f_0 T_0}{g} \mathbf{v}_g \cdot \left(\mathbf{k} \times \frac{\partial \mathbf{v}_g}{\partial z_l} \right) = -\frac{f_0 T_0}{g} \mathbf{k} \cdot \left(\mathbf{v}_g \times \frac{\partial \mathbf{v}_g}{\partial z_l} \right)$$

$$(4.89)$$

Obviously, the geostrophic wind must change its direction with increasing height in order to cause warming or cooling by advection. This can be established more clearly by assuming a thin layer in which the linear approximation

$$\mathbf{v}_g \approx \mathbf{v}_{g0} + \Delta\mathbf{v}_{g\parallel}\left(z_l - \frac{z_1 + z_2}{2}\right) + \Delta\mathbf{v}_{g\perp}\left(z_l - \frac{z_1 + z_2}{2}\right) \quad (4.90)$$

is valid where $\Delta\mathbf{v}_{g\parallel}(\Delta\mathbf{v}_{g\perp})$ denotes the part of the thermal wind which is parallel (perpendicular) to the geostrophic wind $\mathbf{v}_{g0}$. We obtain for the temperature advection

$$-\frac{f_0 T_0}{g}\mathbf{k}\cdot\left(\mathbf{v}_g \times \frac{\partial\mathbf{v}_g}{\partial z_l}\right) = -\frac{f_0 T_0}{g}\mathbf{k}\cdot(\mathbf{v}_{g0} \times \Delta\mathbf{v}_{g\perp}) \quad (4.91)$$

Only the part of the thermal wind that is perpendicular to the mean geostrophic wind contributes to the temperature advection. On the Northern Hemisphere Eq. (4.91) states that a clockwise (anticlockwise) turning of the geostrophic wind with height causes warming (cooling) by horizontal temperature advection. The example illustrated in Fig. 4.3 shows an advective warming since the geostrophic wind turns clockwise. We can also deduce this result easily in this figure by observing that the mean geostrophic wind is directed from the warm towards the cold air. Furthermore, the temperature advection is proportional to the parallelogram area spanned by $\mathbf{v}_g|_{z_1}$ and $\mathbf{v}_g|_{z_2}$ because

$$\mathbf{k}\cdot(\mathbf{v}_{g0} \times \Delta\mathbf{v}_{g\perp}) = \frac{1}{2}\mathbf{k}\cdot\left[(\mathbf{v}_g|_{z_1} + \mathbf{v}_g|_{z_2}) \times \frac{\mathbf{v}_g|_{z_2} - \mathbf{v}_g|_{z_1}}{\Delta z_l}\right]$$

$$= \frac{\mathbf{k}\cdot(\mathbf{v}_g|_{z_1} \times \mathbf{v}_g|_{z_2})}{\Delta z_l}$$

Geostrophic warm and cold advection can be easily diagnosed with weather maps showing geopotential height on different pressure levels since their isolines form streamlines of the geostrophic flow.

Furthermore, it is of importance for the weather forecast if the static stability of the air changes with time. The thermal wind balance equation also provides an answer to this question. The vertical derivative of (4.89) yields

$$\frac{\partial}{\partial t}\left(\frac{\partial T}{\partial z_l}\right) = -\frac{f_0 T_0}{g}\mathbf{k}\cdot\frac{\partial}{\partial z_l}\left(\mathbf{v}_g \times \frac{\partial\mathbf{v}_g}{\partial z_l}\right) = -\frac{f_0 T_0}{g}\mathbf{k}\cdot\left(\mathbf{v}_g \times \frac{\partial^2\mathbf{v}_g}{\partial z_l^2}\right)$$

$$(4.92)$$

Therefore, a vertical change of the vertical shear can alter the static stability. Two effects can be important. First, an increase of clockwise (anticlockwise) vector turning with height leads to stabilization (destabilization) of the atmospheric stratification on the Northern Hemisphere. On the other hand, an increase of wind speed without a change of clockwise (anticlockwise) turning rate yields also stabilization (destabilization). To diagnose a change of static stability it is convenient to choose three equidistant log-pressure height levels, e.g. 1000, 500 and 250 hPa. Then, we obtain approximately for the change of the vertical temperature gradient

$$\frac{\partial}{\partial t}\left(\frac{T_3 - T_1}{z_3 - z_1}\right) \approx -\frac{f_0 T_0}{g\Delta z_l^2}\mathbf{k}\cdot[(\mathbf{v}_g|_{z_2} \times \mathbf{v}_g|_{z_3}) - (\mathbf{v}_g|_{z_1} \times \mathbf{v}_g|_{z_2})]$$

$$(4.93)$$

where the index denotes the selected log-pressure height level. Consequently, the static stability change is proportional to the difference of the parallelogram areas describing the lower- and upper-layer geostrophic temperature advections. Figure 4.4 shows an example where the warm advection increases with height. That is, the static stability rises in the layer bounded by the levels z_1 and z_3.

4.5.2 *Geopotential tendency equation*

The **geopotential tendency equation** is used to predict pressure fall or rise at a certain location. It results from the quasigeostrophic

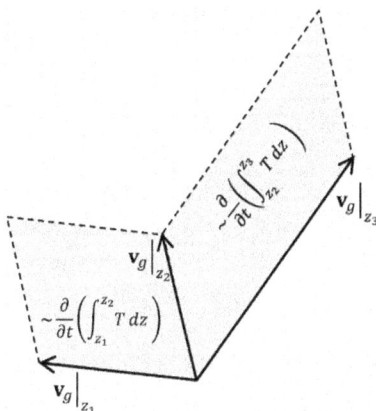

Fig. 4.4. Sketch of the static stability increase due to vertically increasing temperature advection.

potential vorticity equation (4.66) by neglecting heating and turbulent exchange. Then, potential vorticity is conserved and we can write
this equation as

$$\frac{\partial P_g}{\partial t} = -\mathbf{v}_g \cdot \nabla_h P_g \tag{4.94}$$

We see that the local potential vorticity tendency is identical to horizontal potential vorticity advection. The right-hand side of this equation can be split into two parts so that we obtain

$$\frac{\partial P_g}{\partial t} = \underbrace{-\mathbf{v}_g \cdot \nabla_h \zeta_g}_{\substack{\text{vorticity} \\ \text{advection}}} - \underbrace{\frac{\partial}{\partial p}\left[-\frac{R_d f_0}{p \Xi}\mathbf{v}_g \cdot \nabla_h T\right]}_{\substack{\text{differential temperature} \\ \text{advection}}} \tag{4.95}$$

The potential vorticity tendency results from the sum of **vorticity
advection** and **differential temperature advection**. The latter
refers to a nonzero vertical gradient of horizontal temperature advection. With a positive sign, the air column becomes more stable and
the potential vorticity increases. The vorticity advection forms a process that is also included in a barotropic model. On the other hand,
differential temperature advection can only occur within baroclinic
models. It represents a process that is necessary for the baroclinic
development of weather systems while vorticity advection describes
essentially their movement.

Now, the potential vorticity tendency is related to geopotential
tendency by

$$\frac{\partial P_g}{\partial t} = \left[\frac{1}{f_0}\nabla_h^2 + \frac{\partial}{\partial p}\left(\frac{f_0}{\Xi}\frac{\partial}{\partial p}\right)\right]\left(\frac{\partial \phi}{\partial t}\right) \tag{4.96}$$

The differential operator on the left-hand side yields in most cases a
change of sign since synoptic-scale weather systems have a wavelike
structure that can be described by trigonometric functions.[d] Therefore, pressure fall (rise) results usually if the potential vorticity tendency is positive (negative).

Figure 4.5 shows an example for the application of the geopotential tendency equation. Figure 4.5(a) displays 500 hPa geopotential

---

[d]Recall that $\partial^2/\partial x^2(e^{ix}) = -e^{ix}$.

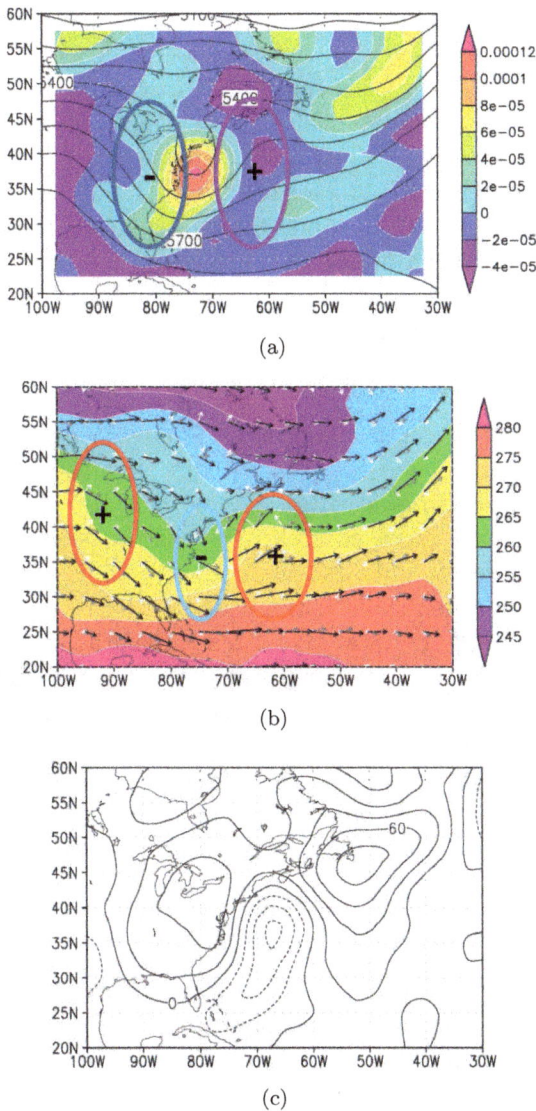

Fig. 4.5.   Analysis of the geopotential tendency equation using NCEP reanalysis data: (a) 500 hPa geopotential height (black isolines, contour interval 100 m) and 500 hPa geostrophic vorticity (colored shadings, interval $2 \times 10^{-5} \text{s}^{-1}$) on January 27, 2008, 18UTC, (b) average temperature of 500 hPa and 1000 hPa (colored shadings, interval 5 K) and geostrophic wind at 1000 hPa (white vectors), 500 hPa (grey vectors) and 250 hPa (black vectors), (c) change of 500 hPa geopotential height in the subsequent 6 hours (contour interval 60 m, negative isolines are dashed). The ovals in (a) and (b) are explained in the text.

height and geostrophic relative vorticity on January 27, 2008, 18 UTC based on NCEP reanalysis. A prominent trough with enhanced vorticity is located at the east coast of North America. It induces a subsequent cyclone development at the surface. The violet and dark blue encircled areas indicate positive and negative vorticity advection, respectively. By this process the trough moves eastward without enhancement. Figure 4.5(b) shows the average of the temperatures on the 1000 hPa and 500 hPa isobaric surfaces while the three arrows in this figure display the geostrophic winds at 1000 hPa (white), 500 hPa (grey) and 250 hPa (black). Red and blue circles indicate areas with enhanced warm advection and cold advection, respectively. The turning of the arrows demonstrates that these advections have larger amplitude in the lower layer between 1000 hPa and 500 hPa compared to the upper layer between 500 hPa and 250 hPa. Therefore, differential temperature advection should lead to pressure rise (fall) in the red and blue encircled areas. By this process, the trough and the adjacent ridges should intensify in amplitude. Figure 4.5(c) shows the difference between the 500 hPa geopotential height 6 hours later and that shown in Fig. 4.5(a). We can see that the trough intensified south of the original position and moved eastward. Furthermore, a significant pressure rise occurs northeast of this trough. Therefore, the geopotential tendency equation can be applied for a qualitative estimate of short-term geopotential height development. However, for a quantitative prediction and longer time scales, a numerical weather forecast model based on the primitive equations is necessary.

4.5.3   *Omega equation*

For the weather analyst, it is of interest to deduce the vertical velocity since it causes adiabatic warming or cooling and upward motion often coincides with areas of large-scale precipitation. At first sight, the continuity equation (2.72) seems suitable for this purpose. However, the measured wind data are too noisy for a reliable determination of the horizontal wind divergence. In contrast, the **omega equation** constitutes a better diagnostic tool for deducing the vertical motion in synoptic-scale weather systems. It can be derived with the quasigeostrophic equations since the geopotential determines both vorticity and temperature. The quasigeostrophic vorticity and temperature tendency equations can be written after neglecting turbulent exchange and diabatic heating as

$$\frac{1}{f_0}\frac{\partial \nabla_h^2 \phi}{\partial t} + \mathbf{v}_g \cdot \nabla_h \zeta_g = f_0 \frac{\partial \omega}{\partial p} \tag{4.97}$$

$$\frac{\partial}{\partial t}\left(\frac{\partial \phi}{\partial p}\right) + \mathbf{v}_g \cdot \nabla_h \left(\frac{\partial \phi}{\partial p}\right) = -\Xi\omega \tag{4.98}$$

Multiplying the vertical derivative of (4.97) with f_0 and applying the horizontal Laplace operator to (4.98) yields

$$\frac{\partial}{\partial t}\left(\nabla_h^2 \frac{\partial \phi}{\partial p}\right) + f_0 \frac{\partial}{\partial p}(\mathbf{v}_g \cdot \nabla_h \zeta_g) = f_0^2 \frac{\partial^2 \omega}{\partial p^2} \tag{4.99}$$

$$\frac{\partial}{\partial t}\left(\nabla_h^2 \frac{\partial \phi}{\partial p}\right) + \nabla_h^2 \left[\mathbf{v}_g \cdot \nabla_h \left(\frac{\partial \phi}{\partial p}\right)\right] = -\Xi\nabla_h^2 \omega \tag{4.100}$$

We see that the tendency terms of both equations are identical. Therefore, we can eliminate them by subtracting Eq. (4.100) from (4.99) which gives the omega equation

$$\left(\Xi\nabla_h^2 + f_0^2 \frac{\partial^2}{\partial p^2}\right)\omega = \underbrace{-f_0 \frac{\partial}{\partial p}(-\mathbf{v}_g \cdot \nabla_h \zeta_g)}_{\substack{\text{differential vorticity} \\ \text{advection}}} + \underbrace{\frac{R_d}{p}\nabla_h^2(\mathbf{v}_g \cdot \nabla_h T)}_{\substack{\text{temperature} \\ \text{advection}}}$$

$$\tag{4.101}$$

Obviously, this equation has a similar structure as the geopotential tendency equation. We can expect that the operator on the left-hand side attains a negative sign. Therefore, upward (downward) motion results if vorticity advection increases (decreases) with height and warm (cold) advection prevails. Note that upward motion is associated with negative ω and the Laplace operator in front of the temperature advection term likely reverses the sign.

Figure 4.6 shows an application of the omega equation to the weather situation already considered in Fig. 4.5. Figure 4.6(a) displays the geopotential height of 400 hPa and 1000 hPa. A weak surface low is found downstream of the prominent trough at the east coast of North America. Such a flow configuration is common for extratropical cyclogenesis resulting from baroclinic instability which will be explained in Section 4.6. The upper-level flow determines

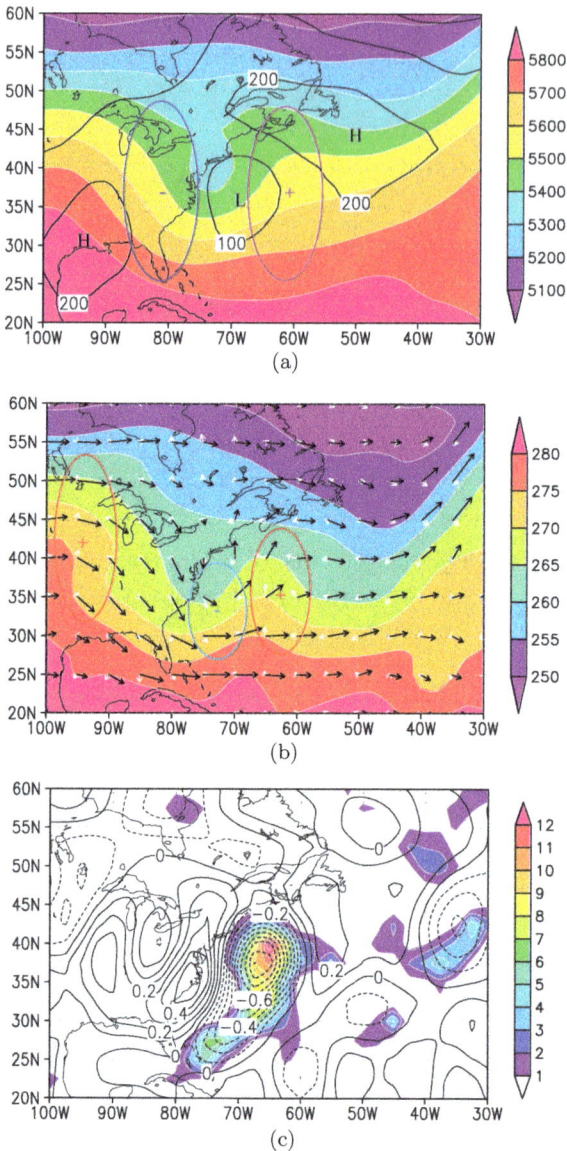

Fig. 4.6.   Analysis of the omega equation using NCEP reanalysis data on January 27, 2008, 18UTC: (a) 400 hPa and 1000 hPa geopotential height (colored shadings and black isolines, interval 100 m), (b) 700 hPa temperature (colored shadings, interval 5 K) and geostrophic wind at 1000 hPa (white vectors) and 400 hPa (black vectors), (c) 700 hPa vertical velocity (black isolines, contour interval 0.1 Pa/s) and 6-h precipitation after the selected date (colored shadings, interval 1 mm). The ovals in (a) and (b) are explained in the text.

the sign of the differential vorticity advection term since the surface geostrophic flow has a much smaller amplitude compared to that at 400 hPa. Therefore, we can infer that positive differential vorticity advection prevails east of the trough (violet encircled area) and negative differential vorticity advection occurs west of the trough (dark blue encircled area). These processes should induce rising (subsiding) vertical motion at 700 hPa east (west) of the trough at 400 hPa. Fig. 4.6(b) shows the 700 hPa level while the two arrows in this figure display the geostrophic winds at 1000 hPa (white) and 400 hPa (black). As in Fig. 4.5(b) we can detect warm and cold advections in the red and blue encircled areas, respectively. Rising (subsiding) motion should prevail in areas of warm (cold) advection. Indeed, Fig. 4.6(c) reveals that a remarkable 700 hPa upward motion occurs in regions with both positive upper layer vorticity advection and 700 hPa warm advection. In contrast, subsiding motion arises upstream where negative upper-layer vorticity advection prevails. Cold advection only coincides with subsiding air in the western part of the blue encircled region. We must admit that the method provides only a rough estimate since a reversion of the sign by applying the operator $\Xi \nabla_h^2 + f_0^2 \partial^2/\partial p^2$ is not always true. Figure 4.6(c) also shows 6-hour precipitation after the selected date. Precipitation mostly occurs in a region with rising motion at 700 hPa and also the amount is related to the strength of vertical velocity as seen in the region east of the trough. Consequently, the omega equation can also be used for short-term prediction of large-scale precipitation.

4.5.4 Q-vector

Warm and cold fronts often develop in synoptic-scale weather systems and there, significant weather events like showers, rain or gusts appear. It is possible to predict this **frontogenesis** with the **Q-vector** which is defined by

$$\mathbf{Q} \equiv \nabla_h \mathbf{v}_g \cdot \nabla_h \left(\frac{\partial \phi}{\partial p} \right) = \frac{\partial \mathbf{v}_g}{\partial x} \cdot \nabla_h \left(\frac{\partial \phi}{\partial p} \right) \mathbf{i} + \frac{\partial \mathbf{v}_g}{\partial y} \cdot \nabla_h \left(\frac{\partial \phi}{\partial p} \right) \mathbf{j}$$

$$\equiv Q_x \mathbf{i} + Q_y \mathbf{j} \tag{4.102}$$

The Q-vector introduced by Hoskins *et al.* (1978) is related to the change of the horizontal temperature gradient in a geostrophic flow.

This can be seen by applying the horizontal gradient operator to the prognostic temperature equation (4.63) which leads to

$$\left(\frac{\partial}{\partial t} + \mathbf{v}_g \cdot \nabla_h\right) \left[\nabla_h \left(\frac{\partial \phi}{\partial p}\right)\right] = -\Xi\nabla_h\omega - \frac{R_d}{pc_p}\nabla_h Q - \mathbf{Q} \quad (4.103)$$

Using temperature T instead of $\partial\phi/\partial p$ gives

$$\left(\frac{\partial}{\partial t} + \mathbf{v}_g \cdot \nabla_h\right)(\nabla_h T) = \frac{p}{R_d}\Xi\nabla_h\omega + \frac{1}{c_p}\nabla_h Q + \frac{p}{R_d}\mathbf{Q} \quad (4.104)$$

This equation states that the horizontal temperature gradient changes by (i) a nonzero horizontal gradient of vertical velocity, (ii) a nonzero horizontal gradient of heating and (iii) a nonzero Q-vector. The latter results because horizontal temperature advection is not uniform due to horizontal variations in the geostrophic wind field. The absolute value of the gradient constitutes a suitable measure for the strength of a front. Therefore, the tendency of this quantity can be used to identify regions where frontogenesis takes place. By multiplying (4.104) with $\nabla_h T$ we get

$$\left(\frac{\partial}{\partial t} + \mathbf{v}_g \cdot \nabla_h\right)\left(\frac{1}{2}|\nabla_h T|^2\right) = \frac{p}{R_d}\Xi\nabla_h T \cdot \nabla_h\omega + \frac{1}{c_p}\nabla_h T \cdot \nabla_h Q$$

$$+ \frac{p}{R_d}\nabla_h T \cdot \mathbf{Q} \quad (4.105)$$

This equation states that fronts can evolve if the horizontal temperature gradient has a component in the direction of (i) the vertical velocity gradient, (ii) the heating gradient or (iii) the Q-vector. Consequently, the Q-vector must be directed toward the warmer air to cause frontogenesis.

There is also a diagnostic relation between the Q-vector and vertical velocity. This can be derived by the thermal wind balance equation in isobaric coordinates:

$$\frac{\partial \mathbf{v}_g}{\partial p} = \frac{1}{f_0}\mathbf{k} \times \nabla_h\left(\frac{\partial \phi}{\partial p}\right) \quad (4.106)$$

which can be used to write (4.103) also in the form

$$\left(\frac{\partial}{\partial t} + \mathbf{v}_g \cdot \nabla_h\right)\left(f_0\mathbf{k} \times \frac{\partial \mathbf{v}_g}{\partial p}\right) = \Xi\nabla_h\omega + \frac{R_d}{pc_p}\nabla_h Q + \mathbf{Q} \quad (4.107)$$

On the other hand, we can recast the quasigeostrophic momentum equation (4.61) for $\boldsymbol{f}_T = 0$ (no turbulent exchange) into the form[e]

$$\left(\frac{\partial}{\partial t} + \mathbf{v}_g \cdot \nabla_h\right)(\mathbf{k} \times \mathbf{v}_g) - f_0 \mathbf{v}_a - \beta y \mathbf{v}_g = 0 \qquad (4.108)$$

Applying the operator $f_0 \partial/\partial p$ yields

$$\left(\frac{\partial}{\partial t} + \mathbf{v}_g \cdot \nabla_h\right)\left(f_0 \mathbf{k} \times \frac{\partial \mathbf{v}_g}{\partial p}\right) - f_0^2 \frac{\partial \mathbf{v}_a}{\partial p} - f_0 \beta y \frac{\partial \mathbf{v}_g}{\partial p} = -\mathbf{Q}$$

$$(4.109)$$

Here, we have used the identity

$$\left[\mathbf{k} \times \nabla_h \left(\frac{\partial \phi}{\partial p}\right)\right] \cdot \nabla_h(\mathbf{k} \times \mathbf{v}_g) = \nabla_h \mathbf{v}_g \cdot \nabla_h\left(\frac{\partial \phi}{\partial p}\right) = \mathbf{Q} \quad (4.110)$$

which does not appear obvious at first sight. It is left to the reader to proof this identity by evaluating these expressions in component form with the rules presented in Section A.3 and noting that $\partial u_g/\partial x = -\partial v_g/\partial x$. Eliminating the time derivative by combining (4.107) and (4.109) leads to

$$\Xi \nabla_h \omega - f_0^2 \frac{\partial \mathbf{v}_a}{\partial p} = f_0 \beta y \frac{\partial \mathbf{v}_g}{\partial p} - \frac{R_d}{pc_p} \nabla_h Q - 2\mathbf{Q} \qquad (4.111)$$

After applying the horizontal divergence, we obtain an alternative form of the omega equation

$$\Xi \nabla_h^2 \omega + f_0^2 \frac{\partial^2 \omega}{\partial p^2} = f_0 \beta \frac{\partial v_g}{\partial p} - \frac{R_d}{pc_p} \nabla_h^2 Q - 2\nabla_h \cdot \mathbf{Q} \qquad (4.112)$$

This equation also includes the contribution of heating to vertical motion which has been neglected in the previous subsection. The equation shows that warming (cooling) usually causes upward (downward) vertical motion. Furthermore, this equation shows that upward (downward) motion prevails in regions with convergence (divergence) of the Q-vector if the β-effect and heating are of minor importance. Therefore, it is much easier to identify regions with vertical motion on

---

[e]The first-order geopotential, $\phi'^{(1)}$, is set to zero for simplicity although it does not play a role for this problem.

weather maps by using the Q-vector. Figure 4.7 shows an application of the Q-vector diagnosis for the same region and date as analyzed in Fig. 4.6. Figure 4.7(a) displays the Q-vector together with temperature and height at the 850 hPa pressure level. We see various regions where the Q-vector is oriented towards the warmer air. The red isolines show the field $(p/R_d)\nabla_h T \cdot \mathbf{Q}$ that measures the tendency of

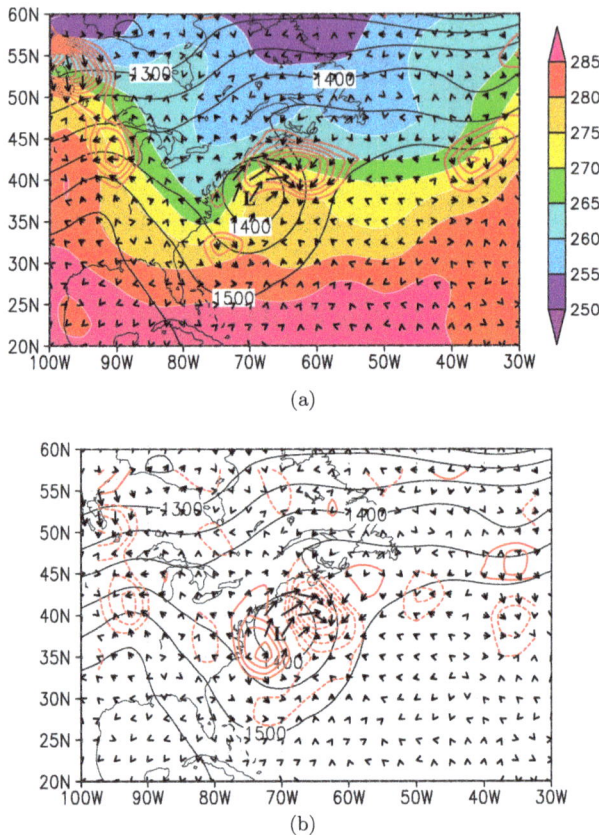

(a)

(b)

Fig. 4.7.   Analysis of the Q-vector field using NCEP reanalysis data on January 27, 2008, 18UTC: (a) 850 hPa temperature (colored shadings, interval 5 K), 850 hPa geopotential height (black isolines, contour interval 100 m), Q-vector (arrows) and frontogenesis tendency $(p/R_d)\nabla_h T \cdot \mathbf{Q}$ (red isolines, contour interval $10^{-10} \mathrm{K}^2/(\mathrm{m}^2 \mathrm{day})$, the zero contour is not displayed), (b) 850 hPa geopotential height (black isolines, contour interval 100 m), Q-vector (arrows) and its divergence (red isolines, contour interval $10^{-13} \mathrm{m}/(\mathrm{kg\, day})$, the zero contour is not displayed and negative values are dashed).

the squared temperature gradient. There is a region with enhanced values in the northern part of the developing low at the east coast of North America where the warm front intensifies. Much weaker tendencies are found for the enhancement of the cold front in the southwestern part of the low. More regions with significant warm and cold frontogenesis occur along the meandering band having a high horizontal temperature gradient. This is the **polar front** where extratropical cyclones typically develop due to baroclinic instability. Figure 4.7(b) shows the divergence of the Q-vector. We detect a convergence at the warm front and divergence at the cold front. This has the consequence that air rises (descends) at warm (cold) fronts. This is in rough agreement with the vertical velocity pattern shown in Fig. 4.6(c). The lifting and lowering of air at these fronts typically accompany the frontogenesis process which will be investigated in more detail in Section 6.5.

4.6 Baroclinic Instability

Baroclinic instability represents the most important process for the development of extratropical synoptic weather systems. Usually, a jet stream as shown in Fig. 2.2b is unstable with respect to baroclinic instability. However, it is not possible to solve the linearized equations analytically for a basic flow of such a shape. Therefore, we consider simpler basic flows that are less realistic but enable a better understanding of the instability. In this section, we first use a simple two-level primitive equation model to explain the physical instability mechanism. Afterwards, the quasigeostrophic approximation is applied and the impact of the β-effect is analyzed. We also present solutions of the vertically continuous quasigeostrophic equations including basic flows having uniform potential vorticity. Finally, instability criteria are derived for baroclinic zonal flows.

4.6.1 *The mechanism of baroclinic instability*

To understand the basic mechanism of baroclinic instability it is sufficient to consider a basic flow on the f-plane with vertical but no horizontal shear. We assume

$$\bar{u} = \Lambda(p_m - p) \tag{4.113}$$

where $p_m = 500 \, \text{hPa}$. Obviously, the basic flow is vertically antisymmetric about the middle level p_m with easterlies at the surface (see Fig. 4.8). This has the advantage of a better understanding because growing perturbations do not travel in this case. However, removing the unrealistic easterlies by adding a uniform barotropic zonal flow would not alter the stability properties. The basic flow is supposed to be in hydrostatic and geostrophic balance. Consequently, the temperature of the basic state becomes

$$\bar{T} = T_r(p) - \frac{p}{R_d} f_0 \Lambda y \qquad (4.114)$$

The temperature decreases in northward direction with the gradient $f_0 \Lambda p / R_d$. With this basic state the linearization of the primitive equations (4.41)–(4.43) yields for $\beta = 0$, $\boldsymbol{f}_T = 0$ and $Q = 0$

$$\left(\frac{\partial}{\partial t} + \bar{u} \frac{\partial}{\partial x} \right) \mathbf{v}_h' - \Lambda \omega' \mathbf{i} + f_0 \mathbf{k} \times \mathbf{v}_h' = -\nabla_h \phi' \qquad (4.115)$$

$$\nabla_h \cdot \mathbf{v}_h' + \frac{\partial \omega'}{\partial p} = 0 \qquad (4.116)$$

$$\left(\frac{\partial}{\partial t} + \bar{u} \frac{\partial}{\partial x} \right) \left(\frac{\partial \phi'}{\partial p} \right) + f_0 \Lambda v' = -\Xi \omega' \qquad (4.117)$$

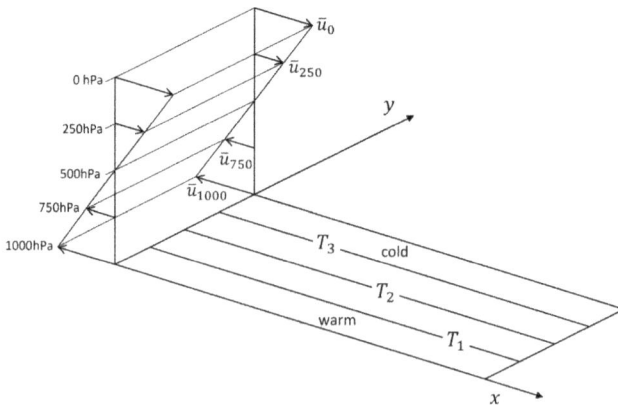

Fig. 4.8.   Sketch of the vertically sheared basic flow. The open arrows show the wind vector of the basic flow and the isolines the temperature field that results from the thermal wind balance equation.

where we have neglected the small term $\Pi R_d \omega' \partial(\bar{\theta}-\theta_r)/\partial p/p$ because it complicates the analysis unnecessarily. To simplify the solution of this problem we approximate the vertical derivative by finite difference expressions and consider the equation only at discrete levels. The simplest formulation is the two-level approximation. Then, the wind vector $\mathbf{v}'_h$ and geopotential ϕ' are evaluated at the pressure levels 750 hPa and 250 hPa while the vertical velocity ω' is only considered at 500 hPa (see Fig. 4.9). We make use of the following approximations:

$$\left(\frac{\partial \omega'}{\partial p}\right)_1 \approx \frac{\omega'_2 - \overbrace{\omega'_0}^{=0}}{\Delta p}, \quad \left(\frac{\partial \omega'}{\partial p}\right)_3 \approx \frac{\overbrace{\omega'_4}^{\approx 0} - \omega'_2}{\Delta p} \qquad (4.118)$$

$$\omega'_1 \approx \frac{\omega'_2 + \overbrace{\omega'_0}^{=0}}{2}, \quad \omega'_3 \approx \frac{\overbrace{\omega'_4}^{\approx 0} + \omega'_2}{2} \qquad (4.119)$$

$$\mathbf{v}'_{h_2} \approx \frac{\mathbf{v}'_{h_1} + \mathbf{v}'_{h_3}}{2}, \quad \left(\frac{\partial \phi'}{\partial p}\right)_2 \approx \frac{\phi'_3 - \phi'_1}{\Delta p} \qquad (4.120)$$

where $\Delta p = 500$ hPa. The indices 0, 1, 2, 3 and 4 refer to the levels 0, 250, 500, 750 and 1000 hPa, respectively. The vertical velocity vanishes at 0 hPa and approximately at 1000 hPa since it is close to the lower boundary.

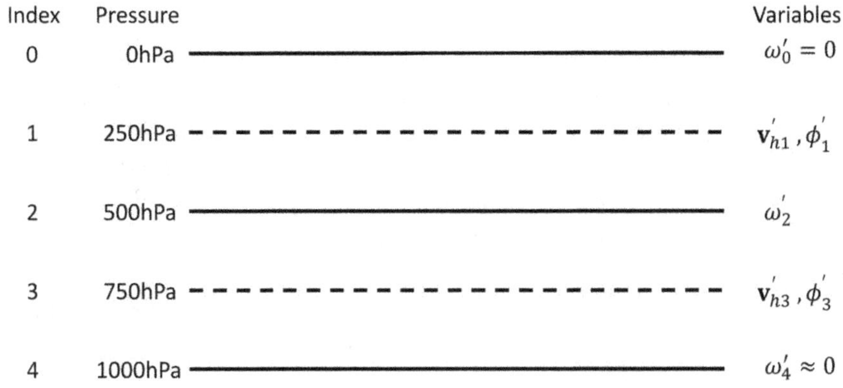

Index	Pressure		Variables
0	0hPa	————————————————————	$\omega'_0 = 0$
1	250hPa	– – – – – – – – – – – – – – – –	$\mathbf{v}'_{h1}, \phi'_1$
2	500hPa	————————————————————	ω'_2
3	750hPa	– – – – – – – – – – – – – – – –	$\mathbf{v}'_{h3}, \phi'_3$
4	1000hPa	————————————————————	$\omega'_4 \approx 0$

Fig. 4.9. Arrangement of the pressure levels in the two-level model.

Using these approximations, we obtain the two-level primitive equations

$$\left(\frac{\partial}{\partial t} + U_T \frac{\partial}{\partial x}\right) \mathbf{v}'_{h_1} - \frac{U_T}{\Delta p}\omega'_2 \mathbf{i} + f_0 \mathbf{k} \times \mathbf{v}'_{h_1} = -\nabla_h \phi'_1 \qquad (4.121)$$

$$\left(\frac{\partial}{\partial t} - U_T \frac{\partial}{\partial x}\right) \mathbf{v}'_{h_3} - \frac{U_T}{\Delta p}\omega'_2 \mathbf{i} + f_0 \mathbf{k} \times \mathbf{v}'_{h_3} = -\nabla_h \phi'_3 \qquad (4.122)$$

$$\nabla_h \cdot \mathbf{v}'_{h_1} + \frac{\omega'_2}{\Delta p} = 0 \qquad (4.123)$$

$$\nabla_h \cdot \mathbf{v}'_{h_3} - \frac{\omega'_2}{\Delta p} = 0 \qquad (4.124)$$

$$\frac{\partial}{\partial t}\left(\frac{\phi'_1 - \phi'_3}{\Delta p}\right) - f_0 \frac{U_T}{\Delta p}(v'_1 + v'_3) = \Xi_2 \omega'_2 \qquad (4.125)$$

where $U_T = \Lambda \Delta p / 2$. The momentum equations can be recast in a more convenient form by taking the sum and difference of Eqs. (4.121) and (4.122):

$$\frac{\partial \mathbf{v}'_{hM}}{\partial t} + U_T \frac{\partial \mathbf{v}'_{hT}}{\partial x} - \frac{U_T}{\Delta p}\omega'_2 \mathbf{i} + f_0 \mathbf{k} \times \mathbf{v}'_{hM} = -\nabla_h \phi'_M \qquad (4.126)$$

$$\frac{\partial \mathbf{v}'_{hT}}{\partial t} + U_T \frac{\partial \mathbf{v}'_{hM}}{\partial x} + f_0 \mathbf{k} \times \mathbf{v}'_{hT} = -\nabla_h \phi'_T \qquad (4.127)$$

where $b_M \equiv (b_1 + b_3)/2$ and $b_T \equiv (b_1 - b_3)/2$ for any field b. We can interpret variables with index M as barotropic while those with index T describe fields that only appear in a baroclinic flow. The analysis and understanding simplify by using vorticity and divergence equations of the barotropic and baroclinic part of the flow. These become

$$\frac{\partial \xi'_M}{\partial t} = -U_T \frac{\partial \xi'_T}{\partial x} + U_T \frac{\partial D'_T}{\partial y} \qquad (4.128)$$

$$\frac{\partial \xi'_T}{\partial t} = -U_T \frac{\partial \xi'_M}{\partial x} - f_0 D'_T \qquad (4.129)$$

$$0 = -2U_T\frac{\partial D'_T}{\partial x} + f_0\xi'_M - \nabla_h^2\phi'_M \qquad (4.130)$$

$$\frac{\partial D'_T}{\partial t} = f_0\xi'_T - \nabla_h^2\phi'_T \qquad (4.131)$$

where we have used the continuity equations (4.123) and (4.124) that lead to the relations $D'_M = (\nabla_h \cdot \mathbf{v}'_{h_1} + \nabla_h \cdot \mathbf{v}'_{h_3})/2 = 0$ and $D'_T = (\nabla_h \cdot \mathbf{v}'_{h_1} - \nabla_h \cdot \mathbf{v}'_{h_3})/2 = -\omega'_2/\Delta p$. Obviously, Eq. (4.130) constitutes a diagnostic relation for the determination of the barotropic geopotential perturbation ϕ'_M. However, this quantity is not needed in the further analysis and, therefore, we can ignore this equation. With the new variables the temperature tendency equation can be written as

$$\frac{\partial \xi'_{gT}}{\partial t} = U_T\frac{\partial \xi'_M}{\partial x} - f_0L_R^2\nabla_h^2 D'_T \qquad (4.132)$$

where $\xi'_{gT} = \nabla_h^2\phi'_T/f_0$ denotes the baroclinic geostrophic vorticity and $L_R = \sqrt{\Xi_2}\Delta p/(\sqrt{2}f_0)$ the **baroclinic Rossby radius** which includes the phase speed of internal instead of external gravity waves as in the shallow water model (cf. Eq. (3.62)). Equations (4.128), (4.129), (4.131) and (4.132) form a closed system of linear differential equations. After making the usual wave ansatz

$$(\xi'_M, \xi'_T, D'_T, \xi'_{gT}) = \Re\{(\hat{\xi}_M, \hat{\xi}_T, \hat{D}_T, \hat{\xi}_{gT})\exp[i(kx+ly-\nu t)]\} \qquad (4.133)$$

we obtain the matrix equation

$$\begin{pmatrix} -i\nu & iU_Tk & iU_Tl & 0 \\ iU_Tk & -i\nu & f_0 & 0 \\ 0 & -f_0 & -i\nu & f_0 \\ -iU_Tk & 0 & -f_0L_R^2K^2 & -i\nu \end{pmatrix} \cdot \begin{pmatrix} \hat{\xi}_M \\ \hat{\xi}_T \\ \hat{D}_T \\ \hat{\xi}_{gT} \end{pmatrix} = 0 \qquad (4.134)$$

The vorticity tilting term $U_T\partial D'_T/\partial y$ in Eq. (4.128) complicates the analysis. However, it vanishes in the case $l = 0$ which is associated with a meridional independence of the perturbations. Although this represents an unrealistic perturbation it suffices to explain the basic mechanism of baroclinic instability.[f] Using this assumption the frequency equation becomes

$$\nu^4 - [f_0^2 + (f_0^2 L_R^2 + U_T^2)k^2]\nu^2 - f_0^2 U_T^2 (k^2 - L_R^2 k^4) = 0 \qquad (4.135)$$

This equation can be written in a more compact form by using the nondimensional frequency $\tilde{\nu} \equiv \nu L_R / U_T$ and nondimensional zonal wavenumber $\tilde{k} = L_R k$:

$$\tilde{\nu}^4 - \left[\frac{1}{\mathrm{Ro}_d^2}(1 + \tilde{k}^2) + \tilde{k}^2\right]\tilde{\nu}^2 - \frac{1}{\mathrm{Ro}_d^2}(\tilde{k}^2 - \tilde{k}^4) = 0 \qquad (4.136)$$

where $\mathrm{Ro}_d = U_T/(f_0 L_R)$ which coincides with the Rossby number if the length scale is $\mathrm{L} = o(L_R)$.[g] This biquadratic equation has the following four frequency solutions

$$\tilde{\nu}_{1,2,3,4} = \pm \sqrt{\frac{1}{2}\left(\frac{1 + \tilde{k}^2}{\mathrm{Ro}_d^2} + \tilde{k}^2\right) \pm \sqrt{\frac{1}{4}\left(\frac{1 + \tilde{k}^2}{\mathrm{Ro}_d^2} + \tilde{k}^2\right)^2 + \frac{(\tilde{k}^2 - \tilde{k}^4)}{\mathrm{Ro}_d^2}}}$$

$$(4.137)$$

It can be shown that the argument of the second root has always a positive sign. Therefore, the frequencies are either real or imaginary. Instability results in the latter case while in the former case east- and westward traveling waves appear. Figure 4.10 shows the real and imaginary parts of $\tilde{\nu}_1$ and $\tilde{\nu}_2$ as a function of wavelength $2\pi/\tilde{k}$ for different Ro_d. The frequencies $\tilde{\nu}_3$ and $\tilde{\nu}_4$ have the same values as $\tilde{\nu}_1$ and $\tilde{\nu}_2$ with negative sign and are not displayed. The frequency $\tilde{\nu}_1$ is associated with inertia-gravity waves. The dimensionless frequency $\tilde{\nu}_1$ increases rapidly with decreasing Ro_d at constant wavelength since the dimensioned frequency of these waves only depends weakly on U_T. The frequency $\tilde{\nu}_2$ belongs to a baroclinic wave. This wave grows exponentially for wavelengths $L_x > 2\pi L_R$ ($\tilde{k} < 1$) but exhibits eastward propagation for $L_x < 2\pi L_R$ ($\tilde{k} > 1$). The transition from instability to neutrality at $L_x = 2\pi L_R$ is referred to as

---

[f]Fraedrich and Frisius (2001) investigated solutions with $l \neq 0$. They found that meridional eddy axes tilt anticlockwise due to vorticity tilting. This induces a southward eddy momentum flux that is absent in our analysis.

[g]We find indeed that $\mathrm{Ro}_d = \mathrm{RoL}/L_R = \mathrm{Ro}/\sqrt{\mathrm{Bu}}$ where $\mathrm{Bu} = (L_R/L)^2$ is the Burger number (cf. Section 4.3).

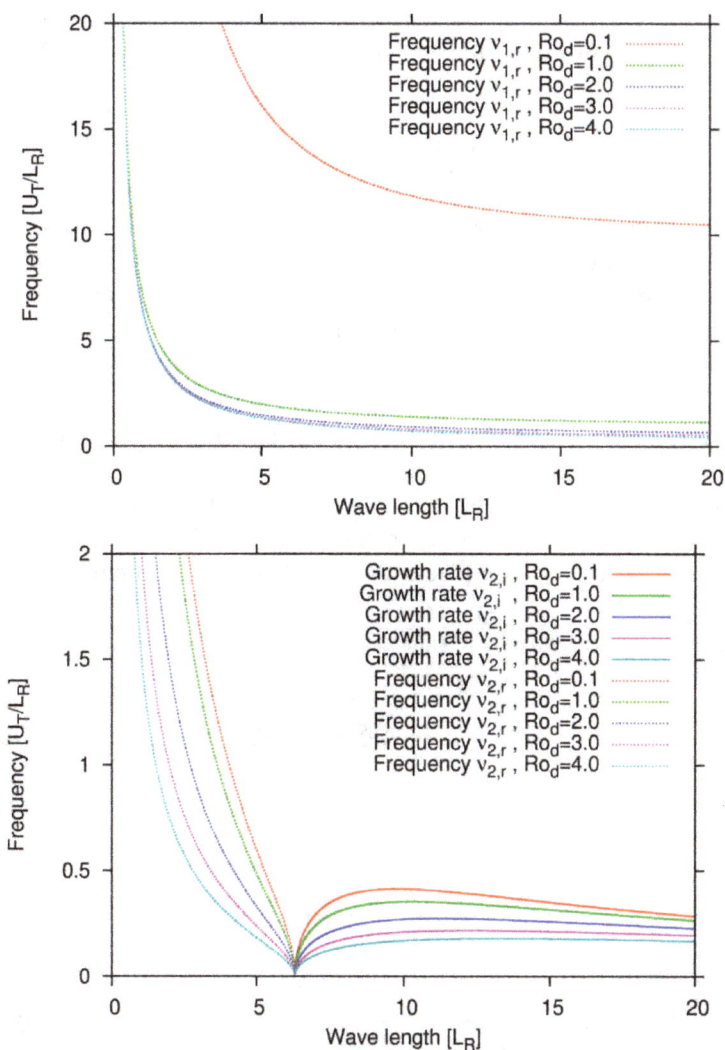

Fig. 4.10. Frequencies $\tilde{\nu}_1$ (upper panel) and $\tilde{\nu}_2$ (lower panel) as a function of wavelength $2\pi/\tilde{k}$ for different Ro_d values. The dashed and solid lines show the real and imaginary parts of the frequency, respectively.

the **short-wave cutoff**. Fastest growth occurs at a certain finite wavelength. It is notable that a similar frequency diagram resulted for barotropic shear instability (cf. Fig. 3.31). However, the instability mechanism of baroclinic and barotropic instability differs substantially as we will see. The dimensionless frequency increases with decreasing Ro_d but approaches — in contrast to the inertia-gravity wave — a finite value. This can be seen by taking the limit $\text{Ro}_d \to 0$:

$$\lim_{\text{Ro}_d \to 0} \tilde{\nu}_{2,4} = \pm \tilde{k} \sqrt{\frac{\tilde{k}^2 - 1}{\tilde{k}^2 + 1}} \tag{4.138}$$

This result is obtained by squaring Eq. (4.137), writing the results as a fraction with Ro_d^2 in the denominator and applying L'Hôpital's rule. This is also the outcome of a corresponding quasigeostrophic two-level model. However, the dimensional frequency becomes zero in this limit for finite $f_0 L_R$ since then $U_T \to 0$. Therefore, the limit (4.138) should rather be interpreted as a suitable approximation for the case $\text{Ro}_d \ll 1$. On the other hand, the inertia-gravity wave frequency has a value which remains finite in the limit $U_T \to 0$. Consequently, the frequency of this wave is much larger than that of the baroclinic wave for $\text{Ro}_d \ll 1$. This is the reason why the inertia-gravity waves are unimportant for synoptic-scale weather development and should be filtered out.

In the case of baroclinic instability ($\tilde{k} < 1$) the wave solution with the growth rate ς becomes

$$\xi'_M = A e^{\varsigma t} \sin(kx) \tag{4.139}$$

$$\xi'_T = \frac{\varsigma}{U_T k} A e^{\varsigma t} \cos(kx) \tag{4.140}$$

$$D'_T = -\frac{U_T^2 k^2 + \varsigma^2}{U_T k f_0} A e^{\varsigma t} \cos(kx) \tag{4.141}$$

$$\xi'_{gT} = \frac{\varsigma}{f_0} \frac{f_0^2 + U_T^2 k^2 + \varsigma^2}{U_T k f_0} A e^{\varsigma t} \cos(kx) \tag{4.142}$$

Figure 4.11 sketches the structure of the growing baroclinic wave. The temperature wave is phase-shifted by 90° with respect to the vorticity wave at level 2 (500 hPa). Anomalous warm and cold air appears east and west of the cyclone, respectively. The divergence D'_T is proportional to the vertical velocity at level 2 and, therefore, upward (downward) motion prevails in warm (cold) air. The fields ξ'_M and ξ'_T together form a vertically westward tilted vorticity wave. Consequently, the upper-level cyclone has a westward phase shift

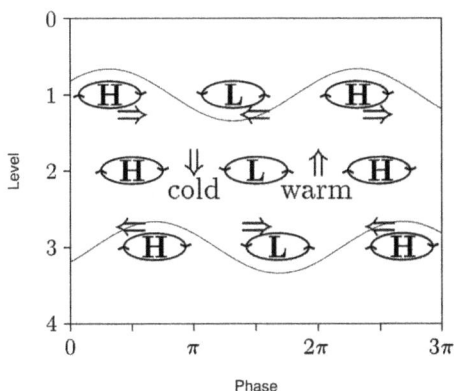

Fig. 4.11. Sketch of the growing baroclinic wave in the two-level model. The symbols L and H indicate the position of the trough and ridge, respectively. The vortex symbols indicate cyclones and anticyclones while the arrows show the divergent circulation in zonal-vertical direction.

with respect to the lower-level cyclone. Finally, the horizontal flow is approximately in geostrophic balance so that lows and highs coincide with the centers of the cyclones and anticyclones, respectively. These features are qualitatively consistent with the fields in the developing cyclone shown in Fig. 4.6.

Figure 4.12 sketches the mechanism of baroclinic instability in this two-level model. The boxes show the four interacting modes of the barotropic vorticity wave ξ'_M, baroclinic vorticity wave ξ'_T, divergence wave D'_T and baroclinic geostrophic vorticity wave ξ'_{gT}. We refer to the latter wave as the temperature wave. These modes combined form a growing baroclinic wave as displayed in Fig. 4.11. The arrows in Fig. 4.12 indicate the various time tendencies which always go from one mode to another mode. The bold arrows having a plus sign characterize the feedback loop of baroclinic instability while the thin arrows with a minus sign represents processes running contrary to this feedback loop. The latter dampen or even terminate the instability. The barotropic vorticity wave induces meridional warm (cold) advection east (west) of the cyclone. Therefore, the temperature wave grows due to this process. On the other hand, zonal vorticity advection by the vertically sheared basic flow reduces the amplitude of the baroclinic vorticity wave so that this process dampens the instability. The enhancement of the temperature wave induces geostrophic imbalance which has the consequence that air flows from

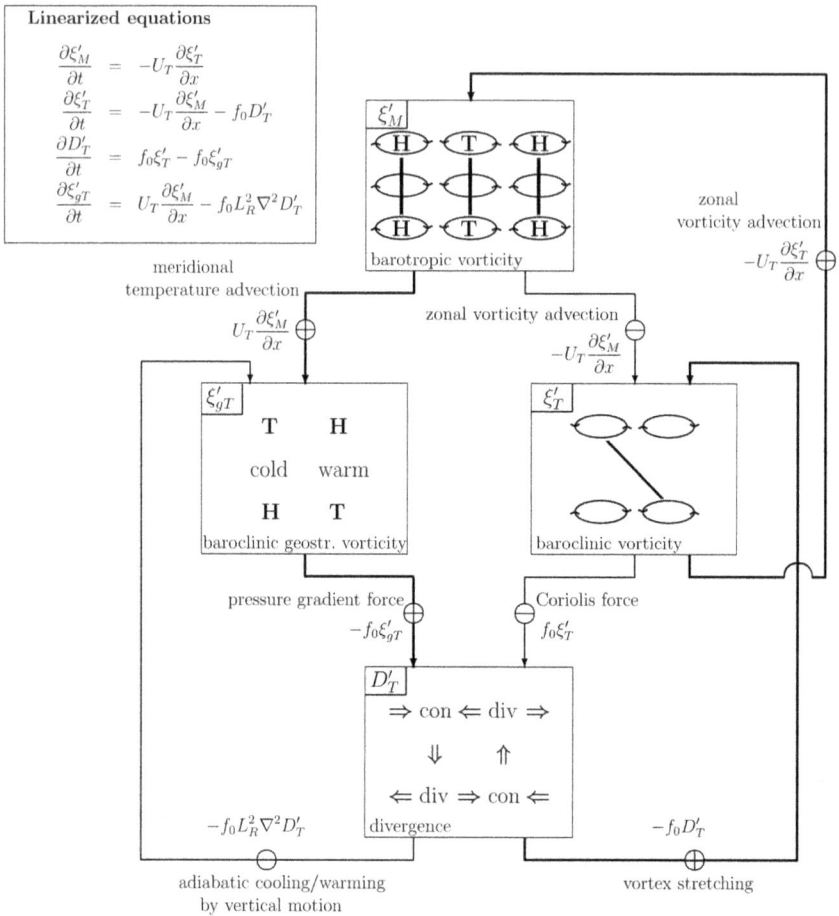

Fig. 4.12.   Sketch of the baroclinic instability mechanism in the two-level model. The growing baroclinic wave is decomposed into four components: barotropic vorticity wave, baroclinic geostrophic vorticity wave, baroclinic vorticity wave and divergence wave. These components (represented as vertical-zonal cross-sections) are linked by the seven tendencies: (I) meridional temperature advection, (II) zonal vorticity advection, (III) pressure gradient force, (IV) Coriolis force, (V) adiabatic cooling/warming by vertical motion, (VI) vortex stretching and (VII) zonal vorticity advection. The arrows indicate tendencies from the originating wave to the wave being affected. The $\oplus$ ($\ominus$) symbol indicates a tendency that reinforces (weakens) the growing baroclinic wave.

the high to the low. This process strengthens the divergence wave. On the other hand, the Coriolis force of the baroclinic vorticity wave leads to a decrease in the divergence wave. The divergence wave weakens the temperature wave by adiabatic cooling (warming) due to upward (downward) vertical motion. On the other hand, vortex stretching amplifies the baroclinic vorticity wave which in turn forces the barotropic vorticity wave by zonal vorticity advection. Note that the barotropic vorticity wave is always in geostrophic balance since the assumption of vanishing vertical velocity at the top and bottom enforces immediate geostrophic adaptation. We can deduce an instability criterion from the schematic figure. In the unstable case, the following three inequalities must be fulfilled:

(i) Meridional temperature advection > Adiabatic cooling of rising air

$$\left| U_T \frac{\partial \xi'_M}{\partial x} \right| > f_0 L_R^2 |\nabla_h^2 D'_T| \qquad (4.143)$$

(ii) Vortex stretching > Zonal vorticity advection

$$f_0 |D'_T| > \left| U_T \frac{\partial \xi'_M}{\partial x} \right| \qquad (4.144)$$

(iii) Pressure gradient force > Coriolis force

$$|\xi'_{gT}| > |\xi'_T| \qquad (4.145)$$

From (4.143) and (4.144) we obtain after inserting the wave ansatz

$$k^2 < L_R^{-2} \qquad (4.146)$$

which reflects the short-wave cutoff. However, this instability criterion does not guarantee that all inequalities are fulfilled. Indeed, these criterions can be violated for a certain time period but the adjustment mechanisms bring the wave to the shown structure. First, let us assume that adiabatic cooling/warming by vertical motion dominates over the meridional temperature advection. Then, the temperature wave weakens and afterwards the divergence wave also decreases in

amplitude due to the weaker pressure gradient. This happens until
the divergent flow is so weak that the inequality (4.143) is fulfilled.
Second, the baroclinic vorticity wave reduces its amplitude in the
case that zonal advection of barotropic vorticity has a larger magnitude than vortex stretching. This has the effect that the divergent
wave amplifies until inequality (4.144) is fulfilled. Finally, the Coriolis force could be larger than the pressure gradient force. Then, the
divergence wave decreases in amplitude which induces a strength-
ening of the temperature wave and a weakening of the baroclinic
vorticity wave. This happens until the inequality (4.145) is fulfilled.
All these restoring processes belong to the geostrophic adjustment
of an unbalanced baroclinic flow that is triggered by the barotropic
wave. The inequality (4.146) leads to a dominance of the temperature wave (mass field) in the adjustment process. Note that we have
found the same result in the case of geostrophic adjustment in the
barotropic shallow water model (see Section 3.6).

4.6.2 *Baroclinic instability in the quasigeostrophic two-level-model*

In this subsection, baroclinic instability is evaluated in a quasigeostrophic two-level model. To derive the equations of this model
we use the same level configuration and vertical discretization as considered in the previous subsection. Therefore, the quasigeostrophic
vorticity equation (4.64) with neglected turbulent exchange is written
at the two levels 1 and 3:

$$\frac{\partial \nabla_h^2 \psi_1}{\partial t} + J_{xy}(\psi_1, \nabla_h^2 \psi_1) + \beta \frac{\partial \psi_1}{\partial x} = f_0 \frac{\omega_2}{\Delta p} \qquad (4.147)$$

$$\frac{\partial \nabla_h^2 \psi_3}{\partial t} + J_{xy}(\psi_3, \nabla_h^2 \psi_3) + \beta \frac{\partial \psi_3}{\partial x} = -f_0 \frac{\omega_2}{\Delta p} \qquad (4.148)$$

where ψ_1 and ψ_3 denote here the geostrophic streamfunction at level
1 and 3, respectively. The quasigeostrophic temperature tendency
equation (4.63) takes the following form in the two-level model

$$\frac{\partial}{\partial t}(\psi_1 - \psi_3) + \frac{1}{2} J_{xy}(\psi_1 + \psi_3, \psi_1 - \psi_3) = \frac{\Delta p \Xi_2}{f_0} \omega_2 \qquad (4.149)$$

Elimination of vertical velocity ω_2 yields the following two potential vorticity equations[h]:

$$\frac{\partial P_{g1}}{\partial t} + J_{xy}(\psi_1, P_{g1}) = 0 \tag{4.150}$$

$$\frac{\partial P_{g3}}{\partial t} + J_{xy}(\psi_3, P_{g3}) = 0 \tag{4.151}$$

where the potential vorticity fields are given by

$$P_{g1} = \nabla_h^2 \psi_1 + \beta y - \frac{1}{2L_R^2}(\psi_1 - \psi_3) \tag{4.152}$$

$$P_{g3} = \nabla_h^2 \psi_3 + \beta y + \frac{1}{2L_R^2}(\psi_1 - \psi_3) \tag{4.153}$$

Therefore, potential vorticity is an individually conserved quantity as in the vertically continuous quasigeostrophic equations. For the stability analysis, we assume a basic flow of the form

$$\bar{\psi}_1 = -U_1 y = -(U_M + U_T)y, \quad \bar{\psi}_3 = -U_3 y = -(U_M - U_T)y \tag{4.154}$$

where $U_T = (U_1 - U_3)/2$ and $U_M = (U_1 + U_3)/2$. We see that the basic flow turns out to be identical to that considered in the previous subsection if $U_M = 0$. The potential vorticity fields of the basic flow become

$$\bar{P}_{g1} = \beta y + \frac{U_T}{L_R^2}y, \quad \bar{P}_{g3} = \beta y - \frac{U_T}{L_R^2}y \tag{4.155}$$

The linearized quasigeostrophic two-level model equations take for this basic flow the form

$$\left(\frac{\partial}{\partial t} + U_1 \frac{\partial}{\partial x}\right)\left(\nabla_h^2 \psi_1' - \frac{\psi_1' - \psi_3'}{2L_R^2}\right) + \left(\beta + \frac{U_T}{L_R^2}\right)\frac{\partial \psi_1'}{\partial x} = 0 \tag{4.156}$$

$$\left(\frac{\partial}{\partial t} + U_3 \frac{\partial}{\partial x}\right)\left(\nabla_h^2 \psi_3' + \frac{\psi_1' - \psi_3'}{2L_R^2}\right) + \left(\beta - \frac{U_T}{L_R^2}\right)\frac{\partial \psi_3'}{\partial x} = 0 \tag{4.157}$$

---

[h]Note that $J_{xy}(\psi_1 + \psi_3, \psi_1 - \psi_3) = J_{xy}(\psi_1, -\psi_3) + J_{xy}(\psi_3, \psi_1) = 2J_{xy}(\psi_3, \psi_1) = 2J_{xy}(\psi_3, \psi_1 - \psi_3) = 2J_{xy}(\psi_1, \psi_1 - \psi_3)$.

Adding and subtracting these equations yield after division by two the barotropic and baroclinic potential vorticity equations:

$$\left(\frac{\partial}{\partial t} + U_M \frac{\partial}{\partial x}\right)\nabla_h^2 \psi_M' + U_T \frac{\partial \nabla_h^2 \psi_T'}{\partial x} + \beta \frac{\partial \psi_M'}{\partial x} = 0 \quad (4.158)$$

$$\left(\frac{\partial}{\partial t} + U_M \frac{\partial}{\partial x}\right)\left(\nabla_h^2 \psi_T' - \frac{\psi_T'}{L_R^2}\right)$$

$$+ U_T \frac{\partial}{\partial x}\left(\nabla_h^2 \psi_M' + \frac{\psi_M'}{L_R^2}\right) + \beta \frac{\partial \psi_T'}{\partial x} = 0 \quad (4.159)$$

Inserting the usual wave ansatz $(\psi_M', \psi_T') = \Re\{(\hat{\psi}_M, \hat{\psi}_T)\exp[i(kx + ly - \nu t)]\}$ leads to the matrix equation[i]

$$\begin{pmatrix} \nu - U_M k + \frac{\beta k}{K^2} & -U_T k \\ U_T k \frac{1 - L_R^2 K^2}{1 + L_R^2 K^2} & \nu - U_M k + \frac{\beta L_R^2 k}{1 + L_R^2 K^2} \end{pmatrix} \cdot \begin{pmatrix} \hat{\psi}_M \\ \hat{\psi}_T \end{pmatrix} = 0 \quad (4.160)$$

By setting the determinant to zero and solving this equation we obtain after some straightforward algebra the frequencies

$$\nu_{1,2} = U_M k - \frac{\beta k(1 + 2L_R^2 K^2) \pm k\sqrt{\beta^2 - 4U_T^2 K^4(1 - L_R^4 K^4)}}{2K^2(1 + L_R^2 K^2)}$$

$$(4.161)$$

We see that the barotropic basic flow U_M just describes a zonal propagation of the wave and has no influence on the stability characteristics. Therefore, neglecting U_M in the previous section was legitimate since it has no relevance for the instability mechanism. On the other hand, the β-effect modifies the stability criterion. The argument of the square root must be negative in the unstable case, i.e.

$$|U_T| > U_{Tc} = \frac{|\beta|}{2K^2\sqrt{1 - L_R^4 K^4}} \quad \text{for} \quad K < 1/L_R \quad (4.162)$$

where U_{Tc} denotes the **critical shear**. Therefore, the vertical shear must exceed a critical value and $K < 1/L_R$. Figure 4.13 displays

---

[i]Note that we have divided the equations by iK^2 and $i(1 + L_R^2 K^2)$, respectively.

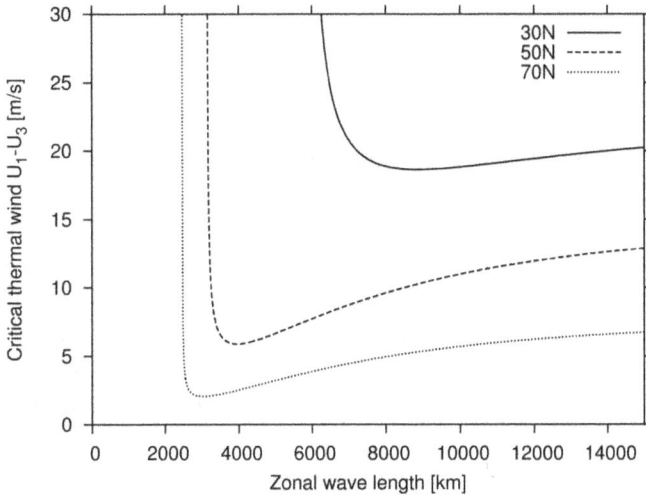

Fig. 4.13. Critical thermal wind $U_1 - U_3$ as a function of zonal wavelength at 30°N (solid line), 50°N (dashed line) and 70°N (dotted line). The basic flow is baroclinically unstable above these lines. For the calculation a static stability parameter $\Xi = 2 \times 10^{-6} \mathrm{m}^2/(\mathrm{Pa}^2\mathrm{s}^2)$ and a meridional wavenumber $l = 1 \times 10^{-6}\mathrm{m}^{-1}$ have been assumed.

the critical thermal wind of the basic flow as a function of zonal wavenumber for realistic values at various latitudes. Obviously, the critical thermal wind has much larger values at low latitudes compared to high latitudes. This happens because the β-effect inhibits baroclinic instability. The latitude has also an effect on the wavelength of the growing baroclinic wave. At low latitude waves are much longer than at high latitudes because the Rossby radius L_R is inversely proportional to the Coriolis parameter. No instability occurs for $K > 1/L_R$ due to the short-wave cutoff. There is also a long-wave cutoff for sufficiently small thermal winds. This additional cutoff solely results from the β-effect. The critical curves in Fig. 4.13 always have a minimum. We can determine the minimum by solving

$$\frac{\partial U_{Tc}^2}{\partial K^4} = \frac{\partial}{\partial K^4}\left[\frac{\beta^2}{4K^4\left(1 - L_R^4 K^4\right)}\right] = 0 \qquad (4.163)$$

Therefore, the minimum critical shear becomes

$$U_{Tcm} = |\beta| L_R^2 \qquad (4.164)$$

at a total wavenumber of

$$K = 1/(\sqrt[4]{2}L_R) \qquad (4.165)$$

From Fig. 4.13, we can deduce that midlatitude baroclinic waves develop in the synoptic scale as specified in Section 2.3. The velocity and length scales found in (4.164) and (4.165) are functions of β, f_0 and $\sqrt{\Xi}\Delta p$ where the latter is proportional to the Brunt–Väisälä frequency. It can be anticipated that the synoptic scale results from these three parameters since baroclinic instability is the dominant development process for synoptic-scale weather systems.

Figure 4.14 shows the growth rate $\varsigma = \nu_i$ at 45°N for different thermal winds $U_1 - U_2$ as a function of dimensionless zonal wavenumber

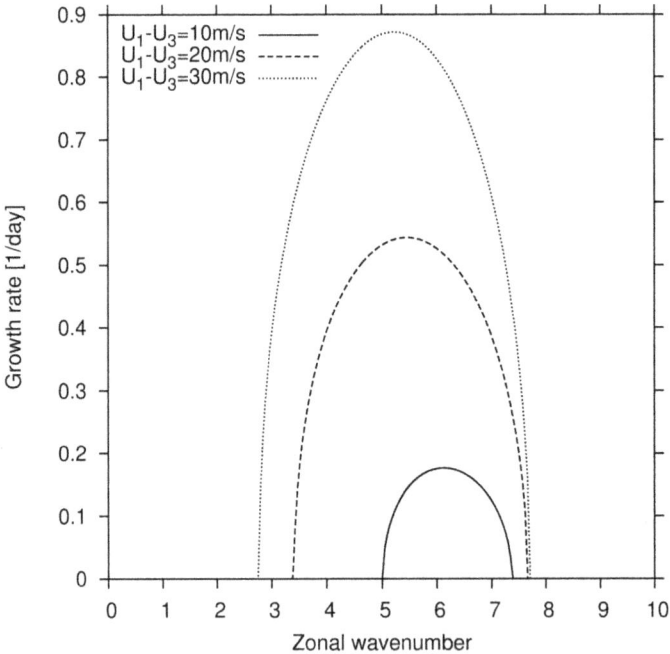

Fig. 4.14. Growth rate of the baroclinic wave at 45°N as a function of nondimensional zonal wavenumber $\tilde{k} = r_E\cos\varphi k$ for different thermal winds $U_1 - U_3$. For the calculation a static stability parameter $\Xi = 2 \times 10^{-6} \mathrm{m^2/(Pa^2s^2)}$ and a meridional wavenumber $l = 2/3k$ have been assumed. With this assumption, the wave eddies have a meridonal extension that is 1.5 times larger than its zonal extension.

$\tilde{k} = r_E \cos\varphi k$ which is an integer because the latitude circle is periodic. We see that instability only appears in a wavenumber range that shrinks with decreasing vertical shear of the basic flow. The maximum growth rate occurs at a certain zonal wavenumber that decreases with increasing U_T. Perturbations having a wavenumber >7 or <3 do not grow for the chosen model parameters.

Figure 4.15 displays a **wavenumber frequency diagram** which shows the **power spectrum** of 500 hPa geostrophic meridional wind at 45°N for winter 2004/2005 as a function of frequency and zonal wavenumber.[j] The power spectrum is proportional to the amplitude squared of the wave having the related wavenumber and frequency. The maxima of the power spectrum reveal weather activity for zonal wavenumbers ranging from 4 to 7. The figure also contains the frequency deduced from Eq. (4.161) for different barotropic basic flows. They show **bifurcations** at high and low wavenumbers which are associated with the short- and long-wave cutoffs, respectively. Obviously, the maxima of the power spectrum occur where the baroclinic basic flow is unstable in the quasigeostrophic model.

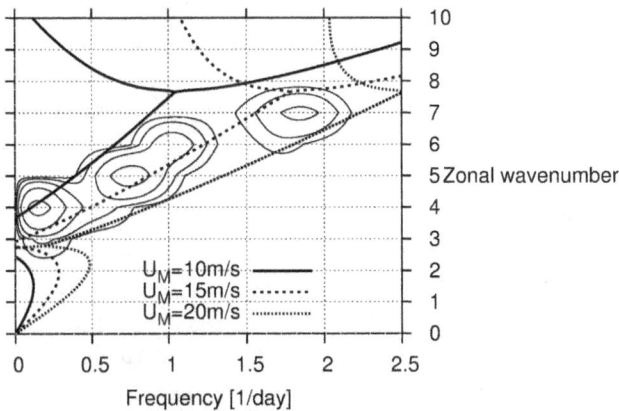

Fig. 4.15. Wavenumber frequency spectrum of 500 hPa geostrophic wind for winter 2004/2005 deduced from NCEP reanalysis data (black isolines). The other lines show the frequency of the baroclinic wave in the quasigeostrophic two-level model for $U_M = 10$ m/s (solid), $U_M = 15$ m/s (dashed) and $U_M = 20$ m/s (short-dashed).

---

[j]For more details of the analysis method see Fraedrich and Boettger (1978).

Furthermore, the frequencies deduced from the model are close to the frequencies with notable observed activity. This result suggests that baroclinic instability causes weather activity in the mid-latitudes. Other processes like barotropic instability, nonlinear interaction or orographic forcing can also be important for triggering extratropical waves but the diagram clearly demonstrates that the important waves are Rossby waves because other atmospheric waves have higher frequencies (see Fig. 4.1).

The best way to determine the structure of the growing wave is to consider the vertically averaged PV equation (4.158). With the wave ansatz inserted we get

$$\left(\nu - U_M k + \frac{\beta k}{K^2} \right) \hat{\psi}_M = U_T k \hat{\psi}_T \tag{4.166}$$

Therefore, the eigensolution of the growing baroclinic wave becomes

$$\psi'_M = A e^{\varsigma t} \sin(ly) \sin(kx - \nu_r t) \tag{4.167}$$

$$\psi'_T = \frac{A}{U_T} e^{\varsigma t} \sin(ly) \left[\frac{\beta}{2(K^2 + L_R^2 K^4)} \sin(kx - \nu_r t) + \frac{\varsigma}{k} \cos(kx - \nu_r t) \right] \tag{4.168}$$

We see that this solution corresponds to the solution (4.139)–(4.140) in the primitive equation two-level model when we set $\beta = U_M = l = 0$. However, the introduction of the β-effect induces a zonal phase shift between the barotropic vorticity and the temperature wave that is smaller than 90°. The decrease of the phase shift weakens the feedback loop explained in Fig. 4.12 so that the growth rate becomes smaller. Furthermore, the β-effect causes a westward migration of the wave relative to the barotropic basic flow U_M. Figure 4.16 shows the fields of a growing baroclinic wave in the quasigeostrophic two-level model. Obviously, the wave has a vertical structure as in the primitive equation two-level model but the superposition of a barotropic basic flow has the effect that the upper level streamfunction pattern has an undular shape without local extrema while at the lower-level isolated highs and lows occur. This is an idealized picture of what is often seen on weather maps, e.g. in Fig. 4.6(a).

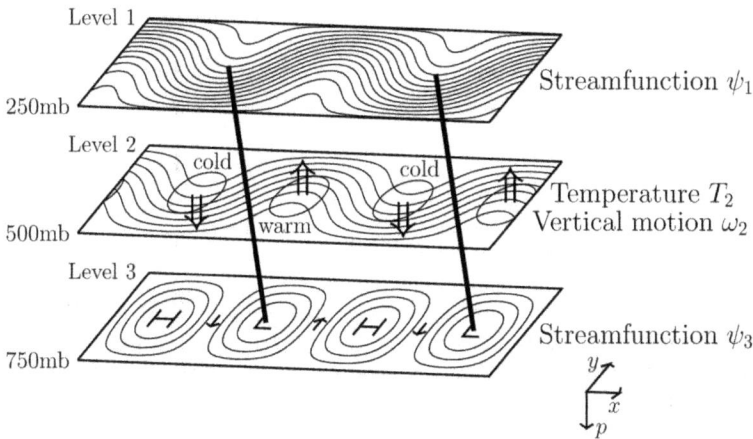

Fig. 4.16. Schematic picture of the growing baroclinic wave in the quasi-geostrophic two-level model.

4.6.3 *Nonmodal growth*

In the previous subsections, we assumed that the growing baroclinic wave takes the form of an eigenvector solution (normal mode) of the linearized equations. However, this is rarely the case in the real atmosphere. Usually, Rossby waves are always present and they sometimes result from other mechanisms than baroclinic instability (e.g. orographic forcing) or they are remainders of previous baroclinic developments. Friction at the surface makes sure that the Rossby waves in the upper part of the troposphere have higher amplitudes. Frequently, baroclinic development of a low-pressure system takes place by an upper-level trough that overtakes a weak surface low which leads to an amplification of both the surface low and the upper-level trough. This kind of baroclinic development is referred to as **type B cyclogenesis** according to Petterssen and Smebye (1971). In contrast, **type A cyclogenesis** refers to the growth of a baroclinic wave at a temperature front. Farrell (1984) suggested that **nonmodal growth** can explain the Type B cyclogenesis. Nonmodal growth results from the superposition of different solutions of the linearized model. These solutions would in our simple quasi-geostrophic two-level model be two eigensolutions. Indeed, perturbation growth can take place even if all eigenvalues have negative real parts. Figure 4.17 sketches this for an example where the eigenvectors

(a) Initial state   (b) After time period t

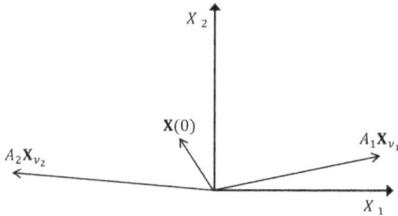

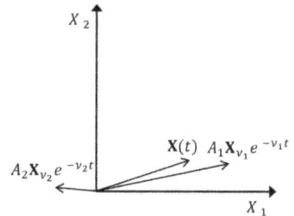

Fig. 4.17. Sketch of nonmodal growth. (a) Initial state $\vec{X}(0)$ resulting from a superposition of two eigenvectors, (b) final state $\vec{X}(t)$ for the case that the decay rate $|\nu_2|$ of the second eigensolution is larger than the decay rate $|\nu_1|$ of the first eigensolution.

form an angle of more than $90°$. The part of the solution $\vec{X}(t)$ that results from the eigenvector $\vec{X}_{\nu_2}$ decays faster with the larger rate ν_2. Therefore, the solution turns into the direction of the other eigenvector $\vec{X}_{\nu_1}$ and, thereby, the length of the vector increases. However, the perturbation decays eventually so that nonmodal growth only happens temporarily in this case. Nonmodal growth is impossible if the eigenvectors form an orthogonal basis.

To determine nonmodal growth, we must search for the perturbation that grows temporarily with the fastest rate. A reasonable measure of perturbation amplitude is the square root of the scalar product

$$\langle \vec{X} | \vec{X} \rangle \equiv \vec{X}^{\mathrm{C}} \cdot \vec{X} \tag{4.169}$$

since it is positive and real. Furthermore, it reduces to the squared vector length if the vector has real components. The perturbation $\vec{X}$ results from the linear differential equation

$$\frac{d\vec{X}}{dt} = \overset{\leftrightarrow}{D} \cdot \vec{X} \tag{4.170}$$

where $\overset{\leftrightarrow}{D}$ is the Jacobi matrix. Consequently, the time tendency of the scalar product $\langle \vec{X} | \vec{X} \rangle$ becomes

$$\frac{d}{dt} \langle \vec{X} | \vec{X} \rangle = \vec{X}^{\mathrm{C}} \cdot \overset{\leftrightarrow}{D} \cdot \vec{X} + \overset{\leftrightarrow}{D}^{\mathrm{C}} \cdot \vec{X}^{\mathrm{C}} \cdot \vec{X}$$

$$= \vec{X}^{\mathrm{C}} \cdot \overset{\leftrightarrow}{D} \cdot \vec{X} + \vec{X}^{\mathrm{C}} \cdot \overset{\leftrightarrow}{D}^{\dagger} \cdot \vec{X} = \vec{X}^{\mathrm{C}} \cdot (\overset{\leftrightarrow}{D} + \overset{\leftrightarrow}{D}^{\dagger}) \cdot \vec{X} \tag{4.171}$$

where $\overset{\leftrightarrow}{D}^\dagger = (\overset{\leftrightarrow}{D}^C)^T$ denotes the **Hermitian conjugate** or **adjoint** of $\overset{\leftrightarrow}{D}$ (the transpose and complex conjugate of $\overset{\leftrightarrow}{D}$). The matrix $\overset{\leftrightarrow}{D}+\overset{\leftrightarrow}{D}^\dagger$ forms a **Hermitian matrix** since $(\overset{\leftrightarrow}{D} + \overset{\leftrightarrow}{D}^\dagger)^\dagger = \overset{\leftrightarrow}{D} + \overset{\leftrightarrow}{D}^\dagger$, i.e. the matrix is equal to its Hermitian conjugate. A Hermitian matrix has real eigenvalues and the eigenvectors are orthogonal. Consequently, maximum instantaneous growth takes place if the initial perturbation forms the eigenvector of $\overset{\leftrightarrow}{D} + \overset{\leftrightarrow}{D}^\dagger$ with the largest real eigenvalue.

To demonstrate the possibility of nonmodal growth we consider baroclinic instability in the quasigeostrophic two-level model. For simplicity, we assume an f-plane and we set $U_M = 0$. Then, the Jacobi-matrix deduced from (4.158) and (4.159) becomes

$$\overset{\leftrightarrow}{D} = \begin{pmatrix} 0 & -ikU_T \\ ikU_T\frac{1-L_R^2K^2}{1+L_R^2K^2} & 0 \end{pmatrix} \tag{4.172}$$

and for the Hermitian matrix $\overset{\leftrightarrow}{D} + \overset{\leftrightarrow}{D}^\dagger$ we obtain

$$\overset{\leftrightarrow}{D} + \overset{\leftrightarrow}{D}^\dagger = \begin{pmatrix} 0 & -\frac{2ikU_T}{1+L_R^2K^2} \\ \frac{2ikU_T}{1+L_R^2K^2} & 0 \end{pmatrix} \tag{4.173}$$

This matrix has the two eigenvalues

$$\varsigma_{I_{1,2}} = \pm\frac{2kU_T}{1 + L_R^2K^2} \tag{4.174}$$

and the associated eigenvectors

$$\vec{X}_{\varsigma_{I_{1,2}}} = (1,\pm\mathrm{i}) \tag{4.175}$$

On the other hand, the normal mode growth rates become in this case

$$\varsigma_{1,2} = \pm U_T k\sqrt{\frac{1 - L_R^2K^2}{1 + L_R^2K^2}} \tag{4.176}$$

and the associated eigenvectors are

$$\vec{X}_{\varsigma_{1,2}} = \left(1,\pm\mathrm{i}\sqrt{\frac{1 - L_R^2K^2}{1 + L_R^2K^2}}\right) = \left(1,\pm\mathrm{i}\frac{\varsigma_1}{U_Tk}\right) \tag{4.177}$$

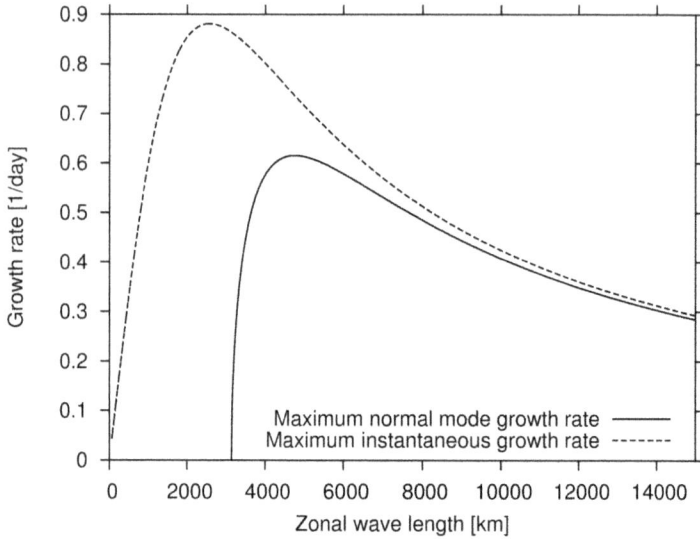

Fig. 4.18.　Maximum normal mode growth rate (solid line) and maximum instantaneous growth rate (dashed line) as a function of zonal wavelength resulting from the baroclinically unstable flow in the quasigeostrophic two-level model on an f-plane. For the calculation the latitude $50°$N, a thermal wind $U_1 - U_3 = 2U_T = 20\,\text{m/s}$, a static stability parameter $\Xi = 2 \times 10^{-6}\text{m}^2/(\text{Pa}^2\text{s}^2)$ and a meridional wavenumber $l = 1 \times 10^{-6}\text{m}^{-1}$ have been assumed.

Figure 4.18 shows the maximum normal mode growth rate ς_1 and the maximum instantaneous growth rate $\varsigma_{I_{1,2}}$ as a function of zonal wavelength for selected parameters. We can deduce that (i) the instantaneous growth rate is larger than the normal mode growth rate, (ii) the maximum instantaneous growth rate appears at a wavelength where the baroclinic flow is not unstable and (iii) the nonmodal perturbation has a larger vertical westward tilt as the normal mode perturbation. We can derive the time evolution of the nonmodal perturbation by projection of the initial state on the eigenvectors. We get for the growing nonmodal perturbation

$$\vec{X}(t) = A \left(\frac{\varsigma_1 + U_T k}{2\varsigma_1} \vec{X}_{\varsigma_1} e^{\varsigma_1 t} + \frac{\varsigma_1 - U_T k}{2\varsigma_1} \vec{X}_{\varsigma_2} e^{-\varsigma_1 t} \right) \tag{4.178}$$

where A is the initial amplitude of the barotropic streamfunction wave. In the case of an imaginary ς_1 it is more convenient to write the solution as

$$\vec{X}(t) = A \left(\frac{\nu_1 - iU_T k}{2\nu_1} \vec{X}_{\varsigma_1} e^{i\nu_1 t} + \frac{\nu_1 + iU_T k}{2\nu_1} \vec{X}_{\varsigma_2} e^{-i\nu_1 t} \right) \quad (4.179)$$

where $\nu_1 = -i\varsigma_1$ denotes the eigenfrequency of the normal modes.

Figure 4.19 shows the time evolution of the amplitude for the nonmodal perturbation with maximum instantaneous growth ($L_x = 2567\,\text{km}$) and close to the short-wave cutoff ($L_x = 3176\,\text{km}$) as well as for the fastest-growing normal mode $L_x = 5286\,\text{km}$. The non-modal perturbation with the fastest instantaneous growth reaches maximum amplitude within 35 h and decays afterwards. In contrast, the longer nonmodal perturbation having a slightly smaller initial growth rate does not exhibit a decline of amplification in the time interval shown. Until 45 h its amplitude is larger as that of the fastest-growing normal mode. Therefore, nonmodal perturbations having wavelengths close to the short-wave cutoff can be relevant for the evolution of real synoptic-scale weather systems since they typically

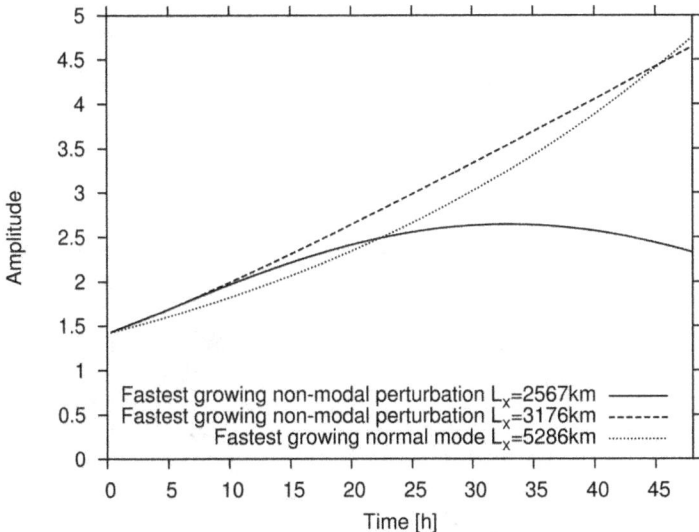

Fig. 4.19. Amplitude $\langle \vec{X} | \vec{X} \rangle^{1/2}$ as a function of time for the fastest-growing non-modal perturbation with the wavelength $L_x = 2567\,\text{km}$ (solid line), the fastest-growing nonmodal perturbation with the wavelength $L_x = 3176\,\text{km}$ (dashed line) and the fastest-growing normal mode with the wavelength $L_x = 5286\,\text{km}$ (dotted line). The model parameters are the same as in Fig. 4.18.

develop within one or two days. However, very short neutral waves
cannot amplify significantly since their frequency is too high.

Figure 4.20 shows the development of upper and lower layer
streamfunction for the nonmodal wave with the wavelength 3176 km.
There is a large phase shift of 90° between the upper and lower layer
at $t = 0\,$h but this phase shift continuously decreases in the course
of time. Meanwhile, the amplitude of the wave has increased to more
than the double of the initial amplitude. In contrast, the normal mode
does not change its structure during amplification and, therefore, the
phase shift remains constant. Observations show that the former is

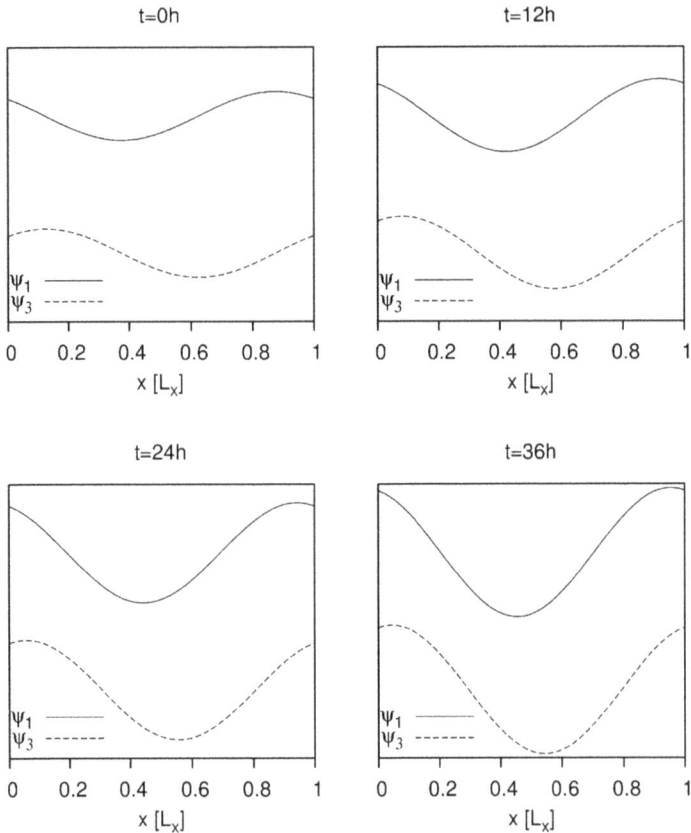

Fig. 4.20. Zonal streamfunction profiles of the nonmodal perturbation with
wavelength $L_x = 3176\,$km (see Fig. 4.19) at different time points. The solid
(dashed) line displays upper (lower) level streamfunction of the quasigeostrophic
two-level model.

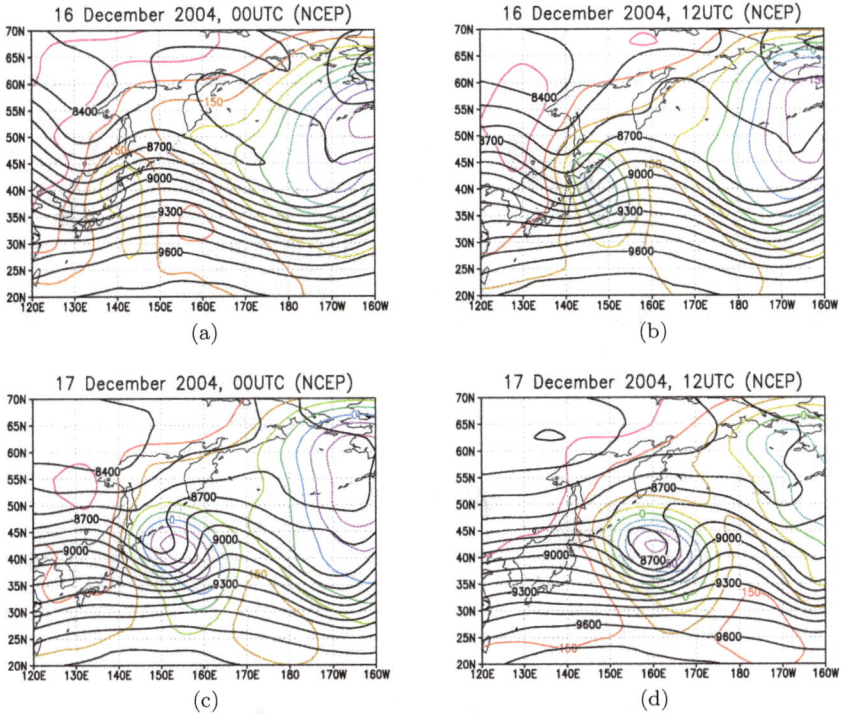

Fig. 4.21. NCEP reanalysis of 1000 hPa geopotential height (colored isolines, contour interval 50 m) and 300 hPa geopotential height (black isolines, contour interval 100 m) on (a) December 16, 2004, 00 UTC, (b) December 16, 2004, 12 UTC, (c) December 17, 2004, 00 UTC and (d) December 17, 2004, 12 UTC.

more typical. Figure 4.21 presents an example of an observed cyclogenesis event. On December 16, 2004, an upper-level trough is located west of Japan while a weak surface low occurs downstream. The system amplifies rapidly and moves eastward. After 36 h the system reaches the **occlusion stage** where the trough axis becomes nearly vertical at the center of the mature low. Although the nonmodal development theory can explain the vertical alignment of the trough axis, it can also result from nonlinear mechanisms to be considered in the next chapter.

The demonstrated nonmodal perturbation analysis has a weakness. In fact, it selects the perturbation with the maximum instantaneous growth rate but this does not mean that the growth lasts very long. For example, in a linearized primitive equation model as considered in Section 4.6.1, such an analysis would yield an inertia-gravity wave because it has temporarily larger time tendencies than those of growing baroclinic waves. However, with the **singular-value decomposition** method, we can circumvent this problem. The method determines modes having the fastest mean growth rate over a chosen time interval. Then, inertia-gravity waves can be filtered out by choosing a sufficiently long time interval. The singular-value decomposition will be illustrated in the following for the simple baroclinic wave example. First, we can write a solution of the linearized equations in the form

$$\vec{X}(t) = A\vec{X}_{\varsigma_1}e^{\varsigma_1 t} + B\vec{X}_{\varsigma_2}e^{-\varsigma_1 t} = \begin{pmatrix} e^{\varsigma_1 t} & e^{-\varsigma_1 t} \\ i\frac{\varsigma_1}{U_T k}e^{\varsigma_1 t} & -i\frac{\varsigma_1}{U_T k}e^{-\varsigma_1 t} \end{pmatrix} \cdot \begin{pmatrix} A \\ B \end{pmatrix}$$

$$= \begin{pmatrix} 1 & 1 \\ i\frac{\varsigma_1}{U_T k} & -i\frac{\varsigma_1}{U_T k} \end{pmatrix} \cdot \begin{pmatrix} e^{\varsigma_1 t} & 0 \\ 0 & e^{-\varsigma_1 t} \end{pmatrix} \cdot \begin{pmatrix} A \\ B \end{pmatrix}$$

$$\equiv \overleftrightarrow{\Lambda} \cdot \overleftrightarrow{I} \cdot \begin{pmatrix} A \\ B \end{pmatrix} \tag{4.180}$$

On the other hand, we can write

$$\begin{pmatrix} A \\ B \end{pmatrix} = \overleftrightarrow{\Lambda}^{-1} \cdot \vec{X}(0) = \begin{pmatrix} \frac{1}{2} & -i\frac{U_T k}{2\varsigma_1} \\ \frac{1}{2} & i\frac{U_T k}{2\varsigma_1} \end{pmatrix} \cdot \vec{X}(0) \tag{4.181}$$

Therefore, we get

$$\vec{X}(t) = \overleftrightarrow{\Lambda} \cdot \overleftrightarrow{I} \cdot \overleftrightarrow{\Lambda}^{-1} \cdot \vec{X}(0) \equiv \overleftrightarrow{P}(t) \cdot \vec{X}(0) \tag{4.182}$$

where $\overleftrightarrow{P}$ constitutes the **propagator matrix** which becomes

$$\overleftrightarrow{P}(t) = \begin{pmatrix} \cosh(\varsigma_1 t) & -i\frac{U_T k}{\varsigma_1}\sinh(\varsigma_1 t) \\ i\frac{\varsigma_1}{U_T k}\sinh(\varsigma_1 t) & \cosh(\varsigma_1 t) \end{pmatrix} \tag{4.183}$$

To measure perturbation growth in the chosen time interval τ we can evaluate the **stretching factor**

$$\sqrt{\frac{\langle \vec{X}(\tau)|\vec{X}(\tau)\rangle}{\langle \vec{X}(0)|\vec{X}(0)\rangle}} = \sqrt{\frac{\vec{X}^C(0) \cdot (\overleftrightarrow{P}^\dagger(\tau) \cdot \overleftrightarrow{P}(\tau)) \cdot \vec{X}(0)}{\vec{X}^C(0) \cdot \vec{X}(0)}} \tag{4.184}$$

The matrix product $\overleftrightarrow{\mathsf{P}}^\dagger(\tau) \cdot \overleftrightarrow{\mathsf{P}}(\tau)$ yields a Hermitian matrix.[k] Therefore, the largest real eigenvalue of this matrix becomes the maximum squared stretching factor and the corresponding eigenvector forms the **optimal perturbation** having the maximum amplitude growth in the considered time interval. We obtain the matrix product

$$\overleftrightarrow{\mathsf{P}}^\dagger(\tau) \cdot \overleftrightarrow{\mathsf{P}}(\tau)$$

$$= \begin{pmatrix} \cosh^2(\varsigma_1\tau) + \frac{\varsigma_1^2}{U_T^2 k^2}\sinh^2(\varsigma_1\tau) & -i\frac{U_T^2 k^2 + \varsigma_1^2}{2\varsigma_1 U_T k}\sinh(2\varsigma_1\tau) \\ i\frac{U_T^2 k^2 + \varsigma_1^2}{2\varsigma_1 U_T k}\sinh(2\varsigma_1\tau) & \cosh^2(\varsigma_1\tau) + \frac{U_T^2 k^2}{\varsigma_1^2}\sinh^2(\varsigma_1\tau) \end{pmatrix}$$

$$(4.185)$$

The real eigenvalues of this matrix become[l]

$$s_{1,2} = \cosh^2(\varsigma_1\tau) + \frac{\varsigma_1^4 + U_T^4 k^4}{2\varsigma_1^2 U_T^2 k^2}\sinh^2(\varsigma_1\tau)$$

$$\pm\sinh^2(\varsigma_1\tau)\sqrt{\left(\frac{\varsigma_1^4 - U_T^4 k^4}{2\varsigma_1^2 U_T^2 k^2}\right)^2 + \left(\frac{\varsigma_1^2 + U_T^2 k^2}{\varsigma_1 U_T k}\right)^2 \coth^2(\varsigma_1\tau)}$$

$$(4.186)$$

A representative growth rate for the optimal perturbation results from the expression

$$\varsigma_0 = \frac{1}{\tau}\ln\left(\sqrt{\frac{\langle \vec{X}(\tau)|\vec{X}(\tau)\rangle}{\langle \vec{X}(0)|\vec{X}(0)\rangle}}\right) = \frac{1}{\tau}\ln(\sqrt{s_1}) \qquad (4.187)$$

In the limit $\tau \to \infty$ this growth rate yields the **Lyapunov exponent** and coincides with the growth rate of the fastest-growing normal mode. In Section 5.5.7 of the next chapter, we will learn more about the Lyapunov exponent. Figure 4.22 shows the optimal growth rate ς_0 for various τ as a function of zonal wavelength. We see that

---

[k]This is true because $(\overleftrightarrow{A}^\dagger \cdot \overleftrightarrow{A})^\dagger = \overleftrightarrow{A}^\dagger \cdot (\overleftrightarrow{A}^\dagger)^\dagger = \overleftrightarrow{A}^\dagger \cdot \overleftrightarrow{A}$.

[l]We assumed a real eigenvalue ς_1. The derivation for an imaginary ς_1 is left to the ambitious reader.

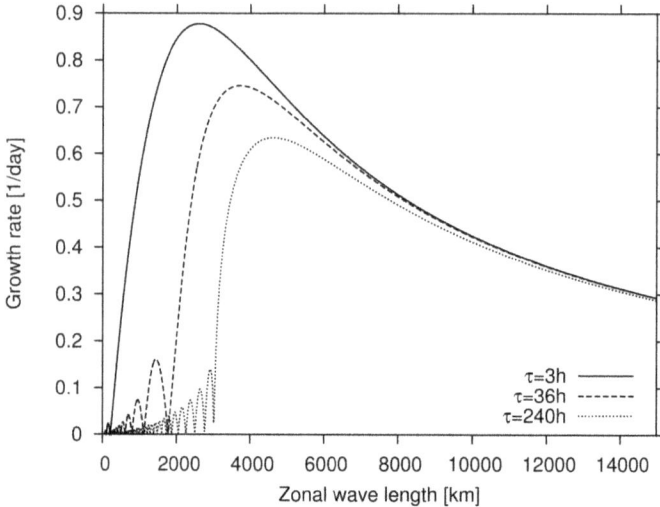

Fig. 4.22.   As in Fig. 4.18 but for the growth rate of the optimal perturbation for $\tau = 3\,\mathrm{h}$ (solid line), $\tau = 36\,\mathrm{h}$ (dashed line) and $\tau = 240\,\mathrm{h}$ (dotted line).

for $\tau = 3\,\mathrm{h}$ and $\tau = 240\,\mathrm{h}$ the optimal growth rate ς_O nearly coincides with ς_{I_1} (fastest instantaneous growth) and ς_1 (fastest normal mode growth), respectively. This is because the limits $\tau \to 0$ and $\tau \to \infty$ approach these values. For the intermediate value $\tau = 36\,\mathrm{h}$ the fastest optimal growth appears close to the wavelength where the short-wave cutoff takes place. This is in agreement with Fig. 4.19 in which we found that such a nonmodal perturbation yields notable amplitude growth within 36 h. Consequently, the singular value decomposition represents a suitable method to determine the perturbations with relevant amplification in a certain time period. The oscillation in the graph results from the periodic solution that occurs below the short-wave cutoff. No amplitude growth can take place in the case that the time period τ is a multiple of the oscillation period $2\pi/\nu_1$. This circumstance applies at discrete zonal wavelengths where we see a vanishing optimal growth rate.

4.6.4   *Baroclinic instability in a uniform potential vorticity flow*

It is complicated to solve the baroclinic instability problem analytically without the two-level approximation. Charney (1947) derived a quasigeostrophic solution for a vertically sheared westerly current on the β-plane. Later, Eady (1949) solved this problem for the f-plane geometry in a channel with an upper lid. Eady's solution was much simpler because he utilized the simplifying properties of a horizontally **uniform potential vorticity flow**. Here, we also consider such a basic flow. This means that the potential vorticity must fulfill

$$\nabla_h P_g = 0 \tag{4.188}$$

This property remains conserved when friction and heating are neglected. Assuming this for a zonally constant basic flow with vertical but no horizontal shear leads to

$$\frac{\partial P_g}{\partial y} = \beta - \frac{\partial}{\partial p}\left(\frac{f_0^2}{\Xi}\frac{\partial \bar{u}_g}{\partial p}\right) = 0, \tag{4.189}$$

We assume in addition that f_0^2/Ξ is constant. Then we get

$$\bar{u}_g = U_M + \Lambda(p_m - p) + \frac{\beta \Xi}{2 f_0^2}(p_m - p)^2, \tag{4.190}$$

where p_m is the pressure at mid-level between the upper and lower boundaries. We see in Fig. 4.23 that the vertical wind profile has a

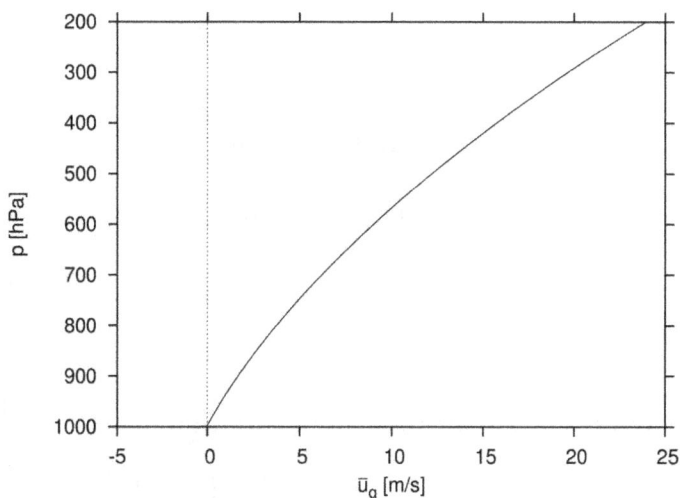

Fig. 4.23. Vertically sheared zonal flow having zero potential vorticity for $\Xi = 2 \times 10^{-6} \text{m}^2/(\text{Pa}^2\text{s}^2)$ at $50°\text{N}$.

convex curvature which balances the meridional planetary vorticity gradient β. This part is associated with an increase (decrease) of the meridional temperature gradient at low (high) levels.

The perturbation must also fulfill (4.188). We can set the perturbation potential vorticity to zero since it cannot grow due to PV conservation. Therefore, we have

$$P'_g = \nabla_h^2 \psi'_g + \frac{f_0^2}{\Xi} \frac{\partial^2 \psi'_g}{\partial p^2} = 0 \qquad (4.191)$$

Inserting the wave ansatz $\psi'_g = \Re\{\hat{\psi}_g(p)\exp[\mathrm{i}(kx + ly - \nu t)]\}$ yields

$$-K^2 \hat{\psi}_g + \frac{f_0^2}{\Xi} \frac{d^2 \hat{\psi}_g}{dp^2} = 0 \qquad (4.192)$$

The general solution of this equation can be written as

$$\hat{\psi}_g = \hat{\psi}_M \frac{\cosh(\Gamma p')}{\cosh(\Gamma \Delta p)} + \hat{\psi}_T \frac{\sinh(\Gamma p')}{\sinh(\Gamma \Delta p)} \qquad (4.193)$$

where $\Gamma = K\sqrt{\Xi}/f_0 = \sqrt{2}KL_R/\Delta p$ and $p' = p - p_m$. The potential vorticity has no use for predicting the development when it is uniformly distributed. Instead, the time-dependent boundary condition is essential for this purpose. In Eady's model, upper and lower boundaries exist where the vertical velocity ω vanishes. The lower boundary at pressure p_b represents the surface while the upper boundary at pressure p_t can be interpreted as the tropopause where the static stability changes suddenly and air parcels experience large restoring forces. At these boundaries, the temperature tendency equation (4.69) yields the boundary conditions which become after linearization:

$$\left[\frac{\partial}{\partial t} + (U_M - \Lambda \Delta p + \beta L_R^2)\frac{\partial}{\partial x}\right]\left(\frac{\partial \psi'_g}{\partial p'}\right) + \left(\Lambda - 2\frac{\beta L_R^2}{\Delta p}\right)\frac{\partial \psi'_g}{\partial x} = 0$$

$$\text{at } p = p_b \qquad (4.194)$$

$$\left[\frac{\partial}{\partial t} + (U_M + \Lambda \Delta p + \beta L_R^2)\frac{\partial}{\partial x}\right]\left(\frac{\partial \psi'_g}{\partial p'}\right) + \left(\Lambda + 2\frac{\beta L_R^2}{\Delta p}\right)\frac{\partial \psi'_g}{\partial x} = 0$$

$$\text{at } p = p_t \qquad (4.195)$$

where $\Delta p = (p_b - p_t)/2$. Inserting solution (4.193) yields

$$-i\Gamma[\nu' + 2U_T k][\hat{\psi}_M \tanh(\Gamma\Delta p) + \hat{\psi}_T \coth(\Gamma\Delta p)]$$

$$+ik\frac{2(U_T - \beta L_R^2)}{\Delta p}[\hat{\psi}_M + \hat{\psi}_T] = 0 \qquad (4.196)$$

$$-i\Gamma[\nu' - 2U_T k][-\hat{\psi}_M \tanh(\Gamma\Delta p) + \hat{\psi}_T \coth(\Gamma\Delta p)]$$

$$+ik\frac{2(U_T + \beta L_R^2)}{\Delta p}[\hat{\psi}_M - \hat{\psi}_T] = 0 \qquad (4.197)$$

where $\nu' = \nu - U_M k - \beta L_R^2 k$ and $U_T = \Lambda\Delta p/2$. Subtracting and adding these equations leads to the matrix equation

$$\begin{pmatrix} \frac{\nu'}{k}\tanh(\Gamma\Delta p) + 2\frac{\beta L_R^2}{\Gamma\Delta p} & 2U_T\left[\coth(\Gamma\Delta p) - \frac{1}{\Gamma\Delta p}\right] \\ 2U_T\left[\tanh(\Gamma\Delta p) - \frac{1}{\Gamma\Delta p}\right] & \frac{\nu'}{k}\coth(\Gamma\Delta p) + 2\frac{\beta L_R^2}{\Gamma\Delta p} \end{pmatrix} \cdot \begin{pmatrix} \hat{\psi}_M \\ \hat{\psi}_T \end{pmatrix} = 0$$

$$(4.198)$$

By setting the determinant of the matrix to zero we obtain the frequencies

$$\nu_{1,2} = U_M k - \frac{\beta k L_R^2}{\Gamma\Delta p}[\tanh(\Gamma\Delta p) + \coth(\Gamma\Delta p) - \Gamma\Delta p]$$

$$\pm k\left\{\frac{\beta^2 L_R^4}{\Gamma^2\Delta p^2}[\tanh(\Gamma\Delta p) - \coth(\Gamma\Delta p)]^2\right.$$

$$\left. + 4U_T^2\left[\tanh(\Gamma\Delta p) - \frac{1}{\Gamma\Delta p}\right]\left[\coth(\Gamma\Delta p) - \frac{1}{\Gamma\Delta p}\right]\right\}^{\frac{1}{2}}$$

$$(4.199)$$

It is worthwhile to compare this result with that of the quasi-geostrophic two-level model. In both cases, we detect baroclinic instability for a sufficiently large vertical shear flow U_T and westward propagation for vanishing midlevel flow U_M. More similarities and differences can be seen in Fig. 4.24 which shows the growth rates and frequencies in both models as a function of zonal wavelength. We see that the eigenvalues depend in a similar way on the wavelength as

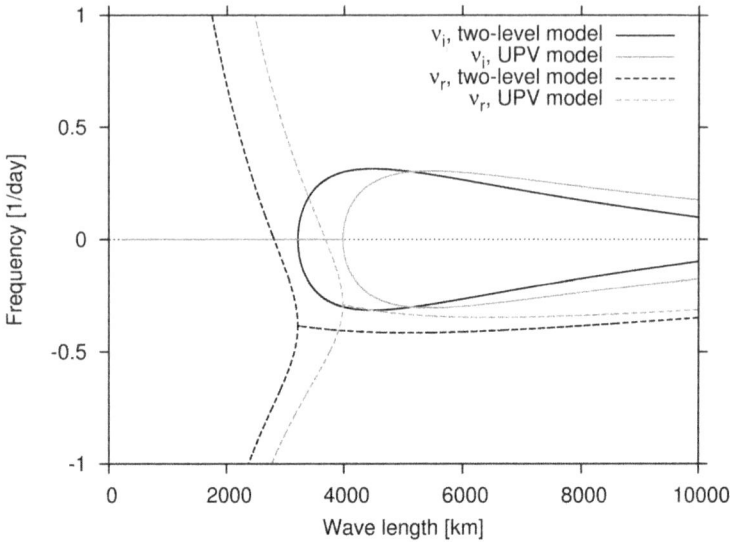

Fig. 4.24.   Real (solid) and imaginary (dashed) parts of the frequency solutions as a function of zonal wavelength for baroclinic instability in the quasigeostrophic two-level model (solid lines) and in the vertically continuous model with uniform potential vorticity (grey lines). The model parameters are the same as assumed for the analysis in Section 4.6.3.

in the two-level model. As differences, we note a longer short-wave cutoff wavelength and a slower westward propagation due to the β-effect because the assumption of uniform potential vorticity requires an additional westerly basic flow to compensate the β-effect. In summary, we can conclude that the crude two-level approximation yields a reasonable qualitative description of baroclinic instability in the vertically sheared basic flow having uniform potential vorticity. A better quantitative agreement could possibly arise if different values for Ξ and U_T were selected for the two-level model.

Figure 4.25 shows the structure of the fastest-growing baroclinic wave perturbation. The wave has as in the two-level model a vertical tilt of the troughs and ridges. Therefore, the trough at the upper boundary appears west of the surface low. Furthermore, the troughs and ridges at the upper boundary have larger amplitudes compared to those at the surface because of the β-effect. This is also the case in the two-level model (cf. Eq. (4.167)–(4.168)). The temperature wave has also a vertical tilt of the warm and cold anomalies but it is in the

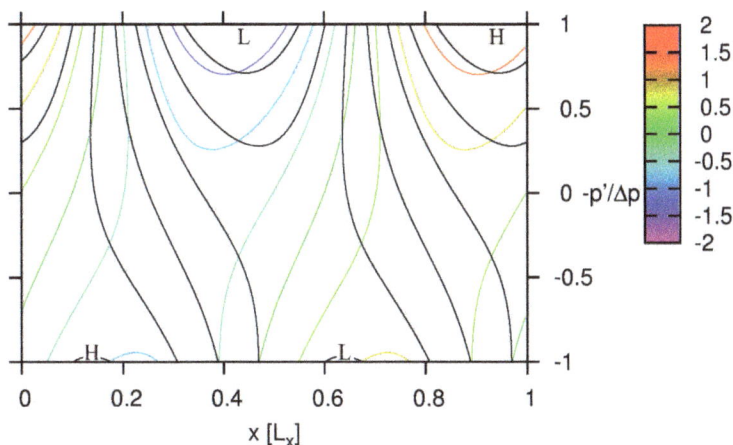

Fig. 4.25. Structure of the fastest growing baroclinic wave in the quasi-geostrophic uniform potential vorticity model. The figure shows anomalies of geopotential (black isolines) and specific volume (colored isolines) in a zonal cross-section. The letters H and L symbolize the local maxima and minima of the geopotential perturbation, respectively.

opposite direction. The consequence is that the warm troughs and cold ridges occur at the surface while at the upper boundary, the troughs and ridges are cold and warm, respectively. The two-level model cannot describe this feature because it has only one level for temperature.

Figure 4.26 shows a zonal cross-section through the extratropical weather system displayed in Fig. 4.21. The observed wave shares some similarities with the analytical result presented in Fig. 4.25. The upper-level trough has a somewhat larger amplitude compared to the surface low and the temperature wave has an eastward tilt with increasing height so that a warm surface low and a cold upper-level trough occur. However, the linearized quasigeostrophic model cannot explain the observed fact that cyclones have larger amplitudes than anticyclones. This is even not possible in linearized models which are not based on the quasigeostrophic approximation and have a more realistic basic flow in the form of a jet. The observed cyclone-anticyclone asymmetry can result from nonlinear dynamics or condensational heating.

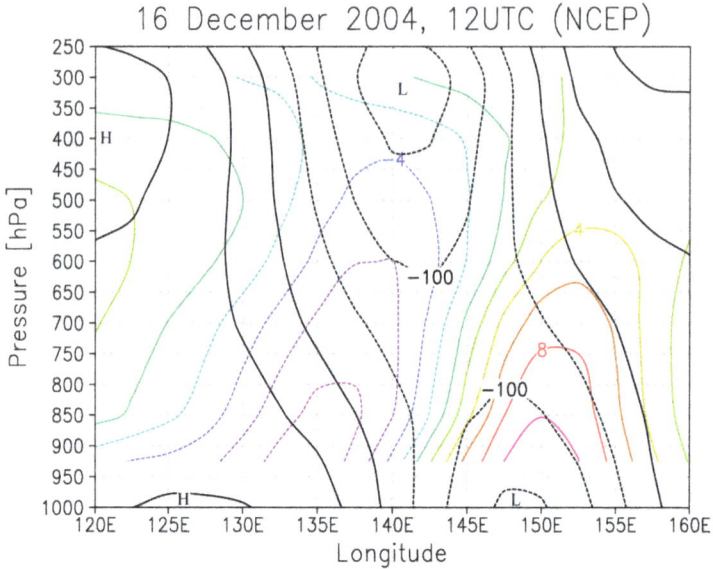

Fig. 4.26.   NCEP reanalysis of the temperature anomaly (colored isolines, contour interval 2 K) and geopotential height (black isolines, contour interval 50 m) in a zonal cross-section at 40 N on December 16, 2004, 12 UTC. The letters H and L symbolize the local maxima and minima of the geopotential perturbation, respectively.

4.6.5   *Criteria for baroclinic stability*

The baroclinic basic flows considered so far have a very simple shape so that analytical solutions can be found but real atmospheric baroclinic flows have a complicated structure as, e.g. seen in the cross-section of Fig. 2.5(b). First, the flow forms a jet with a distinct maximum and second, the static stability is not constant. Some features of baroclinic instability in a meridionally varying flow had been analyzed by Nakamura (1993) in a quasigeostrophic two-level model. However, this analysis is too extensive to include in this book. Instead, we will derive stability criteria for more general baroclinic zonal flows. These can be understood as a generalization of Kuo's criterion for barotropic shear instability (see Section 3.11.3) and have been found by Charney and Stern (1962). We use as the only idealization that the geostrophic basic flow $\bar{u}_g$ is zonally uniform. Then, the linearized quasigeostrophic PV equation for an inviscid and adiabatic atmosphere becomes

$$\left(\frac{\partial}{\partial t} + \bar{u}_g \frac{\partial}{\partial x}\right)\left[\nabla_h^2 \psi_g' + \frac{\partial}{\partial p}\left(\frac{f_0^2}{\Xi}\frac{\partial \psi_g'}{\partial p}\right)\right] + \frac{\partial \bar{P}_g}{\partial y}\frac{\partial \psi_g'}{\partial x} = 0 \quad (4.200)$$

Furthermore, we assume, as in the previous subsection, horizontal boundaries at $p = p_b$ and $p = p_t$. At these boundaries, the following linear temperature tendency equations must be fulfilled:

$$\left(\frac{\partial}{\partial t} + \bar{u}_g \frac{\partial}{\partial x}\right)\left(\frac{\partial \psi_g'}{\partial p}\right) - \frac{\partial \bar{u}_g}{\partial p}\frac{\partial \psi_g'}{\partial x} = 0 \text{ at } p = p_b, p_t \quad (4.201)$$

By inserting the usual wave ansatz $\psi_g' = \Re\{\hat{\psi}_g(y,p)\exp[ik(x - ct)]\}$ in (4.200) and (4.201) we obtain

$$(\bar{u}_g - c)\left[\frac{\partial^2 \hat{\psi}_g}{\partial y^2} + \frac{\partial}{\partial p}\left(\frac{f_0^2}{\Xi}\frac{\partial \hat{\psi}_g}{\partial p}\right) - k^2 \hat{\psi}_g\right] + \frac{\partial \bar{P}_g}{\partial y}\hat{\psi}_g = 0 \quad (4.202)$$

$$(\bar{u}_g - c)\frac{\partial \hat{\psi}_g}{\partial p} - \frac{\partial \bar{u}_g}{\partial p}\hat{\psi}_g = 0 \text{ at } p = p_b, p_t \quad (4.203)$$

After multiplying the PV equation with the complex conjugate $\hat{\psi}_g^C \equiv \hat{\psi}_{g_r} - i\hat{\psi}_{g_i}$ and dividing by $(\bar{u}_g - c)$ we get

$$\hat{\psi}_g^C\left[\frac{\partial^2 \hat{\psi}_g}{\partial y^2} + \frac{\partial}{\partial p}\left(\frac{f_0^2}{\Xi}\frac{\partial \hat{\psi}_g}{\partial p}\right)\right] - k^2|\hat{\psi}_g|^2 + \frac{1}{\bar{u}_g - c}\frac{\partial \bar{P}_g}{\partial y}|\hat{\psi}_g|^2 = 0$$

$$(4.204)$$

Integration over the meridional extension $(y_s < y < y_n)$ of the perturbation (i.e. $\hat{\psi}_g(y_s) = \hat{\psi}_g(y_n) = 0$) and the vertical depth leads to

$$\int_{y_s}^{y_n}\int_{p_t}^{p_p}\left|\frac{\partial \hat{\psi}_g}{\partial y}\right|^2 + \frac{f_0^2}{\Xi}\left|\frac{\partial \hat{\psi}_g}{\partial p}\right|^2 + k^2|\hat{\psi}_g|^2 dp\,dy$$

$$-\int_{y_s}^{y_n}\left[\frac{f_0^2}{\Xi}\hat{\psi}_g^C\frac{\partial \hat{\psi}_g}{\partial p}\right]_{p_t}^{p_p} dy - \int_{y_s}^{y_n}\int_{p_t}^{p_p}\frac{\partial \bar{P}_g}{\partial y}\frac{|\hat{\psi}_g|^2}{\bar{u}_g - c}dp\,dy = 0$$

$$(4.205)$$

With the boundary conditions (4.203) we find that

$$\int_{y_s}^{y_n} \int_{p_t}^{p_p} \left|\frac{\partial \hat{\psi}_g}{\partial y}\right|^2 + \frac{f_0^2}{\Xi} \left|\frac{\partial \hat{\psi}_g}{\partial p}\right|^2 + k^2 |\hat{\psi}_g|^2 dp\, dy$$

$$- \int_{y_s}^{y_n} \left[\frac{\partial \bar{u}_g}{\partial p} \frac{f_0^2 |\hat{\psi}_g|^2}{\Xi(\bar{u}_g - c)}\right]_{p_t}^{p_p} dy - \int_{y_s}^{y_n} \int_{p_t}^{p_p} \frac{\partial \bar{P}_g}{\partial y} \frac{|\hat{\psi}_g|^2}{\bar{u}_g - c} dp\, dy = 0$$

$$(4.206)$$

The imaginary part of this equation becomes

$$-c_i \left\{ \int_{y_s}^{y_n} \left[\frac{\partial \bar{u}_g}{\partial p} \frac{f_0^2 |\hat{\psi}_g|^2}{\Xi|\bar{u}_g - c|^2}\right]_{p_t}^{p_p} dy + \int_{y_s}^{y_n} \int_{p_t}^{p_p} \frac{\partial \bar{P}_g}{\partial y} \frac{|\hat{\psi}_g|^2}{|\bar{u}_g - c|^2} dp\, dy \right\} = 0$$

$$(4.207)$$

From this equation, we can deduce the necessary but not sufficient criteria for instability. Instability requires a nonzero imaginary part of the phase velocity, c_i. In this case, the expression in the curly bracket must vanish. Obviously, only the factors $\partial \bar{u}_g/\partial p$ and $\partial \bar{P}_g/\partial y$ can become negative. First, let us assume that the vertical shear $\partial \bar{u}_g/\partial p$ (meridional temperature gradient) vanishes at the upper and lower boundary. Then, only **internal baroclinic instability** or barotropic shear instability is possible. The latter case happens if the meridional temperature gradient is also zero in the interior of the atmospheric layer. This has the consequence that the PV gradient $\partial \bar{P}_g/\partial y$ reduces to the absolute vorticity gradient $\partial \bar{\zeta}/\partial y$ that must change sign within the considered domain. Consequently, the criteria by Charney and Stern (1962) include the Kuo criterion for barotropic shear instability which we derived in Section 3.11.3. On the other hand, meridional temperature gradients are necessary for internal baroclinic instability. Such instability may occur in the stratosphere where the mid-winter **polar night jet** appears and might break down due to this instability (see Charney and Stern, 1962). However, baroclinic instability in the troposphere usually requires the existence of meridional temperature gradients at the bounding surface. In most cases, the meridional PV gradient is positive as in the jet flow shown in Fig. 2.5(a). Then, $\partial \bar{u}_g / \partial p$ must be negative (positive) at the lower (upper) boundary of the atmospheric layer to compensate for the second integral in (4.207). Charney (1947) investigated a vertically sheared flow in which the PV gradient $\partial \bar{P}_g / \partial y$ reduces to the planetary vorticity gradient β. Baroclinic instability can appear in this case because $\partial \bar{u}_g / \partial p$ is negative at the lower boundary. With a positive $\partial \bar{u}_g / \partial p$ (temperature increase in meridional direction) no boundary baroclinic instability can occur. On the other hand, Eady (1949) considered the same basic flow but for $\beta = 0$. Then, a negative (or positive) meridional temperature gradient must exist at both the lower and upper boundaries. The instability criteria by Charney and Stern are quite unspecific. Therefore, a detailed stability analysis must be conducted to see if a zonal flow is really unstable or not. For realistic jet flows, it is inevitable to do this by solving the model equations with an approximate numerical procedure. In Chapter 5, an example of a numerical model simulation will be presented.

4.7 Atmospheric Energy Cycle

In Section 2.5, we have derived the energy conversions between the various energy forms on the basis of the primitive equations. In this section, we introduce a division in mean and eddy part of the energy and derive the associated energetics. This was first done by Lorenz (1955) in the same article in which he introduced the available potential energy (see Section 2.6). First, we describe the mathematical method to divide into mean and eddy part and then, we derive the **Lorenz energy cycle**. Finally, a simplified energy cycle based on the quasigeostrophic equations is considered.

4.7.1 *Partition in zonal average and its deviation*

When we talk about eddy in this context, we mean a synoptic-scale weather system like a growing baroclinic wave. To separate such a wave from the mean flow it is constructive to subtract the **zonal mean** of the considered field. However, one should keep in mind that parts of a weather system contribute to the zonal mean. For example, a circular cyclone has a nonvanishing zonal mean. The zonal mean of a field b is defined by

$$[b]_\lambda \equiv \frac{1}{2\pi} \int_0^{2\pi} b\, d\lambda \tag{4.208}$$

We denote with **eddy** the deviation from the zonal average

$$b^* \equiv b - [b]_\lambda \tag{4.209}$$

From these definitions, we can derive the following relations in isobaric coordinates:

$$[b^*]_\lambda = 0 \tag{4.210}$$

$$[ab]_\lambda = [a]_\lambda [b]_\lambda + [a^* b^*]_\lambda \tag{4.211}$$

$$\left[\frac{\partial a}{\partial \lambda} \right]_\lambda = 0 \tag{4.212}$$

$$\left[\frac{\partial a}{\partial t} \right]_\lambda = \frac{\partial [a]_\lambda}{\partial t}, \quad \left[\frac{\partial a}{\partial \varphi} \right]_\lambda = \frac{\partial [a]_\lambda}{\partial \varphi}, \quad \left[\frac{\partial a}{\partial p} \right]_\lambda = \frac{\partial [a]_\lambda}{\partial p} \tag{4.213}$$

There is, however, a problem with zonal averaging when the isobaric surface intersects the ground. In this case, one must define how to treat the points below the Earth surface. For this purpose, we assume that the ground is part of the fluid but has quasi-infinite viscosity. Then, the velocity components become zero. To obtain continuous temperature fields we must assume in addition that some heat conduction takes place in the ground. Alternatively, continuous temperature fields can be obtained by assuming a certain vertical lapse rate below the surface and pressure is determined by solving the hydrostatic balance equation. Anyhow, such definitions are somewhat artificial but can be made when its impact is not very large on the overall analysis.

Isobaric coordinates have the advantage of a very simple continuity equation. This can be utilized for deriving zonally averaged equations. We obtain after zonal averaging of Eq. (2.72)

$$\frac{1}{r_E \cos \varphi} \frac{\partial}{\partial \varphi} (\cos \varphi [v]_\lambda) + \frac{\partial [\omega]_\lambda}{\partial p} = 0 \tag{4.214}$$

By subtracting this equation from (2.72) we also find that

$$\frac{1}{r_E \cos \varphi} \frac{\partial u^*}{\partial \lambda} + \frac{1}{r_E \cos \varphi} \frac{\partial}{\partial \varphi}(\cos \varphi v^*) + \frac{\partial w^*}{\partial p} = 0 \qquad (4.215)$$

We consider the following local budget equation in flux form:

$$\frac{\partial b}{\partial t} + \frac{1}{r_E \cos \varphi} \frac{\partial}{\partial \lambda}(ub) + \frac{1}{r_E \cos \varphi} \frac{\partial}{\partial \varphi}(\cos \varphi vb) + \frac{\partial}{\partial p}(wb) = \dot{b}$$

$$(4.216)$$

where $\dot{b}$ also includes the convergence of the microturbulent fluxes. Applying the zonal averaging operator gives

$$\frac{\partial [b]_\lambda}{\partial t} + \frac{1}{r_E \cos \varphi} \frac{\partial}{\partial \varphi}(\cos \varphi [v]_\lambda [b]_\lambda) + \frac{\partial}{\partial p}([\omega]_\lambda [b]_\lambda)$$

$$= [\dot{b}]_\lambda - \frac{1}{r_E \cos \varphi} \frac{\partial}{\partial \varphi}(\cos \varphi [v^* b^*]_\lambda) - \frac{\partial}{\partial p}[\omega^* b^*]_\lambda \quad (4.217)$$

Obviously, the divergence of the zonal advective flux does not contribute to the time development of the zonal mean. The other advective flux components can be split into **mean** and **eddy fluxes**. Baroclinic waves contribute to eddy fluxes while the mean fluxes result from the **meridional overturning circulation**. The latter appears in the atmosphere in the form of the **Hadley, Ferrel** and **polar cells** (cf. Fig. 2).

4.7.2 *Lorenz energy cycle*

The Lorenz energy cycle results from the global budget equations for kinetic and available potential energy, each of which is divided into mean and eddy parts. To derive these, we start with zonal averaging of the momentum equation (2.70) which yields the components

$$\frac{\partial [u]_\lambda}{\partial t} + \frac{[v]_\lambda}{r_E} \frac{\partial [u]_\lambda}{\partial \varphi} + [\omega]_\lambda \frac{\partial [u]_\lambda}{\partial p} - \left(f + \tan \varphi \frac{[u]_\lambda}{r_E}\right)[v]_\lambda$$

$$= [f_{T\lambda}]_\lambda + \tan \varphi \frac{[v^* u^*]_\lambda}{r_E} - \frac{1}{r_E \cos \varphi} \frac{\partial}{\partial \varphi}(\cos \varphi [v^* u^*]_\lambda) - \frac{\partial}{\partial p}[\omega^* u^*]_\lambda$$

$$(4.218)$$

$$\frac{\partial [v]_\lambda}{\partial t} + \frac{[v]_\lambda}{r_E}\frac{\partial [v]_\lambda}{\partial \varphi} + [\omega]_\lambda\frac{\partial [v]_\lambda}{\partial p} + \left(f + \tan\varphi\frac{[u]_\lambda}{r_E}\right)[u]_\lambda$$

$$= -\frac{1}{r_E}\frac{\partial [\phi]_\lambda}{\partial \varphi} + [f_{T\varphi}]_\lambda - \tan\varphi\frac{[u^{*2}]_\lambda}{r_E} - \frac{1}{r_E\cos\varphi}\frac{\partial}{\partial \varphi}(\cos\varphi[v^{*2}]_\lambda)$$

$$- \frac{\partial}{\partial p}[\omega^* u^*]_\lambda \qquad (4.219)$$

Multiplying Eq. (4.218) with $[u]_\lambda$ and Eq. (4.219) with $[v]_\lambda$ yields after the addition of both equations the local budget equation for **zonal kinetic energy (ZKE)**

$$\frac{\partial e_{ZK}}{\partial t} + \frac{1}{r_E\cos\varphi}\frac{\partial}{\partial \varphi}(\cos\varphi[v]_\lambda e_{ZK}) + \frac{\partial}{\partial p}([\omega]_\lambda e_{ZK})$$

$$= -\frac{[v]_\lambda}{r_E}\frac{\partial [\phi]_\lambda}{\partial \varphi} + [\mathbf{v}_h]_\lambda \cdot [\boldsymbol{f}_T]_\lambda - \frac{\tan\varphi}{r_E}[\mathbf{v}_h]_\lambda \cdot \mathbf{k}\times[u^*\mathbf{v}_h^*]_\lambda$$

$$- \frac{[\mathbf{v}_h]_\lambda}{r_E\cos\varphi}\cdot\frac{\partial}{\partial \varphi}(\cos\varphi[v^*\mathbf{v}_h^*]_\lambda) - [\mathbf{v}_h]_\lambda\cdot\frac{\partial}{\partial p}[\omega^*\mathbf{v}_h^*]_\lambda \qquad (4.220)$$

where $e_{ZK} = 1/2[\mathbf{v}_h]_\lambda^2$ denotes the specific zonal kinetic energy. A budget equation for the eddy part of the kinetic energy can be derived by multiplying the components of the momentum equation with the respective eddy wind components and adding the result. This gives

$$\mathbf{v}_h^* \cdot \frac{\partial \mathbf{v}_h}{\partial t} + \frac{u}{r_E\cos\varphi}\mathbf{v}_h^*\cdot\frac{\partial \mathbf{v}_h}{\partial \lambda} + \frac{v}{r_E}\mathbf{v}_h^*\cdot\frac{\partial \mathbf{v}_h}{\partial \varphi} + \omega\mathbf{v}_h^*\cdot\frac{\partial \mathbf{v}_h}{\partial p}$$

$$= -\left(f + \tan\varphi\frac{u}{r_E}\right)\mathbf{v}_h^*\cdot(\mathbf{k}\times\mathbf{v}_h) - \mathbf{v}_h^*\cdot\nabla_h\phi + \mathbf{v}_h^*\cdot\boldsymbol{f}_T$$

$$(4.221)$$

Taking the zonal average leads to

$$\frac{\partial e_{EK}}{\partial t} + \frac{1}{r_E\cos\varphi}\frac{\partial}{\partial \varphi}\left[\cos\varphi v\frac{\mathbf{v}_h^{*2}}{2}\right]_\lambda + \frac{\partial}{\partial p}\left[\omega\frac{\mathbf{v}_h^{*2}}{2}\right]_\lambda$$

$$= -[\mathbf{v}_h^*\cdot\nabla_h\phi^*]_\lambda + [\mathbf{v}_h^*\cdot\boldsymbol{f}_T^*]_\lambda - \frac{\tan\varphi}{r_E}[u^*\mathbf{v}_h^*]_\lambda\cdot(\mathbf{k}\times[\mathbf{v}_h]_\lambda)$$

$$- \frac{1}{r_E}[v^*\mathbf{v}_h^*]_\lambda\cdot\frac{\partial [\mathbf{v}_h]_\lambda}{\partial \varphi} - [\omega^*\mathbf{v}_h^*]_\lambda\cdot\frac{\partial [\mathbf{v}_h]_\lambda}{\partial p} \qquad (4.222)$$

where $e_{EK} = 1/2[\mathbf{v}_h^{*2}]_\lambda$ denotes the specific **eddy kinetic energy (EKE)**. In a similar way, we can derive local budget equations for available potential energy. Starting point is the potential temperature tendency equation

$$\frac{\partial \theta'}{\partial t} + \nabla_h \cdot (\mathbf{v}_h \theta') + \frac{\partial}{\partial p}(\omega \theta') + \omega \frac{\partial \theta_r}{\partial p} = \frac{Q}{c_p \Pi} \qquad (4.223)$$

where θ' is the deviation from the reference temperature $\theta_r = \theta_r(p)$. Zonal averaging and multiplying this equation with $[\theta']_\lambda$ yields

$$\frac{\partial}{\partial t}\left(\frac{[\theta']_\lambda^2}{2}\right) + \frac{1}{r_E \cos\varphi}\frac{\partial}{\partial \varphi}\left(\cos\varphi [v]_\lambda \frac{[\theta']_\lambda^2}{2}\right) + \frac{\partial}{\partial p}\left(\omega \frac{[\theta']_\lambda^2}{2}\right)$$

$$+ [\omega]_\lambda [\theta']_\lambda \frac{\partial \theta_r}{\partial p} = \frac{[\theta']_\lambda [Q]_\lambda}{c_p \Pi} - \frac{[\theta']_\lambda}{r_E \cos\varphi}\frac{\partial}{\partial \varphi}(\cos\varphi [v^*\theta'^*]_\lambda)$$

$$- [\theta']_\lambda \frac{\partial}{\partial p}[\omega^*\theta'^*]_\lambda \qquad (4.224)$$

and an additional multiplication with $\Upsilon \equiv R_d \Pi/(p\partial\theta_r/\partial p)$ leads to the local energy budget equation

$$\frac{\partial e_{ZA}}{\partial t} + \frac{1}{r_E \cos\varphi}\frac{\partial}{\partial \varphi}(\cos\varphi [v]_\lambda e_{ZA}) + \frac{\partial}{\partial p}([\omega]_\lambda e_{ZA})$$

$$= \frac{R_d \Pi}{p}[\omega]_\lambda [\theta']_\lambda + \frac{\Upsilon}{c_p \Pi}[\theta']_\lambda [Q]_\lambda - \Upsilon \frac{[\theta']_\lambda}{r_E \cos\varphi}\frac{\partial}{\partial \varphi}(\cos\varphi [v^*\theta'^*]_\lambda)$$

$$- \Upsilon [\theta']_\lambda \frac{\partial}{\partial p}[\omega^*\theta'^*]_\lambda \qquad (4.225)$$

where $e_{ZA} = \Upsilon [\theta']_\lambda^2/2$ denotes the specific **zonal available potential energy (ZAPE)**. Here, we have ignored the pressure dependence of Υ in the third term on the left-hand side of (4.225). The neglected term includes a triple correlation of anomalies which is small according to Lorenz (1955). Finally, we derive an equation for the eddy part of available potential energy. For this purpose, the temperature tendency equation (4.223) is multiplied with θ'^* and zonally averaged which yields

$$\frac{\partial}{\partial t}\left[\frac{1}{2}\theta'^{*2}\right]_\lambda + \frac{1}{r_E\cos\varphi}\frac{\partial}{\partial\varphi}\left[\cos\varphi v\frac{\theta'^{*2}}{2}\right]_\lambda + \frac{\partial}{\partial p}\left[\omega\frac{\theta'^{*2}}{2}\right]_\lambda$$

$$= -[\omega^*\theta'^*]_\lambda\frac{\partial\theta_r}{\partial p} + \frac{[\theta'^*Q^*]_\lambda}{c_p\Pi} - \frac{1}{r_E}[v^*\theta'^*]_\lambda\frac{\partial[\theta']_\lambda}{\partial\varphi} - [\omega^*\theta'^*]_\lambda\frac{\partial[\theta']_\lambda}{\partial p}$$

$$(4.226)$$

and another multiplication with Υ results in

$$\frac{\partial e_{EA}}{\partial t} + \frac{1}{r_E\cos\varphi}\frac{\partial}{\partial\varphi}\left[\cos\varphi\Upsilon v\frac{[\theta'^{*2}]_\lambda}{2}\right]_\lambda + \frac{\partial}{\partial p}\left[\omega\Upsilon v\frac{[\theta'^{*2}]_\lambda}{2}\right]_\lambda$$

$$= \frac{R_d\Pi}{p}[\omega^*\theta'^*]_\lambda + \frac{\Upsilon}{c_p\Pi}[\theta'^*Q^*]_\lambda - [v^*\theta'^*]_\lambda\frac{1}{r_E}\frac{\partial}{\partial\varphi}(\Upsilon[\theta']_\lambda)$$

$$-[\omega^*\theta'^*]_\lambda\frac{\partial}{\partial p}(\Upsilon[\theta']_\lambda) \qquad (4.227)$$

where $e_{EA} = \Upsilon[\theta'^{*2}]_\lambda/2$ denotes the specific **eddy available potential energy (EAPE)**. Again, the pressure dependence of Υ is ignored. By integrating the local energy equations (4.220), (4.222), (4.225) and (4.227) over the entire atmosphere,[m] we obtain the following global energy equations:

$$\frac{dE_{ZK}}{dt} = C_{E_{EK},E_{ZK}} + C_{E_{ZA},E_{ZK}} - D_{ZK} \qquad (4.228)$$

$$\frac{dE_{EK}}{dt} = -C_{E_{EK},E_{ZK}} + C_{E_{EA},E_{EK}} - D_{EK} \qquad (4.229)$$

$$\frac{dE_{ZA}}{dt} = -C_{E_{ZA},E_{EA}} - C_{E_{ZA},E_{ZK}} + G_{ZA} \qquad (4.230)$$

$$\frac{dE_{EA}}{dt} = C_{E_{ZA},E_{EA}} - C_{E_{EA},E_{EK}} + G_{EA} \qquad (4.231)$$

---

[m]Mountains are included as described in the previous subsection. For the derivation, we must apply many partial integrations like $\int a\partial b/\partial x\,dx = [ab] - \int b\partial a/\partial x\,dx = -\int b\partial a/\partial x\,dx$ where the second identity results because a vanishes at the boundary or the integration interval is periodic. Note that the area integral becomes in spherical coordinates: $\iint_{a_E}\dots da = \int_{-\pi/2}^{\pi/2}\int_0^{2\pi}\dots r_E^2\cos\varphi d\lambda d\varphi$.

where the terms on the right-hand side are:

Transfer of eddy to zonal kinetic energy

$$C_{E_{EK},E_{ZK}} = \iint_{a_E} \int_0^{p_0} \frac{\tan\varphi}{r_E} [u^* \mathbf{v}_h^*]_\lambda \cdot \mathbf{k} \times [\mathbf{v}_h]_\lambda \frac{dp}{g} da$$

$$+ \iint_{a_E} \int_0^{p_0} \frac{1}{r_E} [v^* \mathbf{v}_h^*]_\lambda \cdot \frac{\partial [\mathbf{v}_h]_\lambda}{\partial\varphi} + [\omega^* \mathbf{v}_h^*]_\lambda \cdot \frac{\partial [\mathbf{v}_h]_\lambda}{\partial p} \frac{dp}{g} da \tag{4.232}$$

Transfer of zonal to eddy available potential energy

$$C_{E_{ZA},E_{EA}} = - \iint_{a_E} \int_0^{p_0} \frac{\Upsilon}{r_E} [v^* \theta'^*]_\lambda \frac{\partial [\theta']_\lambda}{\partial\varphi} + [\omega^* \theta'^*]_\lambda \frac{\partial}{\partial p} (\Upsilon [\theta']_\lambda) \frac{dp}{g} da \tag{4.233}$$

Conversion of zonal available potential energy to zonal kinetic energy

$$C_{E_{ZA},E_{ZK}} = - \iint_{a_E} \int_0^{p_0} \frac{R_d \Pi}{p} [\omega]_\lambda [\theta']_\lambda \frac{dp}{g} da \tag{4.234}$$

Conversion of eddy available potential energy to eddy kinetic energy

$$C_{E_{EA},E_{EK}} = - \iint_{a_E} \int_0^{p_0} \frac{R_d \Pi}{p} [\omega^* \theta'^*]_\lambda \frac{dp}{g} da \tag{4.235}$$

Generation of zonal and eddy available potential energy

$$G_{ZA} = \iint_{a_E} \frac{\Upsilon}{c_p \Pi} [\theta']_\lambda [Q]_\lambda \frac{dp}{g} da,$$

$$G_{EA} = \iint_{a_E} \int_0^{p_0} \frac{\Upsilon}{c_p \Pi} [\theta'^* Q^*]_\lambda \frac{dp}{g} da \tag{4.236}$$

Dissipation of zonal and eddy kinetic energy

$$D_{ZK} = - \iint_{a_E} \int_0^{p_0} [\mathbf{v}_h]_\lambda \cdot [\mathbf{f}_T]_\lambda \frac{dp}{g} da,$$

$$D_{EK} = - \iint_{a_E} \int_0^{p_0} [\mathbf{v}_h^* \cdot \mathbf{f}_T^*]_\lambda \frac{dp}{g} da \tag{4.237}$$

The lower boundary pressure p_0 is defined in such a way that the corresponding isobaric surface lies below the ground. Figure 4.27 outlines the energy flows as deduced from the budget equations (4.228)–(4.231). The values for the energy flows (grey boxes) are in W/m^2 and those for the energy amounts (black boxes) are expressed in $10^5 \ J/m^2$. These values have been taken from Peixoto and Oort (1992) who determined them from a ten-year observational global data set. Obviously, the energy flows between the various energy forms describe a cycle. Zonal available energy converts into eddy available potential energy and then into eddy kinetic energy. Eddy kinetic energy is on the other hand transferred into zonal kinetic energy which is finally converted into zonal available potential energy. Furthermore, energy is generated by heating and dissipated by friction. Zonal available potential energy is generated by heating at low latitudes and cooling at high latitudes. Heating results from radiation, surface heat fluxes and release of latent heat. Eddy available potential energy is generated by heating the warm anomalies or cooling of the cold anomalies. A significant contribution stems from the upward warm air flow in extratropical cyclones where latent heat is released. Dissipation of kinetic energy mainly takes place in the boundary layer. Altogether the atmospheric energy cycle describes a thermodynamic heat engine that converts heat into kinetic energy which eventually dissipates. However, the direction of Lorenz energy cycle

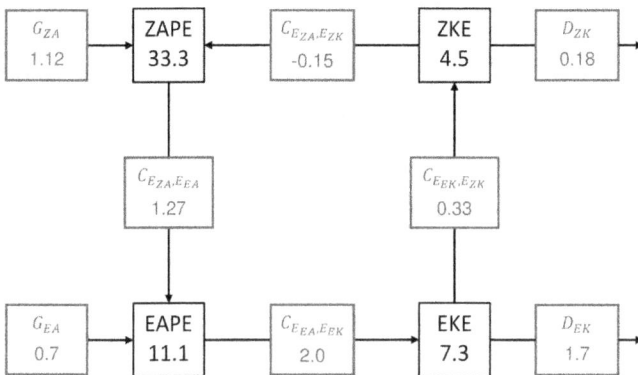

Fig. 4.27.   Lorenz energy cycle: The black boxes show the amount of the various energy forms in $10^5 \ J/m^2$ while the grey boxes display the energy flows in W/m^2. The values result from observations as described in Peixoto and Oort (1992).

is not obvious. It is also conceivable that the cycle runs in a clockwise direction, i.e. zonal available potential energy is converted into zonal kinetic energy which is transferred to eddy kinetic energy and so forth. The observed direction results because baroclinic instability represents the main driver for eddy development. We will investigate this in the next subsection in more detail. Furthermore, the zonally averaged overturning circulation is **thermodynamically indirect**, i.e. this circulation would increase temperature contrasts and build up available potential energy if other processes were absent. The midlatitude **Ferrel cell** describes such a thermodynamically indirect circulation. It is mechanically driven by nonlinear feedbacks of baroclinic waves as we will understand in the next chapter. The tropical **Hadley cell** has in contrast a **thermodynamically direct circulation**. It contributes with a positive $C_{E_{ZA},E_{ZK}}$ to the Lorenz energy cycle.

4.7.3 *Quasigeostrophic energy cycle*

The Lorenz energy cycle can also be derived from the quasigeostrophic equations (4.60)–(4.63). Indeed, we obtain again the set of energy equations (4.228)–(4.231) but the energy flows and energy amounts must be calculated in a different way. We will omit the derivation because it is similar to that in the previous subsection. The reader can do this as an exercise and must bear in mind that $\phi'^{(1)}$ is neglected in (4.61). We obtain the following expressions for the energies:

$$E_{ZK} = \iint_a \int_0^{p_0} \frac{1}{2}[u_g]_x^2 \frac{dp}{g} da \qquad (4.238)$$

$$E_{EK} = \iint_a \int_0^{p_0} \frac{1}{2}[v_g^{*2}]_x \frac{dp}{g} da \qquad (4.239)$$

$$E_{ZA} = \iint_a \int_0^{p_0} \frac{1}{2\Xi} \left[\frac{\partial \phi'}{\partial p}\right]_x^2 \frac{dp}{g} da \qquad (4.240)$$

$$E_{EA} = \iint_a \int_0^{p_0} \frac{1}{2\Xi} \left[\left(\frac{\partial \phi'^*}{\partial p}\right)^2\right]_x \frac{dp}{g} da \qquad (4.241)$$

and the energy flows become

$$C_{E_{EK},E_{ZK}} = \iint_a \int_0^{p_0} [u_g^* v_g^*]_x \frac{\partial [u_g]_x}{\partial y} \frac{dp}{g} da \qquad (4.242)$$

$$C_{E_{ZA},E_{EA}} = -\iint_a \int_0^{p_0} \frac{R_d^2}{p^2 \Xi} [v_g^* T^*]_x \frac{\partial [T']_x}{\partial y} \frac{dp}{g} da \qquad (4.243)$$

$$C_{E_{ZA},E_{ZK}} = -\iint_a \int_0^{p_0} \frac{R_d}{p} [\omega]_x [T']_x \frac{dp}{g} da \qquad (4.244)$$

$$C_{E_{EA},E_{EK}} = -\iint_a \int_0^{p_0} \frac{R_d}{p} [\omega^* T^*]_x \frac{dp}{g} da \qquad (4.245)$$

$$G_{ZA} = \iint_a \int_0^{p_0} \frac{R_d^2}{p^2} \frac{[T']_x [Q]_x}{c_p \Xi} \frac{dp}{g} da,$$

$$G_{EA} = \iint_a \int_0^{p_0} \frac{R_d^2}{p^2} \frac{[T^* Q^*]_x}{c_p \Xi} \frac{dp}{g} da \qquad (4.246)$$

$$D_{ZK} = -\iint_a \int_0^{p_0} [\mathbf{v}_g]_x \cdot [\mathbf{f}_T]_x \frac{dp}{g} da,$$

$$D_{EK} = -\iint_a \int_0^{p_0} [\mathbf{v}_g^* \cdot \mathbf{f}_T^*]_x \frac{dp}{g} da \qquad (4.247)$$

In these equations, a denotes the area and $[\ldots]_x$ the average in x-direction of the considered β-plane domain. The energy flows in the quasigeostrophic system become simpler and it is more convenient to relate these to the energy flows in a growing baroclinic wave. According to the observed Lorenz energy cycle we have a positive transfer of zonal to eddy available potential energy, $C_{E_{ZA},E_{EA}}$. Therefore, the condition

$$[v_g^* T^*]_x \frac{\partial [T']_x}{\partial y} < 0 \qquad (4.248)$$

must be fulfilled in most locations. Therefore, the **meridional eddy temperature flux** is **down-gradient**, i.e. the eddy temperature flux by the wave is on average directed towards the colder air.

In Fig. 4.11, we see a positive correlation between v_g^* and T^* in the unstable case which proves that (4.248) is fulfilled. Indeed, the vertical tilt of the troughs and ridges against the shear is an indicator of potential energy transfer from the zonal mean to the wave (see Fig. 4.11).

The observed positive conversion of eddy available potential energy into eddy kinetic energy, $C_{E_{EA},E_{EK}}$ requires

$$[\omega^* T^*]_x < 0 \tag{4.249}$$

This condition means that warm (cold) temperature anomalies must be connected to rising (sinking) air. Indeed, Fig. 4.11 reveals that a growing baroclinic wave fulfills (4.249). This conversion is proportional to the **vertical eddy temperature flux** and represents the production of kinetic wave energy due to negative work at the expense of available potential energy.

The conversions $C_{E_{EK},E_{ZK}}$ and $C_{E_{ZA},E_{ZK}}$ cannot be explained by the quasigeostrophic baroclinic wave solutions found in Section 1.6 because both conversions remain zero for these solutions. Note that the considered models are linear and that the chosen basic flows have no horizontal shear. Indeed, simulation of baroclinic instability in a jet flow using a numerical nonlinear model yields energy flows consistent with the directions in the observed Lorenz energy cycle as we will see in the next chapter. These results suggest that baroclinic instability plays an important key role for the observed atmospheric energy cycle.

A nice graphical diagnostic tool for the role of eddy-zonal mean flow interaction in the energy cycle can be derived by considering the transfer of zonal mean to eddy energy:

$$C_{E_Z,E_E} = -\iint_a \int_0^{p_0} \frac{R_d^2}{p^2 \Xi} [v_g^* T^*]_x \frac{\partial [T']_x}{\partial y} + [u_g^* v_g^*]_x \frac{\partial [u_g]_x}{\partial y} \frac{dp}{g} da \tag{4.250}$$

By using the thermal wind balance equation (4.86) and transformation (4.70) we get

$$C_{E_Z,E_E} = \iint_a \int_0^\infty \left(\frac{f_0 R_d}{N_l^2 H_s} [v_g^* T^*]_x \frac{\partial [u_g]_x}{\partial z_l} - [u_g^* v_g^*]_x \frac{\partial [u_g]_x}{\partial y} \right) \rho_I dz_l da \tag{4.251}$$

This energy flow can be written in the form

$$C_{EZ,EE} = \iint_a \int_0^\infty \mathbf{E} \cdot \nabla_{z_l}[u_g]_x \, dz_l da \qquad (4.252)$$

where $\mathbf{E}$ denotes the **Eliassen Palm flux** (introduced by Eliassen and Palm, 1961):

$$\mathbf{E} \equiv -\rho_I[u_g^* v_g^*]_x \mathbf{j} + \rho_I \frac{f_0 R_d}{N_l^2 H_s}[v_g^* T^*]_x \mathbf{k} \qquad (4.253)$$

and ∇_{z_l} is the Nabla operator in the log-pressure height coordinate system, i.e.

$$\nabla_{z_l} \equiv \nabla_h + \mathbf{k}\frac{\partial}{\partial z_l} \qquad (4.254)$$

From (4.252) we can state that positive energy transfer from the zonal mean to the eddy takes place if the Eliassen Palm flux $\mathbf{E}$ directs to higher zonal winds. In the opposite case, the transfer becomes negative. Figure 4.28 sketches the Eliassen Palm flux in a meridional cross-section for several cases in a westerly jet flow. In the first case, **barotropic growth** induces the energy transfer. Then, the Eliassen Palm flux is horizontal so that energy transfer results from meridional eddy momentum fluxes. Such a situation indicates barotropic shear instability as studied in Section 3.11.2. The meridional trough and ridge axes lean against the meridional shear of the zonal flow as seen in Fig. 3.32. Then, u_g^* and v_g^* are correlated in such a way that the eddy meridional momentum flux is directed towards lower zonal wind speeds. In the second case, eddy energy rises by **baroclinic growth**. The Eliassen–Palm flux points upward since the meridional eddy temperature flux only contributes to this vector. Baroclinic instability is the likely reason for baroclinic growth. Now, the vertical trough and ridge axes lean against the vertical shear as seen in Fig. 4.11. Also, the opposite situations of **barotropic decay** and **baroclinic decay** are possible. Then, the arrows direct in the opposite direction and energy of the zonal mean flow increases at the expense of eddy energy.

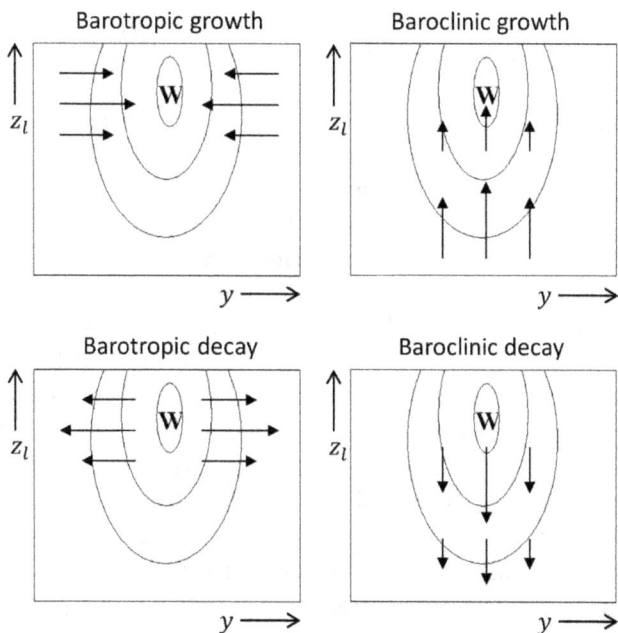

Fig. 4.28. Examples of the orientation of the Eliassen–Palm flux in a meridional cross-section. The lines represent isolines of zonally averaged zonal wind and the W symbolizes the positive maximum of the westerly jet stream.

It is possible to divide the eddy contributions to the energetics into the various zonal wavenumbers. With the Fourier expansion

$$G(x, y, z_l) = \Re \left[\sum_{m=0}^{\infty} \hat{G}_m(y, z_l) \exp(ik_m x) \right] \qquad (4.255)$$

we obtain for the eddy fluxes

$$[u_g^* v_g^*]_x = \Re \left[\frac{1}{2} \sum_{m=1}^{\infty} \widehat{u_g^*}_m (\widehat{v_g^*}_m)^C \right] \qquad (4.256)$$

$$[v_g^* T^*]_x = \Re \left[\frac{1}{2} \sum_{m=1}^{\infty} \widehat{v_g^*}_m (\widehat{T^*}_m)^C \right] \qquad (4.257)$$

$$[\omega^* T^*]_x = \Re \left[\frac{1}{2} \sum_{m=1}^{\infty} \widehat{\omega^*}_m (\widehat{T^*}_m)^C \right] \tag{4.258}$$

Therefore, every eddy energy conversion term can be calculated for each zonal wavenumber separately. The complete eddy energy conversion results from the sum of all contributions. The same is true for all other eddy terms since they all appear as a zonal average of two multiplied eddy quantities. Here, we distinguish **long waves** having zonal wavenumbers from 1 to 5 and **cyclone waves** having zonal wavenumbers larger than 5. The long waves include orographically forced stationary Rossby waves (cf. Section 3.10) and other slowly propagating pressure systems like **blocking highs**. Cyclone waves describe synoptic-scale weather systems that usually result from baroclinic instability and form extratropical cyclones in the mature stage of the baroclinic wave life cycle. Figure 4.29 shows **Eliassen–Palm cross-sections** for January 2008 and July 2008. The Eliassen–Palm flux **E** has been divided by ρ_I to make the direction of **E** at larger heights visible. We see that **E** has a predominantly vertical orientation and the largest amplitude at the surface of the mid-latitudes. At most locations the vertical gradient of zonal wind is positive. Therefore, the baroclinic energy conversions support the eddy perturbations on average. On the other hand, we detect an equatorward turning of **E** in the upper troposphere on the equatorward side of the jet stream. This is more pronounced for cyclone waves and means that the barotropic energy conversion transfers kinetic eddy energy to the zonal mean flow. These results agree with the mean Lorenz energy cycle displayed in Fig. 4.27. Furthermore, we see that the winter hemisphere has more eddy activity than the summer hemisphere. Another interesting feature is the existence of a strong jet stream in the stratosphere of the winter hemisphere. This is the **polar night jet** with superposed long baroclinic waves. These have higher amplitudes in the Northern Hemisphere where a more pronounced wavenumber two orography triggers long Rossby waves which eventually propagate upward into the stratosphere (cf. Section 4.4). The eddy energy conversion of long waves diagnosed with the Eliassen–Palm flux could be beneficial for predicting large-scale weather. For example, a high eddy energy conversion in the long wave range could initiate a blocking high that can dominate the weather situation for more than a week.

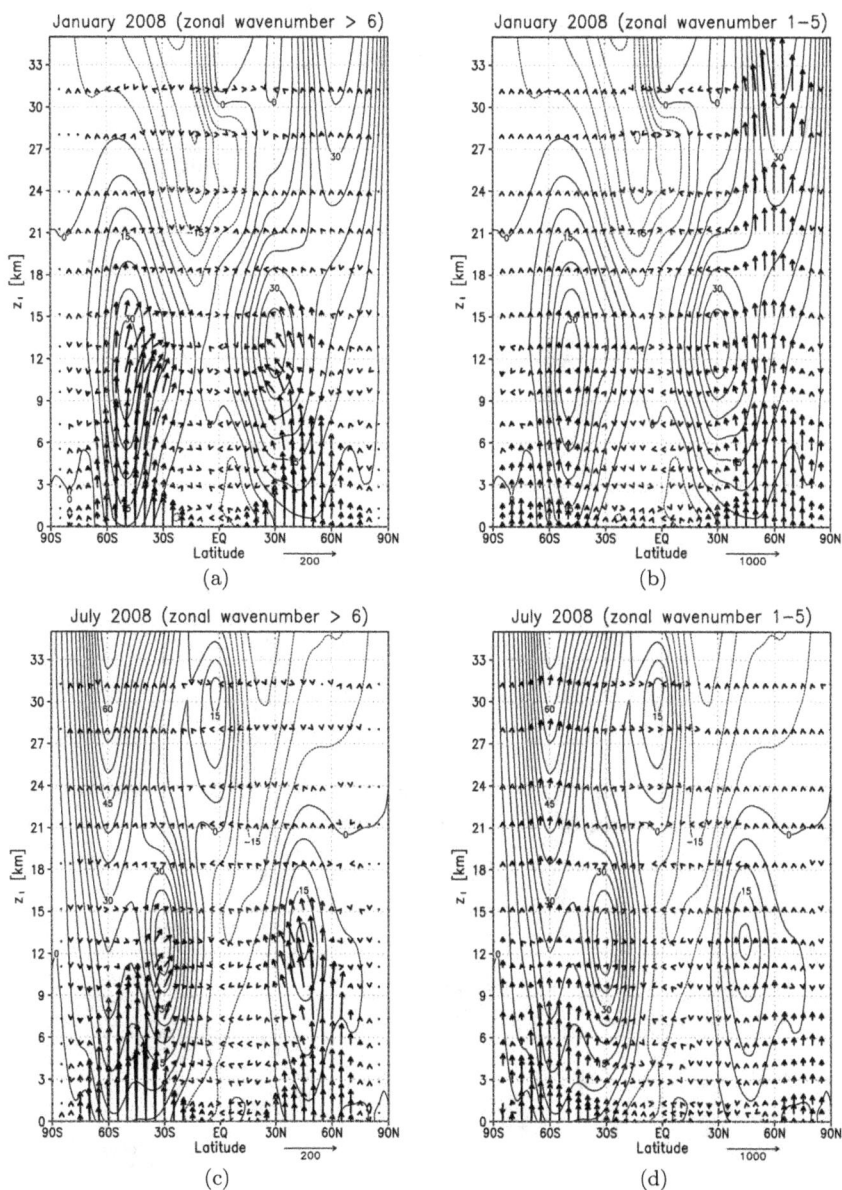

Fig. 4.29. Time averaged Eliassen–Palm flux (arrows) and zonally averaged zonal wind (isolines, contour interval 5 m/s) deduced from NCEP reanalysis in a meridional cross-section in which the vertical axis measures the log-pressure vertical coordinate. The panels display results for (a) cyclone waves in January 2008, (b) long waves in January 2008, (c) cyclone waves in July 2008 and (d) long waves in July 2008.

Finally, we introduce another interesting property of the Eliassen–Palm flux, namely, that it represents a **wave activity flux** in the zonally averaged view as shown by Edmon *et al.* (1980). Eliassen and Palm (1961) derived a similar property on the basis of the linearized quasiqeostrophic equations for stationary waves in an adiabatic and inviscid flow. Here, we only assume the quasigeostrophic approximation and small wave amplitudes. Let us start with the quasistrophic PV equation

$$\left(\frac{\partial}{\partial t} + \mathbf{v}_g \cdot \nabla_h\right) P_g = \dot{P}_g \tag{4.259}$$

where $\dot{P}_g$ denotes the PV tendencies due to diabatic and irreversible processes and microturbulent exchange. Linearization of this equation with respect to the zonal mean flow yields

$$\frac{\partial P_g^*}{\partial t} + [u_g]_x \frac{\partial P_g^*}{\partial x} + v_g^* \frac{\partial [P_g]_x}{\partial y} = \dot{P}_g^* \tag{4.260}$$

Multiplying this equation with P_g^* and zonal averaging leads to

$$\frac{\partial}{\partial t}\left(\frac{1}{2}[P_g^{*2}]_x\right) + [v_g^* P_g^*]_x \frac{\partial [P_g]_x}{\partial y} = [P_g^* \dot{P}_g^*]_x \tag{4.261}$$

To show the relation to the Eliassen–Palm flux we must evaluate the meridional eddy flux of potential vorticity

$$[v_g^* P_g^*]_x = \left[\frac{\partial \psi_g^*}{\partial x}\left[\nabla_h^2 \psi_g^* + \frac{1}{\rho_I}\frac{\partial}{\partial z_l}\left(\rho_I \frac{f_0^2}{N_l^2}\frac{\partial \psi_g^*}{\partial z_l}\right)\right]\right]_x$$

$$= \left[\frac{1}{2}\frac{\partial}{\partial x}(v_g^{*2} - u_g^{*2}) - \frac{\partial}{\partial y}(u_g^* v_g^*) + v_g^* \frac{1}{\rho_I}\frac{\partial}{\partial z_l}\left(\rho_I \frac{f_0^2}{N_l^2}\frac{\partial \psi_g^*}{\partial z_l}\right)\right]_x$$

$$= \left[-\frac{\partial}{\partial y}(u_g^* v_g^*) + \frac{1}{\rho_I}\frac{\partial}{\partial z_l}\left(\rho_I \frac{f_0^2}{N_l^2}v_g^* \frac{\partial \psi_g^*}{\partial z_l}\right) - \frac{f_0^2}{N_l^2}\frac{\partial^2 \psi_g^*}{\partial x \partial z_l}\frac{\partial \psi_g^*}{\partial z_l}\right]_x$$

$$= \frac{1}{\rho_I}\nabla_{z_l} \cdot \mathbf{E} \tag{4.262}$$

Consequently, the meridional eddy potential vorticity flux is proportional to the divergence of the Eliassen–Palm flux. Neglecting slow time variation of $\partial[P_g]_x/\partial y$ leads to the following equation:

$$\frac{\partial A_{P_g^*}}{\partial t} + \nabla_{z_l} \cdot \mathbf{E} = S_{P_g^*} \qquad (4.263)$$

where $A_{P_g^*}$ denotes the **local wave activity** and $S_{P_g^*}$ the irreversible sink (or source) of wave activity. The terms are given by

$$A_{P_g^*} = \frac{\frac{\rho_I}{2}[P_g^{*2}]_x}{\partial[P_g]_x/\partial y}, \quad S_{P_g^*} = \frac{\rho_I[P_g^* \dot{P}_g^*]_x}{\partial[P_g]_x/\partial y} \qquad (4.264)$$

Usually, the meridional gradient of zonally averaged potential vorticity is positive. Hence, wave activity is a positive quantity in this case and proportional to the specific potential eddy enstrophy. Obviously, $\mathbf{E}$ constitutes a wave activity flux in the zonally averaged view as stated above. Looking at Fig. 4.29 we see that wave activity is transported upward on average. Consequently, it must be generated by the wave activity flux passing through the lower boundary. In the upper troposphere, the wave activity flux $\mathbf{E}$ of cyclone waves turns equatorward. Indeed, a numerical simulation of a baroclinic wave life cycle shows the beginning generation of wave activity near the surface, then, upward propagation and finally, propagation towards the equator (as we will see in the next chapter). Long waves propagate farther upward into the stratosphere in winter and reinforce wave activity in the polar night jet. This result is consistent with our finding in Section 4.4 that only long waves can propagate into the stratosphere. Note that the wave activity analysis method is questionable in the case of internal baroclinic instability because $\partial[P_g]_x/\partial y$ must change sign.

Chapter 5

Nonlinear Dynamics

5.1 Introduction

We have learned in the last chapter that hydrodynamic instabilities play an essential role in weather dynamics and this has important consequences for the time behavior of atmospheric fields. Instability results in exponential growth of the perturbation in the underlying linearized model. However, sooner or later the linearized model becomes invalid because the perturbation attains such a large amplitude that nonlinear terms cannot be neglected. This is also evident from another perspective since unlimited growth is impossible due to the limited amount of energy being available. Therefore, nonlinear terms induce usually a decrease of perturbation growth until it eventually halts and the perturbation possibly decays afterward. The nonlinear terms modify the unstable basic flow in such a way that it becomes stable. They can also trigger other perturbations and are responsible for the interaction of different waves. **Chaotic dynamics** and **limited predictability** result in a fluid where hydrodynamic instability is always recovered by external forcing (e.g., solar insulation). We will explain this in more detail later in this chapter. We begin with the study of a baroclinic wave lifecycle in a nonlinear primitive equation model. Afterwards, we derive a solution for a **weakly nonlinear baroclinic wave** in the quasigeostrophic two-level model. Nonlinear models can be formulated as a **dynamical system**. We introduce a method of how to do this and describe how to analyze such a system in order to obtain useful information of the model behavior. Finally, the limited predictability of weather is explained on the basis of the previous considerations.

5.2 The Life Cycle of a Baroclinic Wave

In the last chapter, we analyzed baroclinic instability in simple linearized models. However, we did not consider the question of what happens with the baroclinic wave when the amplitude becomes very large. It is not possible to derive an analytical solution that can give an answer to this question. Instead, we have to solve the model equations with an approximate numerical solution method. Here, the results of the general circulation model Planet Simulator (Fraedrich *et al.*, 2005) are presented.

5.2.1 *Model and experiment description*

The model solves the primitive equations based on the sigma–coordinate. The numerical solution is based on the spectral method, that is, the model variables are expanded in terms of spherical harmonics, and the resulting amplitude equations are time-integrated. The expansion ranges until total wavenumber 170 and 28 vertical sigma levels are employed. This corresponds approximately to a horizontal grid point resolution of about 80 km. The model has **physical parameterization schemes** for microturbulent exchange, cloud microphysics, convection and radiation. However, in the first experiment, we switch them off to obtain an approximate solution of the inviscid and adiabatic primitive equations. More details of the model can be found in Lunkeit *et al.* (2011).

For the experiment, we assume an **aquaplanet**, that is, the whole planet is covered with water. The initial state of the model consists of a zonally constant zonal jet stream in thermal wind balance according to Polvani *et al.* (2004) and a superimposed wavenumber 6 perturbation. Figure 5.1 shows a cross-section of the initial basic flow and the temperature field. The center of the jet stream is at 45°N and $\sigma \approx 0.2$ while the maximum zonal wind amounts to 45 m/s. The zonal wind declines to zero at the top and bottom of the atmosphere which has the consequence that the surface pressure takes a constant value. The temperature field describes a poleward

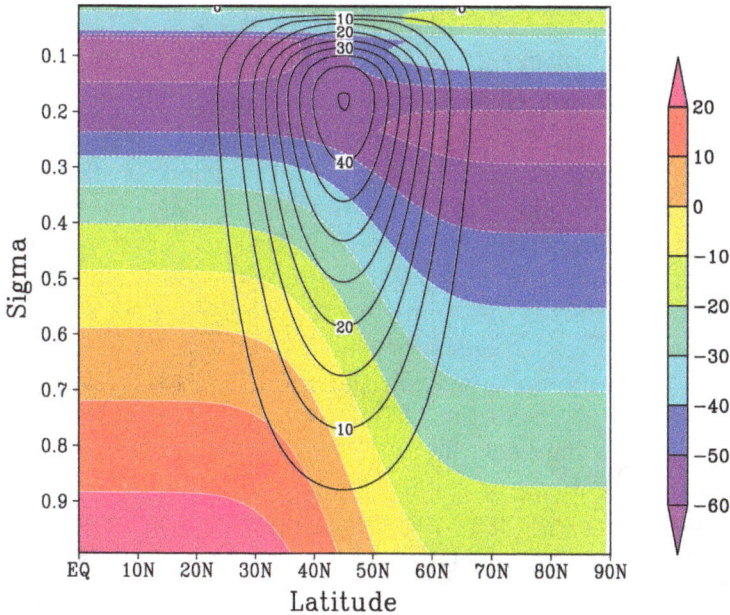

Fig. 5.1.   Initial zonal mean temperature (colored shadings, K) and zonal wind (isolines, contour interval 5 m/s) of the numerical life cycle experiment.

decrease below the jet maximum but the meridional temperature gradient reverses sign above the maximum as a consequence of the thermal wind balance equation. The horizontal mean of the temperature field does not result from this equation and it is assumed that it has the vertical profile of the international standard atmosphere. This has the consequence that the lapse rate below a log-pressure height of 11 km becomes 6.5 K/km and zero above where a stratosphere is assumed. However, the superposition of the temperature anomaly resulting from the thermal wind balance introduces a meridional variation of the tropopause with larger (lower) heights near the equator (pole) which is in agreement with observations. The superimposed wavenumber 6 perturbation is a barotropic wave in geostrophic balance. It can be anticipated that it triggers the fastest growing normal mode having wavenumber 6 because the subharmonic waves (wavenumber 12, 18, etc.) have smaller growth rates. Therefore, this mode dominates the perturbation after a finite time in a linearized model. This can be also true in the nonlinear model if the initial perturbation amplitude is sufficiently small.

5.2.2   *Results*

Figure 5.2 shows the time evolution of several energetic quantities resulting from the Lorenz energy cycle analysis. In Fig. 5.2(a) we see that EKE and EAPE attain appreciable values after day 8 of the model simulation and peak at 11.5 days and 11 days, respectively. Afterward, the eddy energies decline. The growth of eddy energy is accompanied by a decrease in ZAPE. Therefore, the eddy energy grows due to loss of available potential energy of the initial state which indicates that baroclinic instability causes the development. Indeed, Fig. 5.2(b) shows that the conversions $C_{E_{ZA},E_{EA}}$ and $C_{E_{EA},E_{EK}}$ become positive as in the analytical description of baroclinic instability (cf. Section 4.7.3). However, after EAPE peaks a growth of ZKE happens due to the conversion $-C_{E_{ZK},E_{EK}}$ and it equilibrates eventually at a value that is twice as high as the initial one. This describes the **barotropic decay** of the baroclinic wave and is caused by nonlinear terms. A part of the ZKE is converted into ZAPE by $C_{E_{ZA},E_{ZK}}$. Therefore, a thermodynamic indirect cell appears in the zonally averaged circulation. On average the energy conversions have the same signs as in the observed energy cycle (cf. Fig. 4.27). The sum of all energies remains approximately constant during the first 10 days but afterward a slow decrease becomes visible. The absence of total energy conservation can have two reasons. First, the model includes horizontal diffusion to avoid numerical errors due to sharp gradients. Diffusion leads to energy loss and can explain the observed energy decline. On the other hand, the Lorenz energy cycle is based on approximations which have the consequence that total energy conservation must not hold. The energy curves show that there is a growth, a mature, and a decay phase of the baroclinic wave. Any linear model cannot predict this behavior in the case of instability because in such a model the fastest-growing normal model dominates eventually and endless exponential growth takes place.

    A better understanding of the eddy evolution results from Fig. 5.2(c). Here, the ordinate is logarithmically scaled and exponential growth would appear in the form of an upward-sloping straight line. In this graph, the eddy energies approach such a straight line within the first three days. Afterward, exponential energy growth takes place with an approximate growth rate of 0.52/day. Beyond 7 days the growth rate declines (slope of the curve) until it becomes

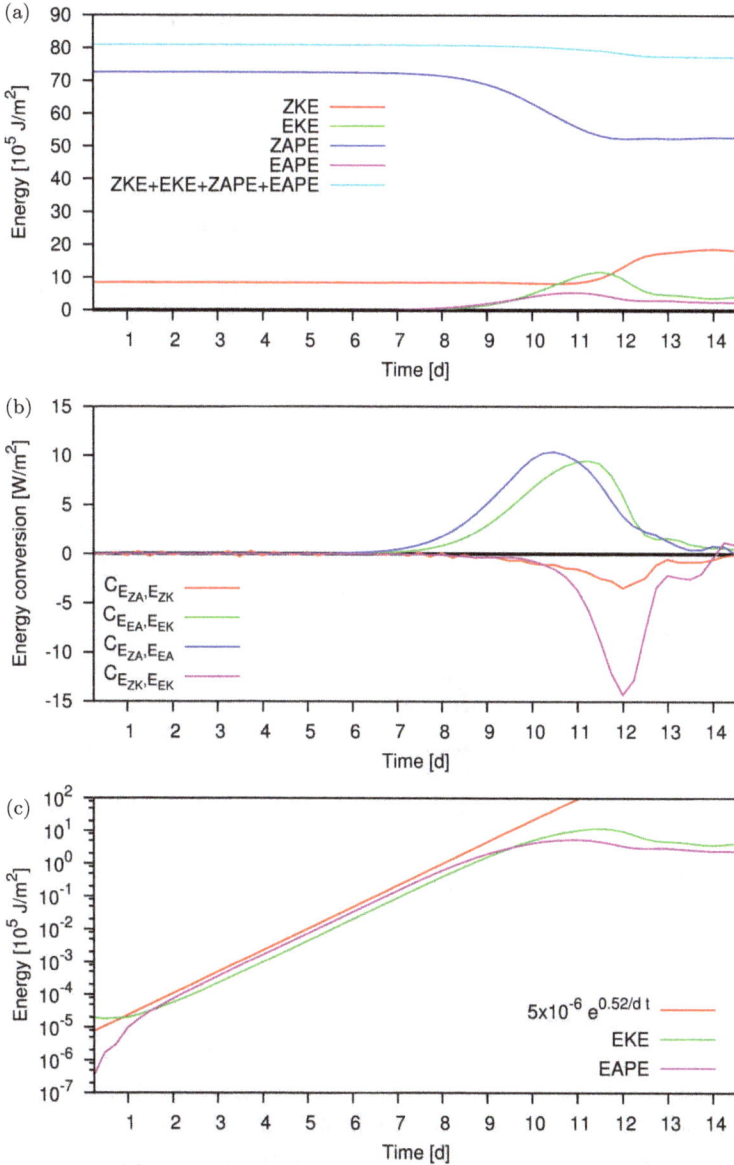

Fig. 5.2. Time evolution of energy quantities of the numerical baroclinic wave life cycle experiment: (a) ZKE (red), EKE (green). ZAPE (blue), EAPE (purple) and total energy (light blue), (b) energy conversions $C_{E_{ZA},E_{ZK}}$ (red), $C_{E_{EA},E_{EL}}$ (green), $C_{E_{ZA},E_{EA}}$ (blue) and $C_{E_{ZK},E_{EK}}$ (purple), (c) EKE and EAPE with logarithmically scaled ordinate. The straight line in (c) describes exponential growth with a growth rate of 0.52/day.

zero at the respective maxima. From the results we can expect that the wave takes approximately the form of the fastest growing normal mode in the period from day 3 to day 7.

Obviously, we have found a cheap method to analyze the stability of a complicated basic flow. We just have to initialize the numerical model with the basic flow and a small perturbation. Nonlinearities can be neglected at the beginning of the simulation and the model behaves as if it were linear. Then, after some time the arbitrary initial perturbation takes the form of the fastest-growing normal mode in the unstable case. Making a full stability analysis is more laborious since it requires the determination of a huge Jacobi matrix and a numerically costly eigenvalue analysis.

Figure 5.3 shows some wave fields at $t = 5$ days when the perturbation grows nearly exponentially. The zonal cross-section at 45°N in Fig. 5.3 reveals a similar baroclinic wave structure as in the quasi-geostrophic uniform potential vorticity model (cf. Fig. 4.25). As differences, we note that the perturbation geopotential is at the surface stronger than at the tropopause and that the temperature wave shows no vertical increase in the upper troposphere. The perturbation geopotential at 500 hPa (Fig. 5.3(b)) has notable amplitude only in the jet stream and the maximum wave amplitude appears northward of the jet stream center. This bias is not understandable with quasigeostrophic theory and leads to a stronger evolution of cyclones in comparison with anticyclones (for more explanation see Snyder et al., 1991). The meridional high and low axes are curved and this has the consequence that the meridional eddy momentum flux $-[u^*v^*]_\lambda$ is directed toward the center of the eddy and the energy flow $C_{E_{ZK}, E_{EK}}$ becomes negative. Consequently, the jet stream's kinetic energy rises due to this process.

So far, we do not know how the nonlinearities can change the time evolution by reducing and eventually stopping growth. The most obvious assumption is that the eddy fluxes of the growing wave alter the zonal mean flow in such a way that it becomes stable. By zonally averaging the zonal momentum and temperature equation we find according to (4.217):

$$\frac{\partial [u]_\lambda}{\partial t} + \frac{[v]_\lambda}{r_E} \frac{\partial [u]_\lambda}{\partial \varphi} + [\omega]_\lambda \frac{\partial [u]_\lambda}{\partial p} - \left(f + \tan \varphi \frac{[u]_\lambda}{r_E} \right) [v]_\lambda$$

$$= -\frac{1}{r_E \cos^2 \varphi} \frac{\partial}{\partial \varphi} (\cos^2 \varphi [v^* u^*]_\lambda) - \frac{\partial}{\partial p} [\omega^* u^*]_\lambda \qquad (5.1)$$

(a)

(b)

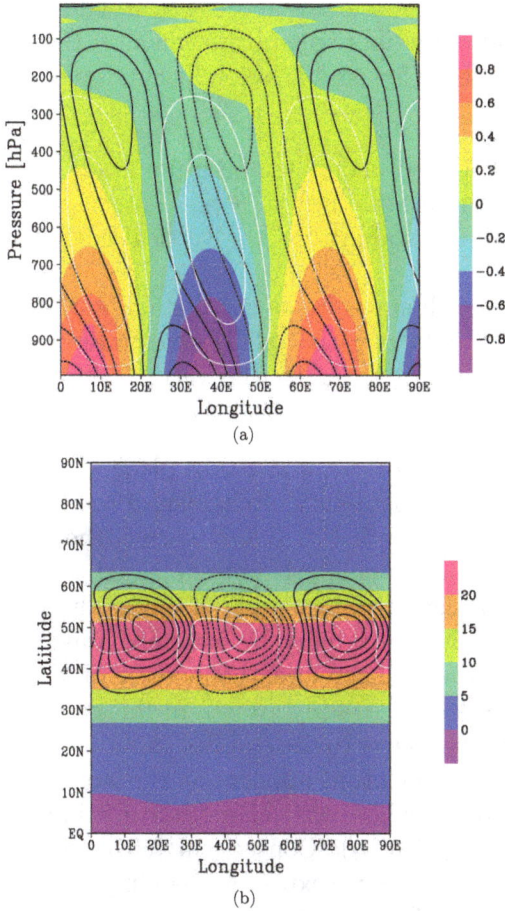

Fig. 5.3. (a) Zonal anomaly of temperature (colored shadings, K), geopotential height (black isolines) and vertical pressure velocity (white isolines) in a zonal cross-section at 45°N after 5 days of the numerical baroclinic wave life cycle experiment. (b) Zonal wind (colored shadings, m/s) and zonal anomaly of geopotential height at 500 hPa (black isolines) and vertical pressure velocity (black isolines) at the same time as in (a). Negative isolines are dashed and the zero contour has been omitted.

$$\frac{\partial [\theta]_\lambda}{\partial t} + \frac{[v]_\lambda}{r_E} \frac{\partial [\theta]_\lambda}{\partial \varphi} + [\omega]_\lambda \frac{\partial [\theta]_\lambda}{\partial p}$$

$$= -\frac{1}{r_E \cos \varphi} \frac{\partial}{\partial \varphi} \left(\cos \varphi [v^* \theta^*]_\lambda \right) - \frac{\partial}{\partial p} [\omega^* \theta^*]_\lambda \qquad (5.2)$$

Therefore, the meridional and vertical eddy fluxes of zonal momentum and potential temperature can alter the zonally averaged basic state. Essentially, stabilization occurs due to the following processes:

1. reduction of the meridional temperature gradient due to the poleward eddy temperature flux $[v^*\theta^*]_\lambda$,
2. increase of static stability by the vertical eddy temperature flux $[\omega^*\theta^*]_\lambda$,
3. increase of meridional shear of zonal wind by the meridional eddy momentum flux $[u^*v^*]_\lambda$ that is directed towards the jet center,
4. decrease of vertical shear by the vertical eddy momentum flux $[\omega^*u^*]_\lambda$.

It is obvious that processes 1 and 2 take place when baroclinic instability constitutes the instability mechanism. Process 3 results because of the eddy shape seen in Fig. 5.3. However, the role of meridional shear for baroclinic instability is not explained so far. James (1987) found with a quasigeostrophic two-level model that meridional shear suppresses baroclinic instability by confining wave perturbations in a small latitude range. Therefore, the increase of meridional shear by momentum fluxes reduces eddy growth in this case. It is not obvious that process 4 contributes to the stabilization of the zonal mean flow. Maximum zonal momentum appears at the northern and southern flanks of the eddies while vertical motion maximizes at the center of the jet (cf. Fig. 5.3(b)). Consequently, the correlation between ω^* and u^* is very low and we expect a small impact on the zonal mean momentum budget. Figure 5.4 shows the various eddy feedbacks on the zonal means of zonal wind and potential temperature. The meridional momentum flux induces an eastward acceleration of more than $0.3\,\text{m/s}$ per day in the jet core and a somewhat weaker westward acceleration at the northern flank of the jet (Fig. 5.4(a)). Higher amplitudes occur at the surface. This process leads to cyclonic shear and stronger cyclones compared to anticyclones. The acceleration amplitude appears small but due to the exponential wave growth, it becomes larger rapidly. The vertical eddy momentum flux has a much smaller impact and the pattern does not convincingly imply a weakening of vertical shear (Fig. 5.4(b)). On the other hand, meridional eddy temperature flux clearly reduces the meridional temperature gradient with maximum impact at the surface (Fig. 5.4(c)). The

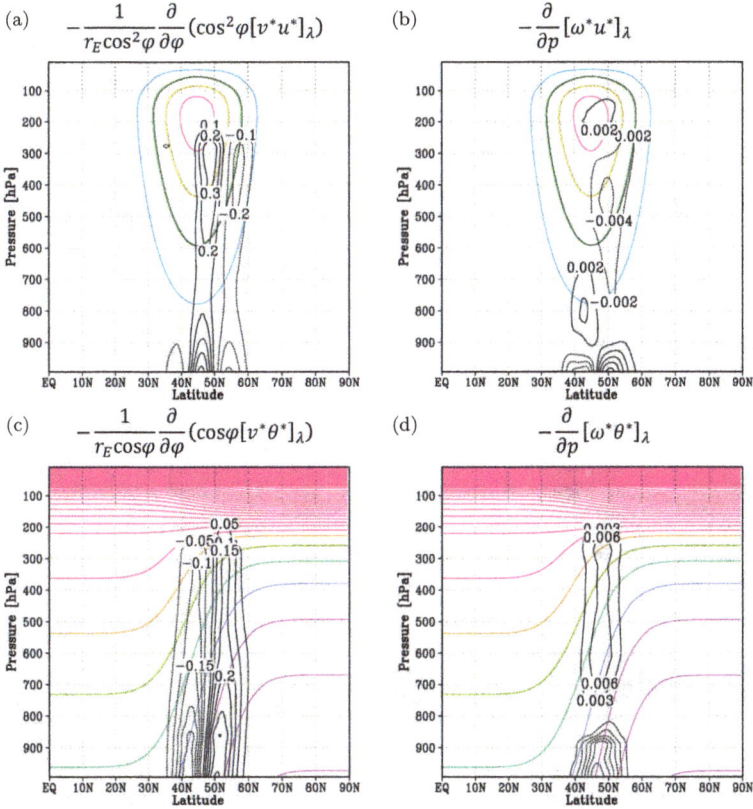

Fig. 5.4. Zonally averaged eddy flux convergence of (a) meridional eddy momentum flux (contour interval $0.1 \, \mathrm{m/s^2}$), (b) vertical eddy momentum flux (contour interval $0.002 \, \mathrm{m/s^2}$), (c) meridional eddy temperature flux (contour interval $0.05 \, \mathrm{K/s}$) and (d) meridional eddy temperature flux (contour interval $0.003 \, \mathrm{K/s}$) on day 5. Negative isolines are dashed and the zero contour has been omitted. The colored isolines in (a) and (b) show zonal mean zonal wind and in (c) and (d) zonal mean potential temperature.

vertical eddy temperature flux enhances the static stability in the latitude range from 40°N to 55°N (Fig. 5.4(d)) but the temperature tendency is smaller than that induced by horizontal eddy temperature fluxes. Consequently, the nonlinear eddy feedback acts in such a way that the zonal mean state becomes stable and this is likely to be the main reason why the eddy energy growth declines and eventually halts.

Figure 5.5 shows the modification of the zonal mean state during the life cycle and the Eliassen–Palm flux. On day 9 the Eliassen Palm flux has the largest amplitude at the surface near 50°N and is primarily directed in vertical direction which indicates baroclinic growth (cf. Fig. 4.28). The temperature gradient at the surface has decreased because the eddy temperature flux warms (cools) the air north (south) of the jet stream. A slight increase of zonal wind at the surface happens. This can result from the meridional eddy momentum

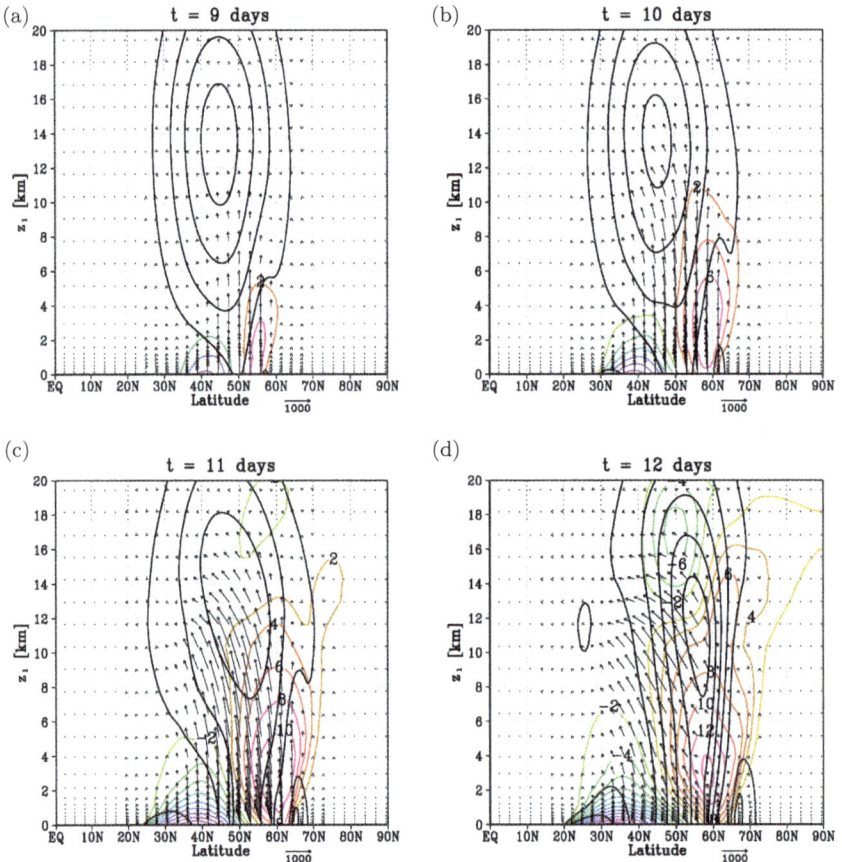

Fig. 5.5. Eliassen–Palm flux (arrows), zonally averaged zonal wind (black isolines, contour interval 10 m/s) and zonally averaged temperature anomaly (colored isolines, contour interval 2 K) in a meridional cross-section on (a) day 9, (b) day 10, (c) day 11 and (d) day 12 of the baroclinic wave lifecycle simulation. Negative isolines are dashed and the zero contour has been omitted.

flux but also from the zonal mean meridional flow that is forced by the geostrophic imbalance due to the modification of the temperature field. On day 10 the Eliassen–Palm flux has gained higher amplitudes at large heights where it turns equatorward. The temperature anomalies increased further and the warm anomaly shifted its maximum upward. Therefore, the static stability rises near the surface which also dampens baroclinic instability. The zonal mean wind at the surface becomes stronger and easterlies appear south and north of the positive maximum. On day 11, the equatorward turning of the Eliassen–Palm flux has proceeded, especially on the southern side of the jet. This signifies the barotropic decay phase (cf. Fig. 4.28). The jet stream has weakened and the zonal mean temperature anomalies attain maximum amplitudes of more than 10 K. Finally on day 12, the eddy activity declines near the surface but at larger heights, the Eliassen–Palm flux established with a significant southward component. This is the time when the barotropic decay is at its maximum (cf. Fig. 5.2(b)). Now we see that the jet stream has a large barotropic component with strong westerlies at the surface which reach velocities of more than 40 m/s. This is not realistic and results from the absence of a surface drag parameterization in this simulation. The remaining jet is possibly unstable with respect to barotropic instability. A longer simulation must be done to check this and it must include longer wave perturbations because wavenumber 6 is possibly too high. The Eliassen–Palm flux also describes the propagation of wave activity in the zonal mean cross-section. Indeed, it can be seen in agreement with the vector field that eddy kinetic energy first increases at higher levels and then on the southern side of the jet (not shown).

Figure 5.6 shows some horizontal fields at various times of the simulated baroclinic wave life cycle. On day 8 we detect alternating highs and lows in the surface pressure field. The low has already a larger intensity than the high and it is deflected northwards while the high appears southward of the center of the zonal anomaly. This happens because zonal mean surface westerlies have modified the pressure field because of geostrophic balance. The low-level temperature wave is also present where the warm (cold) anomaly appears east of the low (high). The 500 hPa geopotential height is phase shifted with respect to the surface pressure wave indicating a growing baroclinic wave. On day 9, the low has intensified further and more rapidly than

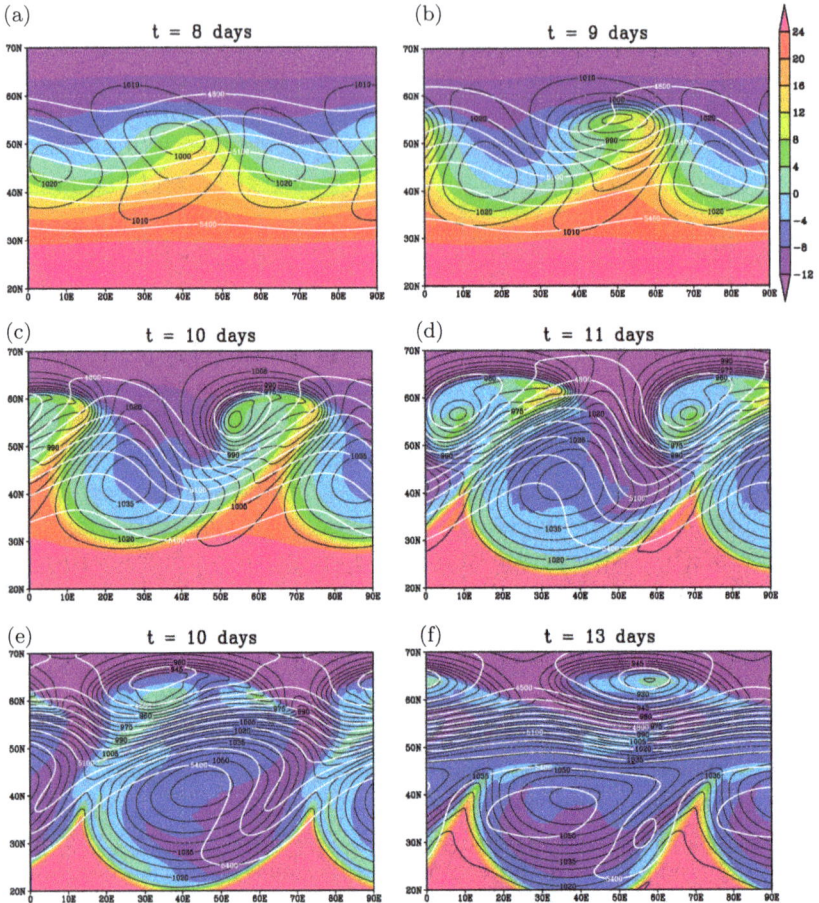

Fig. 5.6. Snapshots of some horizontal fields in the baroclinic wave life cycle experiment: Temperature at the lowest model level $\sigma = 0.995$ (colored shadings, °C), surface pressure (black isolines, contour interval 5 hPa) and 500 hPa geopotential height (white isolines, contour interval 100 m) on (a) day 8, (b) day 9, (c) day 10, (d) day 11, (e) day 12, and (f) day 13.

the high. In the temperature field, we see locally a strong increase in the horizontal temperature gradient although in the zonal average, the opposite happens. This is the frontogenesis that takes place in the nonlinear baroclinic wave development. The temperature field indicates warm and cold fronts. Both seem to merge at the wave cusp that veers towards the low. The region between the two fronts constitutes the **warm sector** of the cyclone. Near the developing cold front (the western boundary of the warm sector) we can detect a strong curvature of the surface isobars indicating high cyclonic vorticity. On day 10, the fronts become more accentuated. Furthermore, we see that the warm sector has shrunk and become very narrow in the northern part. There, one may define the **occlusion front** according to the conceptual Norwegian model by Bjerknes and Solberg (1922) (Fig. 5.7(a)). However, another warm front bends on the northern side around the low. This is a **bent-back warm front** which was introduced in the conceptual model by Shapiro and Keyser (1990) (Fig. 5.7(b)). Now, we see that the upper-level low is nearly at the same position as the surface low indicating the termination of the baroclinic wave development. The warm sector has significantly retreated southwards on day 11 while the surface low has a central pressure of less than 950 hPa. On the other hand, the high has a huge size and a large maximum pressure of nearly 1050 hPa. These intense weather systems result because the damping effect of surface drag is absent in this simulation. On days 12 and 13, the pressure fields attain a marked zonal mean component that describes an intense

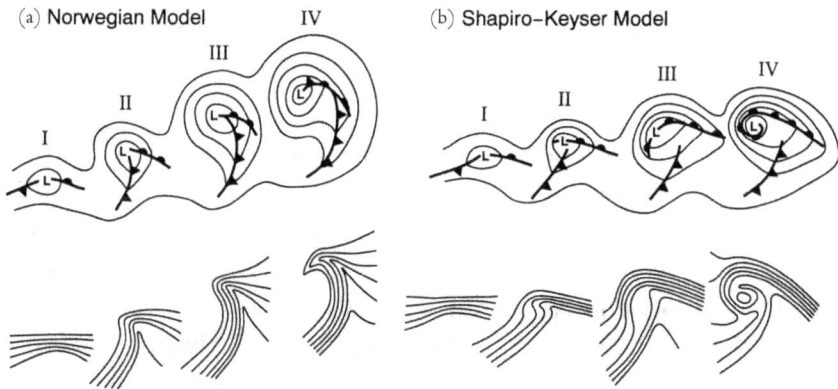

Fig. 5.7. Conceptual models of cyclone evolution showing lower-tropospheric geopotential height and fronts (top), and lower-tropospheric potential temperature (bottom): (a) The Norwegian model by Bjerknes and Solberg (1922): (I) incipient frontal cyclone, (II) and (III) narrowing warm sector, (IV) occlusion, (b) the Shapiro-Keyser model (1990): (I) incipient frontal cyclone, (II) frontal fracture, (III) frontal T-bone and bent-back warm front, (IV) frontal T-bone and warm seclusion. Figure and caption are adapted from Schultz *et al.* (1998), doi: 10.1175/1520-0493(1998)126%3C1767:TEOLSF%3E2.0.CO;2. © American Meteorological Society. Used with permission.

eastward jet. This jet extending to the surface can also be seen in Fig. 5.5(d) and is a consequence of the barotropic decay. The maximum and minimum surface pressures obtain extreme values of 1060 and 920 hPa, respectively, which is unrealistic because the physical parameterizations of the model were switched off. Furthermore, we see that the temperature has significantly decreased at most locations. In the polar front model, it results because the warm sector lifts upward. This is a consequence of vertical eddy temperature fluxes stabilizing the atmosphere and is also evident in the zonal mean picture (see Fig. 5.5(d)).

The numerical experiment has shown that the nonlinear processes transform the simple initial flow (monochromatic wave plus zonal mean flow) into a more complicated flow where different scales play a role. On one hand, wave energy is transferred to the larger scale of the zonal mean and on the other hand to smaller subsynoptic scale where the temperature fronts evolve. The final flow can possibly get unstable again so that another perturbation grows but this must not be of the same kind as the initial instability. It appears that nonlinearities do not only stabilize the unstable flow but they can also introduce a more complicated or even a turbulent flow with many scales of motion. However, this is not always the case as we will see in Section 5.3.

5.2.3   *Impact of microturbulent exchange and condensation*

The numerical model experiment excludes parameterizations for microturbulent exchange in the boundary layer, condensation in clouds and deep convection. All can have an impact on the evolution of the baroclinic wave. In this subsection, we consider these processes by switching on the parameterization schemes in the model. To include moisture, we assume a zonal specific humidity field which has a maximum of 11.25 g/kg at the surface in the tropics. Towards the poles and in the vertical direction it decreases to very low values. Then, the relative humidity is above 60% in a midlatitude belt. The boundary layer parameterization in the model includes the surface transfer of heat, moisture and momentum. Furthermore, it describes vertical transfer of these quantities within the atmosphere (for more details, see Lunkeit *et al.* 2011).

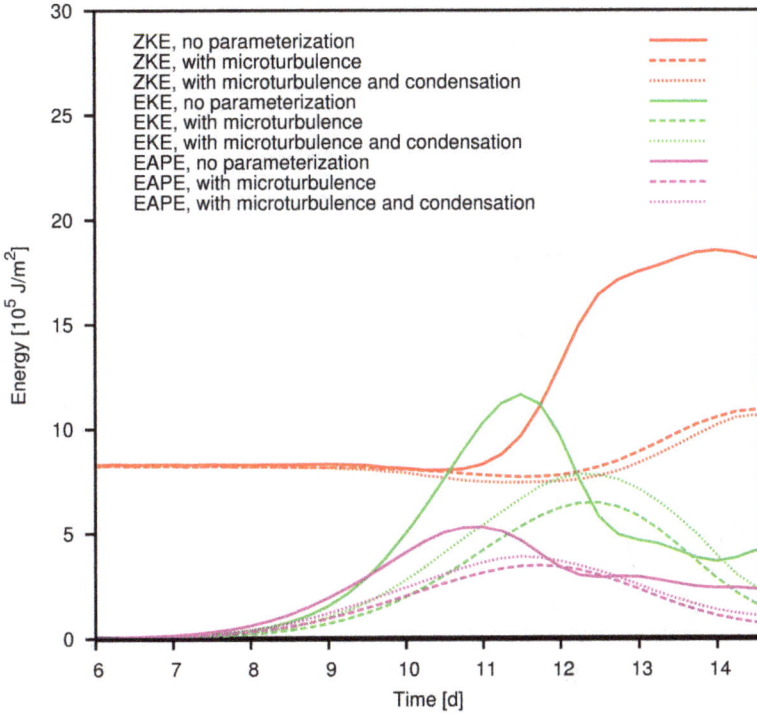

Fig. 5.8.   Time evolution of energy quantities of the numerical baroclinic wave life cycle experiment with no parameterization (solid lines), with parameterization of microturbulence (dashed lines) and with parameterization of microturbulence and condensation in clouds (dotted lines). Shown are ZKE (red), EKE (green) and EAPE (purple).

Figure 5.8 shows the energetics of these sensitivity experiments. Obviously, the inclusion of boundary layer dynamics (represented by the dashed curves) reduces the growth rate of the baroclinic wave and the maximum of EKE and EAPE occurs later at a significantly lower energy level. Consequently, surface drag and heat transfer dampen the baroclinic wave evolution. The zonal kinetic energy also shows a less intense evolution in the wave decay phase. This happens because much kinetic energy is dissipated in the boundary layer. The additional inclusion of moisture, condensation and convection leads to higher eddy energy and a slightly earlier occurrence of its maximum. Therefore, the release of latent heat in regions with ascent supports the baroclinic wave development. We can understand this from the sketch shown in Fig. 4.11. Without condensation, vertical motion cools the warm air due to the stable stratification and this yields negative feedback. Cooling is weaker with condensation because of the latent heat release and the wave can grow faster. However, the supporting effect of condensation does not compensate for the damping effect of boundary layer dynamics so the most vigorous evolution results when no parameterization is switched on.

Figure 5.9(a) shows a snapshot of the surface fields and precipitation in the mature stage of the baroclinic wave for the case with parameterizations of microturbulence, condensation and convection. We see in contrast to the inviscid and dry baroclinic wave (Fig. 5.6) that the fronts have less strong gradients. However, we can still locate the fronts at the positions where the curvature of the isobars is very strong which indicates high geostrophic vorticity. The precipitation pattern forms a "T". This is the **frontal T-bone** in stage III of the life cycle according to the Shapiro-Keyser model (see Fig. 5.7(b)). Along the cold front convective precipitation prevails (as seen by the dotted isolines) while the warm front is accompanied by precipitation induced by large-scale vertical motion. Figure 5.9(b) shows the same fields resulting from NCEP reanalysis of a real mature extratropical marine cyclone. There are some similarities with the simulated cyclone. The frontal T-bone also appears although not as obvious as in the simulated case and the pressure field extrema have similar values. The horizontal temperature gradient at the fronts is more moderate but possibly the sparse observational data are not sufficient for a detailed representation of a front.

The life cycle experiment has shown that a realistic cyclone-anticyclone pair develops when a wave perturbation is added to a zonally constant jet stream. This development can only be reproduced by the full consideration of nonlinear terms and physical parameterizations. This fact makes weather dynamics very complex and, therefore, the forecast of weather is difficult and challenging. In the next sections, we will examine the mathematical properties of nonlinear processes in more detail to obtain a better understanding of its meaning.

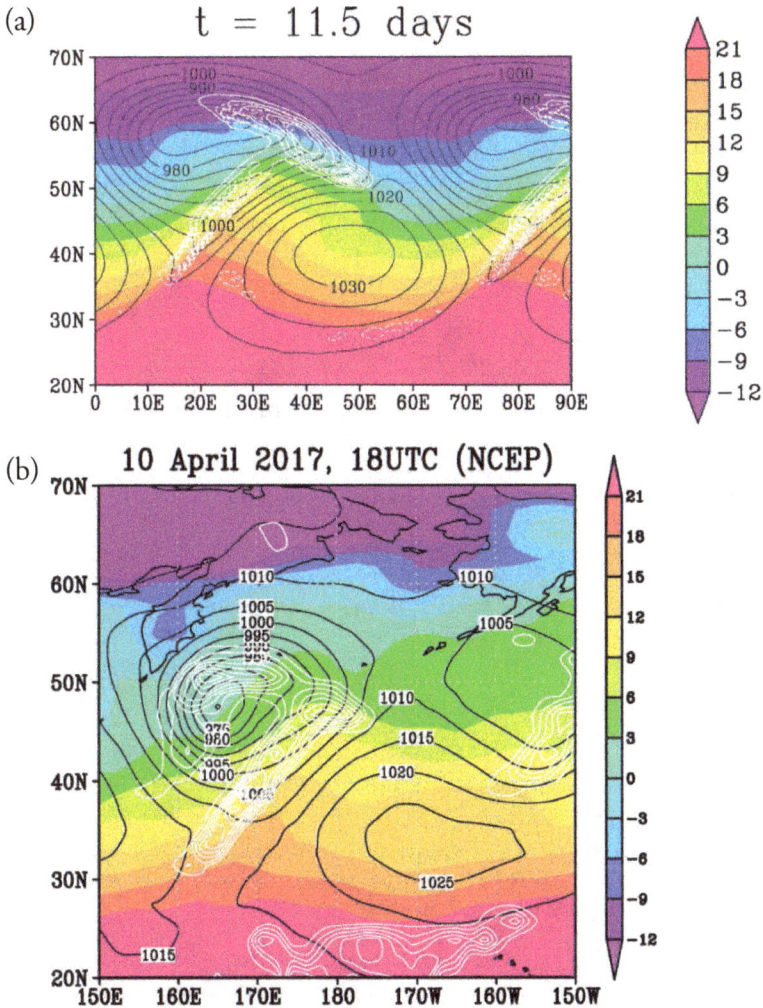

Fig. 5.9. (a) Snapshot of some horizontal fields of the baroclinic wave life cycle experiment with parameterization of microturbulence and condensation in clouds at day = 11.5 days: Temperature at the lowest model level $\sigma = 0.995$ (colored shadings, °C), surface pressure (black isolines, contour interval 5 hPa), precipitation due to the resolved flow (white isolines, contour interval 5 mm/day), precipitation due to convection (white dashed isolines, contour interval 5 mm/day). (b) NCEP reanalysis on April 10, 2017, 18UTC shows the same fields as in (a) except that white isolines picture precipitation of both resolved and convective air motion.

5.3 Dynamics of a Weakly Nonlinear Baroclinic Wave

In this section, we derive an approximate solution of the nonlinear quasigeostrophic two-level model for a **weakly nonlinear baroclinic wave**. Weakly nonlinear means that only the feedback of the first-order nonlinear correction of the zonal mean flow has an impact on the evolution. This feedback is the modification of the zonal mean flow by meridional eddy fluxes. First, we start with the case without frictional dissipation and afterward we include friction in the model.

5.3.1 *Baroclinic wave evolution without frictional dissipation*

The analysis is based on the governing equations (4.150) and (4.151) of the two-level model on an f-plane:

$$\frac{\partial}{\partial t}\left(\nabla_h^2\psi_1 - \frac{\psi_1 - \psi_3}{2L_R^2}\right) + J_{xy}\left(\psi_1, \nabla_h^2\psi_1 - \frac{\psi_1 - \psi_3}{2L_R^2}\right) = 0 \quad (5.3)$$

$$\frac{\partial}{\partial t}\left(\nabla_h^2\psi_3 + \frac{\psi_1 - \psi_3}{2L_R^2}\right) + J_{xy}\left(\psi_3, \nabla_h^2\psi_3 + \frac{\psi_1 - \psi_3}{2L_R^2}\right) = 0 \quad (5.4)$$

Adding and subtracting these equations yield after division by 2:

$$\frac{\partial}{\partial t}\nabla_h^2\psi_M = -\frac{1}{2}\left[J_{xy}\left(\psi_1, \nabla_h^2\psi_1 - \frac{\psi_T}{L_R^2}\right) + J_{xy}\left(\psi_3, \nabla_h^2\psi_3 + \frac{\psi_T}{L_R^2}\right)\right]$$
$$(5.5)$$

$$\frac{\partial}{\partial t}\left(\nabla_h^2\psi_T - \frac{\psi_T}{L_R^2}\right)$$
$$= \frac{1}{2}\left[J_{xy}\left(\psi_3, \nabla_h^2\psi_3 + \frac{\psi_T}{L_R^2}\right) - J_{xy}\left(\psi_1, \nabla_h^2\psi_1 - \frac{\psi_T}{L_R^2}\right)\right] \quad (5.6)$$

where $\psi_M = (\psi_1 + \psi_3)/2$ and $\psi_T = (\psi_1 - \psi_3)/2$ denote the barotropic and baroclinic streamfunction, respectively. With the rules (3.126)–(3.128) we can recast these equations as

$$\frac{\partial}{\partial t}\nabla_h^2\psi_M = -J_{xy}(\psi_M, \nabla_h^2\psi_M) - J_{xy}(\psi_T, \nabla_h^2\psi_T) \quad (5.7)$$

$$\frac{\partial}{\partial t}\left(\nabla_h^2 \psi_T - \frac{\psi_T}{L_R^2}\right) = -J_{xy}\left(\psi_M, \nabla_h^2 \psi_T\right) - J_{xy}\left(\psi_T, \nabla_h^2 \psi_M + \frac{\psi_M}{L_R^2}\right)$$
(5.8)

Now we split into basic flow and perturbation as in the linearized model

$$\psi_M = -U_M y + \psi_M', \quad \psi_T = -U_T y + \psi_T', \tag{5.9}$$

and without loss of generality we can set $U_M = 0$.[a] Inserting (5.9) in the governing equations gives

$$\frac{\partial}{\partial t}\nabla_h^2 \psi_M' = -U_T \frac{\partial}{\partial x}\nabla_h^2 \psi_T' - J_{xy}(\psi_M', \nabla_h^2 \psi_M') - J_{xy}(\psi_T', \nabla_h^2 \psi_T')$$
(5.10)

$$\frac{\partial}{\partial t}\left(\nabla_h^2 \psi_T' - \frac{\psi_T'}{L_R^2}\right) = -U_T \frac{\partial}{\partial x}\left(\nabla_h^2 \psi_M' + \frac{\psi_M'}{L_R^2}\right) - J_{xy}(\psi_M', \nabla_h^2 \psi_T')$$

$$- J_{xy}\left(\psi_T', \nabla_h^2 \psi_M' + \frac{\psi_M'}{L_R^2}\right)$$
(5.11)

By linearizing these equations, we get solutions as discussed in Section 4.6.3. We do not linearize but assume that the instability is weak, i.e. the wavelength is very close to the short-wave cutoff. This can be guaranteed by setting

$$K^2 = \frac{1}{L_R^2(1+\epsilon^2)} \quad \text{with } 0 < \epsilon \ll 1 \tag{5.12}$$

By substituting this in (4.176) we obtain for the linear growth rate

$$\varsigma = U_T k \frac{\epsilon}{\sqrt{2+\epsilon^2}} \tag{5.13}$$

For the analysis, it is necessary to nondimensionalize the equations with the characteristic scales. The time scale is $1/(U_T K \epsilon)$ while for the length scale, we use $1/K$. The scale of the stream function is U_T/K. With these scales the nondimensional equations become

$$\epsilon \frac{\partial}{\partial \tilde{t}}\tilde{\nabla}_h^2 \tilde{\psi}_M' = -\frac{\partial}{\partial \tilde{x}}\tilde{\nabla}_h^2 \tilde{\psi}_T' - J_{\tilde{x}\tilde{y}}\left(\tilde{\psi}_M', \tilde{\nabla}_h^2 \tilde{\psi}_M'\right) - J_{\tilde{x}\tilde{y}}\left(\tilde{\psi}_T', \tilde{\nabla}_h^2 \tilde{\psi}_T'\right)$$
(5.14)

---

[a]A nonzero U_M only describes a zonal propagation of the wave.

$$\varepsilon \frac{\partial}{\partial \tilde{t}} \left[\tilde{\nabla}_h^2 \tilde{\psi}_T' - \left(1 + \varepsilon^2 \right) \tilde{\psi}_T' \right] = -\frac{\partial}{\partial \tilde{x}} \left[\tilde{\nabla}_h^2 \tilde{\psi}_M' + \left(1 + \varepsilon^2 \right) \tilde{\psi}_M' \right]$$

$$- J_{\tilde{x}\tilde{y}} \left(\tilde{\psi}_M' \tilde{\nabla}_h^2 \tilde{\psi}_T' \right) - J_{\tilde{x}\tilde{y}} \left(\tilde{\psi}_T', \tilde{\nabla}_h^2 \tilde{\psi}_M' + \left(1 + \varepsilon^2 \right) \tilde{\psi}_M' \right) \qquad (5.15)$$

Now, we can proceed as we did for the derivation of the quasi-geostrophic approximation (cf. Sections 3.8 and 4.3). We expand the stream functions in a power series of the form

$$\tilde{\psi}_M' = \varepsilon \psi_M^{(1)} + \varepsilon^2 \psi_M^{(2)} + \varepsilon^3 \psi_M^{(3)} + o(\varepsilon^4) \qquad (5.16)$$

$$\tilde{\psi}_T' = \varepsilon \psi_T^{(1)} + \varepsilon^2 \psi_T^{(2)} + \varepsilon^3 \psi_T^{(3)} + o(\varepsilon^4) \qquad (5.17)$$

The zeroth order is not contained in this expansion since it is associated with the basic flow. Equating terms of the order ε in (5.14) and (5.15) yields

$$\frac{\partial}{\partial \tilde{x}} \tilde{\nabla}_h^2 \psi_T^{(1)} = 0, \quad \frac{\partial}{\partial \tilde{x}} \left(\tilde{\nabla}_h^2 + 1 \right) \psi_M^{(1)} = 0 \qquad (5.18)$$

A solution of these equations can be quite complicated because any zonal mean flow fulfills them. Here, we assume a special solution that corresponds to a growing normal mode of the linearized equations, namely

$$\psi_T^{(1)} = 0, \quad \psi_M^{(1)} = A(t) \sin(k\tilde{x}) \sin(l\tilde{y}) \text{ with } k = \sqrt{1 - l^2} \qquad (5.19)$$

The baroclinic stream function $\psi_T^{(1)}$ is zero because it does not have a normal mode contribution of order ε.

Equating terms of order ε^2 leads to

$$\frac{\partial}{\partial \tilde{t}} \tilde{\nabla}_h^2 \psi_M^{(1)} = -\frac{\partial}{\partial \tilde{x}} \tilde{\nabla}_h^2 \psi_T^{(2)} - J_{\tilde{x}\tilde{y}} \left(\psi_M^{(1)}, \tilde{\nabla}_h^2 \psi_M^{(1)} \right) \qquad (5.20)$$

$$\frac{\partial}{\partial \tilde{x}} \left(\tilde{\nabla}_h^2 + 1 \right) \psi_M^{(2)} = 0 \qquad (5.21)$$

The nonlinear term on the right-hand side of (5.20) vanishes because $\tilde{\nabla}_h^2 \psi_M^{(1)} = -(k^2 + l^2)\psi_M^{(1)} = -\psi_M^{(1)}$. The solution of this equation is

$$\psi_T^{(2)} = \frac{1}{k} \frac{dA}{dt} \cos(k\tilde{x}) \sin(l\tilde{y}) + \Psi(\tilde{y}, \tilde{t}) \qquad (5.22)$$

where the zonal mean correction $\Psi(\tilde{y}, \tilde{t})$ will be determined later. Equation (5.21) also allows for nontrivial solutions. However, we will not include them because we are only interested in waves which are directly forced by the primary wave $\psi_M^{(1)}$. Therefore, we set

$$\psi_M^{(2)} = 0 \tag{5.23}$$

For terms of order ε^3 we find

$$\frac{\partial}{\partial \tilde{x}} \tilde{\nabla}_h^2 \psi_T^{(3)} = 0 \tag{5.24}$$

$$\frac{\partial}{\partial \tilde{t}} \left(\tilde{\nabla}_h^2 - 1 \right) \psi_T^{(2)} = -\frac{\partial}{\partial \tilde{x}} \left(\tilde{\nabla}_h^2 + 1 \right) \psi_M^{(3)} - \frac{\partial \psi_M^{(1)}}{\partial \tilde{x}}$$

$$- J_{\tilde{x}\tilde{y}} \left(\psi_M^{(1)}, \tilde{\nabla}_h^2 \psi_T^{(2)} \right) - J_{\tilde{x}\tilde{y}} \left(\psi_T^{(2)}, \tilde{\nabla}_h^2 \psi_M^{(1)} + \psi_M^{(1)} \right) \tag{5.25}$$

By inserting the solutions (5.19) and (5.22) in (5.25) we get

$$-\frac{2}{k} \frac{d^2 A}{d\tilde{t}^2} \cos(k\tilde{x}) \sin(l\tilde{y}) + \frac{\partial}{\partial \tilde{t}} \left(\frac{\partial^2}{\partial \tilde{y}^2} - 1 \right) \Psi$$

$$= -\frac{\partial}{\partial \tilde{x}} \left(\tilde{\nabla}_h^2 + 1 \right) \psi_M^{(3)} - kA \cos(k\tilde{x}) \sin(l\tilde{y})$$

$$+ \frac{A}{k} \frac{dA}{d\tilde{t}} J_{\tilde{x}\tilde{y}}[\sin(k\tilde{x}) \sin(l\tilde{y}), \cos(k\tilde{x}) \sin(l\tilde{y})]$$

$$- Ak \cos(k\tilde{x}) \sin(l\tilde{y}) \frac{\partial^3 \Psi}{\partial \tilde{y}^3} \tag{5.26}$$

Evaluation of the Jacobi-operator yields

$$J_{\tilde{x}\tilde{y}}[\sin(k\tilde{x}) \sin(l\tilde{y}), \cos(k\tilde{x}) \sin(l\tilde{y})]$$

$$= kl[\cos^2(k\tilde{x}) + \sin^2(k\tilde{x})] \sin(l\tilde{y}) \cos(l\tilde{y}) = \frac{kl}{2} \sin(2l\tilde{y}) \tag{5.27}$$

Therefore, collecting all zonally uniform terms leads to the differential equation

$$\frac{\partial}{\partial \tilde{t}} \left(\frac{\partial^2}{\partial \tilde{y}^2} - 1 \right) \Psi = \frac{lA}{2} \frac{dA}{d\tilde{t}} \sin(2l\tilde{y}) \tag{5.28}$$

With the ansatz $\Psi = C(\tilde{t}) \sin(2l\tilde{y})$ the following ordinary differential equation is obtained[b]

$$\frac{lA}{2}\frac{dA}{d\tilde{t}} = -(4l^2 + 1)\frac{dC}{d\tilde{t}} \tag{5.29}$$

The other terms in (5.26) result in

$$-\frac{2}{k}\frac{d^2A}{d\tilde{t}^2}\cos(k\tilde{x})\sin(l\tilde{y}) = -\frac{\partial}{\partial\tilde{x}}(\tilde{\nabla}_h^2 + 1)\psi_M^{(3)} - kA\cos(k\tilde{x})\sin(l\tilde{y})$$

$$- 4kl^3 AC \cos(k\tilde{x})[\sin(l\tilde{y}) - \sin(3l\tilde{y})] \tag{5.30}$$

By equating terms with the spatial pattern $\cos(k\tilde{x})\sin(l\tilde{y})$ we find

$$\frac{2}{k}\frac{d^2A}{d\tilde{t}^2} = kA + 4kl^3 AC \tag{5.31}$$

The remaining terms constitute the third-order streamfunction. However, it does not play a role for the lower orders.[c] With the definition

$$B \equiv \frac{1}{k}\frac{dA}{d\tilde{t}} \tag{5.32}$$

we obtain a closed set of ordinary differential equations of first order:

$$\frac{dA}{d\tilde{t}} = kB \tag{5.33}$$

$$\frac{dB}{d\tilde{t}} = \frac{k}{2}A + 2kl^3 AC \tag{5.34}$$

$$\frac{dC}{d\tilde{t}} = -\frac{kl}{2(4l^2 + 1)}AB \tag{5.35}$$

This equation set constitutes a **dynamical system**. The variables A, B and C represent amplitudes of streamfunction patterns. A

---

[b]The homogeneous solution of (5.28) has the form $\exp(\pm y)$. Such a solution is not forced by the primary wave but it becomes important when real channel boundaries exist.

[c]The streamfunction $\psi_M^{(3)}$ takes the form $\propto \cos(k\tilde{x})\sin(l\tilde{y})$ due to the fourth order equation but this function does not yield a nonzero contribution in (5.30).

and B are the amplitudes of the batrotropic and baroclinic wave streamfunction fields, respectively while C represents the correction of the zonal mean baroclinic flow (see Fig. 5.10). We have drastically simplified the problem with approximations so that only three ordinary differential equations have to be solved instead of two partial

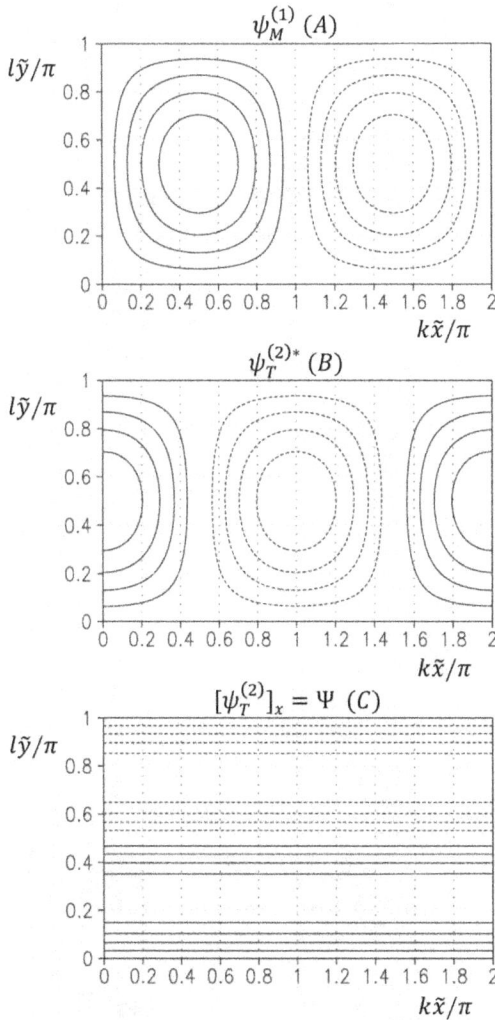

Fig. 5.10.   Interacting streamfunction pattern of the weakly nonlinear baroclinic wave. The associated amplitude variable is specified in brackets. Negative isolines are dashed.

differential equations which depend on the three coordinates x, y and t. The solution of (5.33)–(5.35) can be described as a trajectory in the coordinate system spanned by the model variables A, B and C. This coordinate system is referred to as the **phase space**.

To approach a solution, it is useful to analyze the properties of the dynamical system. The system has two conserved quantities. The first can be obtained by multiplying (5.33) with A and (5.35) with $2(4l^2 + 1)/l$. Then, we obtain by adding these equations

$$\frac{dE_1}{d\tilde{t}} = 0 \text{ with } E_1 = \frac{1}{2}A_r^2 = \frac{1}{2}A^2 + \frac{2(4l^2 + 1)}{l}C \qquad (5.36)$$

where A_r can be understood as a reference amplitude which is identical to A when C becomes zero. The second conserved quantity results by adding (5.34) multiplied with lB and (5.35) multiplied by $2(4l^2 + 1)(2l^3C + 1/2)$

$$\frac{dE_2}{d\tilde{t}} = 0 \text{ with } E_2 = \frac{l}{2}B^2 + 2(4l^2 + 1)\left(l^3C + \frac{1}{2}\right)C \qquad (5.37)$$

With these conserved quantities, we can deduce the trajectories in phase space. For given constants E_1 (or A_r) and E_2 two isosurfaces result in phase space. The line intersecting the two surfaces forms the trajectory of the solution. Two examples are plotted in Fig. 5.11. In the first example, the trajectory is a closed curve where A and B attain both signs. Therefore, the through axis is two times directed against and also with the vertical shear during one cycle. In the second example, two different closed curves result. On each curve, A attains only one sign while B becomes negative and positive.

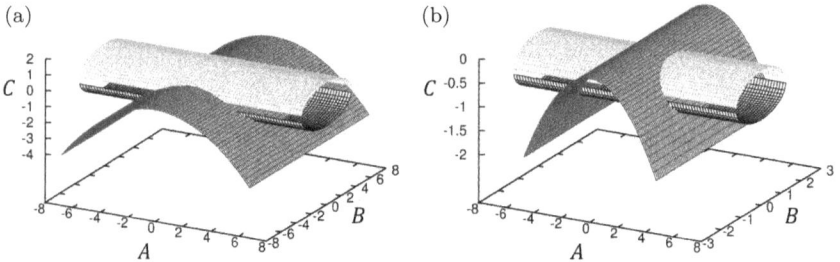

Fig. 5.11.   Isosurfaces $E_1 = \text{const.}$ (dark grey) and $E_2 = \text{const.}$ (light grey) in phase space for (a) $A_r = 0.1$ and $E_2 = 6$ and (b) $A_r = 0.1$ and $E_2 = -1$.

Consequently, the through axis changes its direction only one time during a complete cycle. Obviously, the first example is the more energetic one.

With given A_r we can simplify the dynamical system by eliminating C in (5.34) with (5.36). Then, we obtain

$$\frac{dA}{d\tilde{t}} = kB \tag{5.38}$$

$$\frac{dB}{d\tilde{t}} = \frac{k}{2}A - \frac{kl^4}{2(4l^2+1)}A(A^2 - A_r^2) \tag{5.39}$$

These equations can be written in the form of **Hamiltonian's equations**

$$\frac{dA}{d\tilde{t}} = \frac{\partial \mathcal{H}}{\partial B}, \quad \frac{dB}{d\tilde{t}} = -\frac{\partial \mathcal{H}}{\partial A} \tag{5.40}$$

where the **Hamiltonian** $\mathcal{H}(A, B)$ is given by

$$\mathcal{H} = -\frac{k}{4}A^2 + \frac{k}{2}B^2 + \frac{kl^4}{4(4l^2+1)}\left(\frac{1}{2}A^2 - A_r^2\right)A^2 \tag{5.41}$$

The Hamiltonian represents a "stream function" in phase space spanned by the variables A and B. The streamline $\mathcal{H} = $ const. results by integrating (5.40) in time. Therefore, $\mathcal{H}$ becomes a conserved quantity. Figure 5.12 shows the Hamiltonian $\mathcal{H}$ in phase space and the possible trajectories can be easily deduced by observing the isolines of $\mathcal{H}$. Note that the motion is clockwise around the minima of $\mathcal{H}$ in this case.

Up to now we can only visualize the trajectory of the solution in phase space but do not know anything about the time evolution. To find this we proceed by insertion of (5.38) into (5.39) which leads to

$$\frac{1}{k}\frac{d^2A}{d\tilde{t}^2} - \frac{k}{2}A + \frac{kl^4}{2(4l^2+1)}A(A^2 - A_r^2) = 0 \tag{5.42}$$

This equation corresponds to the equation of motion for a single particle in a conservative force field. By multiplying with dA/dt we get

$$\frac{d}{d\tilde{t}}\left[\frac{1}{2k}\left(\frac{dA}{d\tilde{t}}\right)^2 + \mathcal{V}(A)\right] = \frac{d\mathcal{H}}{d\tilde{t}} = 0 \tag{5.43}$$

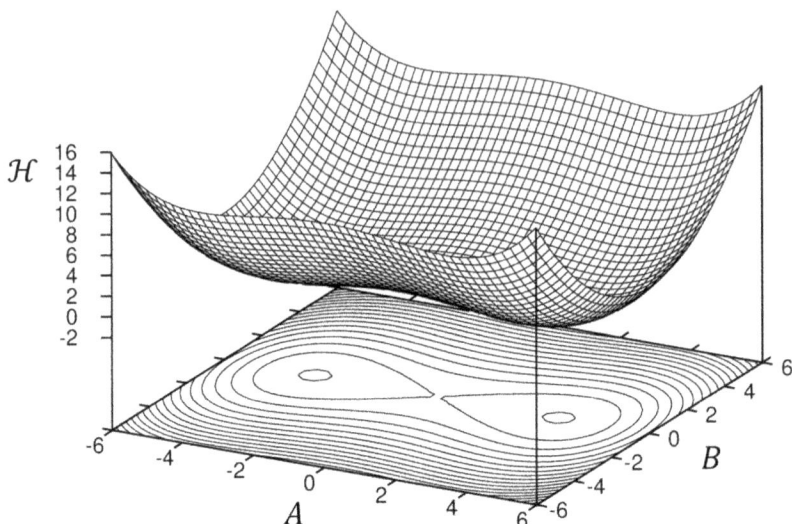

Fig. 5.12.   Hamilton function $\mathcal{H}(A, B)$ as a function of A and B for $k = l = 1/\sqrt{2}$ and $A_r = 0.1$.

where $\mathcal{V}(A)$ denotes a "potential" given by

$$\mathcal{V}(A) = -\frac{k}{4}A^2 + \frac{kl^4}{4(4l^2+1)}\left(\frac{1}{2}A^2 - A_r^2\right)A^2 \qquad (5.44)$$

Therefore, the Hamiltonian $\mathcal{H}$ constitutes as in classical mechanics an "energy" that remains constant. The "potential" $\mathcal{V}$ determines the conservative "force" $-\partial\mathcal{V}/\partial A$ that accelerates the "particle" in this analogy. Figure 5.13 displays the potential $\mathcal{V}$ as a function of A. There are two potential wells and a barrier for $A = 0$. The particle can oscillate from one well to the other if the energy is higher than the potential energy of the barrier. In the other case, the particle is trapped in one of the two potential wells.

With given Hamiltonian $\mathcal{H}$ we find the differential equation

$$\frac{dA}{d\tilde{t}} = \pm\sqrt{2k(\mathcal{H} - \mathcal{V}(A))} \qquad (5.45)$$

where the two signs indicate the **time reversibility** of the system. Inserting of (5.44) and time integration yield

$$\pm(\tilde{t} - \tilde{t}_0) = \int_{A_0}^{A} \frac{dA}{\sqrt{2k\left(\mathcal{H} + \frac{k}{4}A^2 - \frac{kl^4}{4(4l^2+1)}\left(\frac{1}{2}A^2 - A_r^2\right)A^2\right)}} \qquad (5.46)$$

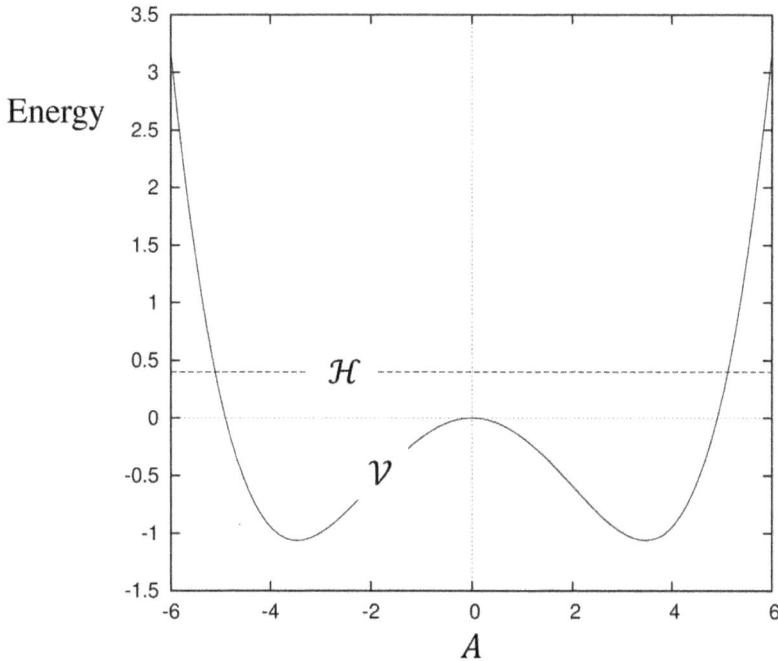

Fig. 5.13. Potential $\mathcal{V}$ as a function of A for $k = l = 1/\sqrt{2}$ and $A_r = 0.1$ (solid line). The dashed line shows the energy level of the "particle" in the described analogy.

where A_0 denotes the amplitude at $\tilde{t} = \tilde{t}_0$. A simple determination of the integral is possible in the special case $\mathcal{H} = 0$. In this case, the wave approaches the growing normal mode solution (4.176)–(4.177) for $\tilde{t} \to -\infty$. We solve the integral (5.46) by using a mathematical handbook

$$\pm(\tilde{t} - \tilde{t}_0) = \frac{1}{\sqrt{c_1}} \left[\ln\left(\frac{A}{A_0}\right) - \ln\left(\frac{\sqrt{c_1 - c_2 A^2} + \sqrt{c_1}}{\sqrt{c_1 - c_2 A_0^2} + \sqrt{c_1}}\right) \right] \quad (5.47)$$

where

$$c_1 = \frac{k^2}{2}\left(1 + \frac{l^4}{4l^2 + 1}A_r^2\right), \quad c_2 = \frac{k^2 l^4}{4(4l^2 + 1)} \quad (5.48)$$

We cannot solve (5.47) for A but it is possible to plot $A(t)$ as shown in Fig. 5.14. After the initial exponential growth phase, the time tendency of A declines, and a maximum is reached. There, the amplitude

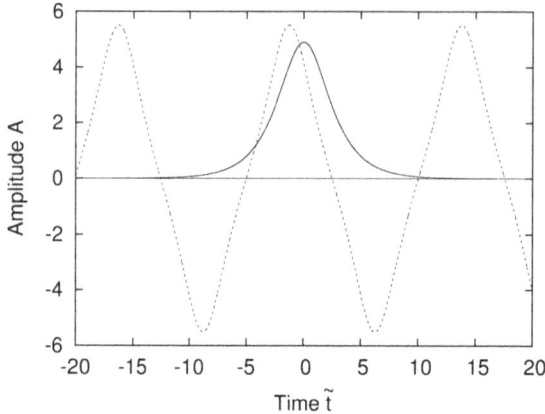

Fig. 5.14.   Time evolution of amplitude A for $k = l = 1/\sqrt{2}$ and $A_r = \mathcal{H} = 0$ (solid line). The dashed line shows a numerical integration of the dynamical system equations (5.33)–(5.35) with the initial conditions $A(-20) = 0.1$, $B(-20) = 2$ and $C(-20) = 0$.

takes the value $\sqrt{c_1/c_2}$ as can be seen by setting the time tendency to zero in Eq. (5.45). Finally, the wave amplitude starts to decline and approaches an exponential decay phase that is associated with the decaying normal mode. A different evolution occurs for the case with a positive $\mathcal{H}$ (dashed curve in Fig. 5.14). Then, the amplitude A executes periodic oscillations and the wave grows much faster at a small amplitude compared to the analytical solution (5.47). This characterizes the nonmodal growth as analyzed in Section 4.6.3. The periodical oscillation makes the system predictable like the ticking of a clock. The oscillation period and the amplitude depend on the initial conditions. Weather does not behave in this way because it is not predictable like a clock. There are many reasons for this discrepancy. One reason is the absence of irreversible processes like turbulent exchange.

5.3.2   *Baroclinic wave evolution with frictional dissipation*

To consider the impact of friction we introduce Ekman layers in the model. We modify the continuity equations by inserting the vertical velocity due to Ekman pumping at the lower and upper boundary (cf. Section 3.9). The latter is unusual because the atmosphere has no boundary layer at the top but it simplifies the analysis significantly. The more realistic case with only one Ekman layer at the bottom was treated by Pedlosky (1983). In the laboratory, it is possible to examine baroclinic waves in a two-layer fluid with two Ekman layers (e.g. Hart, 1985). For the vertical velocities at the top and bottom, we assume:

$$\omega_0 \approx -g\rho_0 w_0 = \frac{\Delta p_{E0}}{2\pi} \nabla_h^2 \psi_1, \quad \omega_4 \approx -g\rho_4 w_4 = -\frac{\Delta p_{E4}}{2\pi} \nabla_h^2 \psi_3$$

(5.49)

where Δp_{E0} and Δp_{E4} are the depths of the upper and lower Ekman layers in isobaric coordinates, respectively. Inserting these in the approximation (4.118) and using the result to substitute the divergence in the quasigeostrophic vorticity equation (4.64) at both levels for $\beta = 0$ gives

$$\frac{\partial \nabla_h^2 \psi_1}{\partial t} + J_{xy}\left(\psi_1, \nabla_h^2 \psi_1\right) = f_0 \frac{\omega_2}{\Delta p} - \frac{\nabla_h^2 \psi_1}{\tau_{E0}}$$

(5.50)

$$\frac{\partial \nabla_h^2 \psi_3}{\partial t} + J_{xy}\left(\psi_3, \nabla_h^2 \psi_3\right) = -f_0 \frac{\omega_2}{\Delta p} - \frac{\nabla_h^2 \psi_3}{\tau_{E4}}$$

(5.51)

where τ_{E0} and τ_{E4} are the corresponding Ekman spin-down times. Setting $\tau_{E0} = \tau_{E4} \equiv \tau_E$ we can derive the following nondimensional equations for the quasigeostrophic two-level model with frictional dissipation

$$\varepsilon \left(\frac{\partial}{\partial \tilde{t}} + \frac{1}{\tilde{\tau}_E}\right) \tilde{\nabla}_h^2 \tilde{\psi}_M'$$

$$= -\frac{\partial}{\partial \tilde{x}} \tilde{\nabla}_h^2 \tilde{\psi}_T' - J_{\tilde{x}\tilde{y}}\left(\tilde{\psi}_M', \tilde{\nabla}_h^2 \tilde{\psi}_M'\right) - J_{\tilde{x}\tilde{y}}\left(\tilde{\psi}_T', \tilde{\nabla}_h^2 \tilde{\psi}_T'\right)$$

(5.52)

$$\varepsilon \frac{\partial}{\partial \tilde{t}} \left[\tilde{\nabla}_h^2 \tilde{\psi}_T' - (1 + \varepsilon^2)\, \tilde{\psi}_T'\right] + \varepsilon \frac{\tilde{\nabla}_h^2 \tilde{\psi}_T'}{\tilde{\tau}_E}$$

$$= -\frac{\partial}{\partial \tilde{x}} \left[\tilde{\nabla}_h^2 \tilde{\psi}_M' + (1 + \varepsilon^2)\, \tilde{\psi}_M'\right]$$

$$- J_{\tilde{x}\tilde{y}}\left(\tilde{\psi}_M' \tilde{\nabla}_h^2 \tilde{\psi}_T'\right) - J_{\tilde{x}\tilde{y}}\left(\tilde{\psi}_T', \tilde{\nabla}_h^2 \tilde{\psi}_M' + (1 + \varepsilon^2)\, \tilde{\psi}_M'\right)$$

(5.53)

where it is assumed that the nondimensional Ekman spin-down time $\tilde{\tau}_E$ is of order unity which corresponds to weak frictional dissipation. First, we expand the stream function in the power of ε. Then, the different orders in ε are considered. The first-order result is identical to the case without frictional dissipation. Therefore,

$$\psi_T^{(1)} = 0, \quad \psi_M^{(1)} = A(t)\sin(k\tilde{x})\sin(l\tilde{y}) \quad \text{with } k = \sqrt{1 - l^2} \quad (5.54)$$

The second-order result experiences a modification since Ekman-pumping in the barotropic vorticity equation appears at this order. We get

$$\psi_T^{(2)} = \frac{1}{k}\left(\frac{dA}{d\tilde{t}} + \frac{A}{\tilde{\tau}_E}\right)\cos(k\tilde{x})\sin(l\tilde{y}) + \Psi(\tilde{y}, \tilde{t}) \quad (5.55)$$

The equation for the zonally uniform stream function is obtained as a result of the third-order consideration

$$\left[\frac{\partial}{\partial\tilde{t}}\left(\frac{\partial^2}{\partial\tilde{y}^2} - 1\right) + \frac{1}{\tilde{\tau}_E}\frac{\partial^2}{\partial\tilde{y}^2}\right]\Psi = \frac{lA}{2}\left(\frac{dA}{d\tilde{t}} + \frac{A}{\tilde{\tau}_E}\right)\sin(2l\tilde{y}) \quad (5.56)$$

Inserting the ansatz $\Psi = C(\tilde{t})\sin(2l\tilde{y})$ in (5.56) yields the ordinary differential equation

$$\frac{lA}{2}\left(\frac{dA}{d\tilde{t}} + \frac{A}{\tilde{\tau}_E}\right) = -(4l^2 + 1)\frac{dC}{d\tilde{t}} - \frac{4l^2}{\tilde{\tau}_E}C \quad (5.57)$$

Collecting all terms proportional to $\cos(k\tilde{x})\sin(l\tilde{y})$ gives

$$\frac{2}{k}\frac{d}{d\tilde{t}}\left(\frac{dA}{d\tilde{t}} + \frac{A}{\tilde{\tau}_E}\right) + \frac{1}{\tilde{\tau}_E k}\left(\frac{dA}{d\tilde{t}} + \frac{A}{\tilde{\tau}_E}\right) = kA + 4kl^3 AC \quad (5.58)$$

finally leading to the dynamical system

$$\frac{dA}{d\tilde{t}} = kB - \frac{A}{\tilde{\tau}_E} \quad (5.59)$$

$$\frac{dB}{d\tilde{t}} = \frac{k}{2}A + 2kl^3 AC - \frac{B}{2\tilde{\tau}_E} \quad (5.60)$$

$$\frac{dC}{d\tilde{t}} = -\frac{kl}{2(4l^2 + 1)}AB - \frac{4l^2}{4l^2 + 1}\frac{C}{\tilde{\tau}_E} \quad (5.61)$$

Now, the system is irreversible in time, i.e. the time-reversed evolution is governed by different equations. This has consequences for the time behavior of the resulting model variables. Figure 5.15 shows a time series of A for different values of Ekman spin-down times $\tilde{\tau}_E$. An infinite $\tilde{\tau}_E$ provides the above-analyzed periodic oscillations. For $\tilde{\tau}_E = 25$, the evolution starts with an aperiodic oscillation until the oscillation becomes periodic. For $\tilde{\tau}_E = 16$ the oscillation remains aperiodic in the time interval shown and for $\tilde{\tau}_E = 5$ the dissipation dampens the oscillation so that finally a steady-state wave arises.

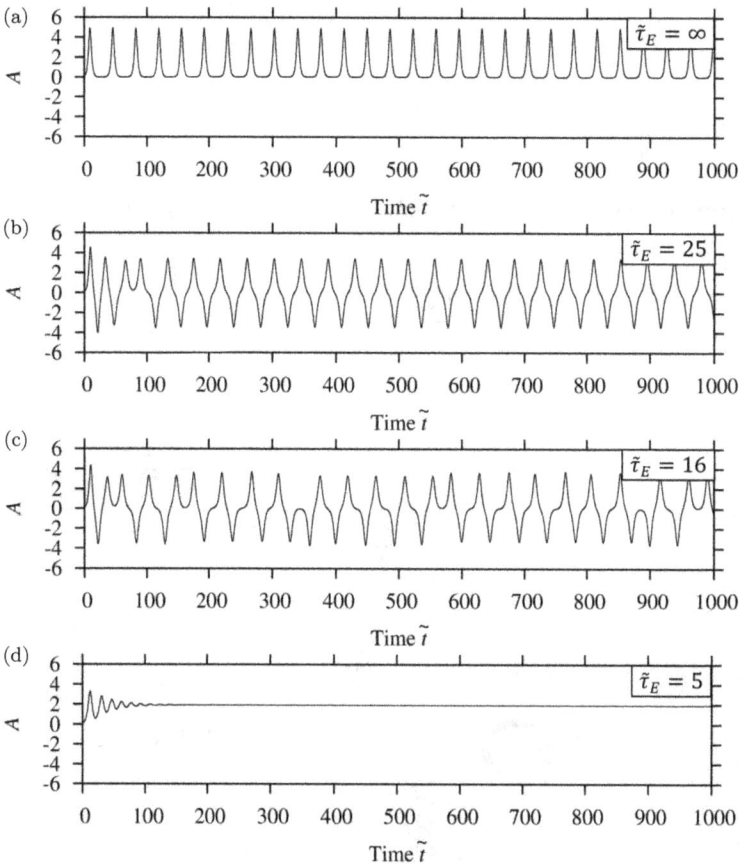

Fig. 5.15. Time evolution of amplitude A for $k = l = 1/\sqrt{2}$ for (a) $\tilde{\tau}_E = \infty$, (b) $\tilde{\tau}_E = 25$, (c) $\tilde{\tau}_E = 16$ and (d) $\tilde{\tau}_E = 5$. The time series results from numerical integrations of the dynamical system equations (5.59)–(5.61) with the initial conditions $A(0) = 0.1$, $B(0) = 0.1/\sqrt{2}$ and $C(0) = 0$.

Figure 5.16 shows these solutions in phase space. The trajectories form eventually closed lines in the cases $\tilde{\tau}_E = \infty$ and $\tilde{\tau}_E = 25$. A spiral curve approaches a point in the case with stronger dissipation. For $\tilde{\tau}_E = 16$, the curve neither approaches a closed line nor a point. Rather, it appears that, with increasing time, the trajectories occupy a surface in phase space. We will see in Section 5.5 that, after a long time, the trajectory forms for finite $\tilde{\tau}_E$ an **attractor** which does not depend on the initial conditions. On the other hand, solutions without frictional dissipation do not approach an attractor. Then, the wave never "forgets" the initial condition.

The dynamical system equations (5.59)–(5.61) can be written in a more concise form by the use of the following definitions:

$$X = \frac{2\tilde{\tau}_E k l^2}{\sqrt{4l^2 + 1}} A, \quad Y = \frac{2\tilde{\tau}_E^2 k^2 l^2}{\sqrt{4l^2 + 1}} B, \quad Z = -4\tilde{\tau}_E^2 k^2 l^3 C \qquad (5.62)$$

and the new timescale $\mathcal{T} = \tilde{t}/(2\tilde{\tau}_E)$. With these definitions, we obtain the dynamical system

$$\frac{dX}{d\mathcal{T}} = \sigma(Y - X) \qquad (5.63)$$

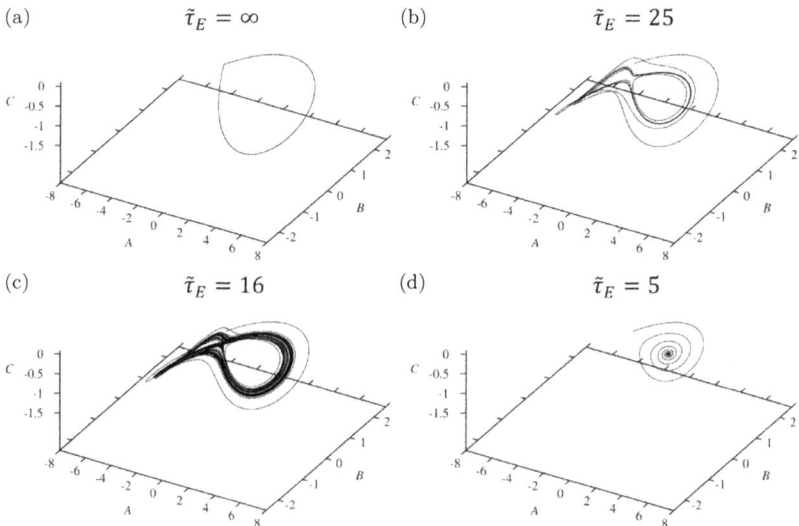

Fig. 5.16.   Phase space trajectories of the numerical experiments, whose time series were shown in Fig. 5.15.

$$\frac{d\mathcal{Y}}{d\mathcal{T}} = \mathcal{R}\mathcal{X} - \mathcal{X}\mathcal{Z} - \mathcal{Y} \tag{5.64}$$

$$\frac{d\mathcal{Z}}{d\mathcal{T}} = \mathcal{X}\mathcal{Y} - \mathcal{B}\mathcal{Z} \tag{5.65}$$

where $\sigma = 2$, $\mathcal{R} = \tilde{\tau}_E^2 k^2$ and $\mathcal{B} = 8l^2/(4l^2 + 1)$. These are the **Lorenz equations** which were originally derived by Lorenz (1963) from the roll convection model by Saltzman (1962). We will examine these equations in detail with dynamical system analysis presented in Section 5.5. For a **square wave** $(k = l = 1/\sqrt{2})$ the parameters become $\sigma = 2$, $\mathcal{R} = \tilde{\tau}_E^2/2$ and $\mathcal{B} = 4/3$.

5.4 Spectral Representation

In the last section, we have derived a dynamical system from the governing equations by assuming weak instability and expansion of the fields in terms of a small parameter. Therefore, this represents a method that applies only under special circumstances. However, a general way to derive dynamical systems from the governing equation exists. It is the **Galerkin method** that transforms the governing equations to a representation in **Hilbert space** as described in the following.

5.4.1 *Spectral method*

The Hilbert space comprises all square-integrable functions in the considered domain which fulfills given boundary conditions. For a rigorous treatment of formalism, we refer to the mathematical literature.[d] We only quote the following properties of Hilbert space of square-integrable functions.

– Every field function $b(\mathbf{r})$ can be represented by an infinite set of complex-valued **orthogonal basis functions**

$$b(\mathbf{r}) = \sum_{n=1}^{\infty} \hat{b}_n f_n(\mathbf{r}) \tag{5.66}$$

---

[d]See e.g. von Neumann (1955).

where the basis functions have the property

$$\frac{1}{V(\mathcal{D})} \int_{V(\mathcal{D})} f_m^C(\mathbf{r}) f_n(\mathbf{r}) dV = \begin{cases} 1 & \text{for } m = n \\ 0 & \text{for } m \neq n \end{cases} \equiv \delta_{m,n} \quad (5.67)$$

In this formula $V(\mathcal{D})$ is the volume of the considered domain $\mathcal{D}$ and $\delta_{m,n}$ the **Kronecker delta**.

– A scalar product exists and is defined by

$$\langle a(\mathbf{r}) \,|\, b(\mathbf{r}) \rangle \equiv \frac{1}{V(\mathcal{D})} \int_{V(\mathcal{D})} a^C(\mathbf{r}) b(\mathbf{r}) dV \qquad (5.68)$$

Obviously, $\langle a(\mathbf{r}) \,|\, a(\mathbf{r}) \rangle$ is positive definite and can be used to define the norm $\|a(\mathbf{r})\| \equiv \sqrt{\langle a(\mathbf{r}) \,|\, a(\mathbf{r}) \rangle}$.

– The representation (5.66) can be interpreted as a vector in a Cartesian coordinate system in which the coordinates measure the coefficients $\hat{b}_n$

$$\vec{b} = \sum_{n=1}^{\infty} \hat{b}_n \vec{e}_n \qquad (5.69)$$

Because of (5.67) we can write the scalar product as

$$\langle a(\mathbf{r}) \,|\, b(\mathbf{r}) \rangle = \sum_{n=1}^{\infty} \hat{a}_n^C \hat{b}_n \equiv \vec{a} \cdot \vec{b} \qquad (5.70)$$

which corresponds to (A.7) in Appendix A but in this case, the sum includes complex numbers and extends until infinity.

– The Hilbert space forms a **vector space**. This means that rules (A.3)–(A.9) in Appendix A apply also for vectors (5.69).

The important step is that the representation can be used to derive an infinite number of ordinary differential equations from the governing equations. Let us assume that the governing equations can be written as

$$\frac{\partial}{\partial t}(\mathcal{L}_j b_j) = \mathcal{F}_j(b_1, b_2, b_3, \ldots, b_J, t) \quad \text{for } j = 1, 2, 3, \ldots, J \qquad (5.71)$$

where $b_j(\mathbf{r}, t)$ is a physical field, $\mathcal{L}_j$ a linear differential operator, and $\mathcal{F}_j$ a nonlinear function of its arguments and of its spatial derivatives.

For every field a representation

$$b_j(\mathbf{r}, t) = \sum_{n=1}^{\infty} \hat{b}_{jn}(t) f_n(\mathbf{r}) \tag{5.72}$$

can be found. The time-dependent $\hat{b}_{jn}$ constitute the **spectral coefficients** of the field $b_j(\mathbf{r}, t)$ and (5.72) is referred to as the **spectral representation**. The orthogonal functions $f_n(\mathbf{r})$ should be eigenfunctions of the operator $\mathcal{L}_j$. This is trivially fulfilled for the primitive equations (2.87)–(2.92) and the shallow water equations (3.10)–(3.11) since $\mathcal{L}_j$ becomes 1 in these cases. $\mathcal{L}_j$ includes the Laplace operator ∇_h^2 in the case of the barotropic potential vorticity equation (3.166) or the quasigeostrophic two-level model equations (5.10)–(5.11). Then, it is possible to find orthogonal eigenfunctions of this operator and we find after the insertion of (5.72) in (5.71)

$$\sum_{n=1}^{\infty} L_{jn} \frac{d\hat{b}_{jn}}{dt} f_n(\mathbf{r}) = \mathcal{F}_j(b_1, b_2, b_3, \dots, b_J, t) \tag{5.73}$$

where L_{jn} denotes the eigenvalue of the operator $\mathcal{L}_j$ applied to the eigenfunction $f_n(\mathbf{r})$. By multiplying this equation with $f_m(\mathbf{r})$ and integrating the result over the domain we get due to the orthogonality relation (5.67)

$$\frac{d\hat{b}_{jm}}{dt} = \frac{1}{L_{jm}} \langle f_m \,|\, \mathcal{F}_j(b_1, b_2, b_3, \dots, b_J, t) \rangle$$

$$\text{for } j = 1, 2, 3, \dots, J \quad \text{and} \quad m = 1, 2, 3, \dots, \infty \tag{5.74}$$

These equations constitute the representation of model dynamics in Hilbert space. However, we cannot work with an infinite number of equations in practice. This problem can be solved by truncating the expansion (5.72) at finite N. Then, the procedure of the Galerkin method is completed and we end up with a dynamical system having $J \times N$ equations. Furthermore, the truncated Hilbert space becomes identical with the complex phase space of the resulting dynamical system. It can be easily associated with the familiar real phase space by considering the real and imaginary parts as independent vector components.

5.4.2 Spectral representation of the quasigeostrophic two-level model

We will demonstrate the Galerkin method for the example of the quasigeostrophic two-level model with frictional dissipation.[e] We consider a domain that is periodic in x direction over the length L_x and that is bounded at $y = 0$ and $y = L_y$. At the meridional boundaries, we assume vanishing meridional geostrophic wind but do not demand this for the ageostrophic wind so there are no real channel boundaries. Then, the barotropic and baroclinic streamfunction can be represented by the Fourier expansions

$$\psi'_M = \sum_{n=1}^{N} \sum_{m=1}^{M} [\Psi_{r,n}^m(t) \cos(k_m x) + \Psi_{i,n}^m(t) \sin(k_m x)] \sin(l_n y)$$

$$+ \sum_{n=0}^{N} \Psi_{r,n}^0(t) \sin(l_n y) \tag{5.75}$$

$$\psi'_T = \sum_{n=1}^{N} \sum_{m=1}^{M} [\Theta_{r,n}^m(t) \cos(k_m x) + \Theta_{i,n}^m(t) \sin(k_m x)] \sin(l_n y)$$

$$+ \sum_{n=0}^{N} \Theta_{r,n}^0(t) \sin(l_n y) \tag{5.76}$$

where $k_m = 2m\pi/L_x$ and $l_n = n\pi/L_y$. We can apply the Galerkin method to this real representation. However, the method is much more efficient by using complex trigonometric functions. A general representation of a two-dimensional field becomes with these functions

$$\psi'_M = \sum_{n=-N}^{N} \sum_{m=-M}^{M} \Psi_n^m(t) e^{ik_m x} e^{il_n y} \tag{5.77}$$

$$\psi'_T = \sum_{n=-N}^{N} \sum_{m=-M}^{M} \Theta_n^m(t) e^{ik_m x} e^{il_n y} \tag{5.78}$$

---

[e]We will use the scaling introduced in Section 5.3.1 but we omit the cumbersome tilde symbol ˜ denoting nondimensional variables.

To obtain a result in agreement with (5.75) the relations

$$\Psi_{-n}^m = -\Psi_n^m, \quad \Psi_{-n}^{-m} = -\Psi_n^{-m}, \quad \Psi_n^{-m} = -\Psi_n^{mC} \tag{5.79}$$

must hold. Then, we calculate

$$\Psi_n^m e^{ik_m x} e^{il_n y} + \Psi_n^{-m} e^{-ik_m x} e^{il_n y}$$

$$+ \Psi_{-n}^m e^{ik_m x} e^{-il_n y} + \Psi_{-n}^{-m} e^{-ik_m x} e^{-il_n y}$$

$$= (\Psi_n^m e^{il_n y} + \Psi_{-n}^m e^{-il_n y}) e^{ik_m x} + (\Psi_n^{-m} e^{il_n y} + \Psi_{-n}^{-m} e^{-il_n y}) e^{-ik_m x}$$

$$= 2i(\Psi_n^m e^{ik_m x} + \Psi_n^{-m} e^{-ik_m x}) \sin(l_n y)$$

$$= 2i(\Psi_n^m e^{ik_m x} - \Psi_n^{mC} e^{-ik_m x}) \sin(l_n y)$$

$$= -4[\Psi_{n\,i}^m \cos(k_m x) + \Psi_{n\,r}^m \sin(k_m x)] \sin(l_n y) \tag{5.80}$$

By equating the coefficients, we find

$$\Psi_{n\,i}^m = -\frac{\Psi_{r,n}^m}{4}, \quad \Psi_{n\,r}^m = -\frac{\Psi_{i,n}^m}{4} \quad \text{for} \quad m, n > 0 \tag{5.81}$$

and for $m = 0$ we get by evaluating $\Psi_n^0 e^{il_n y} + \Psi_{-n}^0 e^{-il_n y}$

$$\Psi_{n\,i}^0 = -\frac{\Psi_{r,n}^0}{2}, \quad \Psi_{n\,r}^0 = 0 \quad \text{for} \quad n > 0 \tag{5.82}$$

Therefore, it is sufficient to predict Ψ_n^m with $m > 0$ and $n > 0$ for determining the fields at a future time point. Note that $\Psi_0^m = 0$ is zero because of the boundary condition. The argumentation is likewise for the spectral coefficients of ψ_T'.

The Fourier functions $F_n^m \equiv e^{ik_m x} e^{il_n y}$ are orthogonal because

$$\langle F_{n_1}^{m_1} | F_{n_2}^{m_2} \rangle = \delta_{m_1,m_2} \delta_{n_1,n_2} \tag{5.83}$$

and the Fourier functions are eigenfunctions of the Laplace operator since

$$\nabla_h^2 F_n^m = -(k_m^2 + l_n^2) F_n^m \equiv -K_n^{m^2} F_n^m, \tag{5.84}$$

Furthermore, the zonal and meridional derivatives of these functions yield

$$\frac{\partial F_n^m}{\partial x} = ik_m F_n^m, \quad \frac{\partial F_n^m}{\partial y} = il_n F_n^m \tag{5.85}$$

and the Jacobi operator of two different Fourier functions becomes

$$J_{xy}(F_{n_1}^{m_1}, F_{n_2}^{m_2}) = (k_{m_2} l_{n_1} - k_{m_1} l_{n_2}) F_{n_1+n_2}^{m_1+m_2} \qquad (5.86)$$

Consequently, the wavenumber sum of the two forcing waves yields the wavenumber of the forced wave. However, this does not mean that the forced wave has a smaller scale because wavenumbers can also be negative in the expansions (5.77) and (5.78). For example, two waves having zonal wavenumbers $|k_{m_1}|$ and $|k_{m_2}|$ trigger waves with wavenumbers $|k_{m_1} + k_{m_2}|$ and $|k_{m_1} - k_{m_2}|$, respectively.

Now, we can derive the spectral equations by inserting (5.77) and (5.78) in (5.52) and (5.53), multiplying this equation with the Fourier function F_n^m and integrating the result over the domain. The result is

$$
\frac{d\Psi_n^m}{dt} = -\frac{ik_m}{\varepsilon}\Theta_n^m - \frac{\Psi_n^m}{\tau_E}
$$

$$
- \sum_{n_1=n-N}^{N} \sum_{m_1=m-M}^{M} \frac{\alpha_{n,n_1}^{m,m_1}}{\varepsilon} \frac{K_{n_1}^{m_1 2}}{K_n^{m2}} (\Psi_{n-n_1}^{m-m_1} \Psi_{n_1}^{m_1} + \Theta_{n-n_1}^{m-m_1} \Theta_{n_1}^{m_1})
$$

$$(5.87)$$

$$
\frac{d\Theta_n^m}{dt} = \frac{ik_m}{\varepsilon} \frac{1 + \varepsilon^2 - K_n^{m2}}{1 + \varepsilon^2 + K_n^{m2}} \Psi_n^m - \frac{K_n^{m2}}{1 + \varepsilon^2 + K_n^{m2}} \frac{\Theta_n^m}{\tau_E}
$$

$$
- \sum_{n_1=n-N}^{N} \sum_{m_1=m-M}^{M} \frac{\alpha_{n,n_1}^{m,m_1}}{\varepsilon} \frac{K_{n_1}^{m_1 2}(\Psi_{n-n_1}^{m-m_1} \Theta_{n_1}^{m_1} + \Theta_{n-n_1}^{m-m_1} \Psi_{n_1}^{m_1})}{1 + \varepsilon^2 + K_n^{m2}}
$$

$$
+ \sum_{n_1=n-N}^{N} \sum_{m_1=m-M}^{M} \frac{\alpha_{n,n_1}^{m,m_1}}{\varepsilon} \frac{(1 + \varepsilon^2)\Theta_{n-n_1}^{m-m_1} \Psi_{n_1}^{m_1}}{1 + \varepsilon^2 + K_n^{m2}} \qquad (5.88)
$$

where α_{n,n_1}^{m,m_1} denotes the **interaction coefficient** that is given by

$$\alpha_{n,n_1}^{m,m_1} = k_1 l_1 (m_1 n - m n_1) \qquad (5.89)$$

As an example, we construct a **low-order model** in which truncation happens already at the low wavenumbers $M = 1$ and $N = 2$. Then, we have to consider 4 coefficients per field. The number of interaction coefficients becomes 48 with values given in Table 5.1. Most of them vanish because $n - n_1 = 0$, $m - m_1 > M$ or $n - n_1 > N$.

Table 5.1. Interaction coefficients α_{n,n_1}^{m,m_1} for the truncation $M = 1$ and $N = 2$.

(m_1,n_1) \ (m,n)	$(0,1)$	$(0,2)$	$(1,1)$	$(1,2)$
$(-1,2)$				
$(-1,-1)$	$-k_1l_1$			
$(-1,1)$		$-2k_1l_1$		
$(-1,2)$	$-k_1l_1$			
$(0,-2)$				
$(0,-1)$	0		k_1l_1	
$(0,1)$		0		$-k_1l_1$
$(0,2)$	0		$-2k_1l_1$	
$(1,-2)$				
$(1,-1)$	k_1l_1		$2k_1l_1$	
$(1,1)$		$2k_1l_1$		k_1l_1
$(1,2)$	k_1l_1		$-k_1l_1$	

Note: The grey shaded boxes have no entries because $n-n_1 = 0$, $m - m_1 > M$ or $n - n_1 > N$.

By evaluating all terms, we obtain the following set of complex differential equations[f]

$$\frac{d\Psi_1^0}{dt} = -\frac{\Psi_1^0}{\tau_E} - \frac{6ik_1l_1}{\varepsilon}\Im\left(\Psi_2^1\Psi_1^{1C} + \Theta_2^1\Theta_1^{1C}\right) \tag{5.90}$$

$$\frac{d\Psi_2^0}{dt} = -\frac{\Psi_2^0}{\tau_E} \tag{5.91}$$

$$\frac{d\Psi_1^1}{dt} = -\frac{ik_1}{\varepsilon}\Theta_1^1 - \frac{\Psi_1^1}{\tau_E} + \frac{k_1l_1}{\varepsilon}\frac{K_1^{0^2} - K_2^{1^2}}{K_1^{1^2}}(\Psi_1^0\Psi_2^1 + \Theta_1^0\Theta_2^1)$$

$$+ \frac{2k_1l_1}{\varepsilon}\frac{K_1^{1^2} - K_2^{0^2}}{K_1^{1^2}}(\Psi_2^0\Psi_1^1 + \Theta_2^0\Theta_1^1) \tag{5.92}$$

$$\frac{d\Psi_2^1}{dt} = -\frac{ik_1}{\varepsilon}\Theta_2^1 - \frac{\Psi_2^1}{\tau_E} - \frac{k_1l_1}{\varepsilon}\frac{K_1^{1^2} - K_1^{0^2}}{K_2^{1^2}}(\Psi_1^0\Psi_1^1 + \Theta_1^0\Theta_1^1) \tag{5.93}$$

---

[f]It is tedious to deduce these equations from (5.87), (5.88) and Table 5.1. Therefore, the reader may proceed without retracing the derivation.

$$\frac{d\Theta_1^0}{dt} = -\frac{K_1^{0^2}}{1+\varepsilon^2+K_1^{0^2}}\frac{\Theta_1^0}{\tau_E} - \frac{1}{\varepsilon}\frac{6ik_1l_1^3}{1+\varepsilon^2+K_1^{0^2}}\Im\left(\Psi_2^1\Theta_1^{1C} + \Theta_2^1\Psi_1^{1C}\right)$$

$$+ \frac{2ik_1l_1}{\varepsilon}\frac{1+\varepsilon^2}{1+\varepsilon^2+K_1^{0^2}}\Im\left(\Psi_2^1\Theta_1^{1C} - \Theta_2^1\Psi_1^{1C}\right) \tag{5.94}$$

$$\frac{d\Theta_2^0}{dt} = -\frac{K_2^{0^2}}{1+\varepsilon^2+K_2^{0^2}}\frac{\Theta_2^0}{\tau_E} - \frac{4ik_1l_1}{\varepsilon}\frac{(1+\varepsilon^2)}{1+\varepsilon^2+K_2^{0^2}}\Im\left(\Psi_1^1\Theta_1^{1C}\right) \tag{5.95}$$

$$\frac{d\Theta_1^1}{dt} = \frac{ik_1}{\varepsilon}\frac{1+\varepsilon^2-K_1^{1^2}}{1+\varepsilon^2+K_1^{1^2}}\Psi_1^1 - \frac{K_1^{1^2}}{1+\varepsilon^2+K_1^{1^2}}\frac{\Theta_1^1}{\tau_E}$$

$$- \frac{k_1l_1}{\varepsilon}\frac{2(K_2^{0^2}-K_1^{1^2})(\Psi_2^0\Theta_1^1 + \Psi_1^1\Theta_2^0) + (K_2^{1^2} - K_1^{0^2})(\Psi_2^1\Theta_1^0 + \Psi_1^0\Theta_2^1)}{1+\varepsilon^2+K_1^{1^2}}$$

$$- \frac{k_1l_1(1+\varepsilon^2)}{\varepsilon}\frac{2(\Psi_1^1\Theta_2^0 - \Psi_2^0\Theta_1^1) + \Psi_1^0\Theta_2^1 - \Psi_2^1\Theta_1^0}{1+\varepsilon^2+K_1^{1^2}} \tag{5.96}$$

$$\frac{d\Theta_2^1}{dt} = \frac{ik_1}{\varepsilon}\frac{1+\varepsilon^2-K_2^{1^2}}{1+\varepsilon^2+K_2^{1^2}}\Psi_2^1 - \frac{K_2^{1^2}}{1+\varepsilon^2+K_2^{1^2}}\frac{\Theta_2^1}{\tau_E}$$

$$- \frac{k_1l_1}{\varepsilon}\frac{(K_1^{1^2} - K_1^{0^2})(\Psi_1^0\Theta_1^1 + \Psi_1^1\Theta_1^0) - (1+\varepsilon^2)(\Psi_1^1\Theta_1^0 - \Psi_1^0\Theta_1^1)}{1+\varepsilon^2+K_2^{1^2}}$$

$$\tag{5.97}$$

These six equations describe the evolution of a baroclinic wave in a vertically sheared basic flow. It is not a very difficult task to improve the model by removing the upper Ekman layer, introducing the β-effect and allowing for a barotropic basic flow. However, it is not evident that the low-order truncation does not introduce large errors. In order to find this out one must compare a low-order model simulation with that of a model having a truncation at high M and N.

We will simplify the equation set (5.90)–(5.97) by assuming that certain waves vanish at the beginning. There is no forcing of Ψ_2^0 by other waves due to the low-order truncation. Consequently, we can set it to zero if it does not play a role in the initial state. Furthermore, starting the model with $\Psi_1^0 = \Theta_1^0 = \Psi_2^1 = \Theta_2^1 = 0$ yields a simplified model of only three complex equations

$$\frac{d\Psi_1^1}{dt} = -\frac{ik_1}{\varepsilon}\Theta_1^1 - \frac{\Psi_1^1}{\tau_E} + \frac{2k_1 l_1}{\varepsilon}(1 - 4l_1^2)\Theta_2^0\Theta_1^1 \qquad (5.98)$$

$$\frac{d\Theta_1^1}{dt} = \frac{ik_1\varepsilon}{2+\varepsilon^2}\Psi_1^1 - \frac{1}{2+\varepsilon^2}\frac{\Theta_1^1}{\tau_E} - \frac{k_1 l_1}{\varepsilon}\frac{2(4l_1^2+\varepsilon^2)\Psi_1^1\Theta_2^0}{2+\varepsilon^2} \qquad (5.99)$$

$$\frac{d\Theta_2^0}{dt} = -\frac{4l_1^2}{4l_1^2+1+\varepsilon^2}\frac{\Theta_2^0}{\tau_E} - \frac{4ik_1 l_1}{\varepsilon}\frac{(1+\varepsilon^2)}{1+\varepsilon^2+4l_1^2}\Im\left(\Psi_1^1\Theta_1^{1C}\right)$$

$$(5.100)$$

Here, we have assumed a length scale such that $K_1^1 = 1$. Further reduction happens when we assume that Ψ_1^1 and Θ_1^1 are initially real and imaginary, respectively. Then, with the notations $A = -4\Psi_{1r}^1/\varepsilon$, $B = -4\Theta_{1i}^1/\varepsilon^2$ and $C = -2\Theta_{2i}^0/\varepsilon^2$ we obtain

$$\frac{dA}{dt} = k_1 B - \frac{A}{\tau_E} + k_1 l_1(1 - 4l_1^2)\varepsilon^2 BC \qquad (5.101)$$

$$\frac{dB}{dt} = \frac{k_1}{2+\varepsilon^2}A - \frac{1}{2+\varepsilon^2}\frac{B}{\tau_E} + k_1 l_1\frac{4l_1^2+\varepsilon^2}{2+\varepsilon^2}AC \qquad (5.102)$$

$$\frac{dC}{dt} = -\frac{4l_1^2}{4l_1^2+1+\varepsilon^2}\frac{C}{\tau_E} - \frac{k_1 l_1}{2}\frac{(1+\varepsilon^2)}{1+\varepsilon^2+4l_1^2}AB \qquad (5.103)$$

Obviously, this dynamical system approaches that investigated in Section 5.3.2 in the limit $\varepsilon \to 0$ (cf. Eqs. (5.59)–(5.61)). Therefore, the same result can be derived with the Galerkin method instead of the technique based on the asymptotic expansions (5.16)–(5.17). However, this approach can also be misleading because we do not know from the outset which is the appropriate truncation. For example, a truncation at $M = 2$ and $N = 1$ would yield an insufficient model even in the case $\varepsilon \ll 1$. To find out the right truncation we must integrate the dynamical system equations with different truncations. We can expect that the solution is proper if the statistics of the outcome do not change with increasing truncation. However, we cannot expect identical trajectories for different truncations because

possible internal instabilities can amplify small errors dramatically as
we will see in Section 5.7. The solid curve in Fig. 5.17 shows the time
series of A from a simulation for $\varepsilon = 0.1$ and $\tau_E = 16$ with the truncation $M = 1$ and $N = 2$ while the dashed curve displays the result for
the truncation $M = 8$ and $N = 20$. Obviously, the curves are nearly
identical for $t < 100$ but later on the developments become different.
This is an indication that the low-order truncation is justified. The
latter difference occurs because of internal instability but the long-
term statistics of the time series seem to be coincident. Figure 5.18
shows the same simulations for $\varepsilon = \sqrt[4]{2}$ which corresponds to the
fastest-growing mode for a fixed eddy **aspect ratio** k_1/l_1 (ratio of

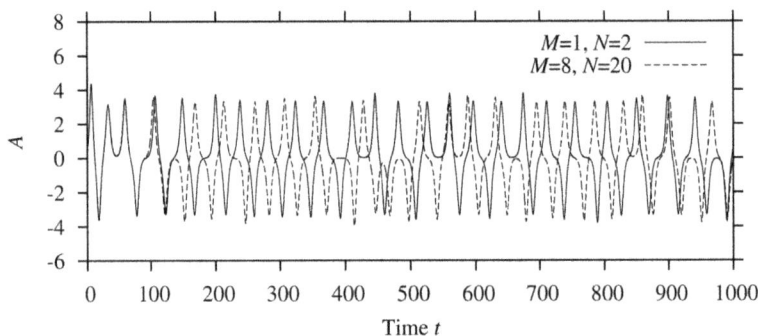

Fig. 5.17.   Time evolution of amplitude A deduced from numerical integrations
of the dynamical system equations (5.87)–(5.88) for $k_1 = l_1 = 1/\sqrt{2}$, $\varepsilon = 0.1$ and
$\tau_E = 16$ using the truncations $M = 1$, $N = 2$ (solid curve) and $M = 8$, $N = 20$
(dashed curve).

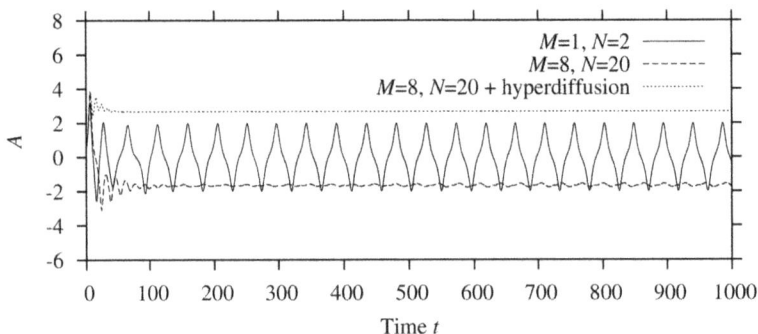

Fig. 5.18.   As in Fig. 5.17 but for $\varepsilon = \sqrt[4]{2}$. The dotted curve shows the higher
resolution result with added second-order hyperdiffusion (see text).

meridional to zonal extension) which is taken to be 1 (square wave). Now, dissimilarities appear much earlier and the long-term statistics differ. With low-order truncation a periodic oscillation of considerable amplitude arises while the higher resolution simulation exhibits small amplitude oscillations with a higher frequency.

Another feature of the simulation with strong baroclinic instability becomes apparent by looking at the actual fields. Figure 5.19 displays potential vorticity and stream function at the lower model

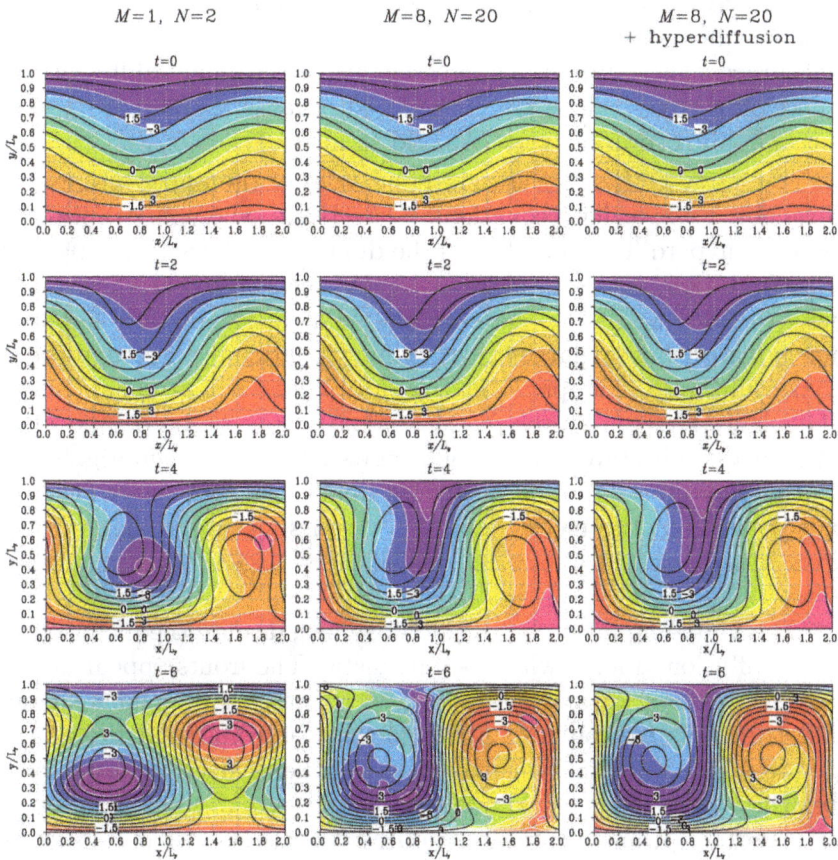

Fig. 5.19. Lower-level potential vorticity (colored shadings) and lower-level stream function (black isolines) deduced from numerical integrations of the dynamical system equations (5.87) and (5.88) at several time points (see plot titles) for a truncation at $M = 1$, $N = 2$ (left panel), at $M = 8$, $N = 20$ (middle panel) and at $M = 8$, $N = 20$ with added hyperdiffusion.

level for selected time points. The low-order model agrees well with the higher-order model until $t = 2$. Afterwards, the low-order model produces cut-off PV anomalies which are not seen in the higher resolution model. However, at $t = 6$ this model exhibits wiggly PV contours which do not appear to be realistic. These perturbations occur due to frontogenesis in the simulation. Even the truncation at high wavenumbers is not sufficient to resolve a sharp PV front. It can be expected that a truncation at a very large wavenumber is necessary to resolve the fronts properly. By the implementation of **hyperdiffusion** we can overcome this problem. The sharpness of fronts is limited by horizontal diffusion of the advected quantity but this process also affects the large-scale fields. With hyperdiffusion the effect on a large-scale is largely reduced because the Laplace operator is applied several times. We add terms $(-1)^{n_h-1} k_h \nabla_h^{2n_h} P_{g1}$ and $(-1)^{n_h-1} k_h \nabla_h^{2n_h} P_{g3}$ to the right-hand sides of the upper and lower-level potential vorticity equations, respectively, where n_h denotes the order of hyperdiffusion and k_h is the diffusion coefficient. In spectral representation, these terms reveal a damping of potential vorticity with the rate $k_h K_n^{m n_h}$. Therefore, the decay at small scales happens much faster than at large scales. The simulation with second-order hyperdiffusion (right panel in Fig. 5.19) does not reveal anymore the wiggly PV contours. Now, the amplitude of the baroclinic wave attains a steady state in the long term (see dotted curve in Fig. 5.18). With hyperdiffusion we parameterize small-scale processes which are not resolved in the model. This is, of course, a rough method but it is widely used in numerical modelling of large-scale atmospheric flows.

The finite-amplitude baroclinic wave has, despite the inclusion of hyperdiffusion, a form which is unrealistic. The fronts appear at the domain boundaries and the flow pattern has a rectangular shape. Many reasons can be responsible for this outcome, e.g. the quasi-geostrophic approximation, the neglect of the β-effect, an unrealistic basic flow, the upper Ekman layer or the absence of spherical geometry. Another problem could be the **flow symmetry** that constrains the development. We already found that only three real coefficients of twelve are nonzero in the system (5.90)–(5.97) for normal mode initial conditions. Similar constraints occur with higher truncation numbers. However, such a flow symmetry can become unstable with respect to shear instability (see Frisius, 1998).

To demonstrate such **symmetry-breaking instability** two additional simulations have been performed. The first one uses the same parameters as in the previous simulation with hyperdiffusion except that an aspect ratio $k_1/l_1 = 2.5$ is prescribed. In the second simulation, a wave with meridional wavenumber two has been added by setting $\Psi_2^1 = i\,0.025$. Then, the barotropic wave axis attains a slight tilt with respect to the meridional direction. The first simulation underlies a flow symmetry which we refer to as **symmetry Z** while in the second simulation, the added perturbation breaks this symmetry giving rise to another symmetry which we refer to as **symmetry A**. Both flow symmetries are sketched in Fig. 5.20. Obviously, symmetry Z has equally shaped cyclones and anticyclones while in symmetry A cyclones or anticyclones can dominate. The quasigeostrophic equations conserve both flow symmetries while the primitive equations only preserve symmetry A (see Frisius, 2003). Figure 5.21 shows the amplitude A evolution in both experiments. It becomes evident that the symmetry breaking dramatically changes the time series. With symmetry Z the wave settles into a steady state characterized by a baroclinic wave with equally strong cyclones and anticyclones (Fig. 5.22(a)). On the other hand, the symmetry-breaking perturbation triggers a breaking of the wave and by the

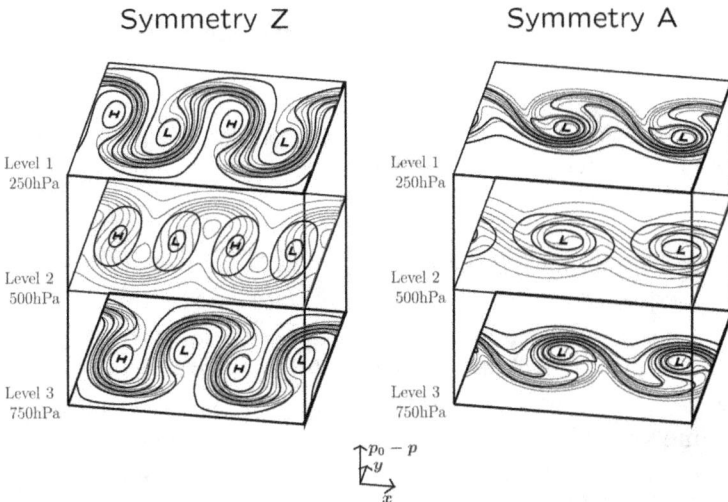

Fig. 5.20. Sketch of flow symmetry Z and A. The black (grey) contours show streamfunction (potential vorticity).

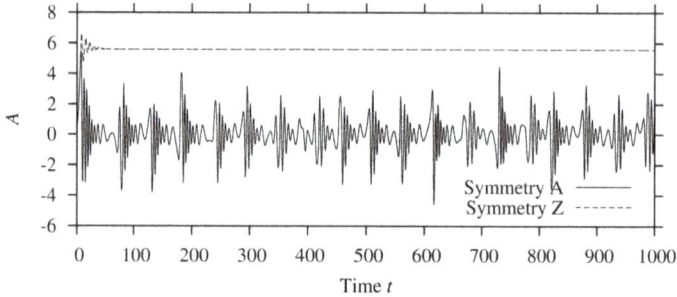

Fig. 5.21. Time evolution of amplitude A deduced from numerical integrations of the dynamical system equations (5.87) and (5.88) for $k_1/l_1 = 2.5$, $\varepsilon = \sqrt[4]{2}$ and $\tau_E = 16$ using the truncation $M = 8$, $N = 20$. The solid (dashed) curve shows the simulation underlying symmetry Z (symmetry A).

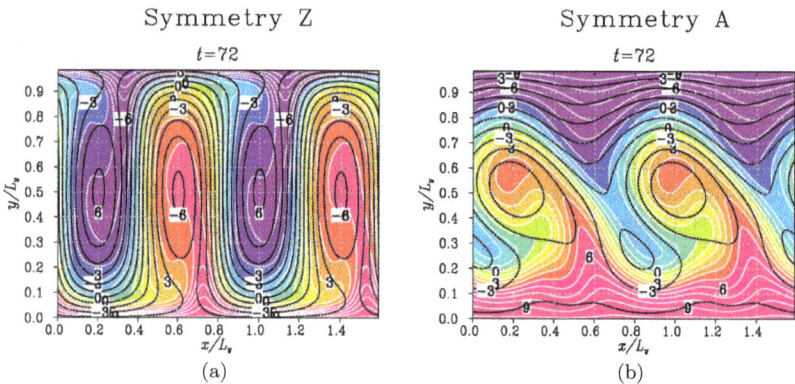

Fig. 5.22. Snap-shot of lower-level potential vorticity (colored shadings) and lower-level stream function (black isolines) at $t = 72$ for (a) symmetry Z and (b) symmetry A simulation.

breaking a cyclonic meridional shear flow is induced (Fig. 5.22(b)). The consequence is that cyclones dominate over anticyclones. Now, the simulated cyclones share much more similarities with those of real extratropical cyclones (see Fig. 5.7) as seen in the snap-shot in Fig. 5.22(b). Furthermore, the symmetry A simulation reveals a vivid time variability. However, there is also a symmetry A solution in which anticyclones dominate. Such a solution results if the initial symmetry-breaking perturbation has the opposite sign. Symmetry A is of course by itself a symmetry that can be broken by other perturbations.

5.4.3 *EOF reduction method*

Sometimes the selected basis functions of the spectral model do not properly describe the physical fields at low truncation although the time evolution suggests a low order description. For example, equatorial waves as introduced in Section 3.7 cannot be represented with a few spherical harmonics which are usually selected as basic functions for flows on the sphere. Therefore, a simple propagation of an equatorial wave must be described with many spectral equations. With **empirical orthogonal functions** (**EOF**) one can simplify the equations in such a case. To determine the EOFs of a system we calculate the time anomaly of each system variable X_j:

$$X'_j = X_j - \bar{X}_j \tag{5.104}$$

Now, we consider real variables X_j. Then, the symmetric **covariance matrix**

$$\overset{\leftrightarrow}{C} \equiv \begin{pmatrix} \overline{X'_1 X'_1} & \overline{X'_1 X'_2} & \cdots & \overline{X'_1 X'_N} \\ \overline{X'_2 X'_1} & \overline{X'_2 X'_2} & & \overline{X'_2 X'_N} \\ \vdots & & \ddots & \vdots \\ \overline{X'_N X'_1} & \overline{X'_N X'_2} & \cdots & \overline{X'_N X'_N} \end{pmatrix} \tag{5.105}$$

yields real eigenvalues and orthogonal eigenvectors. Therefore, the N eigenvectors $\vec{E}_n$ form a new base of the phase space. They constitute the EOFs. With the **orthogonal matrix** $\overset{\leftrightarrow}{O}$ having the $\vec{E}_n$ in the rows, we can calculate the new coordinates Y'_n from the old ones

$$\vec{Y}' = \overset{\leftrightarrow}{O} \cdot \vec{X}' \tag{5.106}$$

The orthogonal matrix $\overset{\leftrightarrow}{O}$ has the property that its inverse is identical to its transpose, i.e. $\overset{\leftrightarrow}{O}^{-1} = \overset{\leftrightarrow}{O}^T$.[g] The new coordinates Y'_n constitute the **principal components**. The scalar product of two vectors is invariant with respect to the orthogonal transformation since

$$\vec{a} \cdot \vec{b} = \vec{a} \cdot (\overset{\leftrightarrow}{O}^{-1} \cdot \overset{\leftrightarrow}{O}) \cdot \vec{b} = \vec{a} \cdot (\overset{\leftrightarrow}{O}^T \cdot \overset{\leftrightarrow}{O}) \cdot \vec{b} = (\overset{\leftrightarrow}{O} \cdot \vec{a}) \cdot (\overset{\leftrightarrow}{O} \cdot \vec{b}) = \vec{a}' \cdot \vec{b}' \tag{5.107}$$

---

[g]In our case this is obvious since the $\vec{E}_n$ form an orthogonal base.

Therefore, we have

$$\sum_{n=1}^{N} \overline{Y'_n Y'_n} = \sum_{n=1}^{N} \overline{X'_n X'_n} \tag{5.108}$$

Hence, the variance of the new phase space coordinates is identical to that of the original ones. The transformed covariance matrix becomes

$$\overset{\leftrightarrow}{C}_O \equiv \overset{\leftrightarrow}{O} \cdot \overset{\leftrightarrow}{C} \cdot \overset{\leftrightarrow}{O}^T \tag{5.109}$$

This is a diagonal matrix since

$$\overset{\leftrightarrow}{O} \cdot \overset{\leftrightarrow}{C} \cdot \overset{\leftrightarrow}{O}^T = \begin{pmatrix} \vec{E}_1 \\ \vdots \\ \vec{E}_N \end{pmatrix} \cdot \overset{\leftrightarrow}{C} \cdot (\vec{E}_1 \cdots \vec{E}_N) = \begin{pmatrix} \vec{E}_1 \\ \vdots \\ \vec{E}_N \end{pmatrix} \cdot (\lambda_1 \vec{E}_1 \cdots \lambda_N \vec{E}_N)$$

$$= \begin{pmatrix} \lambda_1 & 0 & 0 \\ 0 & \ddots & 0 \\ 0 & 0 & \lambda_n \end{pmatrix} \tag{5.110}$$

where λ_n denotes the eigenvalue of $\vec{E}_n$. The eigenvalues represent the variance of the corresponding EOF. This can be seen by the evaluation of the variance

$$\overline{Y'_n Y'_n} = \overline{\left(\sum_{k=1}^{N} O_{n,k} X'_k \right) \left(\sum_{l=1}^{N} O^T_{l,n} X'_l \right)} = C_{On,n} \tag{5.111}$$

where $O_{n,k}$, $O^T_{n,l}$ and $C_{On,n}$ denote the corresponding matrix elements of $\overset{\leftrightarrow}{O}$, $\overset{\leftrightarrow}{O}^T$ and $\overset{\leftrightarrow}{C}_O$, respectively. Therefore, the diagonal elements of $\overset{\leftrightarrow}{C}_O$ are the variances of the EOFs. Usually, the EOFs are ordered in such a way that the first has the highest variance contribution, the second the second highest variance contribution and so on. In some cases, a very large fraction of variance is already described by the first few EOFs. Then, it is meaningful to try an **EOF reduction** of the dynamical system.

We assume that the dynamical system has N equations of the form

$$\frac{dX_n}{dt} = \sum_{n_1=1}^{N} \sum_{n_2=1}^{N} \alpha_{n,n_1,n_2} X_{n_1} X_{n_2} + \sum_{n_2=1}^{N} \beta_{n,n_1} X_{n_1} + F_n(t) \quad (5.112)$$

Indeed, the spectral quasigeostrophic two-level model equations (5.87) and (5.88) can be written in this form. We introduce the shorter **Einstein notation** declaring that indices occurring twice in a term imply a summation from 1 to N. Then, Eq. (5.112) becomes

$$\frac{dX_n}{dt} = \alpha_{n,n_1,n_2} X_{n_1} X_{n_2} + \beta_{n,n_1} X_{n_1} + F_n(t) \quad (5.113)$$

The EOF representation refers to time anomalies. Therefore, we have to reformulate Eq. (5.113) for $X'_n = X_n - \bar{X}_n$:

$$\frac{dX'_n}{dt} = \alpha_{n,n_1,n_2} (X'_{n_1} X'_{n_2} + \bar{X}_{n_1} X'_{n_2} + X'_{n_1} \bar{X}_{n_2} + \bar{X}_{n_1} \bar{X}_{n_2})$$

$$+ \beta_{n,n_1} (X'_{n_1} + \bar{X}_{n_1}) + F_n(t) \quad (5.114)$$

We see that this equation has the same structure as (5.113) because we can write it as

$$\frac{dX'_n}{dt} = \alpha_{n,n_1,n_2} X'_{n_1} X'_{n_2} + \beta'_{n,n_1} X'_{n_1} + F'_n(t) \quad (5.115)$$

where

$$\beta'_{n,n_1} = \beta_{n,n_1} + \alpha_{n,n_1,n_2} \bar{X}_{n_2} + \alpha_{n,n_2,n_1} \bar{X}_{n_2} \quad (5.116)$$

and

$$F'_n(t) = F_n(t) + \alpha_{n,n_1,n_2} \bar{X}_{n_1} \bar{X}_{n_2} + \beta'_{n,n_1} \bar{X}_{n_1} \quad (5.117)$$

Due to Eq. (5.106), the model variables X'_n can be represented by the EOFs

$$X'_n = E_{m,n} Y'_m, \quad (5.118)$$

where m is the index for the EOF running from 1 to $M \leq N$. Inserting this representation in (5.115) yields

$$\frac{dY'_{m_1}}{dt} E_{m_1,n} = \alpha_{n,n_1,n_2} Y'_{m_1} E_{m_1,n_1} Y'_{m_2} E_{m_2,n_2} + \beta'_{n,n_1} Y'_{m_1} E_{m_1,n_1} + F'_n(t) \quad (5.119)$$

Since the EOFs are orthogonal, we obtain them by multiplying with $E_{m,n}$ and summing over all n

$$\frac{dY'_m}{dt} = E_{m,n}\alpha_{n,n_1,n_2}Y'_{m_1}E_{m_1,n_1}Y'_{m_2}E_{m_2,n_2}$$

$$+ E_{m,n}\beta'_{n,n_1}Y'_{m_1}E_{m_1,n_1} + E_{m,n}F'_n(t) \qquad (5.120)$$

Therefore, we can write the dynamical system for the principal components in the form

$$\frac{dY'_m}{dt} = \gamma_{m,m_1,m_2}Y'_{m_1}Y'_{m_2} + \eta'_{m,m_1}Y'_{m_1} + G'_m(t) \qquad (5.121)$$

where

$$\gamma_{m,m_1,m_2} = \alpha_{n,n_1,n_2}E_{m,n}E_{m_1,n_1}E_{m_2,n_2} \qquad (5.122)$$

$$\eta'_{m,m_1} = E_{m,n}\beta'_{n,n_1}E_{m_1,n_1} \qquad (5.123)$$

$$G'_m(t) = E_{m,n}F'_n(t) \qquad (5.124)$$

Consequently, the dynamical system has the same mathematical structure but with different coefficients γ_{m,m_1,m_2}, η'_{m,m_1} and external forcing $G'_m(t)$. It is in some cases possible to truncate the system at $M \ll N$ without a serious error if the relevant dynamics results from the first few EOFs. Then, EOF reduction leads to a simplified dynamical system. Besides EOFs, there are other patterns for model reduction. Hasselmann (1988) introduced the **principal interaction pattern** (PIP). These can be obtained by error minimization. Achatz *et al.* (1995) applied these patterns to reduce a quasigeostrophic two-level model on the sphere. They found that only three PIPs are necessary to simulate repeating life cycles of baroclinic waves.

5.5   Dynamical System Analysis

Dynamical system analysis serves to better understand the behavior of a dynamical system. In this section, we describe the essential steps of the analysis. We begin with a general definition of a dynamical system.

5.5.1 Definition of a dynamical system

We distinguish between **continuous** and **time-discrete dynamical systems**. The former can be written as

$$\frac{d\vec{X}}{dt} = \vec{F}(\vec{X}, c_1, c_2, \ldots, c_P, t) \qquad (5.125)$$

where $\vec{X} = (X_1, X_2, \ldots, X_N)$ is the real **state vector** with N components which describe the state of the considered system and $c_1, c_2, \ldots, c_P$ denote P constant **system parameters**.[h] The system is referred to as **autonomous** if the tendency vector $\vec{F}$ does not explicitly depend upon time. Otherwise, the system becomes **nonautonomous**. Formally, a nonautonomous dynamical system can be easily converted into an autonomous one by introducing an additional component $X_{N+1} \equiv t$ and adding the differential equation

$$\frac{dX_{N+1}}{dt} = 1 \qquad (5.126)$$

Then, the time dependence is described by the new internal variable X_{N+1}. For given parameters, the solution of the autonomous dynamical system can be expressed as

$$\vec{X} = \vec{X}(\vec{X}_0, t) \qquad (5.127)$$

where $\vec{X}_0$ denotes the state vector at $t = t_0$. This function defines the **flow** of the dynamical system. Restricting to a special $\vec{X}_0$ gives an **orbit** in phase space spanned by the axes $X_1, X_2, \ldots, X_N$.

A time-discrete dynamical system takes the form

$$\vec{X}^{j+1} = \vec{F}(\vec{X}^j, c_1, c_2, \ldots, c_P, t_j) \qquad (5.128)$$

where j denotes the integer index for time and $\vec{X}^j$ the state vector at the discrete time t_j. Accordingly, the flow of the time-discrete system is given by

$$\vec{X}^j = \underbrace{\vec{F} \circ \vec{F} \circ \cdots \vec{F}}_{j \times}(\vec{X}^0, t_0) \equiv \vec{F}^j(\vec{X}^0, t_0) \qquad (5.129)$$

---

[h]We restrict our analysis to real state variables. Then, complex numbers in a dynamical system must be split into real and imaginary parts and treated separately.

where ○ denotes the composition of successive functions. This equation shows that it is possible to determine the solution by successive application of $\vec{F}$. This can be done with a computer program.

These definitions bring us to the question of whether complicated atmospheric models like grid point models constitute dynamical systems or not. Indeed, they form at least time-discrete dynamical systems. The phase space of a complex grid point model has a huge number of dimensions since every variable at every grid point in the model represents a system variable. The spatial discretization of the partial differential equations by the introduction of grid points leads to a set of ordinary differential equations which form a continuous dynamical system. Eventually, the necessary discretization of the time coordinate yields a time-discrete dynamical system. It is rarely possible to solve the equations of a nonlinear dynamical system analytically. Therefore, time discretization becomes necessary so that we are practically dealing with time-discrete dynamical systems. However, for dynamical system analysis, it is more convenient to work with continuous systems. The time-discretized solution should approach the unknown continuous solution for a sufficiently short time step.

5.5.2 *Determination of equilibria*

The first step of dynamical system analysis is to find the **equilibria** which form steady-state solutions of the dynamical system. They are also known as **fixed points** and because they do not vary with time, they must fulfill the condition

$$\left.\frac{d\vec{X}}{dt}\right|_{\vec{X}=\vec{X}_e} = \vec{F}(\vec{X}_e, c_1, c_2, \dots, c_P) = 0 \qquad (5.130)$$

where $\vec{X}_e$ is the equilibrium state. Note that nonautonomous dynamical systems have no equilibria since Eq. (5.126) disagrees with (5.130).

In sophisticated models with many equations, it is difficult or even impossible to find equilibria. Sometimes it is possible that the complex system settles into a steady state like the symmetry Z solution shown in Fig. 5.21. Then, the long-term integration does the job of solving (5.130). A requirement for this procedure is that the equilibrium must be stable. A method for finding unstable equilibria in sophisticated models exists (see Dijkstra, 2005). On the other hand, in low-order models, the determination of equilibria is often feasible, at least by a numerical approximation and this is clearly an advantage for the understanding of the system.

Let us reconsider the Lorenz model presented in Section 5.3.2. It has the three system equations

$$\frac{d\mathcal{X}}{dt} = \sigma(\mathcal{Y} - \mathcal{X}) \tag{5.131}$$

$$\frac{d\mathcal{Y}}{dt} = \mathcal{R}\mathcal{X} - \mathcal{X}\mathcal{Z} - \mathcal{Y} \tag{5.132}$$

$$\frac{d\mathcal{Z}}{dt} = \mathcal{X}\mathcal{Y} - \mathcal{B}\mathcal{Z} \tag{5.133}$$

Consequently, the equilibria must fulfill

$$\sigma(\mathcal{Y}_e - \mathcal{X}_e) = 0 \tag{5.134}$$

$$\mathcal{R}\mathcal{X}_e - \mathcal{X}_e\mathcal{Z}_e - \mathcal{Y}_e = 0 \tag{5.135}$$

$$\mathcal{X}_e\mathcal{Y}_e - \mathcal{B}\mathcal{Z}_e = 0 \tag{5.136}$$

To find solutions it is convenient to eliminate as many variables as possible. From (5.134) we find $\mathcal{Y}_e = \mathcal{X}_e$ and by substituting this in (5.136) we get $\mathcal{Z}_e = \mathcal{X}_e^2/\mathcal{B}$. Therefore, we obtain from (5.135) the cubic equation

$$(\mathcal{X}_e^2 - \mathcal{R}\mathcal{B} + \mathcal{B})\mathcal{X}_e = 0 \tag{5.137}$$

It leads to the three solutions

$$\begin{pmatrix} \mathcal{X}_{e_1} \\ \mathcal{Y}_{e_1} \\ \mathcal{Z}_{e_1} \end{pmatrix} = \begin{pmatrix} 0 \\ 0 \\ 0 \end{pmatrix}, \quad \begin{pmatrix} \mathcal{X}_{e_2} \\ \mathcal{Y}_{e_2} \\ \mathcal{Z}_{e_2} \end{pmatrix} = \begin{pmatrix} \sqrt{\mathcal{B}(\mathcal{R}-1)} \\ \sqrt{\mathcal{B}(\mathcal{R}-1)} \\ \mathcal{R}-1 \end{pmatrix},$$

$$\begin{pmatrix} \mathcal{X}_{e_3} \\ \mathcal{Y}_{e_3} \\ \mathcal{Z}_{e_3} \end{pmatrix} = \begin{pmatrix} \sqrt{\mathcal{B}(\mathcal{R}-1)} \\ \sqrt{\mathcal{B}(\mathcal{R}-1)} \\ \mathcal{R}-1 \end{pmatrix} \tag{5.138}$$

The three solutions correspond to (i) the basic state without wave, (ii) the basic state with a steady baroclinic wave and (iii) as in (ii) but the steady baroclinic wave is phase shifted by $180°$. The wave states are only possible if $\mathcal{R} > 1$. Recall that $1/\mathcal{R}$ measures the impact of the Ekman layers. It will be shown later that the baroclinic basic flow becomes unstable only for $\mathcal{R} > 1$. Figure 5.23 shows the three steady-state values of $\mathcal{X}$ as a function of $\mathcal{R}$. At $\mathcal{R} = 1$ one solution is divided into three. The instantaneous appearance of further equilibria is called **bifurcation** which will be further explained in Section 5.5.5. At the bifurcation, the stability of the equilibria changes as indicated in the figure. This will be investigated in Section 5.5.3.

5.5.3 *Stability analysis*

It is of importance for the dynamical system behaviour whether the equilibria are unstable or stable. To find this out we have to do a stability analysis based on the linearized equations. With these equations it can be determined whether small disturbances move towards the equilibrium (stability), move away (instability) or remain at a finite distance (neutrality). In general, the evolution of a perturbation from an equilibrium state $\vec{X}_e$ results from the

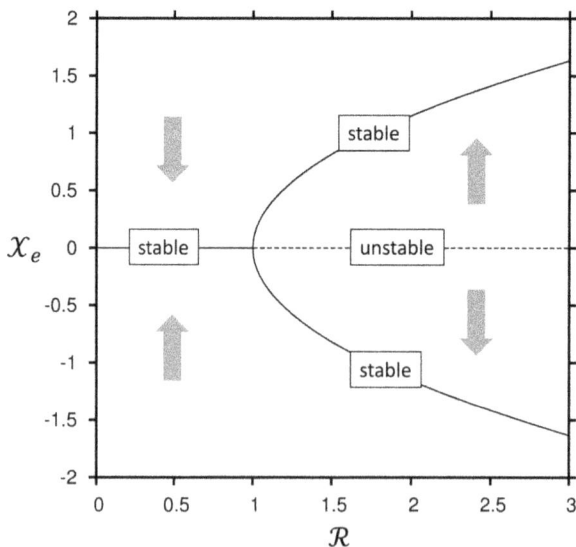

Fig. 5.23. Variable $\mathcal{X}$ of the equilibria in the Lorenz model as a function of model parameter $\mathcal{R}$. The stable (unstable) equilibria are shown as solid (dashed) lines. The grey arrows indicate to which equilibria the solution tends.

equation

$$\frac{d\vec{X}'}{dt} = \vec{F}(\vec{X}_e + \vec{X}') - \vec{F}(\vec{X}_e) \tag{5.139}$$

where $\vec{X}' = \vec{X} - \vec{X}_e$. To first order, we can approximate the right-hand side by

$$\vec{F}(\vec{X}_e + \vec{X}') - \vec{F}(\vec{X}_e) \approx \vec{F}(\vec{X}_e) + \overset{\leftrightarrow}{D} \cdot \vec{X}' - \vec{F}(\vec{X}_e) = \overset{\leftrightarrow}{D} \cdot \vec{X}' \tag{5.140}$$

where $\overset{\leftrightarrow}{D}$ is the Jacobi matrix which is given by

$$\overset{\leftrightarrow}{D} \equiv \begin{pmatrix} \left.\dfrac{\partial F_1}{\partial X_1}\right|_{\vec{X}=\vec{X}_e} & \left.\dfrac{\partial F_1}{\partial X_2}\right|_{\vec{X}=\vec{X}_e} & \cdots & \left.\dfrac{\partial F_1}{\partial X_N}\right|_{\vec{X}=\vec{X}_e} \\[2ex] \left.\dfrac{\partial F_2}{\partial X_1}\right|_{\vec{X}=\vec{X}_e} & \left.\dfrac{\partial F_2}{\partial X_2}\right|_{\vec{X}=\vec{X}_e} & & \left.\dfrac{\partial F_2}{\partial X_N}\right|_{\vec{X}=\vec{X}_e} \\[2ex] \vdots & & \ddots & \vdots \\[2ex] \left.\dfrac{\partial F_N}{\partial X_1}\right|_{\vec{X}=\vec{X}_e} & \left.\dfrac{\partial F_N}{\partial X_2}\right|_{\vec{X}=\vec{X}_e} & \cdots & \left.\dfrac{\partial F_N}{\partial X_N}\right|_{\vec{X}=\vec{X}_e} \end{pmatrix} \tag{5.141}$$

Therefore, the linearized equation becomes

$$\frac{d\vec{X}'}{dt} = \overset{\leftrightarrow}{D} \cdot \vec{X}' \tag{5.142}$$

In most cases, it is sufficient to calculate all eigenvalues of the Jacobi matrix to determine the stability of the equilibrium. The linearized system forms a set of linear ordinary differential equations with constant coefficients. These have special solutions of the form

$$\vec{X} = \vec{X}_\varsigma e^{\varsigma t} \tag{5.143}$$

where ς denotes the eigenvalue and $\vec{X}_\varsigma$ the corresponding eigenvector. Inserting this solution into the linearized system (5.142) results in

$$(\overset{\leftrightarrow}{D} - \varsigma \overset{\leftrightarrow}{I}) \cdot \vec{X} = 0 \tag{5.144}$$

where $\overset{\leftrightarrow}{I}$ denotes the unit matrix. This equation has only nontrivial solutions if the determinant of $\overset{\leftrightarrow}{D} - \varsigma \overset{\leftrightarrow}{I}$ vanishes, i.e. $|\overset{\leftrightarrow}{D} - \varsigma \overset{\leftrightarrow}{I}| = 0$.

This equation yields a polynomial of order N (characteristic polynomial) and we get a spectrum of N or less eigenvalues. Then the eigenvectors can be calculated with Gaussian elimination. Indeed, we already applied this procedure for several perturbation analyses in Chapters 3 and 4. However, here we do not have to assume a spatial structure of the wave since the analysis takes place in phase space.

Now, we analyze the stability of the equilibria in the Lorenz model (5.63)–(5.65). The Jacobi matrix of this system becomes

$$\overset{\leftrightarrow}{D} \equiv \begin{pmatrix} -\sigma & \sigma & 0 \\ \mathcal{R} - \mathcal{Z}_e & -1 & -\mathcal{X}_e \\ \mathcal{Y}_e & \mathcal{X}_e & -\mathcal{B} \end{pmatrix} \tag{5.145}$$

and the associated eigenvalue equation $|\overset{\leftrightarrow}{D} - \varsigma \overset{\leftrightarrow}{I}| = 0$ reads

$$\varsigma^3 + (\mathcal{B} + \sigma + 1)\varsigma^2 + [\sigma(1 + \mathcal{B} + \mathcal{Z}_e - \mathcal{R}) + \mathcal{B} + \mathcal{X}_e^2]\varsigma$$
$$+ \sigma(\mathcal{B} + \mathcal{X}_e^2 + \mathcal{B}\mathcal{Z}_e - \mathcal{B}\mathcal{R} + \mathcal{X}_e\mathcal{Y}_e) = 0 \tag{5.146}$$

First, we study the stability of equilibrium 1, namely the basic flow without wave. Then, (5.146) reduces to

$$\varsigma^3 + (\mathcal{B} + \sigma + 1)\varsigma^2 + [\sigma(1 + \mathcal{B} - \mathcal{R}) + \mathcal{B}]\varsigma + \sigma\mathcal{B}(1 - \mathcal{R}) = 0 \tag{5.147}$$

It is possible to factor out $\varsigma + \mathcal{B}$ in this equation so that

$$(\varsigma + \mathcal{B})[\varsigma^2 + (\sigma + 1)\varsigma + \sigma(1 - \mathcal{R})] = 0 \tag{5.148}$$

Consequently, we obtain the three eigenvalues

$$\varsigma_{1,2} = -\frac{1}{2}(\sigma + 1) \pm \sqrt{\frac{1}{4}(\sigma + 1)^2 - \sigma(1 - \mathcal{R})}, \quad \varsigma_3 = -\mathcal{B} \tag{5.149}$$

The first two eigenvalues $\varsigma_{1,2}$ are associated with the baroclinic instability or stability of the zonal mean basic flow. The third eigenvalue results from the decay of the zonal mean correction $\mathcal{Z}$ due to Ekman pumping. This can be seen by setting $\mathcal{X} = \mathcal{Y} = 0$ in (5.133) so that the solution becomes $\mathcal{Z} = \mathcal{Z}_0 \exp(-\mathcal{B}t)$. Figure 5.24 shows the eigenvalues as a function of $\mathcal{R}$ and it is obvious that for $\mathcal{R} > 1$ the equilibrium becomes unstable with one positive and two negative real eigenvalues. Therefore, the bifurcation also changes the stability properties of the equilibria at $\mathcal{R} = 1$.

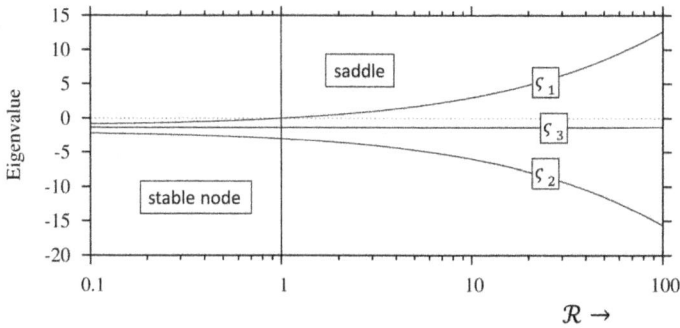

Fig. 5.24. Eigenvalues resulting from the perturbation analysis of the equilibrium 1 in the Lorenz model as a function of $\mathcal{R}$. The names "stable node" and "saddle" will be explained in Section 5.5.4.

It is tedious to calculate the eigenvalues of (5.146) for a nontrivial equilibrium. To prove stability, it is sufficient to show that all eigenvalues have a negative real part. This can be determined with the **Hurwitz criterion** applied to a polynomial of the form

$$a_n \varsigma^n + a_{n-1} \varsigma^{n-1} + \cdots + a_1 \varsigma^1 + a_0 = 0 \qquad (5.150)$$

We can assume $a_n = 1$ without loss of generality. Then, the roots have only negative real parts if and only if the determinants

$$D_1 = a_{n-1}, \quad D_2 = \begin{vmatrix} a_{n-1} & a_{n-3} \\ 1 & a_{n-2} \end{vmatrix}, \quad D_3 = \begin{vmatrix} a_{n-1} & a_{n-3} & a_{n-5} \\ 1 & a_{n-2} & a_{n-4} \\ 0 & 1 & a_{n-3} \end{vmatrix}, \dots,$$

$$D_n = \begin{vmatrix} a_{n-1} & a_{n-3} & a_{n-5} & & 0 & 0 \\ & & & \cdots & & \\ 1 & a_{n-2} & a_{n-4} & & 0 & 0 \\ \vdots & & & \ddots & \vdots & \\ 0 & 0 & 0 & & a_1 & 0 \\ & & & \cdots & & \\ 0 & 0 & 0 & & a_2 & a_0 \end{vmatrix} \qquad (5.151)$$

are positive. Applying this criterion to a polynomial of third order yields the conditions

$$a_2 > 0, \quad a_2 a_1 - a_0 > 0 \quad (a_2 a_1 - a_0) a_0 > 0 \qquad (5.152)$$

All inequality relations can only be fulfilled if $a_1 > 0$ and $a_0 > 0$. Inserting the steady wave solutions 2 or 3 from (5.138) in (5.146) gives

$$\varsigma^3 + (\mathcal{B} + \sigma + 1)\varsigma^2 + (\sigma + \mathcal{R})\mathcal{B}\varsigma + 2\sigma\mathcal{B}(\mathcal{R} - 1) = 0 \qquad (5.153)$$

Obviously, the inequality relations $a_2 > 0$, $a_1 > 0$ and $a_0 > 0$ are fulfilled for defined equilibria ($\mathcal{R} > 1$). It remains to check if $a_2 a_1 - a_0 > 0$. We get

$$a_2 a_1 - a_0 = \mathcal{B}(\mathcal{B} + \sigma + 1)(\sigma + \mathcal{R}) - 2\sigma\mathcal{B}(\mathcal{R} - 1)$$
$$= \mathcal{B}\sigma(\mathcal{B} + \sigma + 3) + \mathcal{B}\mathcal{R}(\mathcal{B} - \sigma + 1) > 0 \quad (5.154)$$

This inequality relation is fulfilled in the case $\mathcal{B} - \sigma + 1 > 0$. Then, the steady wave solutions are always stable. Otherwise,

$$\mathcal{R} < \frac{\sigma(\mathcal{B} + \sigma + 3)}{\sigma - \mathcal{B} - 1} \equiv \mathcal{R}_2 \quad \text{for} \quad \sigma - \mathcal{B} - 1 > 0 \qquad (5.155)$$

must be fulfilled. In this case, the equilibria become unstable when the parameter $\mathcal{R}$ exceeds $\mathcal{R}_2$. The value $\mathcal{B} = 4/3$ resulting in a baro-clinic square wave ($k/l = 1$, cf. Section 5.3.2) satisfies the relation (5.154) for all $\mathcal{R}$ but for an aspect ratio of $k/l > \sqrt{3}$ ($\mathcal{B} < 1$) the steady wave becomes unstable at a sufficiently large $\mathcal{R}_2$. Figure 5.25 shows numerically determined eigenvalues as a function of $\mathcal{R}$ for $\mathcal{B} = 4/3$ ($k/l = 1$) and $\mathcal{B} = 0.711$ ($k/l = 2.5$). Three negative real eigenvalues appear immediately after the bifurcation at $\mathcal{R} = 1$. Two of them change into two complex conjugated eigenvalues at $\mathcal{R} = \mathcal{R}_1$ but their real parts still remain negative. For $k/l = 2.5$ the real parts of the first two eigenvalues become positive at $\mathcal{R} = \mathcal{R}_2 = 39.54$ as predicted by Eq. (5.155).

5.5.4 *Phase portraits in the vicinity of fixed points*

A **phase portrait** is a visualization of the vector field

$$\vec{F} = \vec{F}(\vec{X}, t) \qquad (5.156)$$

in phase space for a specific time t. A readable plot is only possible in a dynamical system with 2 dimensions (or at best with three dimensions). The trajectories are parallel to these vectors in an autonomous

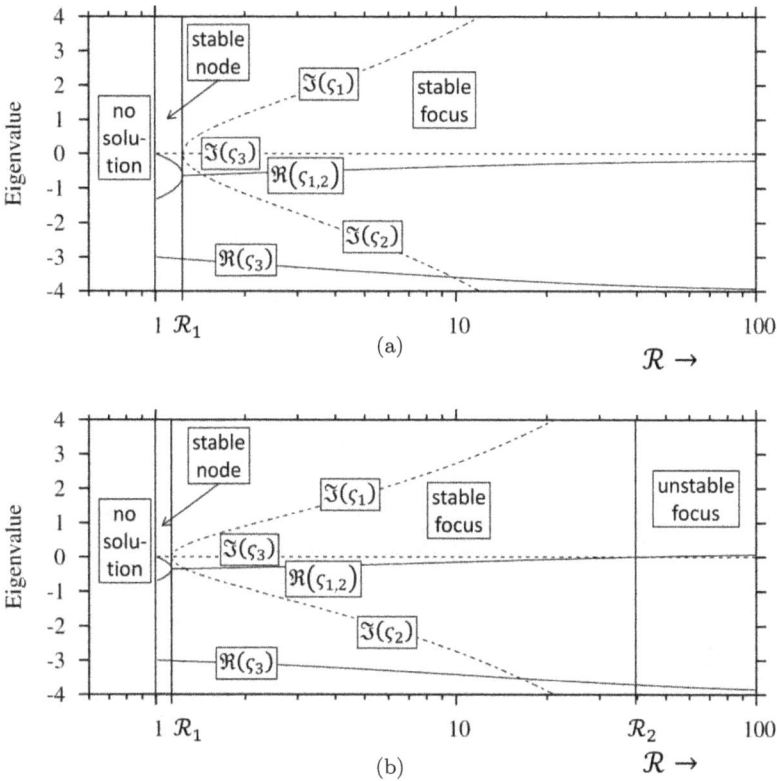

Fig. 5.25. Eigenvalues resulting from the perturbation analysis of the equilibrium 2 and 3 in the Lorenz model as a function of $\mathcal{R}$ for (a) $\mathcal{B} = 4/3$ and (b) $\mathcal{B} = 0.711$. The names "stable node" and "stable focus" and "unstable focus" will be explained in Section 5.5.5.

dynamical system. The eigenvalues resulting from the stability analysis of a fixed point determine the picture of the phase portrait in the vicinity of this fixed point. In this region of phase space, we can calculate the Jacobi matrix $\overset{\leftrightarrow}{D}$ and use the approximation

$$\vec{F} = \overset{\leftrightarrow}{D} \cdot \vec{X}' \tag{5.157}$$

Let $\vec{X} = (\mathcal{X}, \mathcal{Y})$ and $\vec{F} = (u, v)$ then, we get for the Jacobi matrix

$$\overset{\leftrightarrow}{D} = \begin{pmatrix} \dfrac{\partial u}{\partial \mathcal{X}} & \dfrac{\partial u}{\partial \mathcal{Y}} \\ \dfrac{\partial v}{\partial \mathcal{X}} & \dfrac{\partial v}{\partial \mathcal{Y}} \end{pmatrix} = \begin{pmatrix} \dfrac{1}{2}(\mathcal{D} + \mathcal{F}_1) & \dfrac{1}{2}(\mathcal{F}_2 - \mathcal{V}) \\ \dfrac{1}{2}(\mathcal{F}_2 + \mathcal{V}) & \dfrac{1}{2}(\mathcal{D} - \mathcal{F}_1) \end{pmatrix} \tag{5.158}$$

where

$$D = \frac{\partial u}{\partial x} + \frac{\partial v}{\partial y}, \qquad V = \frac{\partial v}{\partial x} - \frac{\partial u}{\partial y}, \tag{5.159}$$

$$\mathcal{F}_1 = \frac{\partial u}{\partial x} - \frac{\partial v}{\partial y}, \qquad \mathcal{F}_2 = \frac{\partial v}{\partial x} + \frac{\partial u}{\partial y} \tag{5.160}$$

The quantities D and V are obviously divergence and vorticity if (u, v) were a two-dimensional fluid flow instead of a vector field in two-dimensional phase space. $\mathcal{F}_1$ and $\mathcal{F}_2$ constitute the **deformations**. The associated vector fields deduced from Eq. (5.157) are displayed in Fig. 5.26. A pure divergent flow (Fig. 5.26(a)) exhibits

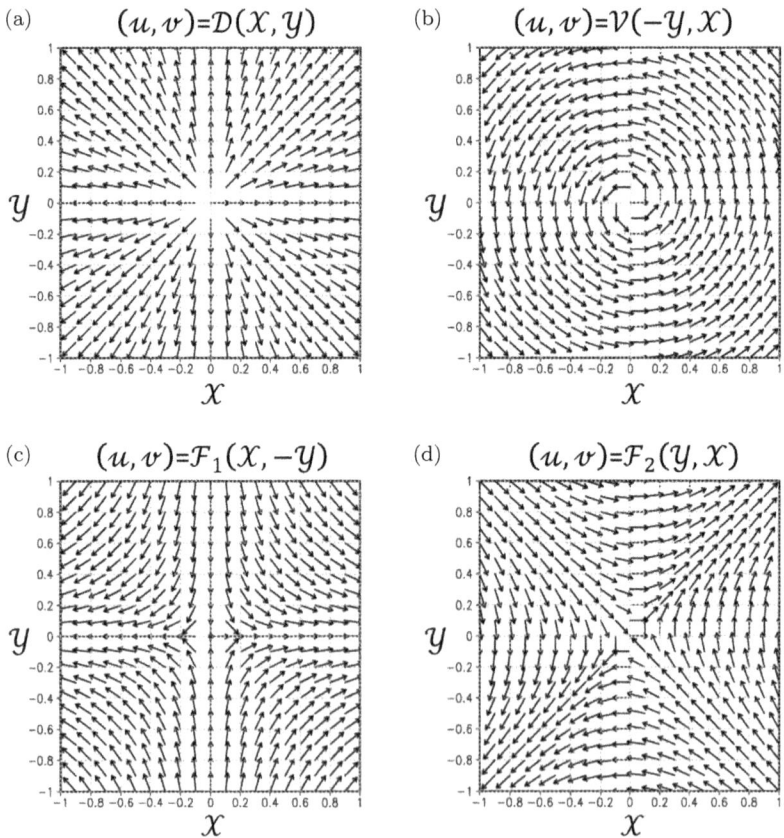

Fig. 5.26.   Phase portrait of the flows described by (a) divergence D, (b) vorticity V, (c) deformation $\mathcal{F}_1$ and (d) deformation $\mathcal{F}_2$.

radiating streamlines. In this case, the fixed point is either stable or unstable depending on the sign of the divergence. The flow resulting from nonzero vorticity (Fig. 5.26(b)) displays circles around the fixed point. Such a flow forms a neutral fixed point where the model variables describe oscillations. Finally, flows resulting from $\mathcal{F}_1$ or $\mathcal{F}_2$ (Figs. 5.26(c) and 5.26(d)) yield curved streamlines which first run toward the fixed point and then run away. Such a fixed point is always unstable. Note that the superposition of $\mathcal{F}_1$ and $\mathcal{F}_2$ leads also to a deformation with another orientation.

The matrix (5.158) has the eigenvalues

$$\varsigma_{1,2} = \frac{\mathcal{D} \pm \sqrt{\mathcal{F}_1^2 + \mathcal{F}_2^2 - \mathcal{V}^2}}{2} \qquad (5.161)$$

This result confirms the statements about the four elementary flows shown in Fig. 5.26. However, in general, the fixed points are associated with a superposition of these elementary flows. To distinguish various forms, it is useful to write (5.161) in the form

$$\varsigma_{1,2} = \frac{\mathcal{D} \pm \sqrt{\mathcal{D}^2 - 4|\overleftrightarrow{D}|}}{2} \qquad (5.162)$$

Obviously, the fixed point is stable or neutral if $\mathcal{D} \leq 0$ and $|\overleftrightarrow{D}| \geq 0$ but it gets unstable otherwise. The diagram in Fig. 5.27 shows for which values of $\mathcal{D}$ and $|\overleftrightarrow{D}|$ the various fixed point types result. Figure 5.28 displays the associated phase portraits. The cases $|\overleftrightarrow{D}| < 0$ yields one positive and one negative eigenvalue. Then, the fixed point is unstable and forms a **saddle**. For $|\overleftrightarrow{D}| > 0$ and $|\overleftrightarrow{D}| < \mathcal{D}^2/4$ (below the parabola) both eigenvalues are real and have the same sign as $\mathcal{D}$. In this case, the fixed point is a **stable** or **unstable node**. These nodes turn into **proper nodes** in the case that $|\overleftrightarrow{D}| = \mathcal{D}^2/4$ is exactly fulfilled. Then, the streamlines become straight. Complex conjugated eigenvalues appear if $|\overleftrightarrow{D}| > \mathcal{D}^2/4$ (above the parabola). Then, the solution exhibits oscillations and again, the sign of $\mathcal{D}$ determines whether the fixed point is stable or not. The associated fixed point constitutes a **stable** or **unstable focus**. For $|\overleftrightarrow{D}| > 0$ and $\mathcal{D} = 0$ the fixed point is neutral and the streamlines are closed. Then, only vorticity $\mathcal{V}$ characterizes the flow where a positive (negative) imaginary part of ς yields anticlockwise (clockwise) rotation.

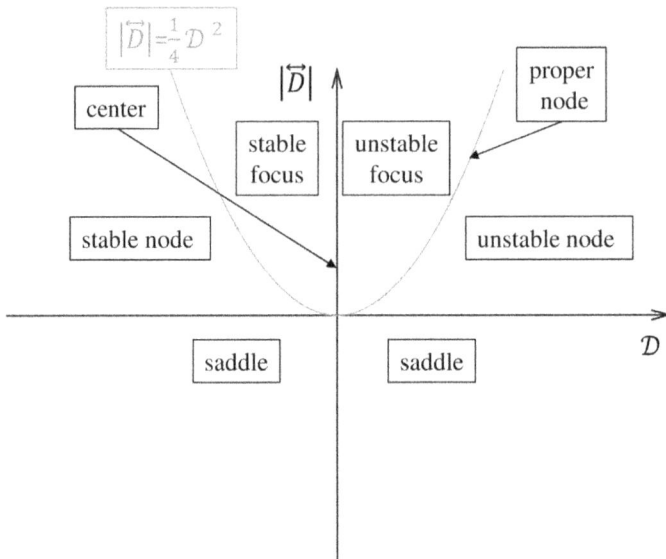

Fig. 5.27. Diagram showing different fixed-point types as a function of $\mathcal{D}$ and $|\overset{\leftrightarrow}{D}|$.

Such a fixed point is called a **center**. The stability analysis of the Lorenz model in Section 5.5.3 reveals that equilibrium 1 can be a stable node or a saddle (see Fig. 5.24) while equilibrium 2 and 3 can take the form of a stable node, stable focus and unstable focus (see Fig. 5.25). However, note that the Lorenz model has three dimensions and, therefore, this fixed-point characterization can only refer to two of the three eigenvalues. Furthermore, there is also the possibility that both eigenvectors have the same direction. Then, they do not form a base anymore and the solution becomes more complicated. In such a case, **improper nodes** arise which are not discussed here (for more information see Dijkstra, 2005).

5.5.5 *Bifurcation analysis*

Number and type of equilibria may change when the system parameters are varied. The bifurcation analysis aims at the detection of those parameter values where such a change occurs. In one-dimensional dynamical systems bifurcations can be classified in a simple way.

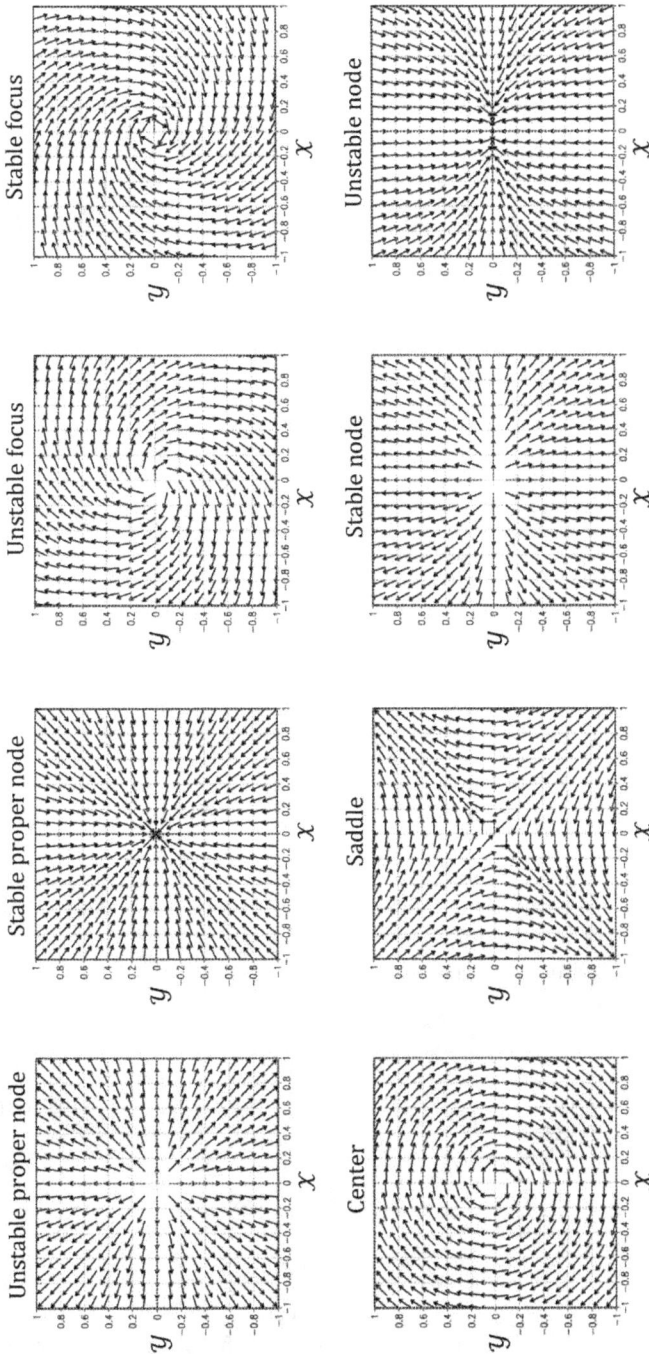

Fig. 5.28.   Phase portrait of the flows described by the different fixed-point types

Then, the system takes the form

$$\frac{d\mathcal{X}}{dt} = F(\mathcal{X}, c) \tag{5.163}$$

where c is a system parameter. The equilibria result from the equation

$$F(\mathcal{X}_e, c) = 0 \tag{5.164}$$

It contains the implicit function $\mathcal{X}_e(c)$. Linearizing (5.163) with respect to one of the equilibria yields

$$\frac{d\mathcal{X}'}{dt} = \left.\frac{\partial F}{\partial \mathcal{X}}\right|_{\mathcal{X}=\mathcal{X}_e} \mathcal{X}' \tag{5.165}$$

Therefore, the eigenvalue is simply $\varsigma = \partial F/\mathcal{X}|_{\mathcal{X}=\mathcal{X}_e}$.

In the following, we analyze basic bifurcation forms. The first one is the **saddle-node bifurcation**. It results from the function

$$F(\mathcal{X}, c) = c + \delta \mathcal{X}^2 \tag{5.166}$$

where $\delta = \pm 1$. The equilibria are

$$\mathcal{X}_{e1,2} = \pm\sqrt{\frac{c}{-\delta}} \tag{5.167}$$

and the associated eigenvalues become

$$\varsigma_{1,2} = \pm 2\delta\sqrt{\frac{c}{-\delta}} \tag{5.168}$$

Obviously, one equilibrium is stable while the other is unstable, if any exists at all. Figure 5.29(a) shows the saddle-node bifurcation for $\delta = -1$ and $\delta = 1$. The bifurcation occurs at $c = 0$ where two equilibria immediately arise ($\delta = -1$) or vanish ($\delta = 1$).

The function

$$F(\mathcal{X}, c) = c\mathcal{X} - \delta\mathcal{X}^2 \tag{5.169}$$

describes the **transcritical bifurcation**. It yields the equilibria

$$\mathcal{X}_{e1} = 0, \quad \mathcal{X}_{e2} = \frac{c}{\delta} \tag{5.170}$$

and the eigenvalues

$$\varsigma_1 = c, \quad \varsigma_2 = -c \tag{5.171}$$

Again, there is one stable and one unstable equilibrium but the equilibria appear for every value of c. Figure 5.29(b) shows that the bifurcation at $c = 0$ is characterized by intersection of the equilibrium curves and a change of the stability characteristics.

A **pitchfork bifurcation** results from the function

$$F(\mathcal{X}, c) = c\mathcal{X} - \delta\mathcal{X}^3 \tag{5.172}$$

It has up to three equilibria given by

$$\mathcal{X}_{e1} = 0, \quad \mathcal{X}_{e2,3} = \pm\sqrt{\frac{c}{\delta}} \tag{5.173}$$

and the associated eigenvalues are

$$\varsigma_1 = c, \quad \varsigma_{2,3} = -2c \tag{5.174}$$

In Fig. 5.29(c) we see that at $c = 0$ one stable and two unstable equilibria turn into one unstable equilibrium for $\delta = -1$. This is a **subcritical pitchfork bifurcation**. On the other hand, for $\delta = 1$ one stable equilibrium turns into one unstable and two stable equilibria, forming a **supercritical pitchfork bifurcation**. We see in Fig. 5.23 that the latter bifurcation appears in the Lorenz model at $\mathcal{R} = 1$.

The **Hopf bifurcation** can only happen in a dynamical system with at least two dimensions. It is described by the following dynamical system:

$$\frac{d\mathcal{X}}{dt} = c\mathcal{X} - \omega\mathcal{Y} - \delta(\mathcal{X}^2 + \mathcal{Y}^2)\mathcal{X} \tag{5.175}$$

$$\frac{d\mathcal{Y}}{dt} = c\mathcal{Y} + \omega\mathcal{X} - \delta(\mathcal{X}^2 + \mathcal{Y}^2)\mathcal{Y} \tag{5.176}$$

To solve the system, it is useful to introduce **polar coordinates** r and α such that $\mathcal{X} = r\cos\alpha$ and $\mathcal{Y} = r\sin\alpha$. Then, we can transform the system equations (5.175) and (5.176) into

$$\frac{dr}{dt} = cr - \delta r^3 \tag{5.177}$$

$$\frac{d\alpha}{dt} = \omega \tag{5.178}$$

The first equation corresponds to (5.172) and, therefore, the dynamics of radius r involves a pitchfork bifurcation. The second equation

(a) Saddle node bifurcation

(b) Transcritical bifurcation

(c) Pitchfork bifurcation

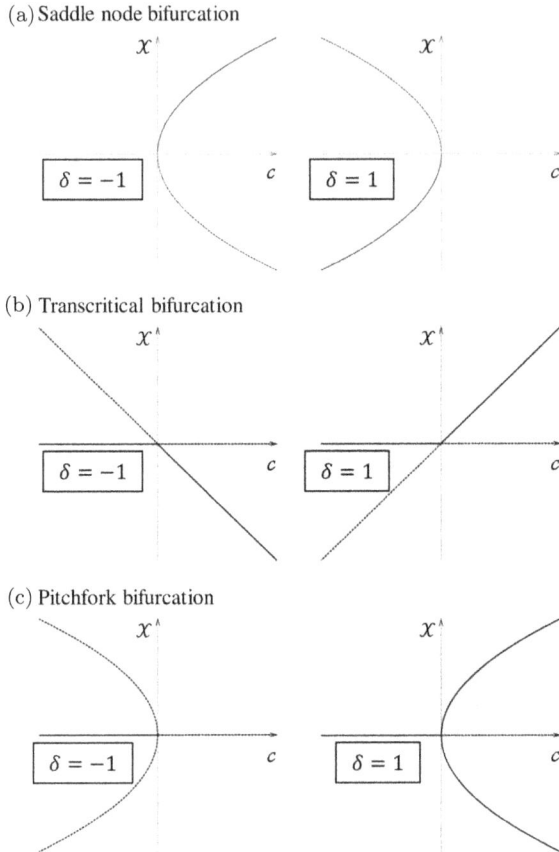

Fig. 5.29. Bifurcation diagrams showing (a) a saddle node bifurcation, (b) a transcritical bifurcation and (c) a pitchfork bifurcation. The solid (dotted) lines display stable (unstable) equilibria.

reveals that the angular coordinate α increases linearly in time. Consequently, a steady-state solution of the radius equation describes a circular path in the two-dimensional phase space. A stable or unstable closed path in phase space constitutes a **limit cycle**. Hence, the Hopf bifurcation consists of the transition from a fixed point to a limit cycle or vice versa which is sketched in Fig. 5.30. Indeed, including a nonzero basic flow U_M in the preceding analysis of a weakly nonlinear baroclinic wave would result in equations that exhibit a Hopf bifurcation instead of the pitchfork bifurcation seen in Fig. 5.23. Then, the resulting baroclinic wave propagation gives rise to a limit cycle in phase space.

Finally, we discuss the **cusp bifurcation** that depends on two model parameters. It results from the function

$$F(\mathcal{X}, c, d) = d - c\mathcal{X} - \mathcal{X}^3 \qquad (5.179)$$

Now, the equilibria become functions of c and d and can be displayed as a surface in the space spanned by c, d and $\mathcal{X}$. For positive c only one solution exists. For negative c three solutions arise if the discriminant $c^3/27 + d^2/4$ becomes negative. Therefore, the criterion for three solutions is in this case

$$|d| < \frac{2}{3}\sqrt{-\frac{c^3}{3}} \qquad (5.180)$$

This is the region inside the wedge shown in Fig. 5.31. The surface describes two folds ranging from the position of the cusp to negative c values along the boundaries of the wedge. The eigenvalues of this system become

$$\varsigma_j = -c - 3\mathcal{X}_{ej}^2 \qquad (5.181)$$

where the index j denotes the considered equilibrium. Obviously, the equilibrium for $c > 0$ is stable. To reveal the stability for the other

(a) Subcritical Hopf bifurcation   (b) Supercritical Hopf bifurcation

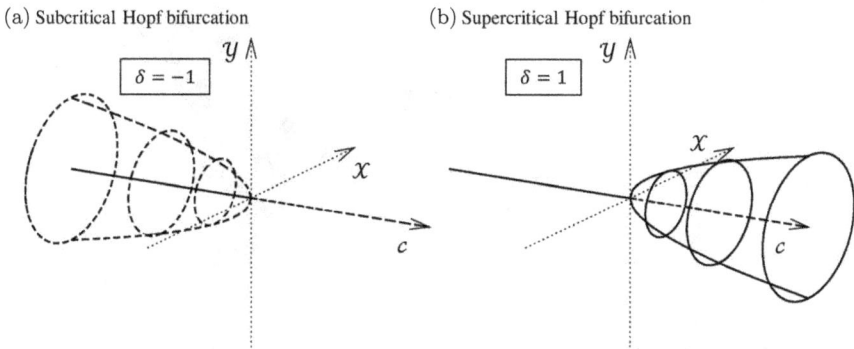

Fig. 5.30. Sketch of (a) a subcritical Hopf bifurcation and (b) a supercritical Hopf bifurcation. The circles are the resulting limit cycles. The solid (dotted) lines display stable (unstable) equilibria.

cases we can write equation (5.163) in terms of a "potential"

$$\frac{d\mathcal{X}}{dt} = -\frac{\partial \mathcal{V}}{\partial \mathcal{X}} \tag{5.182}$$

where

$$\mathcal{V}(\mathcal{X}, c, d) = -d\mathcal{X} + \frac{c}{2}\mathcal{X}^2 + \frac{1}{4}\mathcal{X}^4 \tag{5.183}$$

Maxima and minima in $\mathcal{V}$ yield unstable and stable equilibria, respectively. There must be a maximum inside of two minima (3 equilibria) or only one minimum (one equilibrium) since $\mathcal{V}$ approaches infinity for $\mathcal{X} \to \pm\infty$. Hence, inside the wedge, the middle equilibrium is unstable while the other two are stable. The cusp bifurcation results in a supercritical pitchfork bifurcation for $d = 0$ but a saddle node bifurcation arises for other values of d. Therefore, the singular case $d = 0$ is **structurally unstable** because slight deviations of d alter the bifurcation type. The parameter d describes a slight asymmetry in the model. For example, adding a small heat anomaly in the Lorenz model equation (5.132) alters the equilibria (5.138) with a saddle-node bifurcation appearing instead of a pitchfork bifurcation.

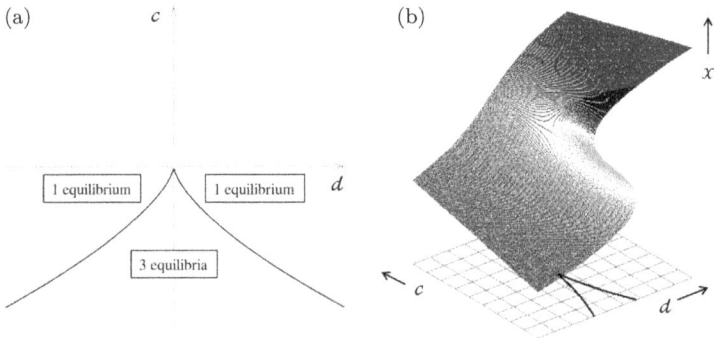

Fig. 5.31.   Sketch of a cusp bifurcation: (a) The lines bounding the wedge in the $d - c$ plane separate the regions with 1 and 3 equilibria, (b) The surface displays the equilibria as a function of d and c. Multiple equilibria occur between the two folds.

5.5.6 *Dissipative dynamical systems and attractors*

For the further analysis, it is important to know whether a dynamical system is dissipative or not. A **dissipative dynamical system** is characterized by a negative divergence of the tendency vector, i.e.

$$\vec{\nabla} \cdot \vec{F} = \sum_{n=1}^{N} \frac{\partial F_n}{\partial X_n} < 0 \tag{5.184}$$

Therefore, a phase space volume shrinks in a dissipative dynamical system. This can be understood by Reynolds transport theorem (1.12) applied to the N dimensional Euclidean space

$$\frac{dV_F}{dt} = \iiint_{V_F} \sum_{n=1}^{N} \frac{\partial F_n}{\partial X_n} dV < 0 \tag{5.185}$$

where V_F is a phase space volume that is deformed by the "velocity" vector field $\vec{F}$. Often, atmospheric models are dissipative dynamical systems because irreversible (mainly volume-shrinking) processes are included.

Conservative dynamical systems, on the other hand, stand out due to zero volume change, i.e.

$$\vec{\nabla} \cdot \vec{F} = 0 \tag{5.186}$$

Such systems often appear in the form of Hamiltonian equations and in this context, volume conservation is known as **Liouville's theorem**. An example for a conservative dynamical system was derived in Section 5.3.1 with the equations (5.33)–(5.35). We can easily check that

$$\vec{\nabla} \cdot \vec{F} = \frac{\partial}{\partial A}\left(\frac{dA}{d\tilde{t}}\right) + \frac{\partial}{\partial B}\left(\frac{dB}{d\tilde{t}}\right) + \frac{\partial}{\partial C}\left(\frac{dC}{d\tilde{t}}\right) = 0 \tag{5.187}$$

For this reason, it was possible to derive Hamiltonian equations. However, we obtain a dissipative dynamical system by including frictional dissipation as done in Section 5.3.2. For the resulting Lorenz model equations (5.131)–(5.133) we get

$$\vec{\nabla} \cdot \vec{F} = \frac{\partial}{\partial \mathcal{X}}\left(\frac{d\mathcal{X}}{dt}\right) + \frac{\partial}{\partial \mathcal{Y}}\left(\frac{d\mathcal{Y}}{dt}\right) + \frac{\partial}{\partial \mathcal{Z}}\left(\frac{d\mathcal{Z}}{dt}\right) = -2 - \mathcal{B} < 0 \tag{5.188}$$

Dissipative dynamical systems have a special property, namely that **attractors** exist in phase space. An attractor is a subset of the phase

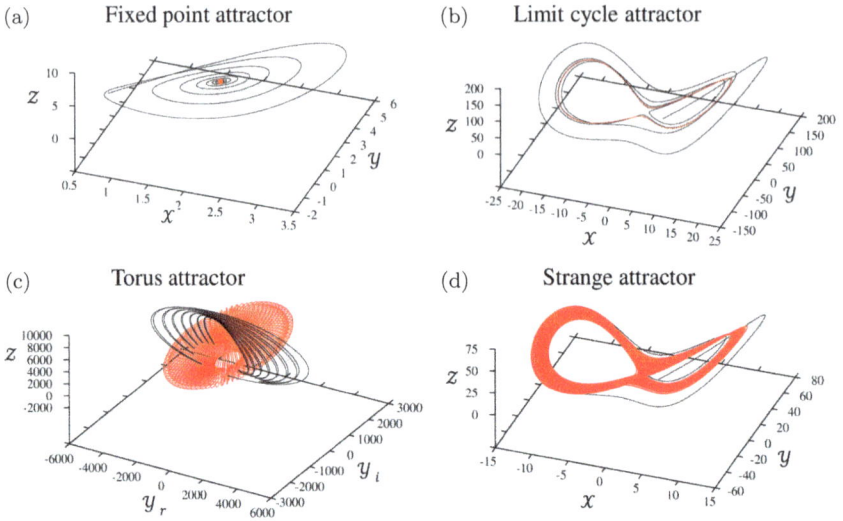

Fig. 5.32.   Attractors in phase space simulated with the Lorenz model: (a) Fixed point attractor, (b) limit cycle attractor, (c) torus attractor and (d) strange attractor. The black lines show the transients approaching the attractor, while the red lines represent the attractor itself.

space with zero volume content. To understand the existence, let us first assume a finite volume in phase space. In the course of time, the volume content decreases to arbitrarily small values when it is deformed by the flow. The volume vanishes for $t \to \infty$ and the initial volume is distributed on one or more attractors. The individual trajectories that end up on an attractor are called **transients**. We distinguish the following attractor types: **fixed point attractor, limit cycle attractor, torus attractor and strange attractor**.

A **fixed-point attractor** is just a stable fixed point. The transients in the vicinity of the fixed point are portrayed for the stable proper node, stable focus and stable node in Fig. 5.28. The volume of a fixed point vanishes since it has no dimensions. Therefore, a one-dimensional dynamical system suffices to describe a fixed-point attractor. Figure 5.32(a) shows the trajectory into a fixed-point attractor occurring in the Lorenz model for $\mathcal{R} = 12.5$. We can only visualize the transient because a fixed-point attractor is just a point in phase space.

A **limit cycle attractor** is a closed line in phase space and, therefore, it describes a periodic solution. The time series can exhibit several frequencies but they must be multiples of the fundamental frequency that results from the periodicity. A limit cycle attractor is one-dimensional and can appear in dynamical systems with at least two dimensions. Figure 5.32(b) shows a limit cycle attractor that results in the Lorenz model for $\mathcal{R} = 312.5$ and $\mathcal{B} = 4/3$. The transient approaches this cycle after several aperiodic oscillations. Note that two fixed points are still stable because Eq. (5.154) is fulfilled. Therefore, we have at least three attractors for these parameters in the Lorenz model.

A **torus attractor** represents an aperiodic solution as a result of several periodic oscillations. Absence of periodicity appears when the ratio of two frequencies is not identical to the ratio of two integers. If the ratio can be written as $\nu_1/\nu_2 = n_1/n_2$, then the two oscillations have the common period $T = 2\pi n_1/\nu_1 = 2\pi n_2/\nu_2$ and the attractor is a limit cycle. Otherwise, there is no periodicity and the attractor becomes a torus. Such a solution is also called to be **quasiperiodic**. An n-**torus** comprises n distinct frequencies and the phase space must have at least $n+1$ dimensions. The Lorenz model does not reveal a torus attractor. However, a modified model is obtained when we add a barotropic basic flow in the derivation presented in Section 5.3.2. Then, the following complex Lorenz equations can be derived which take the form

$$\frac{d\mathcal{X}}{dt} = -i\mathcal{U}X + \sigma(\mathcal{Y} - \mathcal{X}) \tag{5.189}$$

$$\frac{d\mathcal{Y}}{dt} = -i\mathcal{U}\mathcal{Y} + \mathcal{R}\mathcal{X} - \mathcal{X}\mathcal{Z} - \mathcal{Y} \tag{5.190}$$

$$\frac{d\mathcal{Z}}{dt} = \Re(\mathcal{X}\mathcal{Y}^C) - \mathcal{B}\mathcal{Z} \tag{5.191}$$

where $\mathcal{U}$ is a real parameter that is proportional to the strength of the barotropic basic flow. In these equations, $\mathcal{X}$ and $\mathcal{Y}$ are complex variables while $\mathcal{Z}$ is still real. Now, we may have two independent oscillations in the model. The first stems from the zonal advection of the wave while the other results from periodic baroclinic wave lifecycles. We obtain a 2-torus since the resulting frequencies are not related to each other. Figure 5.32(c) shows a torus attractor that appears for $\mathcal{U} = 0.4$ and $\mathcal{R} = 5000$ in the dynamical system (5.189)–(5.191). Note that the system has five real variables and the shown coordinate system displays only three of them. The attractor encloses a finite volume in phase space. It looks like a thread that has been winded up on the surface of the volume. The torus attractor forms a closed surface in the limit $t \to \infty$.

The **strange attractor** is characterized by its aperiodicity and limited predictability. The latter property distinguishes the strange attractor from the torus because it is possible to predict the quasi-periodic solution for arbitrarily long periods. The dynamical system must have at least three dimensions in order to produce a strange attractor. Figure 5.32(d) shows a strange attractor occurring in the Lorenz model for $\mathcal{R} = 128$. It looks like a surface in phase space but the attractor must have a more complicated structure for explaining the limited predictability.

5.5.7   *Lyapunov exponent*

The stability of a trajectory in phase space can be detected by the **Lyapunov exponents** instead of doing a conventional eigenvalue analysis. The latter is not appropriate if the trajectory is not a fixed point. In this case, the linearized system contains differential equations with time-dependent coefficients and the solution cannot be represented as a sum of exponential functions in time. The **maximum Lyapunov exponent** describes the average perturbation growth or decay of nearby trajectories and is given by

$$\varsigma_L = \lim_{t \to \infty} \frac{1}{t} \ln \left[\frac{|\vec{X}'(t)|}{|\vec{X}'(0)|} \right] \tag{5.192}$$

where $\vec{X}'(t)$ is a solution of the linearized dynamical system for which the basic state is given by the orbit $\vec{X}_O(t)$. If $\vec{X}_O(t)$ is on an attractor, the maximum Lyapunov exponent[i] does not depend upon the initial condition since the trajectory reaches every point on the attractor for $t \to \infty$ which is the **ergodic hypothesis**.

The Lyapunov exponent coincides with the maximum real part of the eigenvalues of $\overleftrightarrow{D}$ if $\vec{X}_O(t)$ is a fixed point. We can see this by

---

[i]We omit the adjective maximum in the following.

considering the solution of the linearized system

$$\vec{X}'(t) = \sum_{n=1}^{N} \Re(\vec{X}_n \, e^{\varsigma_n t}) \tag{5.193}$$

So, we obtain the limit of very large t for the norm of the perturbation

$$|\vec{X}'(t)| = \left| \sum_{n=1}^{N} \Re(\vec{X}_n \, e^{\varsigma_n t}) \right| = \left| \sum_{n=1}^{N} e^{\varsigma_{n,r} t} \Re(\vec{X}_n \, e^{i\varsigma_{n,i} t}) \right|$$

$$\approx \left| e^{\varsigma_{1,r} t} \Re(\vec{X}_1 \, e^{i\varsigma_{1,i} t}) \right| = e^{\varsigma_{1,r} t} |\Re(\vec{X}_1 \, e^{i\varsigma_{1,i} t})| \tag{5.194}$$

where $\varsigma_{n,r}$ and $\varsigma_{n,i}$ are the real and imaginary parts of the n-th eigenvalue, respectively. The eigenvalues have been ordered in such a way that $\varsigma_{1,r} \geq \varsigma_{2,r} \geq \cdots \geq \varsigma_{N,r}$.[j] With the approximation (5.189) the Lyapunov exponent becomes

$$\varsigma_L = \lim_{t \to \infty} \frac{1}{t} \ln \left[\frac{|\vec{X}'(t)|}{|\vec{X}'(0)|} \right] = \lim_{t \to \infty} \frac{1}{t} \ln \left[\frac{e^{\varsigma_{1,r} t} |\Re(\vec{X}_1 \, e^{i\varsigma_{1,i} t})|}{|\vec{X}'(0)|} \right]$$

$$= \lim_{t \to \infty} \left\{ \frac{1}{t} \ln(e^{\varsigma_{1,r} t}) + \frac{1}{t} \ln \left[\frac{|\Re(\vec{X}_1 \, e^{i\varsigma_{1,i} t})|}{|\vec{X}'(0)|} \right] \right\} = \varsigma_{r1} \tag{5.195}$$

since $|R(\vec{X}_1 \, e^{i\varsigma_{1,i} t})|$ remains finite in the limit $t \to \infty$. Therefore, a fixed point is stable if $\varsigma_L < 0$ and unstable if $\varsigma_L > 0$.

The stability conclusion can be extended to time-dependent reference trajectories since a continuous growth of distance is associated with a positive Lyapunov exponent. Indeed, this coefficient gives information about the attractor type. Obviously, a fixed-point attractor has a negative ς_L. A limit cycle attractor is associated with $\varsigma_L = 0$. This means that points on adjacent trajectories neither approach nor move away from each other. To understand this let us consider two neighboring points on the limit cycle attractor. After one period the distance between them is the same and this remains true for an arbitrarily large number of periods. An analysis of a torus attractor also yields $\varsigma_L = 0$. The aforementioned explanation is also true

---

[j]In (5.195) we have ignored possible cases like $\varsigma_{r1} = \varsigma_{r2}$ or $\varsigma_{r1} = \varsigma_{r2} = \varsigma_{r3}$. In such cases, the proof is similar.

when the solution is a superposition of two or more periodic oscillations. Finally, a strange attractor is characterized by $\varsigma_L > 0$. For such a system we can state that the trajectories on the attractor are unstable. However, the instability is damped by nonlinearities in the dynamical system so that perturbed solutions stay on the attractor. The positive Lyapunov exponent has the consequence that small initial perturbations grow to a size that is comparable to the diameter of the attractor. Consequently, a prediction of the system for large time is difficult because initial uncertainties have large impacts in the course of time. This is the **butterfly effect**, a term which was introduced by Lorenz (1972). He stated in the context of weather that the flapping of a butterfly can cause significant weather events like a tornado at a distant location. Systems with such behavior are also known to be **chaotic** and form the basis of **chaos theory**.

We have learned that we can use the Lyapunov exponent to characterize the attractor. To determine this exponent, we apply a straightforward method. First, a numerical solution of the dynamical system equations must be found by numerical integration. The solution covers nearly the complete attractor and the initial transient becomes negligible for a large enough time period. In the next step, the linearized equations must be time-integrated to find the evolution of the perturbation $\vec{X}'(t)$. At large t this can be used to calculate ς_L. We can also do everything simultaneously by solving the dynamical system

$$\frac{d\vec{X}}{dt} = \vec{F}(\vec{X}) \tag{5.196}$$

$$\frac{d\vec{X}'}{dt} = \overleftrightarrow{D}(\vec{X}) \cdot \vec{X}' \tag{5.197}$$

and calculating ς_L using $\vec{X}'$. However, there is a practical problem because $\vec{X}'$ might grow or decay exponentially. Then, the computer experiences an overflow or an underflow so that no usable result is produced. This problem can be solved by rescaling the perturbation when it becomes too large or too small. The perturbation remains a solution of the linearized equation after multiplying it with a certain factor. However, we must consider this factor in the calculation of ς_L. Figure 5.33 illustrates the multiple rescaling of $\vec{X}'$ due to exponential growth. Figure 5.34 displays the time series of ς_L for different simulations using the Lorenz model. Obviously, ς_L converges to a

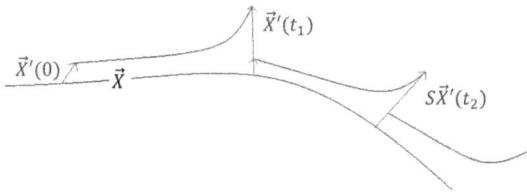

Fig. 5.33. Schematic representation of rescaling the perturbation by multiplying it with a factor S.

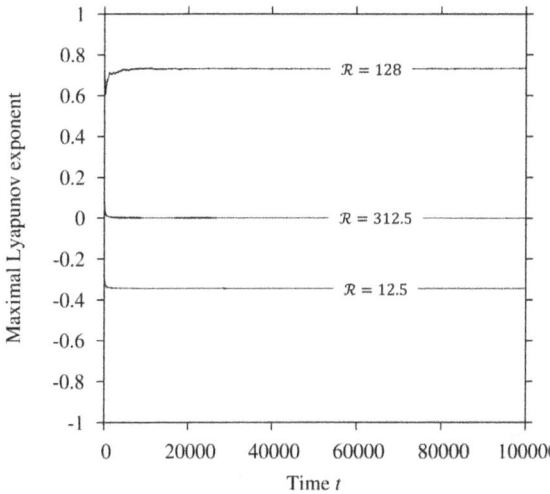

Fig. 5.34. Lyapunov exponent as a function of time deduced from numerical solutions of the Lorenz model for $\mathcal{R} = 12.5$, $\mathcal{R} = 128$ and $\mathcal{R} = 312.5$.

constant value in all cases and, indeed, the value of the final ς_L is in agreement with the attractor type (cf. Figure 5.32). Figure 5.35 shows ς_L as a function of $\mathcal{R}$ for simulations initialized near the equilibrium 1 and equilibrium 2 of the Lorenz model (cf. Eq. (5.138)). The Lyapunov exponents of both cases are negative and coincide up to $\mathcal{R} = 108$. In this range, the attractor is a fixed point and ς_L is identical to the maximum real part of the eigenvalues of the Jacobi matrix. The Lyapunov exponent remains negative for larger $\mathcal{R}$ values when the simulations are initialized close to equilibrium 2. Therefore, the equilibrium is a fixed-point attractor for all $\mathcal{R}$ which is already evident from the eigenvalue analysis. However, the simulations initialized at equilibrium 1 reveal positive Lyapunov exponents

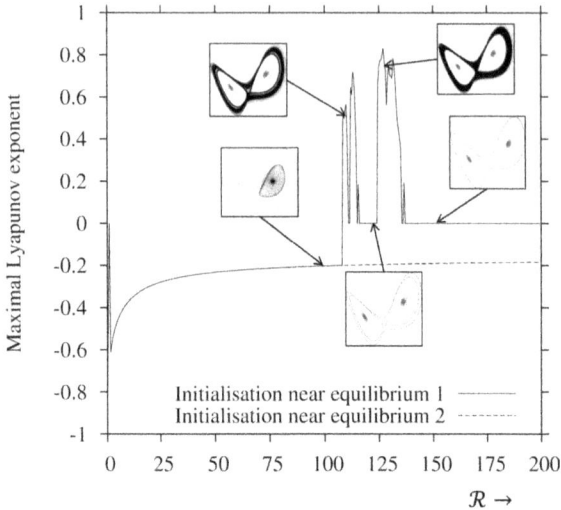

Fig. 5.35.   Lyapunov exponent as a function of $\mathcal{R}$ in numerical integrations of the Lorenz equations initialized near equilibrium 1 (solid line) and near equilibrium 2 (dashed line). Attractors for selected $\mathcal{R}$ values are displayed in the integrated frames.

for $\mathcal{R} > 108$. Consequently, a strange attractor appears beside the two fixed point attractors that coincide with equilibrium 2 and 3. There are intervals where ς_L becomes zero before it takes a positive value again. In these phases, a limit cycle attractor occurs instead of a strange attractor. The transition from a periodic to a chaotic time series happens by sporadically occurring irregular events. This behavior found by Pomeau and Manneville (1980) is called **inter-mittency**. Finally, the limit cycle attractor prevails for $\mathcal{R} > 135$.

In Section 4.6.3, we have introduced a method to find out the optimal perturbation for a certain flow and a chosen time interval τ. We also mentioned that the largest eigenvalue of the analysis becomes the Lyapunov exponent in the limit $\tau \to \infty$. However, the analysis results in a number of eigenvalues and these constitute the **spectrum of Lyapunov exponents** in this limit. This spectrum can be used to further characterize the attractor. An n-torus attractor has n Lyapunov exponents with the value zero while all others are negative. A strange attractor possesses at least one positive Lyapunov exponent and at least one having the value 0 while all others are negative (see Wolf *et al.*, 1985). The atmosphere represents a strange attractor with many positive Lyapunov exponents if we are able to describe it as a dissipative dynamical system. Therefore, predictions are only feasible for a limited time period as we will discuss in more detail in Section 5.7.

5.6   Atmospheric Blocking in the Charney–DeVore Model

The mid-latitude weather evolution is very volatile. Besides growing and decaying baroclinic wave packets we also find a variation of planetary-scale waves and of the zonal mean current. The **zonal index** introduced by Rossby (1939) measures the strength of the zonal mean current. It is given by

$$ZI = [p_{SL35} - p_{SL55}]_\lambda \tag{5.198}$$

where p_{SL35} and p_{SL55} denote the sea level pressure at 35° and 55° latitude, respectively. Figure 5.36 shows a time series of the zonal index for the North Atlantic (zonal average extends from 120W to 40E) in the winter season 2007/2008. We see a large variability in a range between $-8\,$hPa and $22\,$hPa. Often the index changes rapidly from one extreme to the opposite extreme. This behavior is known as the **index cycle** and describes an oscillation between a high-index and a low-index weather state.

Figure 5.37 displays examples for $500\,$hPa geopotential height in the high and low index states. High-index weather is characterized by a zone extending from west to east with a high negative meridional geopotential gradient. The contours have an undulating shape because of the embedded synoptic-scale baroclinic waves. In contrast, the low index state exhibits contours having large displacements in the meridional direction due to the existence of a planetary scale Rossby wave. Obviously, the zonal flow is blocked at 50°W longitude and deflected northwards. Therefore, the huge ridge eastward of 50°W is referred to as a **blocking high**. This high is also called **omega high** because the contours have the shape of the Greek capital letter Ω.

It can be supposed that the blocking high is caused by the orography or the land-sea contrast. Charney and DeVore (1979) used an idealized barotropic model with orography to investigate blocking.

Zonal index for North−Atlantic

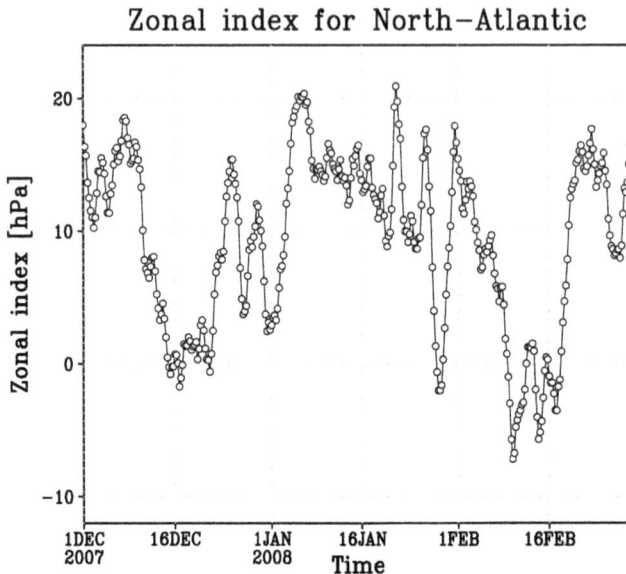

Fig. 5.36. Zonal index (hPa) for a zonal range from 120W to 40E in the winter season 2007/08 derived from NCEP reanalysis.

In this model, two stable equilibria appear which can be associated with the high and low index states, respectively. Here, we will analyze a very similar model. The only difference is that the model is based on different basis functions. In Section 3.9, we derived the quasigeostrophic vorticity equation for a barotropic atmosphere with orography and an underlying Ekman layer. This equation would predict a flow that decays slowly due to Ekman pumping. To overcome the decay Charney and DeVore introduced an artificial forcing term that mimics baroclinic feedbacks and heating. Hence, they tried to parameterize effects which can only be described fully in a baroclinic model. The forcing term in the barotropic vorticity equation has the form $\nabla_h^2 \psi_F / \tau_E$ where ψ_F is the streamfunction pattern that is forced by baroclinic effects. With this additional term, the quasigeostrophic vorticity equation takes the form

$$
\frac{\partial}{\partial t}\left(\nabla_h^2 - \frac{1}{L_R^2}\right)\psi_g + J_{xy}\left(\psi_g, \nabla_h^2\psi_g + f_0\frac{H_s}{\bar{H}_f}\right)
$$

$$
+ \beta\frac{\partial \psi_g}{\partial x} = \frac{\nabla_h^2(\psi_F - \psi_g)}{\tau_E} \tag{5.199}
$$

5 January 2008, 18UTC (NCEP)

(a)

17 February 2008, 6UTC (NCEP)

(b)

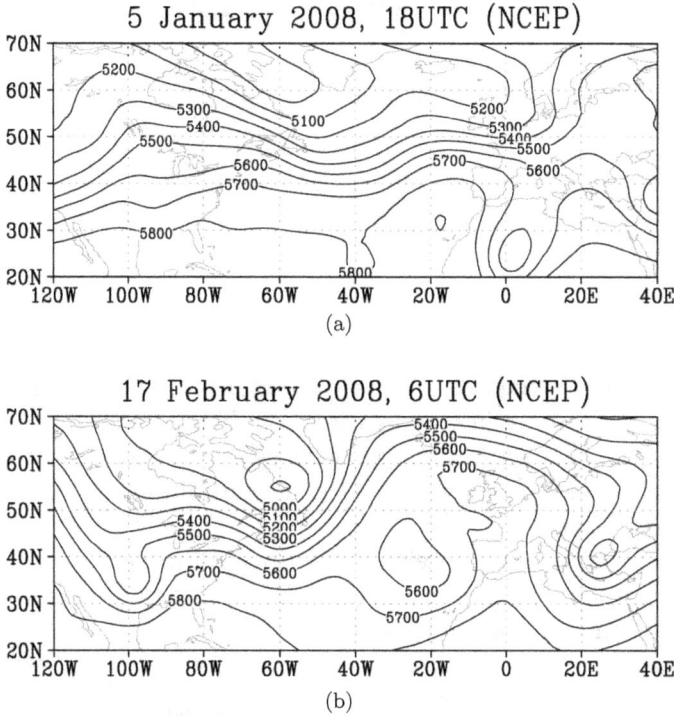

Fig. 5.37. NCEP reanalysis 500 hPa geopotential height (contour interval 100 m) for (a) a high index weather state (January 5, 2008, 18UTC) and (b) a low index weather state (February 17, 2008, 6UTC).

where $\bar{H}_f$ is the mean depth of the free atmosphere. Charney and DeVore neglected the impact of upper surface elevations by setting $L_R \to \infty$ which is equivalent to assuming a rigid lid. We will do it likewise and nondimenionalize the time scale by f_0^{-1}, horizontal length scale by L, stream function by $f_0 L^2$, β by f_0/L and height by $\bar{H}_f$. Then, the nondimensional vorticity equation becomes

$$\frac{\partial}{\partial t}\nabla_h^2\psi + J_{xy}(\psi, \nabla_h^2\psi + h) + \beta\frac{\partial\psi}{\partial x} = \frac{\nabla_h^2(\psi_F - \psi)}{\tau_E} \qquad (5.200)$$

where ψ and h denote the nondimensional stream function and height, respectively. For all other quantities, the tilde symbol denoting nondimensional variables has been omitted. For the baroclinic forcing stream function we assume a meridional profile

$$\psi_F = -U\left[1 - \frac{\pi}{2} - \frac{1}{2}\sin(2y)\right] \qquad (5.201)$$

The associated flow describes a zonal jet stream at $y = \pi/2$ and vanishing wind at $y = 0$ and $y = \pi$ where the model boundaries lie. Orography is prescribed by

$$h = h_0\cos(kx)\sin y \qquad (5.202)$$

where h_0 denotes the height of the mountains. Obviously, mountains and valleys alternate in a zonal direction. This is a somewhat artificial and idealized setup but the purpose of the model is to explain the principal mechanism of blocking instead of making realistic and usable predictions.

In Section 5.4.2, we developed the spectral method which can be directly applied to the barotropic vorticity equation. First, we expand the streamfunction in the form

$$\psi = -Uy + \sum_{n=-N}^{N}\sum_{m=-M}^{M} \Psi_n^m(t)e^{imkx}e^{iny} \qquad (5.203)$$

The coefficients Ψ_n^m are subject to the boundary conditions (5.79). In the next step, we can use the orthogonality of the basis function to derive the ordinary differential equations for the spectral coefficients. Applying this to the governing equation (5.200) yields the dynamical system

$$\frac{d\Psi_n^m}{dt} = -imk\left(U - \frac{\beta}{K_n^{m2}}\right)\Psi_n^m + \frac{\Psi_{Fn}^m - \Psi_n^m}{\tau_E} + i\frac{Umk}{K_n^{m2}}H_n^m$$

$$- \sum_{n_1=n-N}^{N}\sum_{m_1=m-M}^{M}\frac{\alpha_{n,n_1}^{m,m_1}}{K_n^{m2}}(K_{n_1}^{m_1\,2}\Psi_{n-n_1}^{m-m_1}\Psi_{n_1}^{m_1} - \Psi_{n-n_1}^{m-m_1}H_{n_1}^{m_1})$$

$$(5.204)$$

where Ψ_{Fn}^m and $H_{n_1}^{m_1}$ denote the spectral coefficients of ψ_F and h, respectively. Obviously, this is a dissipative dynamical system since the partial derivative of the time tendency with respect to the prognostic variable (real or imaginary part) yields always $-1/\tau_E$. Truncating the system at $M = 1$ and $N = 2$ leads to the following four complex-valued equations

$$\frac{d\Psi_1^0}{dt} = -\frac{\Psi_1^0}{\tau_E} - 6ik\Im\left(\Psi_2^1\Psi_1^{1C}\right) - 2ik\Im\left(\Psi_2^1 H_1^{1C}\right) \quad (5.205)$$

$$\frac{d\Psi_2^0}{dt} = \frac{\Psi_{F2}^0 - \Psi_2^0}{\tau_E} + ik\Im\left(\Psi_1^1 H_1^{1C}\right) \quad (5.206)$$

$$\frac{d\Psi_1^1}{dt} = -i\left(Uk - \frac{\beta k}{1+k^2}\right)\Psi_1^1 - \frac{\Psi_1^1}{\tau_E} - k\frac{3+k^2}{1+k^2}\Psi_1^0\Psi_2^1$$
$$+ 2k\frac{k^2-3}{1+k^2}\Psi_2^0\Psi_1^1 + i\frac{Uk}{1+k^2}H_1^1 - \frac{2k}{1+k^2}\Psi_2^0 H_1^1 \quad (5.207)$$

$$\frac{d\Psi_2^1}{dt} = -i\left(Uk - \frac{\beta k}{4+k^2}\right)\Psi_2^1 - \frac{\Psi_2^1}{\tau_E} - \frac{k^3}{4+k^2}\Psi_1^0\Psi_1^1 + \frac{k}{4+k^2}\Psi_1^0 H_1^1 \quad (5.208)$$

where we have used the assumption that only Ψ_{F2}^0 and H_1^1 are nonzero. We can find a flow symmetry with $\Psi_1^0 = \Psi_2^1 = 0$ and with this symmetry the model can be reduced to the following three real differential equations:

$$\frac{dA}{dt} = \frac{U/2 - A}{\tau_E} + \frac{kh_0}{8}C \quad (5.209)$$

$$\frac{dB}{dt} = -\left(Uk - \frac{\beta k}{1+k^2}\right)C - \frac{B}{\tau_E} - k\frac{k^2-3}{1+k^2}AC \quad (5.210)$$

$$\frac{dC}{dt} = \left(Uk - \frac{\beta k}{1+k^2}\right)B - \frac{C}{\tau_E} + k\frac{k^2-3}{1+k^2}AB - \frac{Ukh_0}{1+k^2} - \frac{kh_0}{1+k^2}A \quad (5.211)$$

where A, B and C are defined to satisfy

$$\psi = -Uy + A(t)\sin 2y + \sin y(B(t)\cos kx + C(t)\sin kx) \quad (5.212)$$

This is the dynamical system describing the interaction of an orographically forced Rossby wave (represented by B and C) with a jet stream (represented by A and U). The first step of dynamical system analysis is to find the equilibria. By setting the tendencies in (5.210) and (5.211) to zero we find after some manipulations

$$C = -\frac{\tau_E k}{1+k^2} \frac{(U+A)h_0}{1 + \tau_E^2 \left(Uk + k\frac{k^2-3}{1+k^2}A - \frac{\beta k}{1+k^2}\right)^2} \qquad (5.213)$$

For $A = 0$ it can be verified that this formula corresponds to the imaginary part of Eq. (3.199) since C represents the sine component of the orographically forced wave. Therefore, the model has also resonance at a certain zonal wavelength. The jet stream strength and shape depend on A. In equilibrium, it is related to C by

$$A = \frac{U}{2} + \tau_E \frac{kh_0}{8} C \qquad (5.214)$$

Therefore, the wave has an impact on the zonal mean flow due to the orography. It can be expected that C is negative so that orography decelerates the zonal flow by the form drag. Equations (5.213) and (5.214) yield functions $C(A)$ and the intersection points constitute the equilibria (combining (5.213) and (5.214) results in a cubic equation). Figure 5.38 shows these functions and it becomes clear that one or three equilibria can appear. One of the equilibria has a high positive A value with small negative C while the other two equilibria are characterized by negative A values and large negative C. The magnitude of the wave part C rises rapidly in Eq. (5.213) when $A = -0.05$ is approached. Therefore, the latter equilibria are close to the wave resonance. Figure 5.39 shows the streamfunction pattern of the three equilibria. The first equilibrium has a strong trough in the lee of the mountain and a strong ridge further downstream. The second equilibrium has also a high amplitude wave but it is phase-shifted by about 180 degrees. The phase shift was already detected in the linear case close to the resonance (cf. Section 3.10) and we assessed a state like the second equilibrium as unrealistic in comparison with observations. Indeed, the second equilibrium is unstable as we will see later. The third equilibrium has only a small amplitude wave. Therefore, we can associate the first equilibrium with the low-index state and the third equilibrium with the high-index state.

The equilibrium values of A as a function of U are shown in Fig. 5.40(a). There are two saddle-node bifurcations which isolate

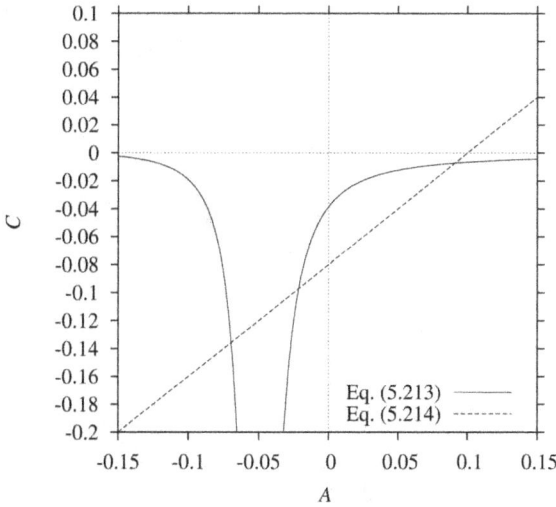

Fig. 5.38. Functions $C(A)$ as given by Eqs. (5.213) and (5.214). For these plots, the parameter values $U = 0.2$, $\beta = 0.5$, $\tau_E = 100$, $h_0 = 0.1$ and $k = 1$ are implemented.

the range where three equilibria appear. The equilibria as a function of U and mountain height h_0 are displayed in terms of a surface in Fig. 5.40(b). In this plot, we see that a cusp bifurcation appears for low U and high h_0. Another one can be detected at high U and intermediate h_0. Consequently, three equilibria arise only in a closed area of the $U - h_0$ plane.

In the next step, the stability of the equilibria is examined. The eigenvalues of the stability analysis are obtained from a cubic equation which is solved numerically. Figure 5.41 shows these eigenvalues for the same parameter values as used for Fig. 5.40(a). Obviously, the analysis reveals that only the equilibrium with intermediate A values is unstable in the range with multiple equilibria. Consequently, we have two fixed point attractors in this range and one fixed point attractor for all other U values. The stable equilibria are stable foci because of the complex conjugated eigenvalues. On the other hand, the unstable fixed point represents an unstable saddle node.

So far it is not clear whether the results of the low-order model are reasonable or rather an artifact of the low-order truncation. To answer this question the spectral model equations (5.204) must be integrated with a truncation at sufficiently large wavenumbers. For

(a)

(b)

(c)

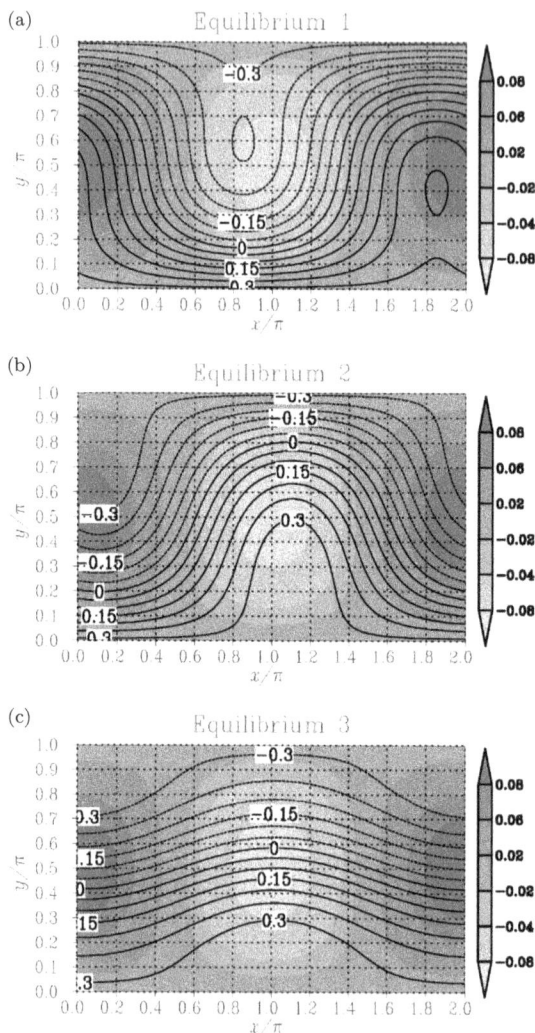

Fig. 5.39.   Streamfunction of the three equilibria resulting from Eqs. (5.213) and (5.214). The model parameters are the same as in Fig. 5.38. The grey shadings indicate the nondimensional height of the orography.

our purpose, a truncation at $M = 8$ and $N = 20$ appears to be sufficient. To damp small-scale numerical perturbations weak horizontal diffusion has been added. In most cases, the numerical model integration leads to a steady state and the final value of A of this integration is displayed as a function of U in Fig. 5.40(a) together

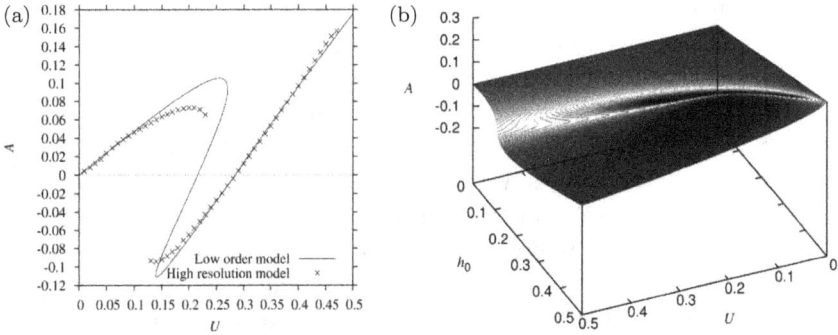

Fig. 5.40. (a) Equilibrium values of A as a function of U. The solid curve shows the result for the low-order model (5.209)–(5.211) and the crosses result from numerical integrations of the spectral model at a high-order truncation. (b) Equilibrium surface of A as a function of U and h_0. The values of the nonvarying parameters are the same as in Fig. 5.38.

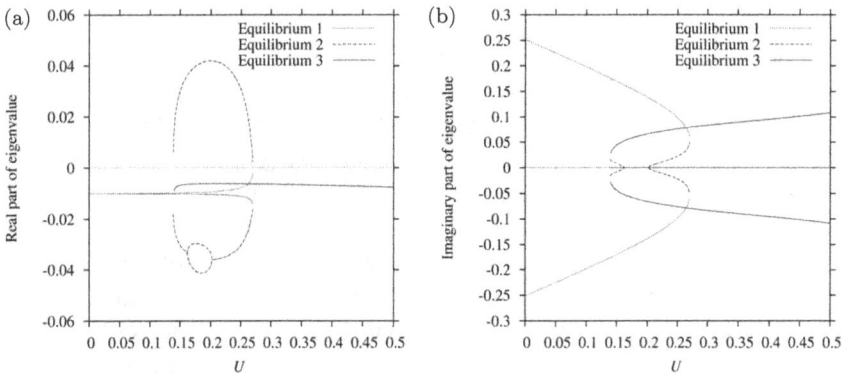

Fig. 5.41. Eigenvalues resulting from the stability analysis of the equilibria shown in Fig. 5.40(a): (a) real part and (b) imaginary part.

with the low-order model result. The agreement is very good for low and high U values but for $U > 0.47$ a limit cycle appears instead of a steady state indicating a Hopf bifurcation. Some discrepancies can be seen for intermediate U where multiple equilibria occur. However, the low-order model agrees qualitatively with the high-resolution numerical model. We cannot determine the unstable intermediate equilibrium in this model but it is likely that such a state exists. With the method of Dijkstra (2005) it might be possible to determine this unstable equilibrium. Comparison of the stream function pattern in the equilibrium state for $U = 0.2$ (shown for the low-order model in Fig. 5.39) reveals little differences. Therefore, the low-order model qualitatively describes what happens in a high-resolution model which should represent the true solution of (5.200) to a high degree. There is also the possibility that a symmetry-breaking instability can alter the results. In the present parameter setup such instability does not appear but Charney and DeVore (1979) found that the blocking state is unstable with respect to a symmetry-breaking disturbance and this led in their model to a limit cycle. However, they used the wavenumber $k = 2$ instead of $k = 1$ as in our case. Indeed, the breaking of the flow symmetry results from a barotropic wave instability which becomes only relevant for meridionally elongated eddies ($k > 1$, see Frisius, 1998).

With the multiple equilibria in this simple model, we could describe qualitatively the existence of high and low-index weather states. However, it was not possible to explain the index cycle, i.e. the repeated switching from one to the other state. It is likely that the absence of time variability in this simple model stems from the neglect of baroclinic instability. Reinhold and Pierrehumbert (1982) extended the Charney-DeVore model by considering the baroclinic two-level equations and truncating at zonal wavenumber 2 to describe the effect of a shorter baroclinic wave. They found 5 equilibria which are all unstable with respect to baroclinic instability. Consequently, the model has no fixed-point attractor and the solution becomes always time dependent. Instead of equilibria they defined quasi-stationary **weather regimes** and found that the model settles aperiodically in a weather regime that can be associated with blocking.

5.7 Weather Prediction

The prediction of weather is limited in time. It is possible to make quite good forecasts of synoptic-scale weather systems for a few days but a forecast beyond one month is likely to fail. In Section 5.5, we have already developed a theory that explains this fact. Synoptic-scale flows are always hydrodynamically unstable in some regions and, therefore, the Lyapunov exponent becomes positive. Consequently, the weather represents a strange attractor in a phase space of the atmospheric dynamical system. Although the system can be regarded as deterministic it is impossible to make an error-free prediction for all future points in time because three errors are inevitable. The first is the **initialization error** that arises because the initial state of the dynamical system is not exactly known. Measurements have errors and the field variables are not measured at every location so some initial values must be guessed. At the same time, we do not know the governing equations of the atmosphere exactly and even if we know them, we cannot solve them without making considerable approximations, especially at small scales. Therefore, there is always a **model error** in the prediction. A third error is the **external error** and stems from the boundary conditions which also cannot be captured without error. Nevertheless, there is also another problem with this error because the boundaries of the atmosphere are also subject to physical laws and change in time. Therefore, it is necessary to predict the surrounding regions to make an error-free weather forecast. However, this is impossible because it means eventually that we have to predict the whole universe. For the atmosphere, the external error is not a big issue since we can assume many things as time independent. An exception is the soil layer in which temperature and moisture can change considerably in the course of a weather forecast. Consequently, we should consider the soil as part of the dynamical system that describes the weather. The total error can also be split into **systematic** and **random** (or **nonsystematic**) **errors**. The random error vanishes after averaging while the systematic error has a certain sign and a nontrivial average. We can state that the model and external errors have more of the character of a systematic error while the initialization error causes more likely a random error. However, this separation is not entirely correct. A model that has no model, external and systematic errors is referred to as a **perfect model**. Such a model does not exist but one can assume it to investigate predictability due to a random initialization error.

In this section, we discuss the prediction of weather. To make such a prediction we must know the initial conditions (of atmosphere and soil) and boundary conditions (e.g. solar insulation and ocean temperature) in as much detail as possible. Since the forecast period is short in comparison to the time scale of the surrounding systems, we can neglect the time variation of the boundary fields (except for the diurnal cycle of solar insulation). In other words, we make a forecast

Weather Dynamics: An Introduction

of an individual trajectory on the attractor of the dynamical weather system with fixed model parameters. This case constitutes a **prediction of the first kind**. On the other hand, the model parameters of the dynamical weather system vary in the course of time because the external systems change. To describe this change, we must predict external systems like the ocean, cryosphere or vegetation but also the feedback of the dynamical weather system on these other systems. In this way, we can only try to predict the statistics of the weather attractor in the future since the timescale is much longer. Therefore, the initial conditions are unimportant while the boundary conditions are essential. This is a **prediction of the second kind** and is carried out in the context of **climate prediction**. Climate has time scales of more than 30 years which is sufficiently long to capture the statistics of the weather attractor. Climate prediction can be quite inaccurate because it is difficult to predict the ocean for which initial data are insufficient and additional uncertainty comes from the coupling with the atmosphere. On the other hand, a significant part of current climate change stems from **anthropogenic greenhouse gas emissions** and this can be described quite well with climate models. However, since the future of these emissions is also not well known we usually use the term **climate projection** for a given **emission scenario** instead of climate prediction.

We have to introduce a measure to assess a weather forecast in comparison to the observed development. With such a measure one can determine the **practical predictability** of the model that comprises non-systematic and systematic errors. Let $\vec{X}(t_m + t')$ be the phase space trajectory deduced from observations (**verification**) and $\vec{X}_m(t_m + t')$ the trajectory deduced from the prediction (**forecast**)[k] where t_m denotes the initial time of forecast m and t' the forecast time. Then, the difference vector $\vec{\varepsilon}_m(t') = \vec{X}_m(t_m + t') - \vec{X}(t_m + t')$ can be used to determine the **mean quadratic error**

$$E(t') = \frac{1}{M} \sum_{m=1}^{M} \frac{\vec{\varepsilon}_m(t') \cdot \vec{\varepsilon}_m(t')}{N} \tag{5.215}$$

---

[k]The phase space variables don't have to be spectral coefficients. It is also possible that the physical quantities at discrete grid points of a weather forecast model span the phase space.

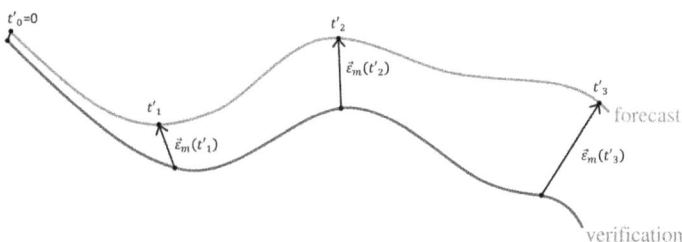

Fig. 5.42. Graphical determination of the forecast error $\vec{\varepsilon}_m(t')$ in phase space. The grey line shows the forecast trajectory and the dark grey line the verification trajectory.

where M denotes the number of forecasts and N is the dimension of the phase space.

Figure 5.42 shows that $|\vec{\varepsilon}_m|$ measures the Euclidean distance between two trajectory points. Usually, $E(t')$ increases with time but it is bounded by the **saturation error** because the attractor has a finite size in phase space. A weather forecast error in the order of the saturation error indicates a useless result and the **limit of predictability** is reached much earlier. A meaningful limit could be the error of the **expectation value** which is just the average state on the attractor. This error becomes

$$E_C(t') = \frac{1}{M} \sum_{m=1}^{M} \frac{\left[\bar{\vec{X}} - \vec{X}(t_m + t')\right] \cdot \left[\bar{\vec{X}} - \vec{X}(t_m + t')\right]}{N} \tag{5.216}$$

where $\bar{\vec{X}}$ denotes the time-averaged state. In the case of a very large M the sum operator approaches the time average. Then, E_C just becomes the **variance**. Practically, E_C yields the error of the climate mean state. In this case, only the long-term average of the considered month is calculated since the annual cycle is significant.

With a very large M we can also find a way to split the error in systematic and non-systematic errors by noting that

$$E(t') = \frac{\overline{\vec{\varepsilon}_m(t') \cdot \vec{\varepsilon}_m(t')}}{N} = \frac{\overline{\vec{\varepsilon}_m(t')} \cdot \overline{\vec{\varepsilon}_m(t')}}{N} + \frac{\overline{\vec{\varepsilon}'_m(t') \cdot \vec{\varepsilon}'_m(t')}}{N} \tag{5.217}$$

where the overbar denotes the average overall t_m and the prime represents the deviation from the average. We can assume that the non-systematic error has no preferred sign so that the time-average

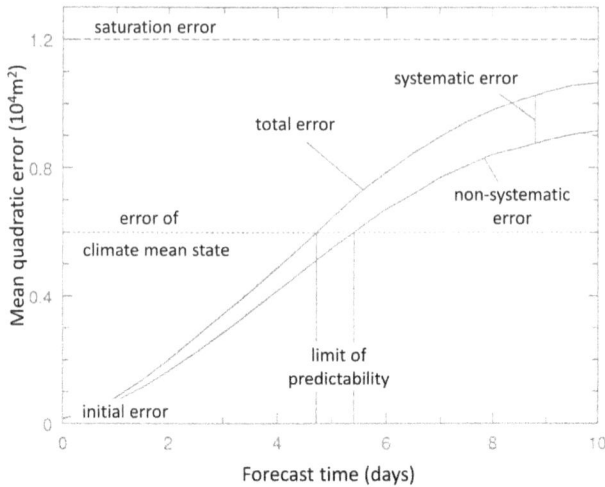

Fig. 5.43. Various forecast errors as a function of forecast time for the ECMWF weather forecast model (adapted from Fraedrich and Ziehmann, 1995, the underlying data is from Dalcher and Kalnay, 1987).

vanishes. Then, the first term on the right-hand side of (5.217) is the systematic and the second one is the non-systematic error. Figure 5.43 shows the various errors as a function of forecast time as deduced from ECMWF 10-day forecasts for 500 hPa geopotential height (for more details, see Dalcher and Kalnay, 1987)). The initial errors grow exponentially at the beginning but on day 4 an inflection point occurs. A little later at around day 5 the limit of predictability is reached because the errors become larger than that of the climate mean state. The error still grows afterwards and approaches the saturation error. We see that the systematic error has only a small contribution to the total error. Therefore, the limit of predictability is only reduced by somewhat more than half a day due to the systematic error. Nowadays, the limit of predictability has been substantially increased by model improvement (Zhang *et al.*, 2019).

Figure 5.44 shows the improvement of ECMWF weather forecasts in terms of the anomaly correlation of 500 hPa height since 1980. There is a continual improvement of short-term prediction but the **anomaly correlation** of 10-day forecasts seems to stagnate in the last 10 years and hardly exceeds 50%. Furthermore, we can see a convergence of northern and Southern Hemisphere forecasts so that

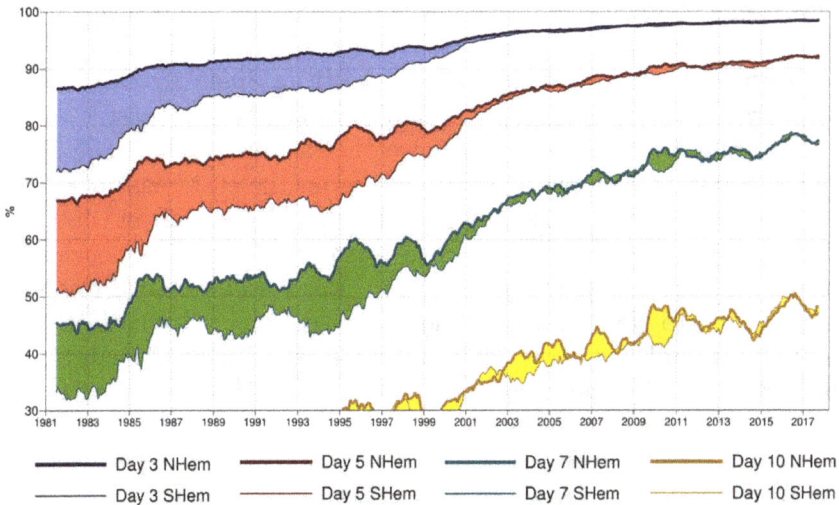

Fig. 5.44. Anomaly correlation of ECMWF 500 hPa height forecasts for various
forecast times and both hemispheres as a function of time. This plot is adapted
from the website: https://charts.ecmwf.int/products/plwww_m_hr_ccaf_adrian_
ts?single_product=_latest

nowadays both hemispheres can be predicted with similar anomaly
correlation.

In the following, we consider an example weather forecast to see
how the model error depends on physical parameterization. These
forecasts are performed with the Planet Simulator (cf. Section 5.2)
with truncation at total wavenumber 85 (T85) and 10 vertical sigma
levels. The Planet Simulator is not a tuned weather forecast model
and it is normally used with lower model resolutions. Therefore, the
forecasts have larger errors than operational weather forecast models
like the Global Forecast System (GFS).[1] The model was initialized
with the GFS analysis on January 25, 2008, 00UTC. Figure 5.45
shows some atmospheric fields at this date for the North Atlantic
region. A low-pressure system forms along a temperature front near
the east coast of North America and a mature low can be found
between Greenland and Iceland. The former has potential for further

---

[1]See https://www.ncei.noaa.gov/products/weather-climate-models/global-
forecast.

development because of the upper-level trough on the upstream side while the latter is fully occluded and seems to decay. A large-scale high-pressure system with its center north of Spain steers the low-pressure system along the jet stream towards the northeast.

Figure 5.46(a) shows the same weather analysis as in Fig. 5.45(a) but 48 h later. The low-pressure system on the east coast of North America migrated northeastwards close to Greenland and intensified significantly. We see developed warm and cold fronts enclosing a warm sector which brings relatively warm weather to Europe. A high moved to Newfoundland and carries cold air southeastward on its eastern flank. The error of a 48-h forecast with the Planet Simulator without physical parameterizations is displayed in Fig. 5.46(b).

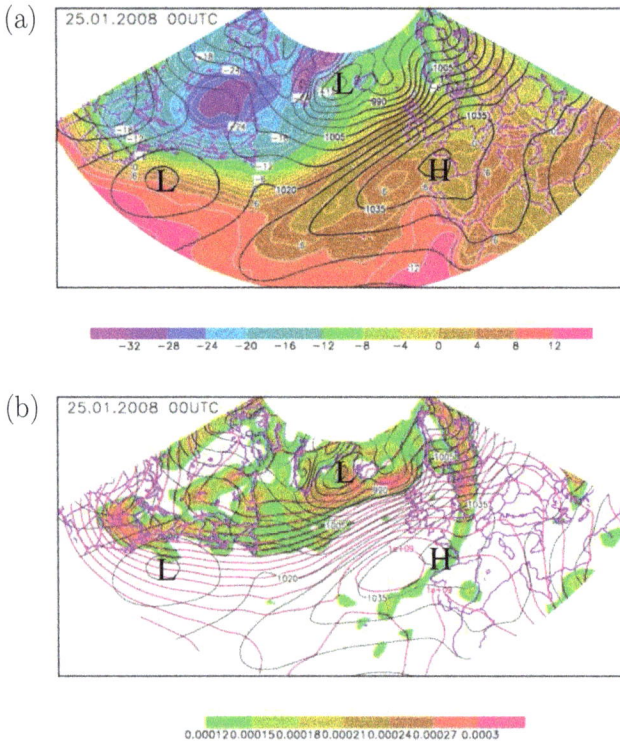

Fig. 5.45. Initial state for the forecast experiment with the Planet Simulator: (a) Temperature at $\sigma = 0.85$ (colored shadings, contour interval 2 K) and sea level pressure (isolines, contour interval 5 hPa), (b) sea level pressure (as in (a)), stream function at $\sigma = 0.55$ (red isolines) and vorticity (colored shadings). The plots were deduced from the GFS analysis on January 25, 2008, 00UTC.

Significant errors in sea level pressure appear near the center of the developed low because baroclinic waves intensify stronger without boundary layer drag as we have seen in Section 5.2. Furthermore, the center of the low in the Planet Simulator is shifted westward in comparison with the observed low. The excessively strong flows associated with the weather system produce large errors in the lower tropospheric temperature field. Adding the boundary layer parameterization reduces the intensity of the cyclone but the position is still wrong so that a remarkable error in sea level pressure and temperature remains (see Fig. 5.45(c)). Adding the cloud physics parameterization leads to a further reduction of the errors (see Fig. 5.45(d)). Now, the error in sea level pressure is so small that only one error isoline is seen in the figure. However, there is still an error of up to 8 K in temperature. Possibly this error can be further reduced by tuning the model parameters or using other physical parameterizations.

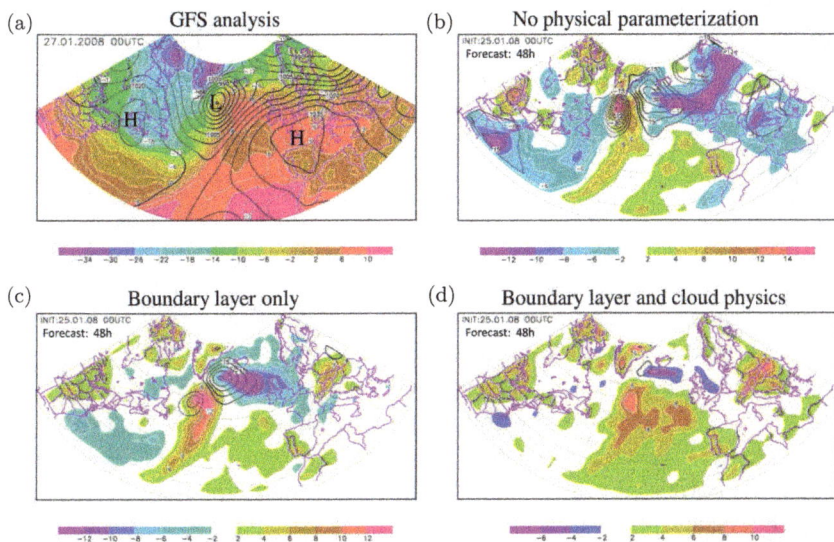

Fig. 5.46. (a) As in Fig. 5.45(a) but 48 hours later. Forecast error (difference between Planet Simulator and GFS analysis) in sea level pressure (contours, 5 hPa) and temperature at $\sigma = 0.85$ (colored shadings, contour interval 2 K) for (b) planet simulator without physical parameterizations, (c) planet simulator with boundary layer parameterization and (d) planet simulator with boundary layer and cloud physics parameterization.

Figure 5.47 shows the root-mean-square error (RMSE) for the various forecast experiments as a function of time. The red line displays the results of the **persistence forecast** in which it is simply assumed that the field does not change in time. The error of the other forecasts must be lower to be considered as a usable forecast. However, the errors of the 6-h forecasts are close to or even above that of persistence. Possibly, the used data assimilation can still be improved since this large error growth does not appear in the GFS model (black line). Later on, the forecasts are much better than persistence except for the forecast experiment without physical parameterizations in which the error curves of sea level pressure and temperature cross the red line. We see that adding physical parameterizations reduces the error remarkably for all three fields. The light blue line shows the error of a forecast in which the inertia-gravity waves are filtered out in the initial fields. This has only a small effect on the forecast error. Therefore, it seems that these waves do not interact significantly with the weather systems which mainly project on Rossby wave modes. The errors of stream function at $\sigma = 0.55$ are smaller and more close to that of the GFS model showing that the prediction of the upper layer flow is less sensitive with respect to the physical parameterization schemes. This circumstance might explain the early success of barotropic models which only predict the upper-level flow and ignore model physics as in the forecast model by Charney *et al.* (1950).

Besides practical predictability experiments we can also investigate the theoretical predictability in **identical twin** experiments. In this case, we assume that we have a **perfect model** without model and external errors. Then, two simulations with slightly different initial states can be compared. The differences between these simulations constitute the forecast error of the perfect model and with many of such simulations, we can determine the **theoretical predictability** of the considered system. A rough estimate can be calculated with the Lyapunov exponent. Perturbations grow by an instability but the growth rate varies in the course of time. The Lyapunov exponent can be regarded as a mean growth rate. Then, the error grows roughly like

$$E(t') = E_0 e^{2\varsigma_L t'} \tag{5.218}$$

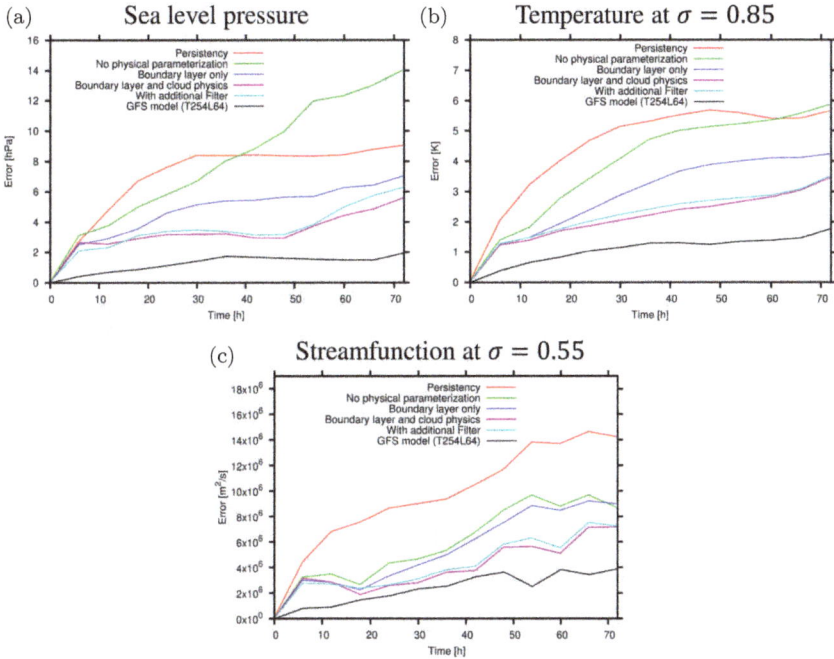

Fig. 5.47. RMSE as a function of time for various forecast experiments with the Planet Simulator (see legend in the plot) and the GFS model (black line). The red line shows the error of the persistence forecast. (a) RMSE of sea level pressure in hPa, (b) RMSE of temperature at $\sigma = 0.85$ in K and (c) RMSE of stream function at $\sigma = 0.55$ in m^2/s.

and the limit of predictability becomes

$$t'_L = \frac{\ln(E_C/E_0)}{2\varsigma_L} \tag{5.219}$$

Indeed, the limit can be extended arbitrarily by reducing the initial error E_0. However, the expansion of predictability increases very slowly with decreasing E_0 because of the logarithm. The method for the calculation of the Lyapunov exponent assumes linearized equations but the forecast error results from the full nonlinear equations. Therefore, one cannot avoid making many forecast experiments for an accurate determination of predictability. As an example, we consider the Lorenz model (5.131)–(5.133) using the parameters which produce a strange attractor (see Fig. 5.32(d)). Then, the Lyapunov exponent is positive and we can calculate t'_L with Eq. (5.219). Figure 5.48(a) shows the time evolution of the mean quadratic error deduced from 98,000 identical twin experiments for different initial errors. At the beginning, the error grows exponentially but later it attains a maximum followed by alternating periods with error decline and growth. This happens because the Lorenz attractor covers regions where the trajectory is stable and the error decreases. However, the error becomes larger in the course of time despite these oscillations and it reaches the "climate mean". Figure 5.48(b) shows the limit of forecast time as a function of initial quadratic error deduced from the numerical simulations and the estimation (5.219). Obviously, this estimation gives too optimistic results regarding the forecast time. However, the slope of the curve is similar to that of the numerical analysis. Therefore, the limit of predictability seems to follow the function in (5.219) but with an included correction factor $f_C < 1$ in front of E_C.

The identical twin experiments suggest that prediction has no limits if the perfect model is known and it is possible to reduce the initial error to arbitrarily small values. These requirements already appear illusory in the context of weather forecasting. Moreover, Lorenz (1969) pointed out another difficulty that stems from the existence of multiple scales in the atmosphere. Let us make a thought experiment by assuming that we have a perfect model and can collect error-free observations on a 10-cm grid that covers the complete atmosphere.

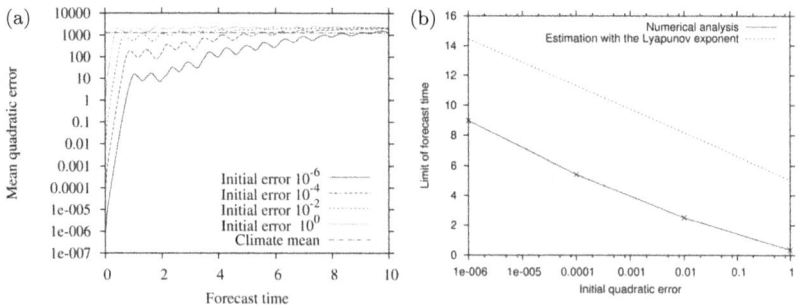

Fig. 5.48. (a) Time evolution of the mean quadratic error as a function of forecast time deduced from 98,000 identical twin experiments with the Lorenz model for different initial errors (see legend in plot). The black horizontal line shows the error of the long-term mean state (climate mean). (b) Limit of forecast time as a function of initial quadratic error deduced from (a) (solid line) and from the estimation (5.219) (dashed line).

Indeed, the initial error is very tiny in this case. Nevertheless, the flow is usually turbulent in the boundary layer and the error grows significantly (some orders of magnitudes) in a few seconds due to shear instabilities in the microturbulent flow. The larger error affects shallow convection by which the error amplifies anew within several minutes. Now, the error is large enough to have an impact on thunderstorms and this leads to another error growth within an hour. Consequently, after one day the error is possibly as large as it would be if we had initialized the model on a much coarser grid, say with a grid point distance of 100 m. This error propagation happens because we have instabilities at all scales in the atmosphere and a determination of the spectrum of Lyapunov exponents would confirm this by finding many positive ones. The thought experiment leads us to the conclusion that weather prediction cannot be extended much beyond that what is possible today.

Besides the improvement of weather models and observations, a new forecast technique has been established, namely the **ensemble prediction**. The idea is to make many forecasts with similar but not identical initial conditions. We can imagine that these initial states cover a small phase space volume which changes in the course of development and the various states in this volume describe several different weather states. It is practicable to assume that each chosen initial state describes the reality with the same probability because of the unavoidable initial error. From the final result, we can calculate the probability of a certain weather event. For example, if 60% of the ensemble forecasts predict that it will rain in London, then we can make the prediction that the rain probability is 60%. On the other hand, with only one forecast based on the observation, it can happen that no rain will be predicted in London. Mathematically, we introduce a **probability density function** $\rho_P(\vec{X}, t)$ (PDF) that describes the probability distribution in phase space. The ensemble of forecasts is used to estimate ρ_P which is illustrated in Fig. 5.49.

There is a striking analogy to quantum mechanics. Quantum mechanics describes the dynamics of elementary particles as, e.g. the electron. Due to Heisenberg's uncertainty principle, it is impossible to determine the state of the electron exactly, namely its position and its momentum.[m] The same thing is true in the case of

---

[m]We ignore the spin in this consideration.

weather forecasting since the initial state cannot be determined without a finite amplitude error due to technical reasons. Usually, the uncertainty rises when the elementary particle moves and interacts with other particles. The same holds for weather forecasts due to the chaotic nature of weather. The increased uncertainty can only be reduced by a new measurement, which applies to both quantum mechanics and weather forecasting. The sketch in Fig. 5.49 could also come from a textbook on quantum mechanics. In this case, it would show imaginable particle paths and the time change of the PDF. The PDF of an elementary particle can be inferred from a prognostic wave equation in quantum mechanics. It is the well-known Schrödinger equation. On the other hand, for a dynamical system the PDF results from the following continuity equation

$$\frac{\partial \rho_P}{\partial t} + \sum_{n=1}^{N} \frac{\partial}{\partial X_n}(\rho_P F_n) = 0 \qquad (5.220)$$

which can be derived from the methods used in continuum physics (cf. Section 1.4). This equation becomes the **Liouville equation** in the special case of a Hamiltonian dynamical system. Since weather forecast models form dynamical systems it is also imaginable to predict the PDF with (5.220). However, due to the huge number of dimensions such an approach appears not to be feasible.

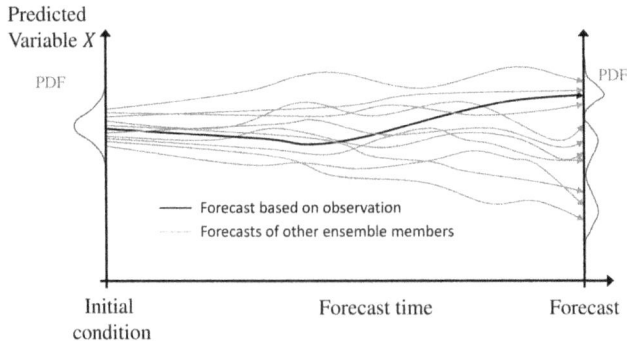

Fig. 5.49.   Schematic representation of ensemble forecasting. The various forecasts start on the left-hand side close to the observed state and yield the PDF shown by the grey curve. The various paths lead to the final forecast result and give rise to the PDF displayed on the right-hand side of this sketch. The black path results from the observation while the grey paths show those of the ensemble members with initially added perturbations.

To make the ensemble predictions more efficient, the different initial states are not chosen randomly. Instead, the method of singular values is used to generate initial perturbations with optimal error growth. Thus, the cases with extreme error growth are captured, so that the range of possible future weather states can be captured even with only a few ensemble members.

Chapter 6

Subsynoptic Weather Systems

6.1 Introduction

So far, our focus was on synoptic-scale weather systems but a lot of weather phenomena appear on **subsynoptic** scales. By subsynoptic we mean that the quasigeostrophic approximation is not valid because the length scale is too small and the Rossby number takes values of one or greater. In this chapter, we will deal with such phenomena, but we must limit ourselves to a selection for reasons of space. We will tackle convection, fronts and tropical cyclones because these subsynoptic phenomena are very frequent or significant. Before starting with convection, we make a scale analysis for such flows and derive another set of governing equations because the primitive equations are not appropriate in this case.

6.2 The Anelastic Approximation

Convection has horizontal length scales ranging from $100\,\mathrm{m}$ to $10\,\mathrm{km}$ and the vertical length scale has the same order of magnitude. Consequently, the hydrostatic approximation becomes invalid. The Euler equations can be used to describe, among other phenomena, the dynamics of sound waves. However, such waves have very high frequencies and have no noticeable influence on convective flows. Therefore, it is useful to introduce an approximation which filters out sound waves. For this purpose, the **anelastic approximation** is suitable, while the quasi-hydrostatic approximation has been proven to be an adequate sound wave filter for synoptic-scale flows. Furthermore, we can neglect spherical geometry and the Coriolis force because of the small horizontal length scale. In the following, we derive the anelastic approximation as described by Ogura and Phillips (1962). This approximation is valid for **shallow convection**, which has a vertical length scale of about 1 km. Then, we address an appropriate approximation for **deep convection** that extends vertically throughout the troposphere.

6.2.1   *Shallow convection*

We start with the Hesselberg-averaged Euler equations (1.210)–(1.213), which can be written with the aforementioned assumptions as

$$\frac{\partial \mathbf{v}_h}{\partial t} + \mathbf{v}_h \cdot \nabla_h \mathbf{v}_h + w \frac{\partial \mathbf{v}_h}{\partial z} = -\frac{1}{\rho} \nabla_h p + \mathbf{f}_{T,h} \tag{6.1}$$

$$\frac{\partial w}{\partial t} + \mathbf{v}_h \cdot \nabla_h w + w \frac{\partial w}{\partial z} = -\frac{1}{\rho} \frac{\partial p}{\partial z} - g + f_{T,z} \tag{6.2}$$

$$\frac{\partial \rho}{\partial t} + \nabla_h \cdot (\rho \mathbf{v}_h) + \frac{\partial}{\partial z}(\rho w) = 0 \tag{6.3}$$

$$\frac{\partial \theta}{\partial t} + \mathbf{v}_h \cdot \nabla_h \theta + w \frac{\partial \theta}{\partial z} = \frac{Q}{c_p \Pi} \tag{6.4}$$

$$\rho = \frac{p}{R_d \, \theta \Pi} \tag{6.5}$$

where $\mathbf{f}_{T,h}$ and $f_{T,z}$ are the horizontal and vertical forces due to turbulent momentum exchange. In these equations, the Reynolds- and Hesselberg-averaging operators are not indicated. This equation set does not include the budget equation for water vapor because it will not be subject to any modification. Substituting pressure by the Exner function, $p = p_0 \Pi^{1/\kappa}$, yields the pressure gradient force

$$-\frac{1}{\rho} \nabla p = -\frac{R_d \, \theta}{p_0} \Pi^{1-\frac{1}{\kappa}} \nabla \left(p_0 \Pi^{\frac{1}{\kappa}} \right) = -c_p \theta \nabla \Pi \tag{6.6}$$

and the equation of state becomes

$$\rho = \frac{p_0 \Pi^{\frac{1}{\kappa}-1}}{R_d \, \theta} \tag{6.7}$$

In the next step, the equations will be nondimensionalized with characteristic scales. The time scale is governed by the Brunt–Väisälä frequency

$$N = \sqrt{\frac{g}{\theta}\frac{\partial\theta}{\partial z}} \tag{6.8}$$

In the case of convective instability, N becomes imaginary but its magnitude still yields the relevant time scale. As a second condition, we assume that the deviation of potential temperature from a constant reference value θ_0 is small,

$$\frac{\theta - \theta_0}{\theta_0} = \frac{\theta^\star}{\theta_0} = o(\epsilon) \ll 1 \tag{6.9}$$

It is obvious that this assumption only holds for subsynoptic scales and a vertical temperature profile close to a dry adiabat. Such conditions only apply for shallow convection extending not much above the boundary layer.

In the next step, we nondimensionalize the equations by assuming the length scale H, time scale $T = (\epsilon g/H)^{-1/2}$, temperature scale θ_0, density scale $p_0/(R_d\theta_0)$ and velocity scale $U = H/T$. Then, the nondimensional equations become

$$\frac{\epsilon g H}{c_p\theta_0}\left(\frac{\partial\tilde{\mathbf{v}}_h}{\partial\tilde{t}} + \tilde{\mathbf{v}}_h\cdot\tilde{\nabla}_h\tilde{\mathbf{v}}_h + \tilde{w}\frac{\partial\tilde{\mathbf{v}}_h}{\partial\tilde{z}}\right) = -\tilde{\theta}\tilde{\nabla}_h\Pi + \frac{\epsilon g H}{c_p\theta_0}\tilde{\mathbf{f}}_{T,h} \tag{6.10}$$

$$\frac{\epsilon g H}{c_p\theta_0}\left(\frac{\partial\tilde{w}}{\partial\tilde{t}} + \tilde{\mathbf{v}}_h\cdot\tilde{\nabla}_h\tilde{w} + \tilde{w}\frac{\partial\tilde{w}}{\partial\tilde{z}}\right) = -\tilde{\theta}\frac{\partial\Pi}{\partial\tilde{z}} - \frac{g H}{c_p\theta_0} + \frac{\epsilon g H}{c_p\theta_0}\tilde{f}_{T,z} \tag{6.11}$$

$$\frac{\partial\tilde{\rho}}{\partial\tilde{t}} + \tilde{\nabla}_h\cdot(\tilde{\rho}\tilde{\mathbf{v}}_h) + \frac{\partial}{\partial\tilde{z}}(\tilde{\rho}\tilde{w}) = 0 \tag{6.12}$$

$$\frac{\partial\tilde{\theta}}{\partial\tilde{t}} + \tilde{\mathbf{v}}_h\cdot\tilde{\nabla}_h\tilde{\theta} + \tilde{w}\frac{\partial\tilde{\theta}}{\partial\tilde{z}} = \frac{\tilde{Q}}{\Pi} \tag{6.13}$$

$$\tilde{\rho} = \frac{\Pi^{\frac{1}{\kappa}-1}}{\tilde{\theta}} \tag{6.14}$$

where the tilde symbol stands for dimensionless variables. We assume that the turbulence and heating terms have the same order of magnitude as the time tendency of the respective quantity. The factor in front of the time derivative in Eqs. (6.10) and (6.11) is a small factor because

$$\frac{\epsilon g H}{c_p \theta_0} = \epsilon \frac{H}{H_a} \ll 1 \tag{6.15}$$

where $H_a = c_p \theta_0 / g$ denotes the height of the isentropic atmosphere (cf. Eq. (3.5)) and has a typical value of 30 km. Obviously, the height scale H must be smaller than H_a. We expand the variables as a power series in ϵ so that each variable is written as

$$\tilde{a} = a^{(0)} + a^{(1)} \epsilon + a^{(2)} \epsilon^2 + o\left(\epsilon^3\right) \tag{6.16}$$

Inserting this expansion in the governing equations yields the various orders in ϵ. For the zeroth order we obtain

$$\theta^{(0)} = 1 \tag{6.17}$$

$$\tilde{\nabla}_h \Pi^{(0)} = 0 \tag{6.18}$$

$$\frac{\partial \Pi^{(0)}}{\partial \tilde{z}} = -\frac{H}{H_a} \tag{6.19}$$

$$\rho^{(0)} = \left(\Pi^{(0)}\right)^{\frac{1}{\kappa}-1} \tag{6.20}$$

$$\tilde{\nabla}_h \cdot \left(\rho^{(0)} \mathbf{v}_h^{(0)}\right) + \frac{\partial}{\partial \tilde{z}}\left(\rho^{(0)} w^{(0)}\right) = 0 \tag{6.21}$$

$$Q^{(0)} = 0 \tag{6.22}$$

Equations (6.18) and (6.19) lead to the solution

$$\Pi^{(0)} = \Pi_s - \frac{H}{H_a}\tilde{z} \tag{6.23}$$

where Π_s is the constant Exner function at $\tilde{z} = 0$. With this result, the continuity equation reduces to

$$\tilde{\nabla}_h \cdot \mathbf{v}_h^{(0)} + \frac{1}{\rho^{(0)}} \frac{\partial}{\partial \tilde{z}}(\rho^{(0)} w^{(0)}) = 0 \tag{6.24}$$

Collecting first-order terms yields the following equations:

$$\frac{H}{H_a}\left(\frac{\partial \mathbf{v}_h^{(0)}}{\partial \tilde{t}} + \mathbf{v}_h^{(0)} \cdot \nabla_h \mathbf{v}_h^{(0)} + w^{(0)} \frac{\partial \mathbf{v}_h^{(0)}}{\partial \tilde{z}}\right)$$

$$= -\tilde{\nabla}_h \Pi^{(1)} + \frac{H}{H_a} \mathbf{f}_{T,h}^{(0)} \tag{6.25}$$

$$\frac{H}{H_a}\left(\frac{\partial w^{(0)}}{\partial \tilde{t}} + \mathbf{v}_h^{(0)} \cdot \tilde{\nabla}_h w^{(0)} + w^{(0)} \frac{\partial w^{(0)}}{\partial \tilde{z}}\right)$$

$$= -\frac{\partial \Pi^{(1)}}{\partial \tilde{z}} + \frac{H}{H_a} \theta^{(1)} + \frac{H}{H_a} f_{T,z}^{(0)} \tag{6.26}$$

$$\frac{\partial \theta^{(1)}}{\partial \tilde{t}} + \mathbf{v}_h^{(0)} \cdot \tilde{\nabla}_h \theta^{(1)} + w^{(0)} \frac{\partial \theta^{(1)}}{\partial \tilde{z}} = \frac{Q^{(1)}}{\Pi^{(0)}} \tag{6.27}$$

With the definitions

$$\mathbf{v}_h = \frac{H}{T}\mathbf{v}_h^{(0)}, \quad w = \frac{H}{T}w^{(0)}, \quad \theta^\star = \epsilon \theta_0 \theta^{(1)},$$

$$\Pi_0 = \Pi^{(0)}, \quad \Pi^\star = \epsilon \Pi^{(1)} \tag{6.28}$$

$$\rho_0 = \frac{p_0}{R_d \theta_0}\rho^{(0)}, \quad Q = \epsilon \frac{c_p \theta_0}{T}Q^{(1)},$$

$$\mathbf{f}_{T,h} = \frac{H}{T^2}\mathbf{f}_{T,h}^{(0)}, \quad f_{T,z} = \frac{H}{T^2}f_{T,z}^{(0)} \tag{6.29}$$

we obtain the following dimensioned equations:

$$\frac{\partial \mathbf{v}_h}{\partial t} + \mathbf{v}_h \cdot \nabla_h \mathbf{v}_h + w\frac{\partial \mathbf{v}_h}{\partial z} = -c_p \theta_0 \nabla_h \Pi^\star + \mathbf{f}_{T,h} \tag{6.30}$$

$$\frac{\partial w}{\partial t} + \mathbf{v}_h \cdot \nabla_h w + w\frac{\partial w}{\partial z} = -c_p \theta_0 \frac{\partial \Pi^\star}{\partial z} + g\frac{\theta^\star}{\theta_0} + f_{T,z} \tag{6.31}$$

$$\frac{\partial \theta^\star}{\partial t} + \mathbf{v}_h \cdot \nabla_h \theta^\star + w\frac{\partial \theta^\star}{\partial z} = \frac{Q}{c_p \Pi_0} \tag{6.32}$$

$$\nabla_h \cdot \mathbf{v}_h + \frac{1}{\rho_0} \frac{\partial}{\partial z}(\rho_0 w) = 0 \tag{6.33}$$

$$\Pi_0 = \Pi_s - \frac{z}{H_a} \tag{6.34}$$

$$\rho_0 = \frac{p_0}{R_d \, \theta_0} \Pi_0^{\frac{1}{\kappa} - 1} \tag{6.35}$$

These are the **anelastic equations** and they filter sound waves out. For shallow convection we have $H \ll H_a$ and this justifies further simplification by setting $\Pi_0 = const$. Then, the density cancels out in the continuity equation (6.33) so that it becomes identical to that of an incompressible fluid. The resulting equation set forms the widely used **Boussinesq equations** which have proved to be useful to describe the dynamics of the planetary boundary layer and shallow convection. Note that the Boussinesq equations differ from those of a fluid with constant density only in the **buoyancy** term $g\theta^\star/\theta_0$, which constitutes the driving force of convection. It appears when the air has less density than in the adjacent environment and vertical acceleration results because the vertical pressure gradient force becomes larger than downward directed gravity. Buoyancy can of course also become negative which leads to a downward acceleration of air.

Microturbulent exchange cannot be neglected in shallow convection. To consider this process we make a flux-gradient ansatz as in (1.247) and (1.248). Note that the Reynolds stress tensor (1.247) reduces to $-K_m(\nabla\mathbf{v} + \nabla\mathbf{v}^T)$ due to vanishing divergence. With this parameterization, we get for the Boussinesq equations

$$\frac{\partial \mathbf{v}}{\partial t} + \mathbf{v} \cdot \nabla \mathbf{v} = -c_p\theta_0\nabla\Pi^\star + g\frac{\theta^\star}{\theta_0}\mathbf{k} + \nabla \cdot \left[K_m\left(\nabla\mathbf{v} + \nabla\mathbf{v}^T\right)\right] \tag{6.36}$$

$$\frac{\partial \theta^\star}{\partial t} + \mathbf{v} \cdot \nabla\theta^\star = \frac{Q_L + Q_R}{c_p\Pi_0} + \nabla \cdot (K_\theta\nabla\theta^\star) \tag{6.37}$$

$$\nabla \cdot \mathbf{v} = 0 \tag{6.38}$$

where Q_L and Q_R denote the heating due to latent heat release and radiation, respectively. In this equation set, we have merged the horizontal and vertical momentum equations into one equation.

6.2.2 *Deep convection*

Deep convection has a vertical length scale that is not very small in comparison with H_a. Therefore, the approximation $\rho_0 \approx$ const. is not valid anymore. Furthermore, the potential temperature deviates significantly from a constant value θ_0 since the temperature profile in the troposphere is not a dry adiabat. Instead, a reversible moist adiabat as shown in Fig. 1.8 is a much better representation of the vertical temperature profile and deep convection can only appear by latent heat release in clouds. Therefore, deep convection is always moist convection and we can only assume that the potential temperature deviates slightly from a reference profile $\theta_r(z)$. This means that the Brunt–Väisälä frequency N does not give rise to a long timescale as demanded in the scale analysis by Ogura and Phillips (1962). To circumvent this issue, we argue that the time scale of deep convection is governed by $N_{es} = g/\theta_r \partial \theta_{es}/\partial z$. The saturated equivalent potential temperature θ_{es} (cf. Eq. (1.146)) has in comparison to θ_r a smaller vertical change (i.e. $|N_{es}| < N$) because the latent heat release (removal) in clouds increases (decreases) the buoyancy. Outside of the cloud, high-frequency internal gravity waves may occur but these have typically not such a large vertical scale and they are not so important for the evolution of a deep convective cloud. Therefore, we can again start with the nondimensional equations (6.10)–(6.14) but find for the zeroth order

$$\theta^{(0)} = \tilde{\theta}_r(\tilde{z}) \tag{6.39}$$

$$\tilde{\nabla}_h \Pi^{(0)} = 0 \tag{6.40}$$

$$\frac{\partial \Pi^{(0)}}{\partial \tilde{z}} = -\frac{H}{H_a \tilde{\theta}_r} \tag{6.41}$$

$$\rho^{(0)} = \frac{(\Pi^{(0)})^{\frac{1}{\kappa}-1}}{\tilde{\theta}_r} \tag{6.42}$$

$$\tilde{\nabla}_h \cdot (\rho^{(0)} \mathbf{v}_h^{(0)}) + \frac{\partial}{\partial \tilde{z}}(\rho^{(0)} w^{(0)}) = 0 \tag{6.43}$$

$$\frac{Q^{(0)}}{\Pi^{(0)}} = w^{(0)} \frac{\partial \tilde{\theta}_r}{\partial \tilde{z}} \tag{6.44}$$

and for the first-order terms

$$\frac{H}{H_a}\left(\frac{\partial \mathbf{v}_h^{(0)}}{\partial \tilde{t}} + \mathbf{v}_h^{(0)} \cdot \tilde{\nabla}_h \mathbf{v}_h^{(0)} + w^{(0)} \frac{\partial \mathbf{v}_h^{(0)}}{\partial \tilde{z}}\right)$$
$$= -\tilde{\theta}_r \tilde{\nabla}_h \Pi^{(1)} + \frac{H}{H_a} \mathbf{f}_{T,h}^{(0)} \tag{6.45}$$

$$\frac{H}{H_a}\left(\frac{\partial w^{(0)}}{\partial \tilde{t}} + \mathbf{v}_h^{(0)} \cdot \tilde{\nabla}_h w^{(0)} + w^{(0)} \frac{\partial w^{(0)}}{\partial \tilde{z}}\right)$$
$$= -\tilde{\theta}_r \frac{\partial \Pi^{(1)}}{\partial \tilde{z}} + \frac{H}{H_a}\frac{\theta^{(1)}}{\tilde{\theta}_r} + \frac{H}{H_a} f_{T,z}^{(0)} \tag{6.46}$$

$$\frac{\partial \theta^{(1)}}{\partial \tilde{t}} + \mathbf{v}_h^{(0)} \cdot \tilde{\nabla}_h \theta^{(1)} + w^{(0)} \frac{\partial \theta^{(1)}}{\partial \tilde{z}} = \frac{Q^{(1)}}{\Pi^{(0)}} \tag{6.47}$$

We have in contrast to the shallow convection case a vertically dependent factor in front of the gradient of $\Pi^{(1)}$ in the momentum equations and the zeroth order heating $Q^{(0)}$ is nonzero as seen in Eq. (6.44). The solution of this equation becomes a reversible moist adiabat if $Q^{(0)}$ results solely from latent heat release. The analysis is strictly speaking not valid outside of clouds. However, here we can assume that vertical velocities are small so that no large deviations from $\tilde{\theta}_r$ occur.

With the definition

$$Q = \frac{c_p \theta_0}{T}\left(Q^{(0)} + \epsilon Q^{(1)}\right) \tag{6.48}$$

and the remaining ones in (6.28) and (6.29) we find the anelastic equations valid for deep convection.

$$\frac{\partial \mathbf{v}_h}{\partial t} + \mathbf{v}_h \cdot \nabla_h \mathbf{v}_h + w\frac{\partial \mathbf{v}_h}{\partial z} = -c_p \theta_r \nabla_h \Pi^\star + \mathbf{f}_{T,h} \tag{6.49}$$

$$\frac{\partial w}{\partial t} + \mathbf{v}_h \cdot \nabla_h w + w\frac{\partial w}{\partial z} = -c_p \theta_r \frac{\partial \Pi^\star}{\partial z} + g\frac{\theta^\star}{\theta_r} + f_{T,z} \tag{6.50}$$

$$\frac{\partial \theta^\star}{\partial t} + \mathbf{v}_h \cdot \nabla_h \theta^\star + w\frac{\partial \theta^\star}{\partial z} + w\frac{\partial \theta_r}{\partial z} = \frac{Q}{c_p \Pi_0} \tag{6.51}$$

$$\nabla_h \cdot (\rho_0 \mathbf{v}_h) + \frac{\partial}{\partial z}(\rho_0 w) = 0 \tag{6.52}$$

$$\Pi_0 = \Pi_s - \int_{z_s}^{z} \frac{g}{c_p \theta_r} dz' \tag{6.53}$$

$$\rho_0 = \frac{p_0}{R_d \theta_r} \Pi_0^{\frac{1}{\kappa} - 1} \tag{6.54}$$

To write this equation set in a more convenient form, we make use of the approximation

$$-c_p \theta_r \nabla \Pi^\star \approx -\frac{1}{\rho_0} \nabla p^\star \tag{6.55}$$

Then, we can use the continuity equation (6.52) to obtain the following flux form of the anelastic equations

$$\frac{\partial}{\partial t}(\rho_0 \mathbf{v}) + \nabla \cdot (\rho_0 \mathbf{v}\mathbf{v})$$

$$= -\nabla p^\star + g\frac{\rho_0 \theta^\star}{\theta_r}\mathbf{k} + \nabla \cdot \left[\rho_0 K_m \left(\nabla \mathbf{v} + \nabla \mathbf{v}^T\right)\right] \tag{6.56}$$

$$\frac{\partial}{\partial t}(\rho_0 \theta^\star) + \nabla \cdot (\rho_0 \mathbf{v}\theta^\star) + \rho_0 w \frac{\partial \theta_r}{\partial z}$$

$$= \rho_0 \frac{Q_L + Q_R}{c_p \Pi_0} + \nabla \cdot (\rho_0 K_\theta \nabla \theta^\star) \tag{6.57}$$

$$\nabla \cdot (\rho_0 \mathbf{v}) = 0 \tag{6.58}$$

Here, we have again used the flux-gradient closures (1.247) and (1.248) and have neglected the three-dimensional divergence in (1.247).

6.3 Shallow Convection

Shallow convection appears when the planetary boundary layer is heated from below and the potential temperature decreases with height up to a stable layer where convection reaches its upper boundary. Such a phenomenon often happens on sunny days. Shallow convection starts immediately in the presence of convective instability because microturbulence provides the necessary perturbation. Condensation and the associated latent heat release may occur in the updraft with the consequence that a cumulus cloud develops. However, the latent heat is not the essential driver so that we can neglect it in this section. First, we derive the instability in a linearized set of Boussinesq equations and afterward, we consider the nonlinear evolution.

6.3.1   *The convective instability problem*

We assume two-dimensional roll convection as sketched in Fig. 3.27 to solve the instability problem and direct the y-axis parallel to the rolls. Such a situation occurs indeed in the form of **cloud streets** (see, e.g. Kuettner, 1971) but it requires the existence of a horizontal basic flow which we do not consider here. On the other hand, three boundary layer rolls with different orientations can be superimposed in such a way that convection forms **hexagonal cells**, which are also sometimes visible on satellite pictures (see, e.g. Agee *et al.*, 1973). For simplicity, we assume constant values for the turbulent exchange coefficients. With these assumptions the equations (6.36)–(6.38) can be written in the form

$$\frac{\partial u}{\partial t} + u\frac{\partial u}{\partial x} + w\frac{\partial u}{\partial z} = -c_p\theta_0\frac{\partial \Pi^\star}{\partial x} + K_m\nabla^2 u \qquad (6.59)$$

$$\frac{\partial w}{\partial t} + u\frac{\partial w}{\partial x} + w\frac{\partial w}{\partial z} = -c_p\theta_0\frac{\partial \Pi^\star}{\partial z} + g\frac{\theta^\star}{\theta_0} + K_m\nabla^2 w \qquad (6.60)$$

$$\frac{\partial \theta^\star}{\partial t} + u\frac{\partial \theta^\star}{\partial x} + w\frac{\partial \theta^\star}{\partial z} = \frac{Q_L + Q_R}{\Pi_0} + K_\theta\nabla^2\theta^\star \qquad (6.61)$$

$$\frac{\partial u}{\partial x} + \frac{\partial w}{\partial z} = 0 \qquad (6.62)$$

where the relation $K_m\nabla\cdot(\nabla\mathbf{v} + \nabla\mathbf{v}^T) = K_m\nabla^2\mathbf{v}$ has been used due to vanishing divergence $\nabla\cdot\mathbf{v}$. The continuity equation (6.62) allows us to define a stream function ψ such that

$$u = -\frac{\partial\psi}{\partial z}, \quad w = \frac{\partial\psi}{\partial x} \qquad (6.63)$$

With this definition, we can derive from (6.59) and (6.60) a prognostic equation for the y-component of the vorticity vector[a]

$$\frac{\partial}{\partial t}\nabla^2\psi + J_{xz}(\psi, \nabla^2\psi) = \frac{g}{\theta_0}\frac{\partial\theta^\star}{\partial x} + K_m\nabla^2(\nabla^2\psi) \qquad (6.64)$$

Convective instability requires a vertical decrease of potential temperature with height as we will see. Therefore, it is useful to prescribe the following basic state:

$$\overline{\psi} = 0, \quad \overline{\theta}^\star = \theta_s^\star - \Delta\theta\frac{z}{H} \qquad (6.65)$$

It is easy to show that this basic state fulfills the governing equations (6.64) and (6.61) when there are no heat sources, i.e. $Q_L + Q_R$. Adding horizontal rigid boundaries at $z = 0$ and $z = H$ yields the **Rayleigh–Bénard convection** problem that has been solved by Rayleigh (1916). The described phenomenon of convection has been observed much earlier by Bénard (1900). Before doing the analysis, it is useful to nondimensionalize the governing equations in the following way with characteristic scales

$$\tilde{x} = \frac{x}{H}, \quad \tilde{z} = \frac{z}{H}, \quad \tilde{t} = \frac{K_\theta t}{H^2}, \quad \tilde{\psi} = \frac{\psi}{K_\theta}, \quad \tilde{\theta}^\star = \frac{\theta^\star}{\Delta\theta} \qquad (6.66)$$

The time has been nondimensionalized with the time scale of turbulent heat conduction. Therefore, it is assumed that convection does not have a much smaller time scale which is reasonable when the lower boundary is not intensively heated. Inserting these nondimensionalized variables in (6.64) and (6.61) yields

$$\frac{\partial}{\partial\tilde{t}}\tilde{\nabla}^2\tilde{\psi} + J_{\tilde{x}\tilde{z}}\left(\tilde{\psi}, \tilde{\nabla}^2\tilde{\psi}\right) = \mathrm{PrRa}\frac{\partial\tilde{\theta}'}{\partial x} + \mathrm{Pr}\tilde{\nabla}^2\left(\tilde{\nabla}^2\tilde{\psi}\right) \qquad (6.67)$$

$$\frac{\partial\tilde{\theta}'}{\partial\tilde{t}} + J_{\tilde{x}\tilde{z}}(\tilde{\psi}, \tilde{\theta}') - \frac{\partial\tilde{\psi}}{\partial\tilde{x}} = \nabla^2\tilde{\theta}' \qquad (6.68)$$

where $\tilde{\theta}'$ is the nondimensional potential temperature deviation from the basic state, $\mathrm{Pr} = K_m/K_\theta$ the **Prandtl number** (cf. Section 1.13)

---

[a]Strictly speaking, $-\nabla^2\psi$ is the y component of the vorticity vector but the incorrect sign is not of importance for the subsequent analysis.

and Ra is the **Rayleigh number** given by

$$\mathrm{Ra} = \frac{gH^3\Delta\theta}{K_\theta K_m \theta_0} \qquad (6.69)$$

Linearizing the equations (6.67) and (6.68) with respect to the basic state (6.65) leads to

$$\frac{\partial}{\partial \tilde{t}}\tilde{\nabla}^2\tilde{\psi} = \mathrm{PrRa}\frac{\partial \tilde{\theta}'}{\partial x} + \mathrm{Pr}\tilde{\nabla}^2\left(\tilde{\nabla}^2\tilde{\psi}\right) \qquad (6.70)$$

$$\frac{\partial \tilde{\theta}'}{\partial \tilde{t}} - \frac{\partial \tilde{\psi}}{\partial \tilde{x}} = \nabla^2\tilde{\theta}' \qquad (6.71)$$

Before solving these equations, we must settle the boundary conditions. Rayleigh (1916) considered the simplest case with free-slip boundaries at constant temperature. We will also use these conditions although they might be somewhat unrealistic. However, we do this for the sake of simplicity and this assumption is not more problematic than that of a constant exchange coefficient K_m. For a discussion of other boundary conditions, we refer to Drazin and Reid (1981) and Sparrow *et al.* (1964). With rigid free-slip boundaries, the vertical velocity and the vertical turbulent momentum flux must vanish there. Consequently, we set

$$\tilde{\psi} = \frac{\partial^2 \tilde{\psi}}{\partial \tilde{z}^2} = 0 \quad \text{at } \tilde{z} = 0,1 \qquad (6.72)$$

On the other hand, the condition of constant temperature at the boundaries leads to

$$\tilde{\theta}' = 0 \quad \text{at } \tilde{z} = 0,1 \qquad (6.73)$$

These conditions justify the following perturbation ansatz:

$$\tilde{\psi} = \Re\left[\hat{\psi}e^{\varsigma\tilde{t}}e^{ik\tilde{x}}\sin(n\pi\tilde{z})\right], \quad \tilde{\theta}' = \Re\left[\hat{\theta}e^{\varsigma\tilde{t}}e^{ik\tilde{x}}\sin(n\pi\tilde{z})\right] \qquad (6.74)$$

where the vertical wavenumber n is an integer. By inserting this ansatz, we obtain the following matrix equation:

$$\begin{pmatrix} \varsigma + \mathrm{Pr}(\delta_h k^2 + \delta_v \pi^2 n^2) & \mathrm{PrRa}\dfrac{ik}{\delta_{nh}k^2 + \pi^2 n^2} \\ -ik & \varsigma + \delta_h k^2 + \delta_v \pi^2 n^2 \end{pmatrix} \cdot \begin{pmatrix} \hat{\psi} \\ \hat{\theta} \end{pmatrix} = 0 \qquad (6.75)$$

where δ_{nh}, δ_h and δ_v are switches for nonhydrostatic terms, horizontal diffusion and vertical turbulent diffusion, respectively. For nontrivial solutions the growth rate ς becomes

$$\varsigma_{1,2} = -\frac{1+\text{Pr}}{2}(\delta_h k^2 + \delta_v \pi^2 n^2)$$

$$\pm\sqrt{\left(\frac{1-\text{Pr}}{2}\right)^2 (\delta_h k^2 + \delta_v \pi^2 n^2)^2 + \frac{\text{PrRa}k^2}{\delta_{nh}k^2 + \pi^2 n^2}} \quad (6.76)$$

Obviously, the two eigenvalues are real in the case of unstable stratification (Ra > 0) and at least one eigenvalue is negative. We see that the fastest growth is associated with the largest vertical mode $n = 1$. Therefore, convection is likely producing cells, which extend vertically over the whole depth of the unstable layer. Figure 6.1 shows the growth rate of the growing mode as a function of nondimensional horizontal wavelength for Ra = 1200, Pr = 0.74 and $n = 1$.

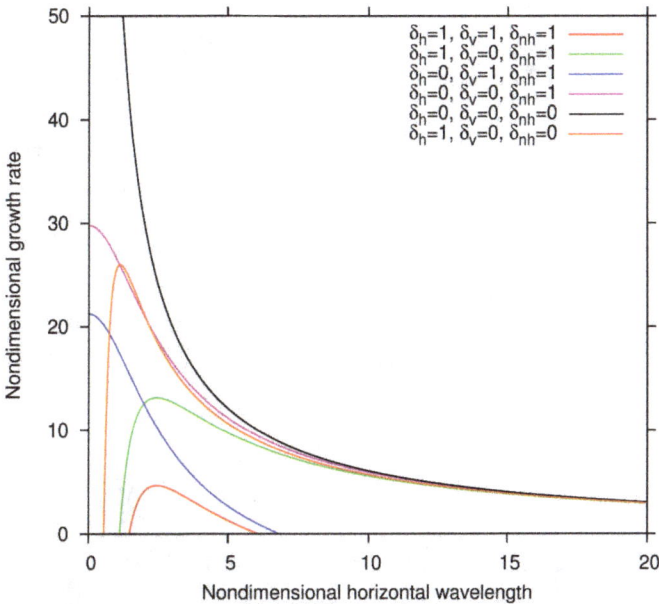

Fig. 6.1. Nondimensional growth rate of the Rayleigh–Bénard solution for Ra = 1200, Pr = 0.74 and $n = 1$. The curves refer to various combinations of switches δ_h, δ_v and δ_{nh} (see legend in the plot).

Results are displayed for various combinations of the switches δ_h, δ_v and δ_{nh}. With all switches set to one, instability occurs only in a limited wavelength range where the growth rate maximizes at a certain wavelength. Consequently, there is a preferred horizontal scale at which convection develops. We obtain the largest growth rate with all switches set to zero. In fact, the growth rate approaches infinity with decreasing wavelength and it is the quasihydrostatic approximation which is responsible for this unrealistic result. However, leaving the nonhydrostatic terms as they are and neglecting viscosity and heat conduction ($\delta_h = \delta_v = 0$ and $\delta_{nh} = 1$) yields finite growth rates but the maximum growth rate appears at zero wavelength. This leads to a problem in numerical modeling because it means that convection is excited on the grid-point scale which causes instability in the numerical solution scheme. Therefore, primitive equation models usually adopt convection parameterization schemes to overcome this problem. Another possibility is to include horizontal diffusion which shifts the wavelength of maximum growth to a finite value as seen by the green and orange curves. However, a huge value for the exchange coefficient must be implemented to control convection in an atmospheric model with coarse resolution and the quasihydrostatic approximation makes this problem even more dramatic (orange curve). An instability cutoff at a long wavelength does not occur when only horizontal diffusion is implemented. On the other hand, a short-wave cutoff is absent when only vertical diffusion is present (blue curve). These results indicate the importance of irreversible processes for the dynamics of convection.

Instability can only occur when the square root in (6.76) is larger than the absolute value of the first term. By setting $\varsigma = 0$ in (6.76) we obtain the instability criterion (assuming $\delta_h = \delta_v = \delta_{nh} = 1$):

$$\text{Ra} > \text{Ra}_c \equiv \frac{(k^2 + \pi^2 n)^3}{k^2} \tag{6.77}$$

where Ra_c is the **critical Rayleigh number**. It attains its minimum for $n = 1$ and $k = \pi/\sqrt{2}$. Then, Ra_c takes the value 657.5 and the aspect ratio of the overturning cell (ratio of vertical to horizontal extension) turns out to be $k/\pi = 1/\sqrt{2}$. Therefore, the expansion of the cell is by a factor $\sqrt{2}$ higher in the horizontal direction than in the vertical direction.

The eigensolution takes the form

$$\tilde{\psi} = Ae^{\varsigma\tilde{t}}\sin(k\tilde{x}+\alpha)\sin(n\pi\tilde{z}) \qquad (6.78)$$

$$\tilde{\theta}' = \frac{k}{\varsigma + \delta_h k^2 + \delta_v \pi^2 n^2} Ae^{\varsigma\tilde{t}}\cos(k\tilde{x}+\alpha)\sin(n\pi\tilde{z}) \qquad (6.79)$$

where α is an arbitrary phase shift. Therefore, the growing mode has the form of overturning cells in which the positive and negative potential temperature anomalies correlate with the updrafts and downdrafts, respectively (see Fig. 6.2). The instability mechanism is rather simple: An initial updraft advects warm air from below upwards. Then, the lifted air has a higher potential temperature than in the adjacent environment. Consequently, the buoyancy is positive and the updraft accelerates. This yields more vertical advection of warm air and more buoyancy so that a positive feedback loop results. The same reasoning holds for downdrafts which advect cold air downward and produce negative buoyancy.

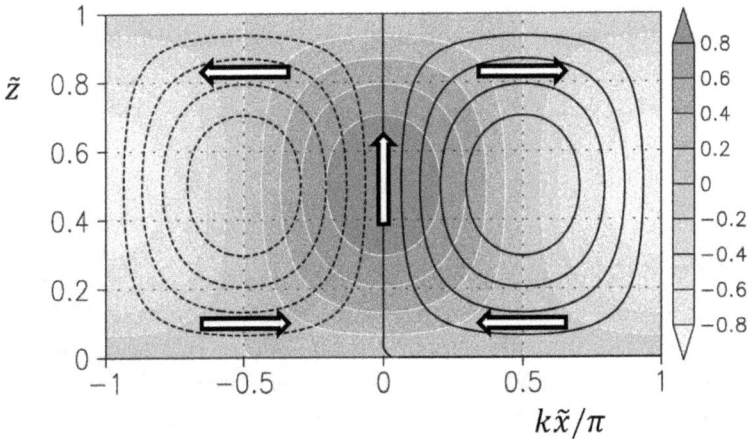

Fig. 6.2. Potential temperature anomaly (grey shadings) and stream function (isolines) of the growing mode calculated from the linear perturbation analysis of the Rayleigh-Bénard convection problem. The arrows show the direction of the flow.

Three solutions of the form (6.77)–(6.78) can be superimposed to form a hexagonal cell. For example, the potential temperature field

$$\tilde{\theta}' = \hat{\theta} e^{\varsigma \tilde{t}} \left\{ \cos(k\tilde{y}) + \cos\left[\frac{k}{2}(\sqrt{3}\tilde{x} - \tilde{y})\right] \right.$$
$$\left. + \cos\left[\frac{k}{2}(\sqrt{3}\tilde{x} + \tilde{y})\right] \right\} \sin(\pi\tilde{z}) \qquad (6.80)$$

yields a hexagonal convection cell. The vertical velocity field has the same structure due to Eqs. (6.63) and (6.78).

$$\tilde{w} = \hat{w} e^{\varsigma \tilde{t}} \left\{ \cos(k\tilde{y}) + \cos\left[\frac{k}{2}(\sqrt{3}\tilde{x} - \tilde{y})\right] \right.$$
$$\left. + \cos\left[\frac{k}{2}\left(\sqrt{3}\tilde{x} + \tilde{y}\right)\right] \right\} \sin(\pi\tilde{z}) \qquad (6.81)$$

We can describe the horizontal flow with a velocity potential $\tilde{\chi}$ because the vertical component of vorticity vanishes. Then, we get from the continuity equation

$$\tilde{\nabla}_h^2 \tilde{\chi} = -\frac{\partial \tilde{w}}{\partial \tilde{z}} \qquad (6.82)$$

Therefore, the velocity potential of the hexagonal cell becomes

$$\tilde{\chi} = \pi \frac{\hat{w}}{k^2} e^{\varsigma \tilde{t}} \left\{ \cos(k\tilde{y}) + \cos\left[\frac{k}{2}\left(\sqrt{3}\tilde{x} - \tilde{y}\right)\right] \right.$$
$$\left. + \cos\left[\frac{k}{2}\left(\sqrt{3}\tilde{x} + \tilde{y}\right)\right] \right\} \cos(\pi\tilde{z}) \qquad (6.83)$$

and the wind components can be deduced from the relation $\tilde{\mathbf{v}}_h = \tilde{\nabla}_h \tilde{\chi}$. Figure 6.3 shows a top view of the hexagonal convection cell. In this case, the updraft appears at the center of the cell while the vertical motion is downward with weaker amplitude at the edges of the cell. Air flows towards the center of the cell near the lower boundary but near the upper boundary, the flow is exactly in the opposite direction. Such a flow is called **closed cellular convection**. On the other hand, **open cellular convection** is characterized by downdrafts at the cell centers. Both types of convection can often be seen on satellite pictures as in Fig. 6.4 but the flow patterns are not

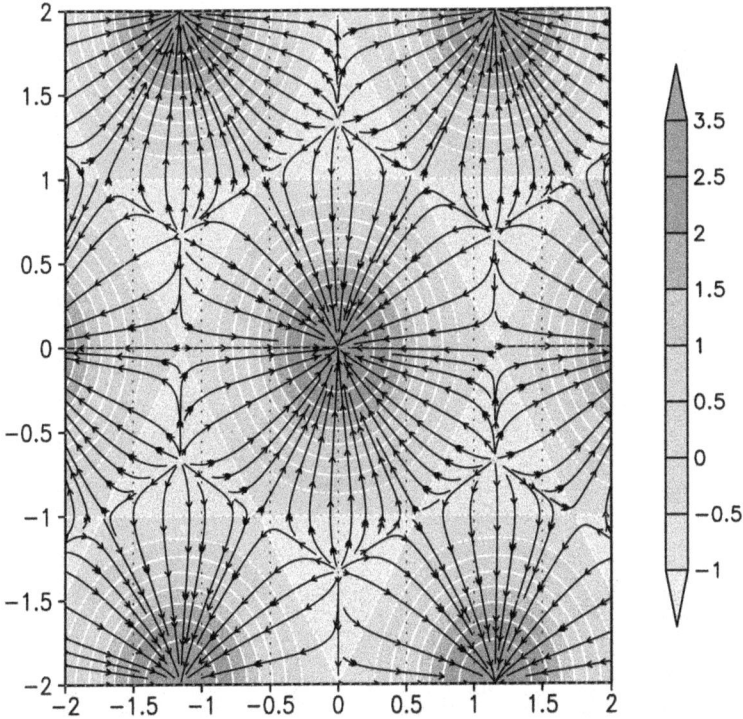

Fig. 6.3. Top view of a hexagonal convection cell: The grey shadings display nondimensional vertical velocity at $\tilde{z} = 0.5$ while the white lines show the streamlines at the bottom of the cell.

(a) Closed cellular convection   (b) Open cellular convection

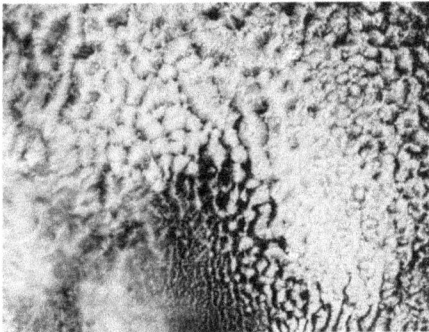

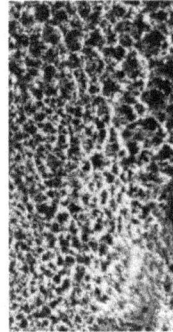

Fig. 6.4. Satellite images providing detail of clouds in (a) closed cellular convection. This MODIS Terra image was taken southeast of South Africa. (*Source*: https://commons.wikimedia.org/wiki/File:Closed Cellular Convection Cloud.png) and in (b) open cellular convection (*Source*: https://commons.wiki media.org/wiki/File:Open_Cellular_Convection.JPG).

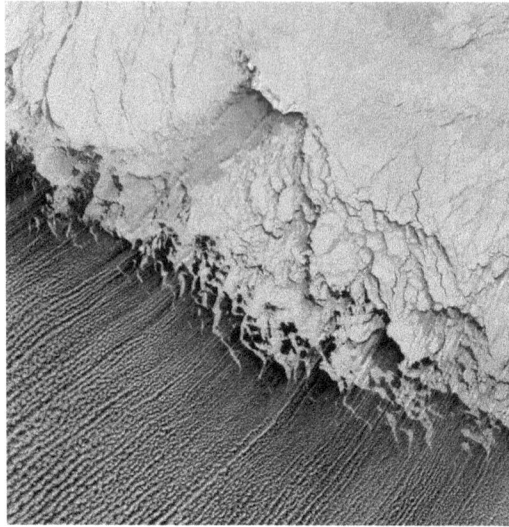

Fig. 6.5. Horizontal convective rolls producing cloud streets over the Bering Sea *Source*: https://commons.wikimedia.org/wiki/File:CloudStreets_BeringSea_20060120.jpg).

nearly as ideal as in theory and the aspect ratio of the observed cells is much smaller (see Agee *et al.*, 1973).

Shallow convection is sometimes organized in cloud streets. In this case a horizontal wind is present which is directed approximately parallel to the roll axis. Cloud streets often appear during **cold air outbreaks** in polar latitudes. Typically, air flows from regions covered with sea ice to open oceans. Convection evolves rapidly because the sea surface temperature is much higher than of the air coming from the sea ice region. Figure 6.5 shows a satellite picture of cloud streets formed at the southern boundary of the sea ice region in the Bering Sea. It is seen that the rolls broaden with increasing distance from the sea ice boundary. Likely, the increasing depth of the boundary layer provides the explanation. Real cloud streets are — in contrast to our simple model — created by a combination of shear and convective instability (e.g. LeMone, 1973).

6.3.2 *Nonlinear development of convective instability*

The linear instability analysis does not reveal the behavior of the convective flow when it reaches finite amplitude. It can possibly oscillate or settle into a steady state. Furthermore, the maximum amplitude of the convection cell can also not be determined with the linearized model. That is, growth comes to a halt when the vertical temperature fluxes reduce the vertical potential temperature gradient so that convective instability vanishes.

To approach this issue, we make use of the spectral method presented in Section 5.4.2 and apply it to the governing equations (6.67) and (6.68) of two-dimensional convection. Then, we represent the fields as follows:

$$\tilde{\psi} = \sum_{n=-N}^{N} \sum_{m=-M}^{M} \Psi_n^m(\tilde{t}) e^{ik_m \tilde{x}} e^{il_n \tilde{z}} \tag{6.84}$$

$$\theta' = \sum_{n=-N}^{N} \sum_{m=-M}^{M} \Theta_n^m(\tilde{t}) e^{ik_m \tilde{x}} e^{il_n \tilde{z}} \tag{6.85}$$

where $k_m = 2m\pi/\tilde{L}_x$ and $l_n = n\pi$. To fulfill all boundary conditions the following relations must hold (cf. Section 5.4.2)

$$\Psi_{-n}^m = -\Psi_n^m, \quad \Psi_{-n}^{-m} = -\Psi_n^{-m}, \quad \Psi_n^{-m} = -\Psi_n^{mC} \tag{6.86}$$

$$\Theta_{-n}^m = -\Theta_n^m, \quad \Theta_{-n}^{-m} = -\Theta_n^{-m}, \quad \Theta_n^{-m} = -\Theta_n^{mC} \tag{6.87}$$

Therefore, it is sufficient to predict only the coefficients with positive m and n. Note that the coefficients with $n = 0$ vanish. Inserting the Fourier expansions (6.84) and (6.85) in (6.67) and (6.68) yields after applying the scalar product with the various Fourier functions the following dynamical system equations:

$$\frac{d\Psi_n^m}{d\tilde{t}} = -i\mathrm{PrRa}\frac{k_m}{K_n^{m2}}\Theta_n^m - \mathrm{Pr}K_n^{m2}\Psi_n^m$$

$$- \sum_{n_1=n-N}^{N} \sum_{m_1=m-M}^{M} \alpha_{n,n_1}^{m,m_1} \frac{K_{n_1}^{m_1 2}}{K_n^{m2}} \Psi_{n-n_1}^{m-m_1} \Psi_{n_1}^{m_1} \tag{6.88}$$

$$\frac{d\Theta_n^m}{d\tilde{t}} = ik_m\Psi_n^m - K_n^{m2}\Theta_n^m$$

$$- \sum_{n_1=n-N}^{N} \sum_{m_1=m-M}^{M} \alpha_{n,n_1}^{m,m_1} \Psi_{n-n_1}^{m-m_1} \Theta_{n_1}^{m_1} \tag{6.89}$$

where $K_n^{m^2} = k_m^2 + l_n^2$ is the total wavenumber and $\alpha_{n,n_1}^{m,m_1} = k_1 l_1 (m_1 n - m n_1)$ the interaction coefficient (cf. Section 5.4.2). These equations were originally derived by Saltzman (1962) and provided the basis for the Lorenz model. The mathematical similarity with the spectral equations of the quasigeostrophic two-level model (5.87) and (5.88) is evident. In both models, two linear interaction terms are responsible for the instability and in each equation, a linear damping term appears. The nonlinear terms have also a similar form except that an additional nonlinearity appears in the spectral vorticity equation (5.87). This similarity is the reason why model reduction leads to similar low-order equations. Truncating the equations at $M = 1$ and $N = 2$ and considering only terms that remain nonzero after initialization with the growing normal mode yield the low-order dynamical system

$$\frac{d\Psi_1^1}{d\tilde{t}} = -i\mathrm{PrRa}\frac{k_1}{K_1^{1^2}}\Theta_1^1 - \mathrm{Pr}K_1^{1^2}\Psi_1^1 \tag{6.90}$$

$$\frac{d\Theta_1^1}{d\tilde{t}} = ik_1\Psi_1^1 - K_1^{1^2}\Theta_1^1 - 2k_1 l_1 \Psi_1^1 \Theta_2^0 \tag{6.91}$$

$$\frac{d\Theta_2^0}{d\tilde{t}} = -4l_1^2\Theta_2^0 - 4ik_1 l_1 I\left(\Psi_1^1 \Theta_1^{1^C}\right) \tag{6.92}$$

According to (6.78) and (6.79) we assume that Ψ_1^1 and Θ_1^1 are initially real and imaginary, respectively. Then, we get with the notations $A = -4\Psi_{1_r}^1$, $B = -4\Theta_{1_i}^1$ and $C = -2\Theta_{2_i}^0$

$$\frac{dA}{dt} = \mathrm{PrRa}\frac{k_1}{K_1^{1^2}}B - \mathrm{Pr}K_1^{1^2}A \tag{6.93}$$

$$\frac{dB}{dt} = k_1 A - K_1^{1^2}B + k_1 l_1 AC \tag{6.94}$$

$$\frac{dC}{dt} = -4l_1^2 C - \frac{k_1 l_1}{2}AB \tag{6.95}$$

This system can be converted into the Lorenz equations (cf. Section 5.3.2)

$$\frac{d\mathcal{X}}{d\mathcal{J}} = \sigma(\mathcal{Y} - \mathcal{X}) \tag{6.96}$$

$$\frac{d\mathcal{Y}}{d\mathcal{J}} = \mathcal{R}\mathcal{X} - \mathcal{X}\mathcal{Z} - \mathcal{Y} \tag{6.97}$$

$$\frac{d\mathcal{Z}}{d\mathcal{J}} = \mathcal{X}\mathcal{Y} - \mathcal{B}\mathcal{Z} \tag{6.98}$$

using the following definitions:

$$\mathcal{X} = \frac{k_1 l_1}{\sqrt{2}K_1^{1^2}}A, \quad \mathcal{Y} = \mathrm{Ra}\frac{k_1^2 l_1}{\sqrt{2}K_1^{1^6}}B, \quad \mathcal{Z} = -\mathrm{Ra}\frac{k_1^2 l_1}{K_1^{1^6}}C \tag{6.99}$$

$$\tau = K_1^{1^2}\tilde{t}, \quad \sigma = \mathrm{Pr}, \quad \mathcal{R} = \frac{\mathrm{Ra}}{\mathrm{Ra}_c}, \quad \mathcal{B} = \frac{4l_1^2}{K_1^{1^2}} \tag{6.100}$$

Lorenz (1963) derived these equations in a similar way and simulated the **Lorenz attractor** with the parameter values $\sigma = 10$, $\mathcal{R} = 28$ and $\mathcal{B} = 8/3$. The latter value is based on the aspect ratio $k_1/l_1 = 1/\sqrt{2}$ where convection becomes unstable at $\mathrm{Ra} = \mathrm{Ra}_c$ (cf. Eq. (6.77)). However, there is a problem with this solution because $\mathcal{R} = 28$ means that the Rayleigh number Ra is a factor 28 larger than its critical value. Hence, the evolving convection is not weakly nonlinear. This is in contrast to the baroclinic instability problem treated in Section 5.3 where the Lorenz equations must not disagree with weak nonlinearity for every $\mathcal{R}$ value. The reason is that convection can only be made weakly nonlinear by dissipative processes while baroclinic instability can also be weakly nonlinear without dissipation (cf. Section 5.3). The consequence of this problem is demonstrated in Fig. 6.6 showing the solution of the Lorenz equations in phase space for the aforementioned parameters and that of the spectral equations (6.88)–(6.89) for a truncation at $M = N = 20$. A strange attractor can be seen for the low

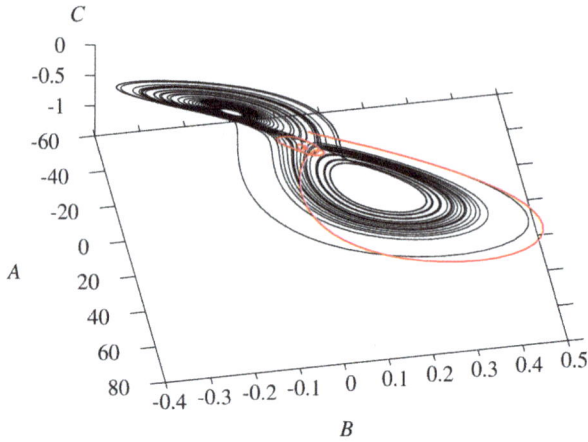

Fig. 6.6. Lorenz attractor obtained by numerical integration of (6.93)–(6.95) for the parameters $Ra = 18410$, $Pr = 10$, $k_1 = \pi/\sqrt{2}$ and $l_1 = \pi$ (black trajectory). The red curve shows the phase space trajectory that results from an integration of the spectral equations (6.88) and (6.89) for truncation at $M = N = 20$ and the same parameter values.

order truncation while the high-resolution simulation reveals a transient flow that approaches a stable fixed point. Figure 6.7 shows the potential temperature and stream function fields at $\tilde{t} = 50$ for both solutions. The low-order solution resulting from the Lorenz equations is an arbitrary snapshot (Fig. 6.7(a)) where upwelling and positive temperature anomalies appear at the center. However, the anomalies have the opposite signs at a later time point. The solution of the high-resolution spectral equations ($M = N = 20$) already settled into a steady state (Fig. 6.7(b)). The horizontal extension of the overturning cells halved in contrast to the low-order solution. Furthermore, we see that updrafts and downdrafts become narrower than those of the normal mode. This result is even more emphasized for higher values of the Rayleigh number (not shown). Updrafts in the convective atmospheric boundary layer are indeed narrow and intense (see, e.g. Fig. 11.19 in Stull, 1988). These are known as **thermals** and glider pilots use such flows to ascend. However, there is no corresponding downdraft counterpart as it appears in Fig. 6.7(b). The reason is that the assumption of a constant temperature at the top of the boundary layer is not realistic and causes a thin unstable layer immediately below. In the atmospheric boundary layer, thermal stratification is rather stable in the upper part since upward temperature fluxes warm

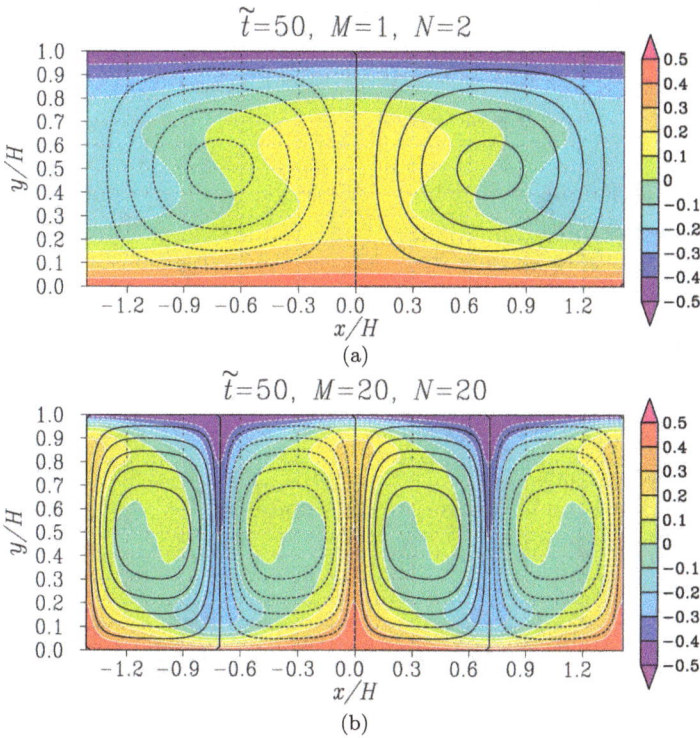

Fig. 6.7. Potential temperature (colored shadings) and stream function (isolines) at $\tilde{t} = 50$ from numerical integrations of (6.88) and (6.89) for a truncation at (a) $M = 1, N = 2$ and (b) $M = N = 20$. The used model parameters are Ra $= 18410$, Pr $= 10$, $k_1 = \pi/\sqrt{2}$ and $l_1 = \pi$.

the air and the cooling by radiation is too weak to compensate for this warming.

Although the Lorenz equations are not valid for high Rayleigh numbers, they can provide excellent solutions for cases in which the Rayleigh number is slightly higher than its critical value. In this case, we can calculate the amplitude of the steady convection cells by using the equilibrium solution Eq. (5.138). With this solution, we obtain for the steady-state vertical velocity and potential temperature anomaly (assuming $k_1/l_1 = 1/\sqrt{2}$)

$$w = \frac{\partial \psi}{\partial x} = \frac{3\pi K_\theta}{\sqrt{2}H} \sqrt{\frac{8}{3}\left(\frac{\text{Ra}}{\text{Ra}_c} - 1\right)} \cos\left(\frac{\pi x}{\sqrt{2}H}\right) \sin\left(\frac{\pi z}{H}\right) \quad (6.101)$$

$$\theta^* = \theta - [\theta]_x = \frac{\sqrt{2}\mathrm{Ra}_c\Delta\theta}{\pi\mathrm{Ra}}\sqrt{\frac{8}{3}\left(\frac{\mathrm{Ra}}{\mathrm{Ra}_c} - 1\right)}\cos\left(\frac{\pi x}{\sqrt{2}H}\right)\sin\left(\frac{\pi z}{H}\right)$$

$$(6.102)$$

where $[\cdot\cdot]_x$ denotes the average in x direction and the superindex $*$ the corresponding anomaly. We have chosen arbitrarily that the phase of the wave is as in Fig. 6.7(a). With (6.101) and (6.102) the vertical temperature flux by convection becomes

$$[\theta^* w^*]_x = \frac{4K_\theta\Delta\theta}{H}\left(1 - \frac{\mathrm{Ra}_c}{\mathrm{Ra}}\right)\sin^2\left(\frac{\pi z}{H}\right) \qquad (6.103)$$

The factor $K_\theta\Delta\theta/H$ in this equation would be the vertical temperature flux by microturbulence if no convection were present. The ratio of both temperature fluxes constitutes the **Nusselt number** Nu:

$$\mathrm{Nu} = \frac{[\theta^* w^*]_x}{K_\theta\Delta\theta/H} = 4\left(1 - \frac{\mathrm{Ra}_c}{\mathrm{Ra}}\right)\sin^2\left(\pi\frac{z}{H}\right) \qquad (6.104)$$

The Nusselt number is a measure for the intensity of convection. Figure 6.8 shows the Nusselt number at $z = H/2$ as a function of the Rayleigh number deduced from (6.104) and from steady-state solutions of the Saltzman model with appropriate truncation for Pr $= 0.74$. For low Rayleigh numbers Ra $< 2\mathrm{Ra}_c$ there is a good agreement between both models. Then, the Saltzman model becomes weakly nonlinear and the low-order truncation $M = 1$ and $N = 2$ is acceptable. For higher Rayleigh numbers the low-order model approaches 4 while the high-resolution Saltzman model reveals approximately a Ra$^{1/3}$ power law scaling (dotted straight line in Fig. 6.8) which is in agreement with other model studies on convection (see, e.g. Iyer *et al.*, 2020).

We already know from the analysis in Section 5.5.3 that the equilibrium (5.138) is always stable for Pr $= \sigma = 0.74 < 1$ (cf. Eq. (5.154)). Furthermore, the Saltzman model settles even at very high Rayleigh number into a steady state as seen in Fig. 6.8 and different initializations lead to the same result. We can estimate Ra

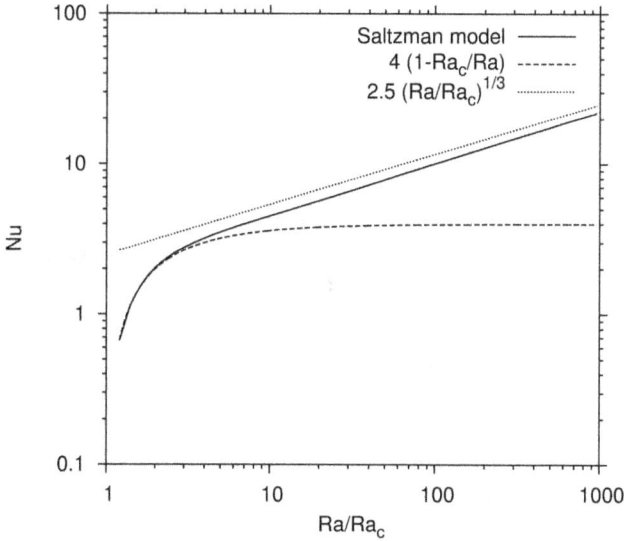

Fig. 6.8. Nusselt number Nu at $z = H/2$ as a function of Ra/Ra_c deduced from numerically determined steady-state solutions of the Saltzman model at appropriate truncation for $\text{Pr} = 0.74$ (solid line) and from the lower order model solution $4(1 - \text{Ra}_c/\text{Ra})$ (dashed line). The dotted line displays the $\text{Ra}^{1/3}$ power law scaling.

for the atmospheric convective boundary layer using some typical values

$$\text{Ra} = \frac{gH^3\Delta\theta}{K_\theta K_m \theta_0} = \frac{9.81\frac{\text{m}}{\text{s}^2}(1000\,\text{m})^3 2\text{K}}{1.35(10\,\text{m}^2/\text{s})^2 300\,\text{K}} \approx 4.36 \times 10^5 \qquad (6.105)$$

At this value, stable convective rolls develop with a maximum vertical velocity of $10\,\text{m/s}$ which is very large but not completely unrealistic. However, the Saltzman model has several significant deficiencies in the context of the atmospheric boundary layer. The boundary conditions, the two-dimensional model configuration and the assumption of a constant turbulent exchange coefficient are not appropriate. To overcome these issues, one must adopt a high-resolution **large eddy simulation** model as, e.g. used by Müller and Chlond (1996) to study convective cells during a cold air outbreak. Such a model resolves large turbulent eddies, but parameterization is still needed for smaller eddies.

6.4 Deep Convection

Deep convection differs from shallow convection by several properties in addition to its greater vertical extent. It always includes phase transformation of water vapor as an essential process and precipitation is always formed by the convective cell. In addition, a trigger must be present to release deep convection. The latter property has the consequence that **convective available potential energy (CAPE)** can build up if no trigger is present. This is the reason why deep convective cells can form vigorous **thunderstorms** due to a high amount of accumulated CAPE being released. Before analyzing the deep convective instability, we explain CAPE and associated quantities because these are important indicators for the potential development of deep convection. Finally, a numerical simulation of an axisymmetric thunderstorm is presented.

6.4.1 *Convective available potential energy*

We have already learned in Section 2.6 that the atmosphere can contain available potential energy which is convertible into kinetic energy. Then, the temperature must vary in the horizontal direction. However, there is also the possibility that available potential energy can exist in a horizontally homogeneous atmosphere. This case happens with unstable stratification that was excluded in Section 2.6. It is tedious to derive an expression for available potential energy that also includes the latter case. Instead, we derive the work of an imaginary air parcel that is lifted in the atmosphere. CAPE exists if the integrated work is positive. In the inviscid case, CAPE will be completely converted into kinetic energy and this amount of energy quantifies CAPE. For a global climatology of CAPE, we refer to Riemann-Campe *et al.* (2009). To derive CAPE, we start with the inviscid vertical momentum equation of the parcel

$$\frac{Dw_p}{Dt} = -\frac{1}{\rho_p}\frac{\partial p}{\partial z} - g \tag{6.106}$$

where the index p denotes a property of the air parcel. We can assume that the atmosphere is in hydrostatic balance so that

$$\frac{Dw_p}{Dt} = g\frac{\rho - \rho_p}{\rho_p} \tag{6.107}$$

Therefore, the parcel is accelerated vertically by the buoyancy. Multiplication of this equation with w_p and integration in time yields an equation for the specific kinetic energy of the parcel

$$e_{Kp} = \int_0^t g \frac{\rho - \rho_p}{\rho_p} w_p dt \qquad (6.108)$$

We can convert the integral by the substitution $dz = w_p dt$ and obtain

$$e_{Kp} = \int_{z_s}^{z_t} g \frac{\rho - \rho_p}{\rho_p} dz \qquad (6.109)$$

where z_s and z_t are the heights of the bottom and the top of the considered atmospheric layer, respectively. This expression constitutes the **surface-based CAPE (SBCAPE)**. It is more convenient to replace density with virtual temperature using the equation of state (1.80). Then, SBCAPE becomes

$$e_{Kp} = \int_{z_s}^{z_t} g \frac{T_{v,p} - T_v}{T_v} dz = \int_{p_t}^{p_s} \frac{T_{v,p} - T_v}{\rho T_v} dp$$

$$= \int_{\ln(p_t/p_0)}^{\ln(p_s/p_0)} R_d (T_{v,p} - T_v) d\ln(p/p_0) \qquad (6.110)$$

where we have assumed that the pressures of the parcel and the environment are identical. The virtual temperature of the parcel, $T_{v,p}$ results in the idealized case from the reversible moist adiabat (see Section 1.10) and T_v is the actual virtual temperature that can be deduced from an atmospheric sounding. For specifying the reversible moist adiabat one must define the initial temperature T_I and initial specific humidity m_{vI} of the parcel. It is suitable to prescribe values observed in the boundary layer. Then, the conservation of potential temperature of moist air (1.144) can be assumed so that

$$T_p(p) = T_I \left(\frac{p}{p_s} \right)^{\frac{R_d + (R_v - R_d) m_{vI}}{c_{pd} + (c_{pv} - c_{pd}) m_{vI}}} \qquad (6.111)$$

This assumption holds until the parcel reaches the **lifting condensation level** which appears at the pressure level where condensation happens for the first time. Then, the identity

$$T_{d,p}(p, m_{v0}) = T_I \left(\frac{p}{p_s} \right)^{\frac{R_d + (R_v - R_d)m_{vI}}{c_{pd} + (c_{pv} - c_{pd})m_{vI}}} \tag{6.112}$$

holds where $T_{d,p}$ is the dew point of the parcel (cf. Eq. (1.103)). T_{dp} results from the Magnus formula (1.98) and Eq. (1.100)

$$T_{d,p}(p, m_{v0}) = \frac{T_0 - 30\,\text{K}}{17.625\ln\left\{ \dfrac{pm_{vI}}{p_{vs}(T_0)\left[\frac{R_d}{R_v}(1-m_{vI})+m_{vI}\right]} \right\}^{-1} - 1} + T_0 \tag{6.113}$$

where $p_{vs}(T_0)$ denotes the saturation vapor pressure at the freezing point temperature T_0. Equation (6.112) can only be solved with a numerical approximation method. The determined pressure p_{lcl} at the lifting condensation level, the dew point $T_{d,p}$ and the specific humidity m_{vI} constitute the initial values for the ascent in saturated air. In this part, the saturated equivalent potential temperature (1.146) is conserved if the condensed water remains in the parcel which thereby describes a reversible process. To obtain the parcel temperature $T_p(p)$, we must solve the equation

$$\theta_{es}(T_{dp}, p_{lcl}) = T_p \left(\frac{p_0}{p - p_{vs}} \right)^{\frac{R_d(1-m_{vI})}{\tilde{c}_p}} \exp\left[\frac{L_v m_{vs}}{\tilde{c}_p T_p} \right] \tag{6.114}$$

where $\tilde{c}_p = c_{pd} + (c_l - c_{pd})m_{vI}$. Note that m_{vs} and p_{vs} are functions of temperature T_p and pressure p. Again, a numerical approximation method must be used to solve this equation. With this method, we ignore the ice phase and precipitation. The former yields a somewhat larger value for CAPE due to the higher latent heat release. Precipitation leads to an increase in CAPE because the air becomes lighter with the loss of liquid water. To estimate this impact, we prescribe a **pseudoadiabatic ascent**. Then, all condensed water leaves the air parcel so that the process is irreversible. To determine the **pseudoadiabat** we use the approximated first law of thermodynamics in differential form

$$c_p dT_p = \frac{R_d T_p}{p} dp - L_v dm_{vs} \tag{6.115}$$

Therefore, the complete amount of latent heat released warms the air that remains in the parcel while the condensed water attains no heat. Saturation specific humidity m_{vs} is a function of T_p and p, as already mentioned. Hence, we get by using Eqs. (1.95) and (1.100)

$$dm_{vs} = \frac{\partial m_{vs}}{\partial T_p} dT_p + \frac{\partial m_{vs}}{\partial p} dp = \frac{p}{p - \left(1 - \frac{R_d}{R_v}\right) p_{vs}} \frac{L_v m_{vs}}{R_v T_p^2} dT_p$$

$$- \frac{m_{vs}}{p - \left(1 - \frac{R_d}{R_v}\right) p_{vs}} dp \qquad (6.116)$$

With (6.115) and (6.116), we get for the vertical temperature gradient of the pseudoadiabat

$$\frac{dT_p}{dp} = \frac{\frac{R_d T_p}{p} + \frac{L_v m_{vs}}{p - (1 - R_d/R_v) p_{vs}}}{c_p + \frac{p}{p - (1 - R_d/R_v) p_{vs}} \frac{L_v^2 m_{vs}}{R_v T_p^2}} \qquad (6.117)$$

Multiplying this expression with $gp/(R_d T_p)$ yields the **pseudoadiabatic lapse rate** $\Gamma_m \equiv -dT_p/dz$ which is always smaller than the dry adiabatic lapse rate $\Gamma_d = g/c_p$ due to the latent heat release. Integrating Eq. (6.117) leads to the temperature of the pseudoadiabat. Figure 6.9 shows a sounding from Springfield (USA) on May 22, 2011 (12UTC) together with the corresponding reversible moist adiabat and pseudoadiabat in a **T-log-p diagram**. The abscissa and ordinate of this thermodynamic diagram are virtual temperature and log-pressure height. Obviously, SBCAPE can be determined graphically by the area between the sounding and the corresponding adiabat. A positive contribution to SBCAPE results for $T_{v,p} > T_v$ (the adiabat is located on the right side of the observed profile) and a negative for $T_{v,p} < T_v$ (the adiabat is located on the left side of the observed profile). The positive contribution (dark grey area in Fig. 6.9) constitutes CAPE which is the energy that can potentially be converted in the updraft. The negative contribution (grey area in Fig. 6.9) yields the **convective inhibition energy (CIN)**. This energy must be invested by another mechanism to release convection. Therefore, deep convection requires in most cases a trigger that overcomes the CIN. A trigger to overcome CIN is often a low-level convergence of the large-scale horizontal flow. Then, the vertical motion

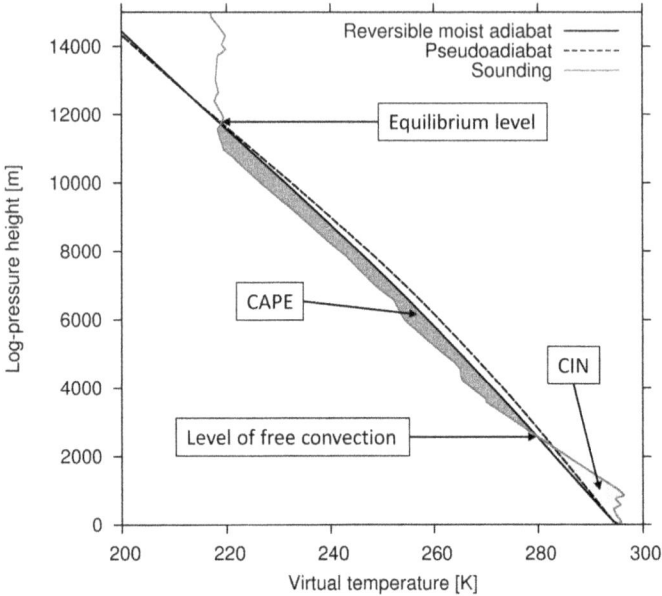

Fig. 6.9. Reversible moist adiabat (black solid line), pseudoadiabat (black dashed line) and the corresponding sounding (grey line) in a T log-p diagram. The latter was compiled on May 22, 2011 (12UTC) in Springfield (USA). The area in dark grey is proportional to CAPE while the area in grey yields the CIN. The data on the sounding was provided by the University of Wyoming (https://weather.uwyo.edu/upperair/sounding.html).

resulting from mass continuity lifts the air to the **level of free convection (LFC)** where buoyancy becomes positive for the first time and convection can start. Another trigger is the heating of the surface by which the boundary layer becomes convective and extents in the vertical until the LFC is reached. Finally, air could also be lifted to the LFC when it approaches a mountain. Convection is decelerated above the **equilibrium level (EL)** where the virtual temperature of the parcel becomes identical with that of the environment again.

Using the aforementioned information, we can derive the following relations between the various energies:

$$\underbrace{\int_{z_s}^{z_t} g \frac{T_{v,p} - T_v}{T_0} dz_l}_{\text{SBCAPE}} = -\underbrace{\int_{z_s}^{z_f} g \frac{T_v - T_{v,p}}{T_0} dz_l}_{\text{CIN}} + \underbrace{\int_{z_f}^{z_t} g \frac{T_{v,p} - T_v}{T_0} dz_l}_{\text{CAPE}}$$

$$(6.118)$$

where $z_l = -H_S \ln(p/p_s)$ is the log-pressure height, z_s the surface level, z_f the LFC and z_t the EL. The data used for Fig. 6.9 yields for CAPE 1170 J/kg and for CIN 305 J/kg. A vertical velocity of 48 m/s would result if CAPE were converted completely into kinetic energy. Using the pseudoadiabat instead of the reversible moist adiabat in (6.118) leads to a higher CAPE (1741 J/kg) and a lower CIN (243 J/kg). This is understandable because the air without liquid water has a higher virtual temperature (see the dashed line in Fig. 6.9).

Deep convection must not be released by air originating from the surface. It is possible that the CIN is much lower when an air parcel of a higher level is lifted vertically. Indeed, Fig. 6.10(a) shows the adiabats for air from 894 hPa instead from the surface level at 962 hPa. Now, the CIN is smaller with a value of only 69 J/kg and CAPE becomes larger attaining 2763 J/kg. These values can be explained by the high specific humidity in the upper part of the boundary layer and it can be expected that air close to this level initiates convection. At a higher level (843 hPa) CAPE vanishes as seen in Fig. 6.10(b) because the air parcel starts above the boundary layer where the specific humidity is much smaller.

Convective inhibition is impossible in the case of dry convective instability because the fluid is always unstable at some height if there is some CAPE. This is the main reason why dry convective cells cannot be as severe as a thunderstorm.

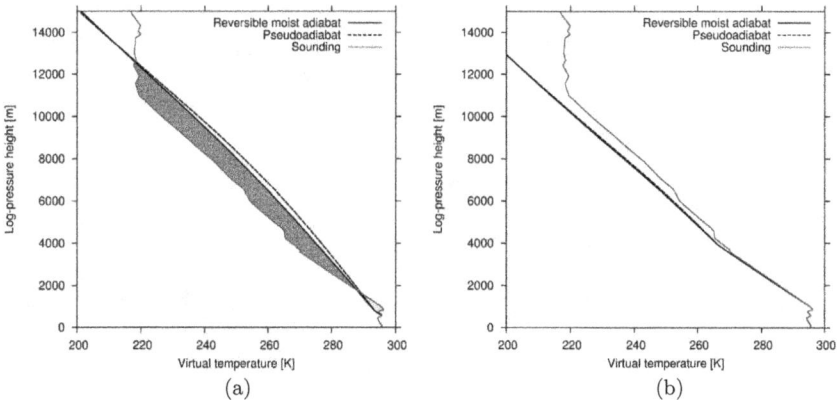

Fig. 6.10.   As Fig. 6.9 but the adiabats originate at (a) 894 hPa and (b) 843 hPa.

6.4.2 *Conditional instability*

The instability that leads to deep convection is referred to as **conditional instability** because a condition must be fulfilled to initiate the instability. The condition is just that CIN vanishes at least at one location. This happens in most cases by saturation of air at the LFC where latent heat release forces the updraft. In the environment, however, air remains unsaturated and no release of latent heat takes place. This **conditional heating** requires a different instability analysis in contrast to dry convective instability. For the analysis, it is convenient to assume an axisymmetric cumulus cloud that is created by the instability. Then, the use of **cylindrical coordinates** is appropriate. The transformation from Cartesian to cylindrical coordinates only concerns the horizontal coordinates since the vertical coordinate does not change. The cylindrical coordinates are given by[b]

$$r = \sqrt{x^2 + y^2}, \quad \alpha = \tan^{-1}\left(\frac{y}{x}\right), \quad z = z \tag{6.119}$$

while the inverse transformation can be obtained by

$$x = r\cos\alpha, \quad y = r\sin\alpha, \quad z = z \tag{6.120}$$

We see that cylindrical coordinates coincide with polar coordinates in the horizontal plane (cf. Section 5.5.5). With (6.119) we can determine the Lamé-coefficients

$$h_r = \frac{1}{|\nabla r|} = 1, \quad h_\alpha = \frac{1}{|\nabla \alpha|} = r, \quad h_z = \frac{1}{|\nabla z|} = 1 \tag{6.121}$$

and the unit vectors

$$\mathbf{e}_r = \frac{\nabla r}{|\nabla r|} = \cos\alpha\,\mathbf{i} + \sin\alpha\,\mathbf{j},$$

$$\mathbf{e}_\alpha = \frac{\nabla \alpha}{|\nabla \alpha|} = -\sin\alpha\,\mathbf{i} + \cos\alpha\,\mathbf{j}, \quad \mathbf{e}_z = \mathbf{k} \tag{6.122}$$

---

[b]We use for the radius coordinate r the same notation as for radius in spherical coordinates. From now on r refers to cylindrical coordinates.

Obviously, these unit vectors form an orthogonal base. We get for
the nonzero derivatives of the unit vector

$$\frac{\partial \mathbf{e}_r}{\partial \alpha} = -\sin \alpha \mathbf{i} + \cos \alpha \mathbf{j} = \mathbf{e}_\alpha,$$

$$\frac{\partial \mathbf{e}_\alpha}{\partial \alpha} = -\cos \alpha \mathbf{i} - \sin \alpha \mathbf{j} = -\mathbf{e}_r \qquad (6.123)$$

In cylindrical coordinates, the velocity vector is represented by

$$\mathbf{v} = v_r\,\mathbf{e}_r + v_\alpha\,\mathbf{e}_\alpha + v_z\,\mathbf{e}_z \qquad (6.124)$$

where v_r is the **radial wind**, v_α the **tangential wind** and v_z the vertical wind. With this information, it is possible to derive the anelastic
equations for deep convection (6.56)–(6.58) in cylindrical coordinates
with radiation being neglected. These take the form

$$\frac{\partial V_r}{\partial t} + V_r\frac{\partial v_r}{\partial r} + \frac{V_\alpha}{r}\frac{\partial v_r}{\partial \alpha} + V_z\frac{\partial v_r}{\partial z} - \frac{v_\alpha V_\alpha}{r}$$
$$= -\frac{\partial p^\star}{\partial r} + \rho_0 f_{T,r} \qquad (6.125)$$

$$\frac{\partial V_\alpha}{\partial t} + V_r\frac{\partial v_\alpha}{\partial r} + \frac{V_\alpha}{r}\frac{\partial v_\alpha}{\partial \alpha} + v_z\frac{\partial v_\alpha}{\partial z} + \frac{v_r V_\alpha}{r}$$
$$= -\frac{1}{r}\frac{\partial p^\star}{\partial \alpha} + \rho_0 f_{T,\alpha} \qquad (6.126)$$

$$\frac{\partial V_z}{\partial t} + V_r\frac{\partial v_z}{\partial r} + \frac{V_\alpha}{r}\frac{\partial v_z}{\partial \alpha} + V_z\frac{\partial v_z}{\partial z}$$
$$= -\frac{\partial p^\star}{\partial z} + B + \rho_0 f_{T,z} \qquad (6.127)$$

$$\frac{\partial B}{\partial t} + \frac{1}{r}\frac{\partial}{\partial r}(rv_r B) + \frac{1}{r}\frac{\partial(v_\alpha B)}{\partial \alpha} + \frac{1}{\theta_r}\frac{\partial(v_z\theta_r B)}{\partial z} + N_r^2 V_z$$
$$= \frac{\rho_0 g Q_L}{c_p \theta_r \Pi_0} + \frac{g\rho_0}{\theta_r}D_\theta \qquad (6.128)$$

$$\frac{1}{r}\frac{\partial}{\partial r}(rV_r) + \frac{1}{r}\frac{\partial V_\alpha}{\partial \alpha} + \frac{\partial V_z}{\partial z} = 0 \qquad (6.129)$$

where $V_r = \rho_0 v_r$, $V_\alpha = \rho_0 v_\alpha$, $V_z = \rho_0 v_z$, $B = g\rho_0\theta^\star/\theta_r$, $N_r^2 = g/\theta_r\partial\theta_r/\partial z$. Note that the new variable B is a measure for buoyancy. In the following, we omit all derivatives with respect to α because of the assumed axisymmetry. Furthermore, the tangential velocity v_α is set to zero. The representation of turbulence exchange terms in cylindrical coordinates requires special attention. For the turbulent exchange of heat, we assume:

$$\frac{g\rho_0}{\theta_r} D_\theta = K_\theta \nabla^2 B = K_\theta \nabla \cdot (\nabla B)$$

$$= K_\theta \left[\frac{1}{r}\frac{\partial}{\partial r}\left(r\frac{\partial B}{\partial r}\right) + \frac{\partial^2 B}{\partial z^2} \right] \qquad (6.130)$$

Therefore, a constant exchange coefficient is prescribed, and the height-dependent factor $g\rho_0/\theta_r$ has been shifted behind the Nabla operator (cf. Eq. (6.57)). Although this representation of turbulence is crude, it greatly simplifies the analysis. For the turbulent momentum exchange, we get with the same assumptions:

$$\rho_0 f_T = K_M \nabla^2 (\rho_0 \mathbf{v})$$

$$= K_M \nabla \cdot \left[\left(\mathbf{e}_r \frac{\partial}{\partial r} + \mathbf{e}_\alpha \frac{1}{r}\frac{\partial}{\partial \alpha} + \mathbf{e}_z \frac{\partial}{\partial z} \right) (V_r \mathbf{e}_r + V_z \mathbf{e}_z) \right]$$

$$= K_M \nabla \cdot \left(\frac{\partial V_r}{\partial r}\mathbf{e}_r\mathbf{e}_r + \frac{\partial V_z}{\partial r}\mathbf{e}_r\mathbf{e}_z + \frac{\partial V_r}{\partial z}\mathbf{e}_z\mathbf{e}_r \right.$$

$$\left. + \frac{\partial V_z}{\partial z}\mathbf{e}_z\mathbf{e}_z + \frac{V_r}{r}\mathbf{e}_\alpha\mathbf{e}_\alpha \right)$$

$$= K_M \left[\left(\frac{\partial^2 V_r}{\partial r^2} + \frac{1}{r}\frac{\partial V_r}{\partial r} - \frac{V_r}{r^2} + \frac{\partial^2 V_r}{\partial z^2} \right) \mathbf{e}_r \right.$$

$$\left. + \left(\frac{\partial^2 V_z}{\partial r^2} + \frac{1}{r}\frac{\partial V_z}{\partial r} + \frac{\partial^2 V_z}{\partial z^2} \right) \mathbf{e}_z \right] \qquad (6.131)$$

The derivative $\partial/\partial\alpha$ has not been discarded in this derivation because the unit vector $\mathbf{e}_r$ depends on α. We see that the Laplace operator can in curvilinear coordinates not just be applied to the components. Instead, we have to evaluate the divergence of a tensor leading to the additional term $-K_M V_r/r^2\, \mathbf{e}_r$.

The governing equations must be linearized for the stability analysis. For the basic state, we assume a horizontally uniform atmosphere at rest. Then, the linearized equations become

$$\frac{\partial V_r'}{\partial t} = -\frac{\partial p^{*\prime}}{\partial r} + K_M \left[\frac{1}{r}\frac{\partial}{\partial r}\left(r\frac{\partial V_r'}{\partial r}\right) - \frac{V_r'}{r^2} + \frac{\partial^2 V_r'}{\partial z^2} \right] \qquad (6.132)$$

$$\frac{\partial V_z'}{\partial t} = -\frac{\partial p^{*\prime}}{\partial z} + B' + K_M \left[\frac{1}{r}\frac{\partial}{\partial r}\left(r\frac{\partial V_z'}{\partial r}\right) + \frac{\partial^2 V_z'}{\partial z^2} \right] \qquad (6.133)$$

$$\frac{\partial B'}{\partial t} + N_r^2 V_z' = \frac{\rho_0 g Q_L}{c_p \theta_r \Pi_0} + K_\theta \left[\frac{1}{r}\frac{\partial}{\partial r}\left(r\frac{\partial B'}{\partial r}\right) + \frac{\partial^2 B'}{\partial z^2} \right] \qquad (6.134)$$

$$\frac{1}{r}\frac{\partial}{\partial r}(rV_r') + \frac{\partial V_z'}{\partial z} = 0 \qquad (6.135)$$

To eliminate the pressure anomaly $p^{*\prime}$ we derive a vorticity equation by applying $\partial/\partial r$ to Eq. (6.133) and $\partial/\partial z$ to Eq. (6.132) before subtracting these equations. Then, we get

$$\frac{\partial \Lambda'}{\partial t} = \frac{\partial B'}{\partial r} + K_M \left[\frac{1}{r}\frac{\partial}{\partial r}\left(r\frac{\partial \Lambda'}{\partial r}\right) - \frac{\Lambda'}{r^2} + \frac{\partial^2 \Lambda'}{\partial z^2} \right] \qquad (6.136)$$

where $\Lambda' = \partial V_z'/\partial r - \partial V_r'/\partial z$ is the horizontal vorticity component of the mass-weighted flow. Because of (6.135) we can define a mass stream function Ψ of the **toroidal overturning circulation** such that

$$rV_r' = -\frac{\partial \Psi'}{\partial z}, \quad rV_z' = \frac{\partial \Psi'}{\partial r} \qquad (6.137)$$

Then, the vorticity Λ' becomes

$$\Lambda' = \frac{\partial}{\partial r}\left(\frac{1}{r}\frac{\partial \Psi'}{\partial r}\right) + \frac{1}{r}\frac{\partial^2 \Psi'}{\partial z^2} \qquad (6.138)$$

We still have to express the latent heat Q_L as a function of known variables. We assume that the atmosphere is saturated and, therefore, the linearized equation for specific humidity takes the form

$$\frac{\partial m_v'}{\partial t} = -v_z'\frac{\partial m_{vs}}{\partial z} + S_{l,v} \qquad (6.139)$$

where we have neglected the turbulent exchange. Upward motion leads to a positive tendency because $\partial m_{vs}/\partial z < 0$. Condensation takes place to keep the relative humidity at 100% and since this process is very fast, we can set the left-hand side of (6.139) to zero. On the other hand, downward motion decreases the relative humidity below 100% and no phase transition happens because liquid water is absent. Consequently, latent heat release only appears in ascending air and it can be written as

$$Q_L = -L_v S_{l,v} = \begin{cases} -v_z' L_v \dfrac{\partial m_{vs}}{\partial z} & \text{for } v_z' > 0 \\ 0 & \text{for } v_z' \le 0 \end{cases} \tag{6.140}$$

Hence, heating is conditional and this complicates the analysis. Strictly speaking, the heating term is nonlinear because it depends not only on vertical velocity but also on the sign of vertical velocity. It is not possible to superpose different solutions since the addition of two different vertical velocity fields yields a new vertical velocity field leading to a heating that does not coincide with the sum of the two primary heating fields. It is convenient to combine the heating term with the vertical advection term in the case of ascending motion:

$$V_z' N_r^2 - \frac{\rho_0 g Q_L}{c_p \theta_r \Pi_0} = V_z' \left(\frac{g}{\theta_r} \frac{\partial \theta_r}{\partial z} + \frac{g L_v}{c_p T_r} \frac{\partial m_{vs}}{\partial z} \right)$$

$$\approx V_z' \frac{g}{\theta_{es}} \frac{\partial \theta_{es}}{\partial z} \equiv V_z' N_m^2 \tag{6.141}$$

where θ_{es} denotes the approximated saturated equivalent potential temperature (cf. Eq. (1.150)) given by

$$\theta_{es} = \theta_r \exp\left(\frac{L_v m_{vs}}{c_p T_r} \right) \tag{6.142}$$

Instability can only occur if θ_{es} decreases with height. For simplicity, we assume that N_r^2 and N_m^2 are constants. Furthermore, we will make use of the same boundary conditions as in the dry convective instability problem treated in Section 6.3.1. Then, we can prescribe the vertical dependence by

$$\Lambda' = \hat{\Lambda}(r,t) \sin\left(\pi \frac{z}{H} \right), \qquad B' = \hat{B}(r,t) \sin\left(\pi \frac{z}{H} \right),$$

$$\Psi' = \hat{\Psi}(r,t) \sin\left(\pi \frac{z}{H} \right) \tag{6.143}$$

where H denotes the depth of the convective layer. Inserting this solution ansatz in (6.134), (6.136) and (6.138) yields

$$\frac{\partial \hat{\Lambda}}{\partial t} = \frac{\partial \hat{B}}{\partial r} + K_M \left[\frac{1}{r} \frac{\partial}{\partial r} \left(r \frac{\partial \hat{\Lambda}}{\partial r} \right) - \left(\frac{1}{r^2} + \frac{\pi^2}{H^2} \right) \hat{\Lambda} \right] \quad (6.144)$$

$$\frac{\partial \hat{B}}{\partial t} = K_\theta \left[\frac{1}{r} \frac{\partial}{\partial r} \left(r \frac{\partial \hat{B}}{\partial r} \right) - \frac{\pi^2}{H^2} \hat{B} \right] - \begin{cases} N_m^2 \dfrac{\partial \hat{\Psi}}{\partial r} & \text{for} \quad \dfrac{\partial \hat{\Psi}}{\partial r} > 0 \\[2ex] N_r^2 \dfrac{\partial \hat{\Psi}}{\partial r} & \text{for} \quad \dfrac{\partial \hat{\Psi}}{\partial r} \leq 0 \end{cases}$$

$$(6.145)$$

$$\hat{\Lambda} = \frac{\partial}{\partial r} \left(\frac{1}{r} \frac{\partial \hat{\Psi}}{\partial r} \right) - \frac{1}{r} \frac{\pi^2}{H^2} \hat{\Psi} \quad (6.146)$$

It is tedious to find analytical solutions of these equations. Instead, we present approximate numerical solutions of them by using a grid point discretization technique and integrating the equations until the growing normal mode has been established. The grid point distance is 50 m in radial direction and Eq. (6.146) is inverted with Richtmyer's method (Richtmyer, 1957). A solid lateral boundary is assumed at $r = 120$ km and for the depth of the convection cloud, we choose $H = 10$ km. The Brunt-Väisälä frequency N_r of the air outside the cloud is 10^{-2}s^{-1} while inside the cloud we assume $N_m^2 = -0.3 \cdot 10^{-4} \text{s}^{-2}$. Furthermore, we set $K_\theta = K_M$ ($\text{Pr} = 1$) and vary K_M. The model selects the fastest-growing normal mode and it is not possible to determine other normal modes with smaller growth rates as in the analysis presented in the previous section. Figure 6.11 shows the growth rate and the radius of the cloud as a function of turbulent exchange coefficient K_M (see the solid curves). The growth rate declines while the cloud radius becomes large with increasing K_M and for $K_M > 17000 \text{ m}^2/\text{s}$ instability vanishes. The diameter of the cloud shrinks to zero in the frictionless limit $K_M = 0$ so that it does not appear to be appropriate to neglect turbulent exchange. The growth rate can reach nearly 20/h for weak turbulent exchange. This corresponds to an e-folding time of 3 minutes and emphasizes the fast development of deep convective flows compared to large-scale phenomena like baroclinic instability. The diagrams in Fig. 6.11 also show results for **slab-symmetry** (dashed curves).

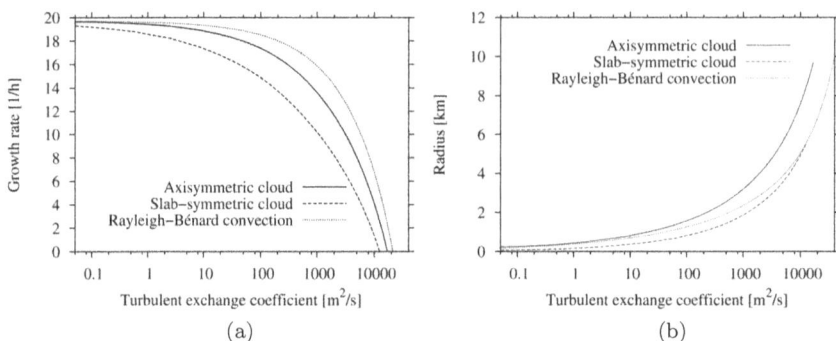

Fig. 6.11. (a) Growth rate of the fastest growing mode of the axisymmetric and slab-symmetric conditional instability problems (solid and dashed curves, respectively) and of the corresponding Rayleigh–Bénard instability problem (dotted curve) as a function of the turbulent exchange coefficient K_M, (b) as in (a) but for the cloud radius.

With this symmetry all quantities do not depend on y and the cloud takes the form of a rectilinear strip instead of a cylinder. Note that the same symmetry was assumed in the Rayleigh–Bénard instability problem treated in the previous subsection. The slab-symmetric instability has similar properties compared to its axisymmetric counterpart. However, the growth rate and the radius (half-width of the cloud strip) are smaller. We can show that the slab-symmetric deep convection problem is mathematically equivalent to the Rayleigh–Bénard instability problem when conditional heating is discarded by allowing corresponding cooling in downdrafts. Figure 6.11 displays the growth rate and quarter wavelength of the fastest-growing mode of the corresponding Rayleigh–Bénard instability problem (dotted curves). The growth rate of this mode is larger than for the conditional heating solutions. The scale is for weak turbulent exchange comparable to that of the axisymmetric case while for high turbulent exchange it is similar to the slab-symmetric case. It seems that all three solutions have very similar growth rates in the inviscid limit.

Figure 6.12(a) shows the toroidal mass stream function and the mass-weighted potential temperature anomaly of the fastest-growing axisymmetric mode for $K_M = 32.5 \, \mathrm{m^2/s}$. Inside of $r \approx 1 \, \mathrm{km}$ the vertical motion is upward and this region is associated with the cloud. Here, the temperature is increased due to the latent heat release that exceeds the adiabatic cooling by vertical motion. The

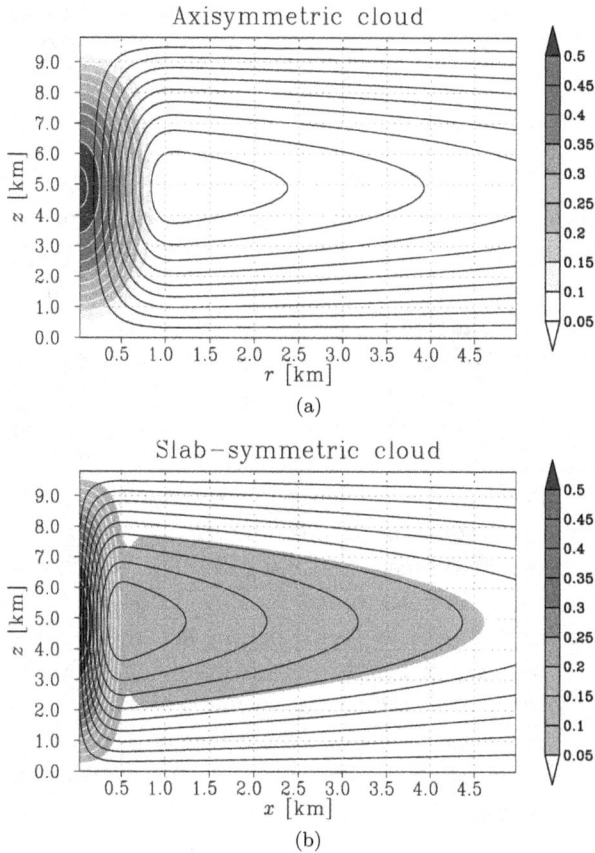

Fig. 6.12. (a) Mass stream function of the toroidal circulation (black isolines) and mass-weighted temperature anomaly (grey shadings) of the fastest growing normal mode of the axisymmetric conditional instability problem, (b) as (a) but for the slab-symmetric instability.

maximum mass-weighted temperature anomaly appears at the center and mid-level where the mass-weighted vertical velocity has also its maximum. The downward motion outside of the cloud is very weak and therefore, the corresponding adiabatic warming leads only to a weak temperature increase which is not visible in this figure. Figure 6.12(b) displays the fastest-growing normal mode for the slab-symmetric setup and the same K_M. The cloud strip has only a width equal to about half the diameter of the axisymmetric cloud. Now, the downwelling outside the cloud is stronger because the horizontal mass fluxes in and out of the cloud are larger due to the different symmetry. This leads to a notable warming in the region adjacent to the cloud.

6.4.3 *Numerical simulation of a cumulonimbus cloud*

In this subsection, a numerical simulation of a **cumulonimbus cloud** with the axisymmetric cloud model HURMOD (Frisius and Wacker, 2007) is presented. Such a cloud develops when conditions for deep convection are fulfilled and it is associated with a thunderstorm. The model HURMOD is based on governing equations similar to (1.174)–(1.179) with a turbulence closure scheme that also includes horizontal turbulent exchange (in contrast to Eqs. (1.256)–(1.258)). The model is formulated in cylindrical coordinates and axisymmetry is prescribed. Therefore, it has only the two spatial coordinates radius r and height z. Turbulence is parameterized with a scheme based on the approach introduced by Lilly (1962) in which the turbulent exchange coefficient is determined by Eq. (1.254). The cloud microphysical parameterization is based on the scheme by Kessler (1969) which will be described briefly in the following.[c] This scheme distinguishes **cloud water** (subscript c) and **rainwater** (subscript r). The former floats in the air while the latter falls downward with the **terminal velocity** W_r. The budget equations for the various water categories become

$$\frac{Dm_v}{Dt} = -S_{v,c} + S_{r,v} - \frac{1}{\rho}\frac{\partial J_{m_v}}{\partial z} + D_{m_v} \qquad (6.147)$$

$$\frac{Dm_c}{Dt} = S_{v,c} - S_{c,r} - \frac{1}{\rho}\frac{\partial J_{m_c}}{\partial z} + D_{m_c} \qquad (6.148)$$

$$\frac{Dm_r}{Dt} = -S_{r,v} + S_{c,r} - \frac{1}{\rho}\frac{\partial J_{m_r}}{\partial z} + D_{m_r} \qquad (6.149)$$

where

$$J_{m_v} = \rho m_v m_r W_r, \quad J_{m_c} = \rho m_c m_r W_r,$$
$$J_{m_r} = -\rho(1 - m_r)m_r W_r \qquad (6.150)$$

---

[c]See Frisius and Wacker (2007) for a more detailed description of HURMOD.

are the diffusive fluxes. Note that the barycenter of the air parcel moves slowly downward due to rain-water sedimentation. Therefore, the diffusive fluxes J_{m_v} and J_{m_c} are positive. The terminal velocity is parameterized by

$$W_r = 12.63 \, \text{m/s} \left(\frac{\rho m_r}{\rho_{00}} \right)^{\frac{1}{8}} \tag{6.151}$$

where $\rho_{00} = 1 \, \text{kg/m}^3$. Hence, the terminal velocity is always below $12.63 \, \text{m/s}$ since ρm_r can hardly exceed $1 \, \text{kg/m}^3$. However, for the realistic value $\rho m_r = 5 \, \text{g/m}^3$ the terminal velocity reaches $6.51 \, \text{m/s}$.

To apply the scheme, it is necessary to express the conversions $S_{v,c}$, $S_{r,v}$ and $S_{c,r}$ in terms of predicted variables. The cloud microphysical scheme in HURMOD distinguishes the five processes **condensation, autoconversion, accretion, evaporation of cloud drops** and **evaporation of rain**. The implementation of these conversions in the Kessler scheme is sketched in Fig. 6.13. Condensation is expressed in many models with a **saturation adjustment technique** to keep the relative humidity in clouds at 100%. In HURMOD, however, the condensation is described with a relaxation term so that

$$S_{v,c} = \frac{m_v - m_{vs}}{\tau_{\text{cond.}}} \quad \text{for } m_v \geq m_{vs} \tag{6.152}$$

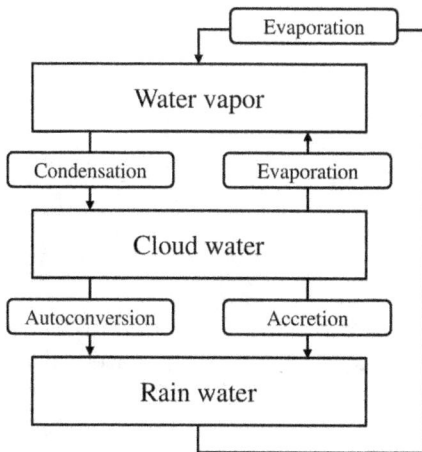

Fig. 6.13.   Sketch of the conversions in the Kessler scheme.

where τ_{cond} is the timescale of condensation. Cloud air might get undersaturated by mixing or warming. Then, cloud droplets evaporate which, in HURMOD, is parameterized by

$$S_{v,c} = \frac{m_v - m_{vs}}{\tau_{\text{evap}}} \frac{m_c}{m_{vs}} \quad \text{for } m_v < m_{vs} \qquad (6.153)$$

where τ_{evap} is the timescale of cloud drop evaporation. Cloud droplets convert to rain by **coagulation** if the density of cloud droplets is sufficiently large. This process constitutes the autoconversion and is parameterized by

$$S_{\text{aut}} = \begin{cases} \dfrac{m_c - m_{c0}}{\tau_{\text{aut}}} & \text{for } m_c \geq m_c 0 \\ 0 & \text{for } m_c < m_c 0 \end{cases} \qquad (6.154)$$

where m_{c0} is the threshold cloud water mass fraction for autoconversion and τ_{aut} its timescale. Raindrops also grow by the collection of cloud droplets during downward sedimentation. This process constitutes the **accretion** and is parameterized by

$$S_{\text{accr}} = 1.7242 \text{s}^{-1} m_c \left(\frac{\rho m_r}{\rho_{00}} \right)^{\frac{7}{8}} \qquad (6.155)$$

Finally, the evaporation of rain in undersaturated air results from

$$S_{r,v} = 0.007 \, \text{s}^{-1} (m_{vs} - m_v) \sqrt{\frac{\rho m_r}{\rho_{00}}}$$

$$\times \frac{1 + 9.1737 \sqrt{\frac{\rho}{\rho_{00}}} \left(\frac{\rho m_r}{\rho_{00}} \right)^{\frac{3}{16}}}{1 + 0.000921 \frac{\text{Ks}^2}{\text{m}} {}^2 \frac{L_{v0}^2 \rho m_{vs}}{\rho_{00} R_v T^2}} \qquad (6.156)$$

This formula is quite complicated because the ventilation of falling rain drops is taken into account. Usually, rain formation includes also the ice phase. However, a cloud microphysical scheme that considers the ice phase is much more complicated and is, therefore, not explained here.

The initial state of the simulation consists of a horizontally homogeneous atmosphere and a superimposed warm moist bubble at the center. The vertical temperature profile of the base state has a lapse rate of $6.5 \, \text{K/km}$ below the tropopause at $z = 12 \, \text{km}$ and zero above

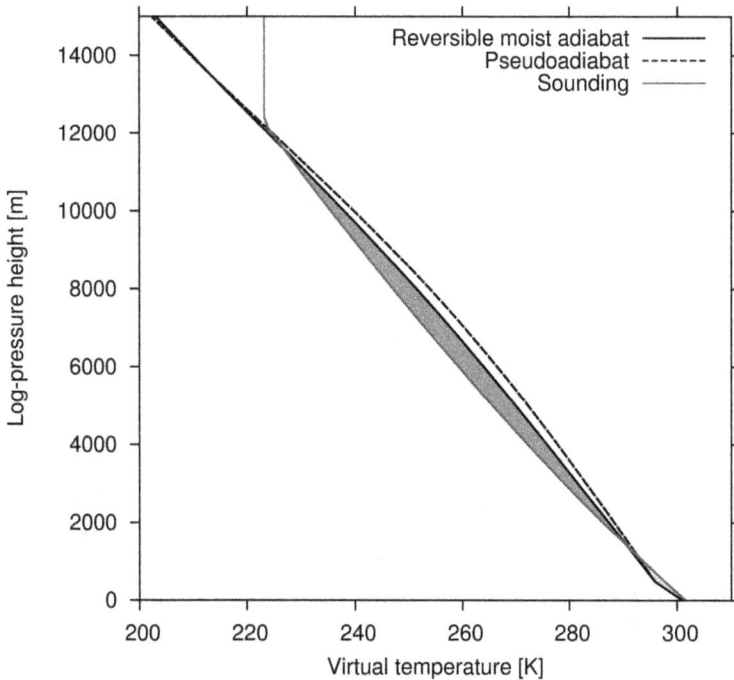

Fig. 6.14. As in Fig. 6.9 but for the base state of the cumulonimbus cloud simulation.

this level. The surface temperature is 28°C and the relative humidity of the tropopause amounts to 70%. Figure 6.14 shows the vertical profile of the virtual temperature together with the corresponding adiabats. The CAPE resulting from this profile is 1151 J/kg for a moist adiabatic ascent and 1876 J/kg for a pseudoadiabatic ascent. These values correspond to hypothetical maximum vertical velocities of 48 m/s and 61 m/s, respectively. The real maximum vertical velocity must be lower because the turbulent mixing of heat, moisture and momentum dampens the updraft. The CIN of this base state is 51 J/kg and must be overcome by the initial bubble. The gridpoint distance is 100 m for this experiment and the model is time integrated for two hours. The parameters for the cloud microphysical scheme are $\tau_{\text{cond}} = \tau_{\text{evap}} = 1\,\text{s}$, $m_{c0} = 1\,\text{g/kg}$ and $\tau_{\text{aut}} = 1000\,\text{s}$.

Figure 6.15 shows snapshots of cloud water (colored shadings), rainwater (blue isolines) and toroidal mass stream function (black isolines). At $t = 20$ minutes the warm air bubble has excited the development of a cumulus cloud as seen by the cloud water field and

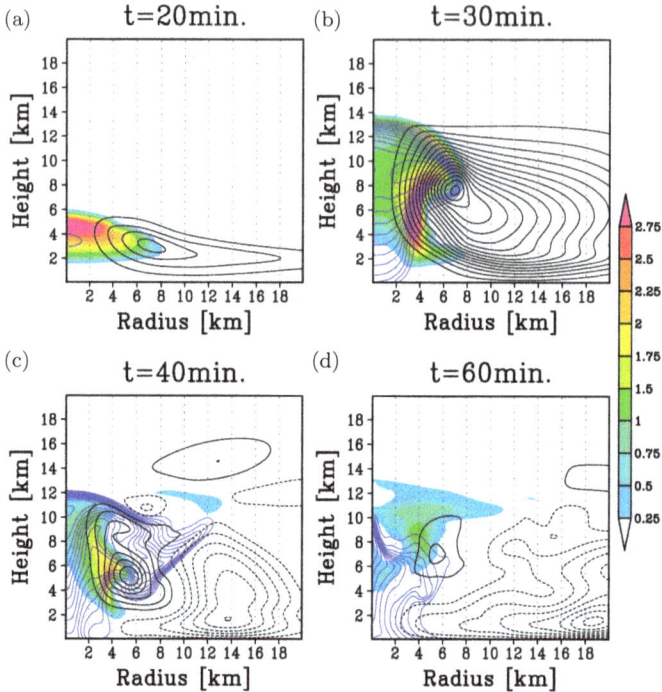

Fig. 6.15. Mass fractions of cloud water (colored shadings, g/kg), rain water (blue isolines, contour interval 1 g/kg) and the toroidal mass stream function (black isolines, contour interval $10^8\,\mathrm{kgm^{-1}s^{-1}}$) at (a) $t = 20$ min., (b) $t = 30$ min, (c) $t = 40$ min and (d) $t = 60$ min of the cumulonimbus cloud simulation. Negative isolines are dashed.

the toroidal overturning. At this time the cloud undergoes the **towering cumulus stage**. The amount of rainwater is very low in this developing cloud but it increases by autoconversion since the cloud water fraction exceeds the critical value 1 g/kg inside of the cloud. At $t = 30$ minutes the cumulonimbus cloud is in its **mature stage** and the vertical velocity attains its maximum value at $29.63\,\mathrm{m/s}$. Now, the cloud extends vertically through the whole troposphere and the overturning circulation has intensified. The cloud has produced a lot of rainwater which reached the surface. The cloud water mass fraction is close to its critical value m_{c0} in the interior of the cloud. Therefore, most of the cloud water has been converted to rain by autoconversion and accretion. At $t = 40$ minutes the **dissipation stage** is initiated. The overturning circulation has weakened and heavy precipitation appears at the surface because the updraft cannot impede the fall of the raindrops. At 8 km radius a downdraft developed due to the high mass density of air with a high rain mass fraction. At $t = 60$ minutes the dissipation stage is advanced since the updraft has vanished completely in the lower part of the troposphere where all rain falls to the surface. Near the tropopause an **anvil dome** developed which can be seen in the cloud water field. In reality, this would be cloud ice because of the low temperatures below the freezing point. The experiment shows that neither the assumptions for moist-adiabat nor for the pseudoadiabat are true. However, the moist-adiabatic ascent appears more realistic since most of the condensed liquid water remains within the cloud until its mature stage.

Figure 6.16 shows the time evolution of hydrological quantities during the life cycle of the simulated cumulonimbus cloud. In the

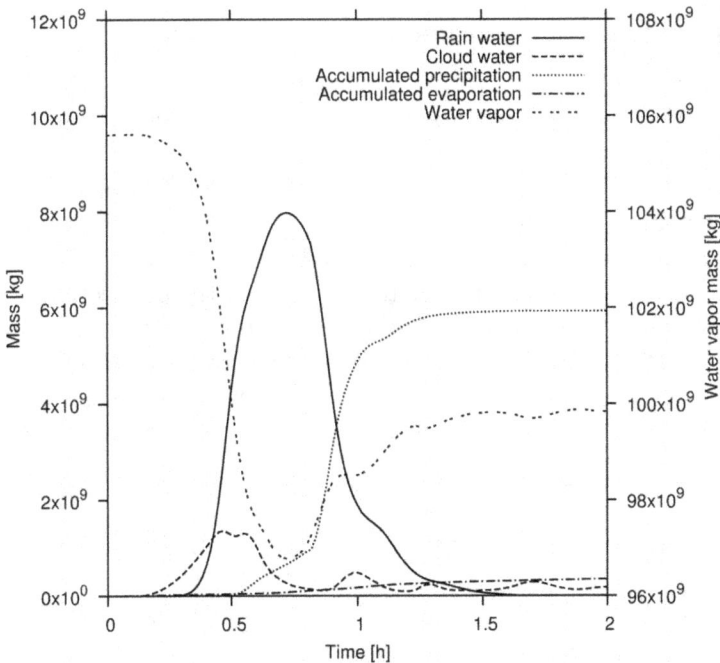

Fig. 6.16. Time evolution of the hydrological cloud properties: Rain water mass (solid line), cloud water mass (dashed line), accumulated mass by precipitation (dotted line), accumulated mass by evaporation at the surface (dot-dashed line) and water vapor mass (double-dashed line, ordinate at the right-hand side) deduced from the cumulonimbus cloud simulation.

early cumulus stage, only cloud water mass is generated by condensation but rain water mass increases shortly afterwards. The initiation of rainwater generation takes place by autoconversion but this process is too slow to explain the rapid increase. Therefore, accretion is the main process for the conversion of cloud water into rain. Water vapor mass decreases by condensation and reaches a minimum at about 40 min. Afterwards, it increases again by rain evaporation. Precipitation starts after half an hour and it becomes very strong in the last 10 minutes of the first simulation hour. Finally, rainwater and precipitation decline due to the dissipation phase of the cloud. Evaporation at the surface increases the water vapor mass. It results from winds at the surface but this process cannot compensate for the water vapor loss due to condensation. Therefore, atmospheric water mass is decreased by the cumulonimbus cloud.

6.4.4 *Convection parameterization schemes*

Weather forecasts and climate models have usually a horizontal resolution that is too coarse to resolve convection.[d] Therefore, convection must be parameterized in order to avoid numerical problems. We will describe briefly two simple schemes. The first one is a **moisture budget scheme** developed by Kuo (1965) and the second one is an **adjustment scheme** developed by Betts and Miller (Betts, 1986; Betts and Miller, 1986). In both cases, the convection scheme only introduces additional tendencies in temperature and specific humidity. Therefore, momentum tendencies due to convective flows are neglected. The modified equations in isobaric coordinates for temperature and specific humidity take after the implementation of such a scheme the form

$$\frac{\partial T}{\partial t} + \mathbf{v}_h \cdot \nabla_h T + \Pi \frac{\partial \theta}{\partial p} \omega = \frac{Q}{c_p} + \frac{Q_C}{c_p} \tag{6.157}$$

$$\frac{\partial m_v}{\partial t} + \mathbf{v}_h \cdot \nabla_h m_v + \omega \frac{\partial m_v}{\partial p} = S_{l,v} + D_{m_v} + S_C \tag{6.158}$$

---

[d]This will change in the future because the gridpoint resolution will increase due to increasing computing power.

where Q_C is the heating and S_C the specific humidity tendency due to convection. The fields predicted by these equations are understood to comprise only the large-scale part of the flow while the convective motions are filtered out by a suitable average (in analogy to the treatment of microturbulence in atmospheric models).

6.4.4.1 *Kuo scheme*

Kuo (1965) considered an atmospheric air column which contains many cumulus cells in the case of deep convection but is small compared to the horizontal scale of the large-scale flow. Then, the change of water vapor mass in this column is proportional to

$$I = -\int_0^{p_s} \nabla_h \cdot (\mathbf{v}_h m_v) \frac{dp}{g} \tag{6.159}$$

Kuo (1965) assumed that deep convection can only take place if SBCAPE and the moisture convergence I are positive. With the latter condition, it is implicitly assumed that large-scale vertical motion is strong enough that the inhibiting CIN is overcome. The moisture supply by I is completely used by forming cumulus clouds in the case of deep convection. These clouds dissolve after its lifecycle and the cloud air mixes with the environment. In the cloud, the profile follows a pseudoadiabat $T_p(p)$. Then, the moisture needed for the cloud becomes:

$$\delta m_{v1} = m_{vs}(T_p, p) - m_v \tag{6.160}$$

However, additional moisture is consumed by the latent heat release which is given by[e]

$$\delta m_{v2} = \frac{c_p}{L_v}(T_p - T) \tag{6.161}$$

This part of the moisture addition falls out as convective precipitation because of the pseudoadiabatic ascent. Consequently, the total moisture mass that must be added turns out to be

$$\delta M_v = \int_{p_t}^{p_l} (\delta m_{v1} + \delta m_{v2}) \frac{dp}{g} \tag{6.162}$$

---

[e]With this equation, it is implicitly assumed that the adiabatic cooling in the updraft is compensated by adiabatic warming of a downdraft in the cloud.

where p_l denotes the lifting condensation level and p_t the EL. The ratio

$$a = \frac{I\Delta t}{\delta M_v} \tag{6.163}$$

can be interpreted as the area fraction occupied by cumulus clouds during the time step Δt. Then, mixing of cloud air with environmental air leads to the following modification of the mean values:

$$T_n = (1-a)T + aT_p = T + a(T_p - T) \tag{6.164}$$

$$m_{vn} = (1-a)m_v + am_{vp} = m_v + a(m_{vp} - m_v) \tag{6.165}$$

By assuming a sufficiently small time step that ensures $a < 1$ we get for the time tendencies due to convection

$$\frac{Q_C}{c_p} = \frac{a}{\Delta t}(T_p - T) \quad \text{for} \quad a < 1 \tag{6.166}$$

$$S_C = \frac{a}{\Delta t}(m_{vp} - m_v) + \mathbf{v}_h \cdot \nabla_h m_v + \omega \frac{\partial m_v}{\partial p} \quad \text{for} \quad a < 1 \tag{6.167}$$

The tendencies are applied for $p_t \leq p \leq p_l$. At other levels, only the correction for water vapor advection must be considered in the water vapor budget equation to fulfill mass consistency.

The Kuo scheme has the advantage that convection is only triggered in the typical case of large-scale moisture convergence. However, the criterion for convection could also be misleading in certain situations. We have already learned that the air must overcome CIN by upward motion but this is not guaranteed despite large-scale moisture convergence. An example is a uniform horizontal flow which embeds a moist anomaly. In this case, moisture convergence appears on the downstream side of the anomaly although the vertical velocity is zero. With this information in mind, it is not difficult to construct an example in which moisture convergence correlates with downward motion. Using positive vertical velocity in the lower troposphere as an additional criterion could resolve this issue. Another problem could be that the temperature below the cloud base remains unaffected by the Kuo scheme. Deep convection can create downdrafts by ice melting and rain evaporation which can penetrate the ground and cool the boundary layer air. Using the surface as the lower boundary in integral (6.162) and applying (6.166) and (6.167) also in the subcloud layer could partly resolve this issue.

6.4.4.2   Betts–Miller scheme

The subsequent description of the Betts–Miller scheme follows closely that given by Frierson (2007). The scheme relaxes temperature and humidity towards a reference profile with the time scale τ_{BM} when conditional instability is present, i.e.

$$\frac{Q_C}{c_p} = \frac{T_r - T}{\tau_{BM}} \tag{6.168}$$

$$S_C = \frac{m_{vr} - m_v}{\tau_{BM}} \tag{6.169}$$

for positive SBCAPE. With these tendencies the associated water mass losses in the column become

$$P_T = \int_{p_t}^{p_s} \frac{c_p}{L_v} \frac{T_r - T}{\tau_{BM}} \frac{dp}{g} \tag{6.170}$$

$$P_{m_v} = -\int_{p_t}^{p_s} \frac{m_{vr} - m_v}{\tau_{BM}} \frac{dp}{g} \tag{6.171}$$

For $P_T > 0$ and $P_{m_v} > 0$ the masses fall out as precipitation. Then, the scheme describes deep convection. Furthermore, it is assumed that convection and subsequent mixing does not change enthalpy. Therefore, the relation

$$\int_{p_t}^{p_s} c_p T_r + L_v m_{vr} \frac{dp}{g} = \int_{p_t}^{p_s} c_p T + L_v m_v \frac{dp}{g} \tag{6.172}$$

or $P_T = P_{m_v}$ must hold. The pseudoadiabat T_p is used as a first guess for T_r. This temperature does, however, not necessarily fulfill (6.172) for a given m_{vr} profile. Then, the reference temperature is corrected by

$$T_r = T_p - \frac{1}{c_p(p_s - p_t)} \int_{p_t}^{p_s} c_p(T_p - T) + L_v(m_{vr} - m_v) \frac{dp}{g} \tag{6.173}$$

Therefore, the reference temperature must not coincide with an adiabat. The difference to the adiabat can be explained by the mixing with the environment after completion of the cumulus life cycle. This is in contrast to the Kuo scheme in which nonconservation of enthalpy is caused by the moisture import due to the large-scale flow.

The case $P_{m_v} < 0$ leads to shallow convection in the Betts–Miller scheme. Then, the atmosphere is too dry to produce deep precipitating convection and a new lower top-level p_{t2} is calculated such that

$$\int_{p_{t2}}^{p_s} \frac{m_{vr} - m_v}{\tau_{BM}} \frac{dp}{g} = 0 \qquad (6.174)$$

Consequently, no precipitation is produced. In this case, enthalpy conservation requires that

$$T_r = T_p - \frac{1}{c_p(p_s - p_{t2})} \int_{p_{t2}}^{p_s} c_p(T_p - T)\frac{dp}{g} \qquad (6.175)$$

The Betts–Miller scheme depends sensitively on the reference profile for moisture. Frierson (2007) made the simple assumption that the specific humidity profile yields a constant specific relative humidity $\mathcal{H}_{BM}$ with respect to the pseudoadiabat T_p.

6.4.4.3 *Comparison of the schemes*

Figure 6.17 shows the temperature and specific humidity changes that result from the two convection parameterization schemes when the initial state of the numerical cumulonimbus simulation presented in Section 6.4.3 is prescribed. The figure also displays the 2-h change of horizontally averaged temperature and specific humidity deduced from the HURMOD simulation. Note that the magnitude of the change depends on the moisture convergence I in the Kuo scheme and on the timescale τ_{BM} in the Betts–Miller scheme. Therefore, the shown high values of the temperature and moisture changes due to the schemes are not representative. We see that the Betts–Miller scheme predicts less warming than the Kuo scheme because of the assumed enthalpy conservation. Both schemes reveal cooling at low levels but the Kuo scheme yields no temperature change below the cloud base. Consequently, the temperature profile becomes discontinuous. This is another reason why it should be better to include also

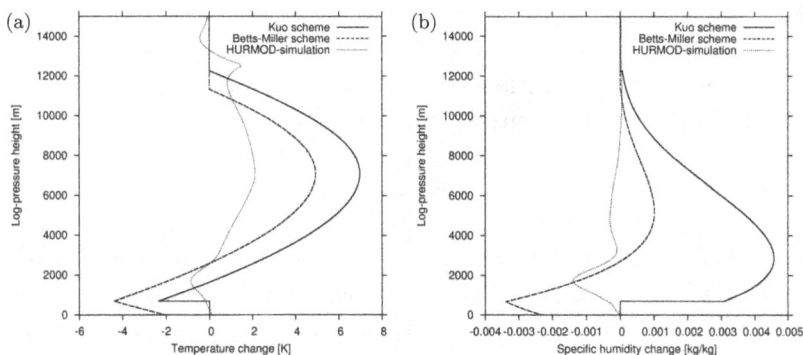

Fig. 6.17. (a) Vertical profiles of 2-h temperature change in K resulting from the Kuo scheme (black solid line), the Betts–Miller scheme (black dashed line) and the HURMOD simulation (dotted line). The area fraction $a = 1$ is assumed for the Kuo scheme while $\tau_{BM} = 2$ h and $\mathcal{H}_{BM} = 0.6$ are used for the determination of change by the Betts–Miller scheme. (b) as in (a) but for the specific humidity change in kg/kg.

changes below the cloud base. HURMOD also simulates warming and cooling at upper and lower levels, respectively but the magnitude is much smaller. The reason is likely that only one cumulonimbus cell was simulated in a cylinder having a diameter of 50 km. The moisture changes of the two schemes deviate significantly from each other. The Kuo scheme reveals only moistening while the Betts–Miller scheme predicts drying at lower levels. Moreover, HURMOD shows drying at all levels so that the Betts–Miller scheme is in better agreement with the simulation. However, it must be noted that the HURMOD experiment starts with an artificial warm bubble at the beginning, and no moisture convergence has been prescribed. Therefore, a direct comparison with the Kuo scheme is misleading. Indeed, without a warm bubble, nothing would happen in HURMOD and this would be consistent with the Kuo scheme because the moisture convergence vanishes. For a more suitable evaluation of the two schemes, one should adopt a weather forecast model and compare the predictions with observations for several characteristic cases.

6.5 Fronts

We learned in Section 5.1 that warm and cold fronts develop during a nonlinear baroclinic wave life cycle. They almost form as discontinuities in the life cycle simulation without irreversible processes being included (see Fig. 5.6(d)). Indeed, in this section, we will present a front solution which evolves into a discontinuity after a finite time. Frontogenesis can already be described by quasigeostrophic equations but there are some important details missing which are investigated in this section. We start with the derivation of a developing front solution in a quasigeostrophic model and use this solution as a basis for the more elaborate **semigeostrophic model**. It yields a more realistic picture of the front. Finally, we present a static front model that also describes features of the jet stream at the polar front.

6.5.1 *Quasigeostrophic frontogenesis*

Frontogenesis happens on the basis of quasigeostrophic equations when the Q-vector (4.102) has the same direction as the horizontal temperature gradient (cf. Section 4.5.4). A characteristic idealized flow that leads to quasigeostrophic frontogenesis is sketched in Fig. 6.18. It is a simple deformation flow described by the geostrophic stream function

$$\psi_g = F_1 xy \tag{6.176}$$

By assuming a y-independent temperature field we obtain for the Q-vector

$$\mathbf{Q} = -\frac{R_d}{p}\frac{\partial u_g}{\partial x}\frac{\partial T}{\partial x}\mathbf{i} - \frac{R_d}{p}\frac{\partial u_g}{\partial y}\frac{\partial T}{\partial x}\mathbf{j} = F_1 \frac{R_d}{p}\frac{\partial T}{\partial x}\mathbf{i} \tag{6.177}$$

Obviously, $\mathbf{Q}$ directs towards the warm air and, therefore, this flow configuration causes frontogenesis. Such a flow pattern appears in extratropical cyclones as seen in Fig. 5.9(b).

We prescribe the irrotational deformation flow (6.176) and consider a frictionless and adiabatic uniform potential vorticity flow on an f-plane (cf. Section 4.6.4). The prescribed flow (6.176) does not represent a steady state solution of the quasigeostrophic system but its slow time-dependency will be ignored in the following which is a common simplification in studies on frontogenesis (e.g. Hoskins and Bretherton, 1972). In the quasigeostrophic model it is more

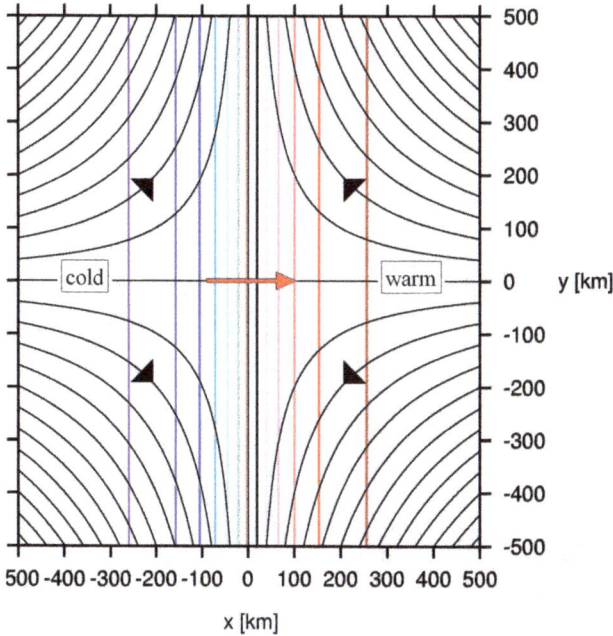

Fig. 6.18. Sketch of a frontogenetic flow showing streamlines (black contours) and isotherms (colored contours). The red arrow at the center displays the direction of the Q-vector.

convenient to work with the specific volume $v = R_d T/p = -\partial\phi/\partial p$ instead of temperature. We prescribe the initial anomaly of specific volume at the upper and lower boundaries by

$$v_b'|_{t=0} = v_t'|_{t=0} = \hat{v} \sum_{j=0}^{\frac{n-1}{2}} \binom{n}{j} \frac{1}{n-2j} \sin[(n-2j)kx] \qquad (6.178)$$

where v_b' and v_t' denote the specific volume anomalies at the pressure levels p_b and p_t, respectively. This sum with odd n is the integral of the expansion (3.80) and it describes a zone with a zonal temperature increase in the vicinity of $x = 0$. There are distant mirror zones with the same profile but opposite gradients. These have no impact on the front evolution at $x = 0$ for a sufficiently large n.

To find the solution to this problem, we consider the temperature tendency equation (4.69) with neglected heating at the lower and upper boundaries

$$\left(\frac{\partial}{\partial t} + \mathbf{v}_g \cdot \nabla_h\right)\left(\frac{\partial \phi'}{\partial p}\right) = -\left(\frac{\partial}{\partial t} + \mathbf{v}_g \cdot \nabla_h\right)v' = 0 \quad \text{at } p = p_b, p_t$$

(6.179)

If the initial temperature field is y-independent, it remains so forever because the zonal geostrophic wind is $u_g = -\partial \psi_g/\partial y = -F_1 x$ and, therefore, also y-independent. In this case, we get

$$\frac{\partial v'}{\partial t} = F_1 x \frac{\partial v'}{\partial x} \quad \text{at } p = p_b, p_t$$

(6.180)

The solution of this equation takes the form

$$v_b'(x, t) = v_t'(x, t) = F(x e^{F_1 t})$$

(6.181)

where $F(x e^{F_1 t})$ is a differentiable function of the argument. By substitution in (6.180), it can be easily checked that (6.181) constitutes a solution. Consequently, a special solution of (6.180) can be obtained when we make the wave number k time-dependent in (6.178) and prescribe it by

$$k(t) = k_0 e^{F_1 t}$$

(6.182)

Uniform potential vorticity requires that

$$\nabla_h^2 \phi' + \frac{f_0^2}{\Xi} \frac{\partial^2 \phi'}{\partial p^2} = 0$$

(6.183)

where the stability parameter Ξ is assumed to be constant (as in Section 4.6.4). Then, the boundary conditions (6.178) yield the geopotential anomaly

$$\phi' = -\hat{v} \sum_{j=0}^{\frac{n-1}{2}} \binom{n}{j} \frac{1}{\Gamma_j(n-2j)} \frac{\sinh(\Gamma_j p')}{\cosh(\Gamma_j \Delta p)} \sin(k_j x)$$

(6.184)

where $k_j = (n - 2j)k, \Gamma_j = k_j \sqrt{\Xi}/f_0, p' = p - (p_b + p_t)/2$ and $\Delta p = (p_b - p_t)/2$.

It is of interest to compute the vertical velocity field of the developing front since this is relevant for the formation of precipitation that often occurs at fronts. We can use the omega equation (4.101) for this purpose but it is more convenient to adopt the temperature tendency equation (4.98) for the case of an already-known solution. With this equation we get

$$\omega = -\frac{1}{\Xi}\left(\frac{\partial}{\partial t} + \mathbf{v}_g \cdot \nabla_h\right)\left(\frac{\partial \phi'}{\partial p}\right) = \frac{F_1 \hat{v}}{\sqrt{\Xi} f_0} k_0 e^{F_1 t} \sum_{j=0}^{\frac{n-1}{2}} \binom{n}{j}$$

$$\times \left(\frac{\sinh(\Gamma_j p')}{\cosh(\Gamma_j \Delta p)} p' - \frac{\cosh(\Gamma_j p')}{\cosh(\Gamma_j \Delta p)} \tanh(\Gamma_j \Delta p) \Delta p\right) \sin(k_j x)$$

(6.185)

The ageostrophic zonal velocity u_a can be deduced from the mass continuity equation (4.62) which justifies to use a stream function ψ_a of the zonal overturning circulation such that

$$u_a = -\frac{\partial \psi_a}{\partial p}, \quad \omega = \frac{\partial \psi_a}{\partial x} \tag{6.186}$$

Due to symmetry, we have $u_a|_{x=-\pi/(2k)} = 0$ and, therefore, the stream function ψ_a becomes

$$\psi_a = \int_{-\pi/(2k)}^{x} \omega \, dx = -\frac{F_1 \hat{v}}{\sqrt{\Xi} f_0} k_0 e^{F_1 t} \sum_{j=0}^{\frac{n-1}{2}} \frac{1}{k_j} \binom{n}{j}$$

$$\times \left(\frac{\sinh(\Gamma_j p')}{\cosh(\Gamma_j \Delta p)} p' - \frac{\cosh(\Gamma_j p')}{\cosh(\Gamma_j \Delta p)} \tanh(\Gamma_j \Delta p) \Delta p\right) \cos(k_j x)$$

(6.187)

Now, it is straightforward to compute u_a using (6.186).

Figure 6.19 shows snapshots of the quasigeostrophic frontogenesis based on the above-mentioned solution. The lines of constant specific volume anomaly converge remarkably at the top and bottom when $F_1 t$ increases from zero to three. On the other hand, the midlevel increase of the zonal gradient happens at a much smaller rate. The reason is that the vertical motion causes adiabatic cooling and warming in warm and cold air, respectively, due to the stable stratification and this yields a reduction of the zonal gradient. The vertical motion field with ascent and descent in warm and cold air, respectively, can be diagnosed with the divergence of

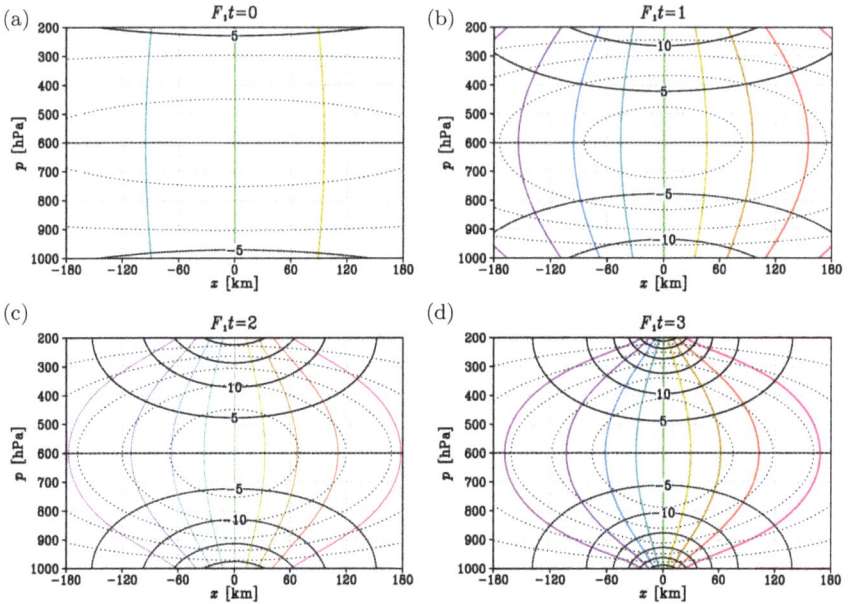

Fig. 6.19. Specific volume anomaly v' (colored isolines, contour interval $0.005\,\mathrm{m^3/kg}$), geostrophic meridional wind anomaly v'_g (black isolines, contour interval $5\,\mathrm{m/s}$) and stream function ψ_a (dotted isolines, contour interval $10{,}000\,\mathrm{Pa\,m/s}$) of the quasigeostrophic frontogenesis solution for (a) $F_1 t = 0$, (b) $F_1 t = 1$, (c) $F_1 t = 2$ and (d) $F_1 t = 3$. The parameters used are $f_0 = 10^{-4}\mathrm{s^{-1}}$, $\Xi = 2 \times 10^{-6}\mathrm{m^2/(Pa^2 s^2)}$, $k_0 = 0.5 \times 10^{-7}\mathrm{m^{-1}}$ and $F_1 = 1.15 \times 10^{-5}\mathrm{s^{-1}}$. The parameter $\hat{v}$ has been calculated such that the v' difference across the front becomes $0.05\,\mathrm{m^3/kg}$.

the Q-vector (6.177) (cf. Eq. (4.112)). The ageostrophic zonal overturning circulation results from the geostrophic adjustment of the geostrophic meridional wind field which is characterized by northward and southward jets at the top and bottom of the developing front, respectively. With increasing time, the jets strengthen and become narrower. At observed fronts, only one jet appears in the upper troposphere which is seen in Fig. 2.5(b). The absence of the jet at the lower boundary results from the surface drag, which is neglected in the present quasigeostrophic model. The frontal zone in the quasigeostrophic solution has no tilt with respect to the vertical direction. This is in contrast with observations where the warm air moves above the cold air leading to a tilt of the front (see Fig. 2.5(b)). This shortcoming stems obviously from the neglection of advection by the ageostrophic flow in the quasigeostrophic equations. The advection by the overturning circulation seen in Fig. 6.19 would cause the missing tilt in the results. This issue will be resolved with a more elaborated equation set that will be introduced in the following.

6.5.2 *Geostrophic momentum approximation and semigeostrophic frontogenesis*

The flow along fronts and jet streams has small curvature but the gradients in cross-stream direction can be very large. The geostrophic balance is still proper in such a flow although the quasigeostrophic approximation is not valid because the small cross-stream scale results in a large Rossby number. The **geostrophic momentum approximation** introduced by Hoskins (1975) takes these aspects into account and yields a suitable equation set for the dynamics of fronts and jet streams. To derive this approximation, we realign the Cartesian coordinates in such a way that the y-axis corresponds to the tangent along the curved streamline described by the geostrophic wind (see Fig. 6.20). Consequently, the geostrophic wind has only a

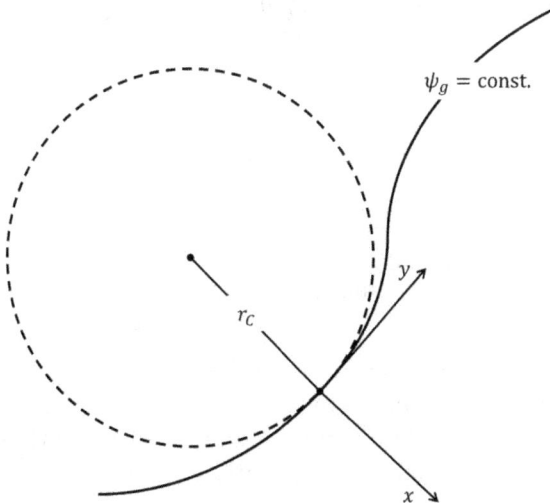

Fig. 6.20. Sketch of a geostrophic streamline and the local Cartesian coordinate system in which the y-axis parallels the streamline at the selected point. The dashed circle with a radius r_C has the same curvature as the streamline at the selected point.

meridional component. Let r_c be the local curvature radius of the streamline. Then, it is suitable for fronts and jet streams that the Rossby number calculated by $U/(f_0 \, r_C)$ is still very small because of the small curvature. However, the radial and vertical gradients of the geostrophic wind and other quantities can be very large so that the associated scale L results in a large Rossby number. Such a scaling gives rise to the following nondimensional momentum equations on the f-plane:

$$\mathrm{Ro}\left(\frac{\partial \tilde{u}}{\partial \tilde{t}} + \tilde{v}\frac{\partial \tilde{u}}{\partial \tilde{y}}\right) + \tilde{u}\frac{\partial \tilde{u}}{\partial \tilde{x}} + \tilde{\omega}\frac{\partial \tilde{u}}{\partial \tilde{p}} - \tilde{v} = -\frac{\partial \tilde{\phi}'}{\partial \tilde{x}} + \mathrm{Ro}\tilde{f}_{T,x} \quad (6.188)$$

$$\mathrm{Ro}\left(\frac{\partial \tilde{v}}{\partial \tilde{t}} + \tilde{v}\frac{\partial \tilde{v}}{\partial \tilde{y}}\right) + \tilde{u}\frac{\partial \tilde{v}}{\partial \tilde{x}} + \tilde{\omega}\frac{\partial \tilde{v}}{\partial \tilde{p}} + \tilde{u} = \mathrm{Ro}\tilde{f}_{T,y} \quad (6.189)$$

The scaling in these equations is based on $\tilde{x} = \mathrm{Ro}\mathrm{L}x, \mathrm{W} = \mathrm{H}\mathrm{U}/(\mathrm{Ro}\mathrm{L}), \tilde{\phi}' = \phi'/(f_0\mathrm{Ro}\mathrm{L}\mathrm{U})$ instead of those adopted in Section 4.3. These equations should be compared with Eq. (4.45) used for deriving the quasigeostrophic approximation. Inserting the expansions (4.51)–(4.53) yields in the zeroth order

$$\tilde{u}^{(0)}\frac{\partial \tilde{u}^{(0)}}{\partial \tilde{x}} + \tilde{\omega}^{(0)}\frac{\partial \tilde{u}^{(0)}}{\partial \tilde{p}} - \tilde{v}^{(0)} = -\frac{\partial \tilde{\phi}'^{(0)}}{\partial \tilde{x}} \quad (6.190)$$

$$\tilde{u}^{(0)}\frac{\partial \tilde{v}^{(0)}}{\partial \tilde{x}} + \tilde{\omega}^{(0)}\frac{\partial \tilde{v}^{(0)}}{\partial \tilde{p}} + \tilde{u}^{(0)} = 0 \quad (6.191)$$

The only realistic solution is

$$\tilde{u}^{(0)} = \tilde{\omega}^{(0)} = 0, \quad \tilde{v}^{(0)} = \frac{\partial \tilde{\phi}'^{(0)}}{\partial \tilde{x}} \quad (6.192)$$

since the ageostrophic wind does not have a magnitude that is comparable to the geostrophic wind, even at fronts. Furthermore, it is not clear if it is possible to find a useful solution with nonzero $\tilde{u}^{(0)}$ and $\tilde{\omega}^{(0)}$ that conforms to the continuity equation (4.38). For the first order, we get

$$\frac{\partial \tilde{v}^{(0)}}{\partial \tilde{t}} + \tilde{v}^{(0)}\frac{\partial \tilde{v}^{(0)}}{\partial \tilde{y}} + \tilde{u}^{(1)}\frac{\partial \tilde{v}^{(0)}}{\partial \tilde{x}} + \tilde{\omega}^{(1)}\frac{\partial \tilde{v}^{(0)}}{\partial \tilde{p}} + \tilde{u}^{(1)} = \tilde{f}_{T,y}^{(0)} \quad (6.193)$$

The higher orders can be neglected if Ro $\ll$ 1.[f] All other primitive equations do not suffer any simplifications and, therefore, the geostrophic momentum approximation yields the following dimensioned equations

$$\mathbf{v}_g = \frac{1}{f_0}\mathbf{k} \times \nabla_h \phi'$$ (6.194)

$$\frac{\partial \mathbf{v}_g}{\partial t} + \mathbf{v}_h \cdot \nabla_h \mathbf{v}_g + \omega \frac{\partial \mathbf{v}_g}{\partial p} + f_0 \mathbf{k} \times \mathbf{v}_h = -\nabla_h \phi' + \mathbf{f}_T$$ (6.195)

$$\nabla_h \cdot \mathbf{v}_a + \frac{\partial \omega}{\partial p} = 0$$ (6.196)

$$\frac{\partial}{\partial t}\left(\frac{\partial \phi'}{\partial p}\right) + \mathbf{v}_h \cdot \nabla_h\left(\frac{\partial \phi'}{\partial p}\right) - \Pi \frac{R_d}{p}\frac{\partial \theta}{\partial p}\omega = \frac{R_d Q}{p c_p}$$ (6.197)

We see that the additional consideration of advection by the ageostrophic and vertical velocity components constitutes the only difference to the quasigeostrophic equations (4.60)–(4.63). In the following, turbulent exchange and diabatic warming are neglected as in the previous subsection.

Now, we derive the **semigeostrophic equations** by conducting a coordinate transformation to a **semigeostrophic space**. We assume for simplicity a y-independent front embedded in a deformation flow as in the previous subsection and also analyzed by Hoskins and Bretherton (1972). We will briefly consider the full transformation without a prescribed symmetry later.

In the assumed deformation flow, the solution takes the form

$$u = -F_1 x + u_a(x, p, t), \quad v = F_1 y + v_g{'}(x, p, t)$$ (6.198)

To get a y-independent meridional wind anomaly v_g', the geopotential must become

$$\phi = \phi_0(p) + f_0 F_1 xy - \frac{1}{2}F_1{}^2 y^2 + \phi'(x, p, t)$$ (6.199)

---

[f] $\tilde{v}^{(1)}$ is possibly nonzero but it has no impact on $\tilde{v}^{(0)}$ due to the assumed axis orientation.

Then, the meridional component of the approximated momentum equation (6.195) turns into

$$\frac{\partial v'_g}{\partial t} + u\frac{\partial v'_g}{\partial x} + \omega\frac{\partial v'_g}{\partial p} + F_1 v_g' + f_0 u_a = 0 \qquad (6.200)$$

The **geostrophic coordinates** used for the transformation are given by

$$x_g = x + \frac{v_g'}{f_0}, \quad p_g = p, \quad t_g = t \qquad (6.201)$$

The new x-coordinate fulfills the following relation:

$$\frac{Dx_g}{Dt_g} = \frac{Dx}{Dt} + \frac{1}{f_0}\frac{Dv_g'}{Dt} = u_a - F_1 x - F_1\frac{v_g'}{f_0} - u_a = -F_1 x_g \qquad (6.202)$$

because the material time derivative does not depend on the used coordinate system, i.e.

$$\frac{D}{Dt} = \frac{D}{Dt_g} = \frac{\partial}{\partial t_g} + \frac{Dx_g}{Dt_g}\frac{\partial}{\partial x_g} + \frac{Dp_g}{Dt_g}\frac{\partial}{\partial p_g}$$

$$= \frac{\partial}{\partial t_g} - F_1 x_g\frac{\partial}{\partial x_g} + \omega\frac{\partial}{\partial p_g} \qquad (6.203)$$

We get for the transformation of the spatial derivatives

$$\begin{pmatrix}\dfrac{\partial}{\partial x}\\[2ex]\dfrac{\partial}{\partial p}\end{pmatrix} = \begin{pmatrix}\dfrac{\partial x_g}{\partial x} & \dfrac{\partial p_g}{\partial x}\\[2ex]\dfrac{\partial x_g}{\partial p} & \dfrac{\partial p_g}{\partial p}\end{pmatrix}\cdot\begin{pmatrix}\dfrac{\partial}{\partial x_g}\\[2ex]\dfrac{\partial}{\partial p_g}\end{pmatrix} = \begin{pmatrix}1+\dfrac{1}{f_0}\dfrac{\partial v_g'}{\partial x} & 0\\[2ex]\dfrac{1}{f_0}\dfrac{\partial v_g'}{\partial p} & 1\end{pmatrix}\cdot\begin{pmatrix}\dfrac{\partial}{\partial x_g}\\[2ex]\dfrac{\partial}{\partial p_g}\end{pmatrix}$$

$$(6.204)$$

Let **A** be the transformation matrix, then the inverse becomes

$$\mathbf{A}^{-1} = \frac{1}{\det(\mathbf{A})}\begin{pmatrix}a_{22} & -a_{12}\\ -a_{21} & a_{11}\end{pmatrix} = \frac{1}{J}\begin{pmatrix}1 & 0\\[2ex] -\dfrac{1}{f_0}\dfrac{\partial v_g'}{\partial p} & J\end{pmatrix} \qquad (6.205)$$

where J denotes the determinant of the transformation matrix $\mathbf{A}$. It is proportional to geostrophic vorticity because

$$J = 1 + \frac{1}{f_0} \frac{\partial v_g'}{\partial x} = \frac{\zeta_g}{f_0} \qquad (6.206)$$

With the definition

$$\phi_g \equiv \phi' + \frac{1}{2} v_{g'}^2 \qquad (6.207)$$

we calculate

$$\frac{\partial \phi_g}{\partial x_g} = \frac{1}{J} \frac{\partial \phi_g}{\partial x} = \frac{1}{J} \left(\frac{\partial \phi'}{\partial x} + v_g' \frac{\partial v_g'}{\partial x} \right)$$

$$= \frac{1}{J} \left(\frac{\partial \phi'}{\partial x} + \frac{1}{f_0} \frac{\partial \phi'}{\partial x} \frac{\partial v_g'}{\partial x} \right) = \frac{\partial \phi'}{\partial x} \qquad (6.208)$$

$$\frac{\partial \phi_g}{\partial p_g} = -\frac{1}{J f_0} \frac{\partial v_g'}{\partial p} \frac{\partial \phi_g}{\partial x} + \frac{\partial \phi_g}{\partial p}$$

$$= -\frac{1}{J f_0} \frac{\partial v_g'}{\partial p} \left(\frac{\partial \phi'}{\partial x} + v_g' \frac{\partial v_g'}{\partial x} \right) + \frac{\partial \phi'}{\partial p} + v_g' \frac{\partial v_g'}{\partial p}$$

$$= -\frac{1}{f_0} \frac{\partial v_g'}{\partial p} \frac{\partial \phi'}{\partial x} + \frac{\partial \phi'}{\partial p} + \frac{1}{f_0} \frac{\partial v_g'}{\partial p} \frac{\partial \phi'}{\partial x} = \frac{\partial \phi'}{\partial p} \qquad (6.209)$$

Therefore, the geostrophic and hydrostatic balance equations in semigeostrophic space simply become

$$v_g' = \frac{1}{f_0} \frac{\partial \phi_g}{\partial x_g}, \frac{\partial \phi_g}{\partial p_g} = -\frac{R_d T'}{p} \qquad (6.210)$$

The meridional momentum equation (6.200) can be written as

$$\frac{D v_g'}{D t_g} = \frac{\partial v_g'}{\partial t_g} - F_1 x_g \frac{\partial v_g'}{\partial x_g} + \omega \frac{\partial v_g'}{\partial p_g} = -F_1 v_g' - f_0 u_a \qquad (6.211)$$

With the modified ageostrophic velocity

$$u_{a,sg} = u_a + \frac{\omega}{f_0} \frac{\partial v_g'}{\partial p_g} \qquad (6.212)$$

Equation (6.211) takes the form

$$\frac{\partial v_g'}{\partial t_g} - F_1 x_g \frac{\partial v_g'}{\partial x_g} = -F_1 v_g' - f_0 u_{a,sg} \tag{6.213}$$

It is more elaborate to derive the continuity equation in semi-geostrophic space. First, we take the zonal derivative of $u_{a,sg}$

$$\frac{\partial u_{a,sg}}{\partial x_g} = \frac{\partial u_a}{\partial x_g} + \frac{\omega}{f_0} \frac{\partial^2 v_g'}{\partial p_g \partial x_g} + \frac{1}{f_0} \frac{\partial \omega}{\partial x_g} \frac{\partial v_g'}{\partial p_g} \tag{6.214}$$

Furthermore, we get

$$\frac{\partial u_a}{\partial x_g} = \frac{1}{J} \frac{\partial u_a}{\partial x} = -\frac{1}{J} \frac{\partial \omega}{\partial p} = -\frac{1}{J} \left(\frac{\partial \omega}{\partial p_g} + \frac{1}{f_0} \frac{\partial v_g'}{\partial p} \frac{\partial \omega}{\partial x_g} \right) \tag{6.215}$$

and

$$\frac{1}{f_0} \frac{\partial v_g'}{\partial p_g} = \frac{1}{f_0} \left(\frac{\partial v_g'}{\partial p} - \frac{1}{f_0 J} \frac{\partial v_g'}{\partial p} \frac{\partial v_g'}{\partial x} \right)$$

$$= \frac{1}{f_0} \frac{\partial v_g'}{\partial p} \left(1 - \frac{J-1}{J} \right) = \frac{1}{f_0 J} \frac{\partial v_g'}{\partial p} \tag{6.216}$$

Inserting these results in (6.214) yields

$$\frac{\partial u_{a,sg}}{\partial x_g} = -\frac{1}{J} \left(\frac{\partial \omega}{\partial p_g} + \frac{1}{f_0} \frac{\partial v_g'}{\partial p} \frac{\partial \omega}{\partial x_g} \right) + \frac{\omega}{f_0} \frac{\partial^2 v_g'}{\partial p_g \partial x_g} + \frac{1}{f_0 J} \frac{\partial \omega}{\partial x_g} \frac{\partial v_g'}{\partial p}$$

$$= -\frac{1}{J} \frac{\partial \omega}{\partial p_g} + \frac{\omega}{f_0} \frac{\partial}{\partial p_g} \left(\frac{1}{J} \frac{\partial v_g'}{\partial x} \right) = -\frac{1}{J} \frac{\partial \omega}{\partial p_g} + \omega \frac{\partial}{\partial p_g} \left(\frac{J-1}{J} \right)$$

$$= -\frac{1}{J} \frac{\partial \omega}{\partial p_g} - \omega \frac{\partial}{\partial p_g} \left(\frac{1}{J} \right) = -\frac{\partial}{\partial p_g} \left(\frac{\omega}{J} \right) \tag{6.217}$$

With the modified vertical velocity

$$\omega_{a,sg} = \frac{\omega}{J} \tag{6.218}$$

the transformed continuity equation becomes

$$\frac{\partial u_{a,sg}}{\partial x_g} + \frac{\partial \omega_{a,sg}}{\partial p_g} = 0 \tag{6.219}$$

Now, we can write the semigeostrophic equations for the y-independent frontogenesis problem as

$$v'_g = \frac{1}{f_0}\frac{\partial \phi_g}{\partial x_g} \tag{6.220}$$

$$\frac{\partial v'_g}{\partial t_g} - F_1 x_g \frac{\partial v'_g}{\partial x_g} = -F_1 v'_g - f_0 u_{a,sg} \tag{6.221}$$

$$\frac{\partial u_{a,sg}}{\partial x_g} + \frac{\partial \omega_{a,sg}}{\partial p_g} = 0 \tag{6.222}$$

$$\frac{\partial}{\partial t_g}\left(\frac{\partial \phi_g}{\partial p_g}\right) - F_1 x_g \frac{\partial}{\partial x_g}\left(\frac{\partial \phi_g}{\partial p_g}\right) - \omega_{a,sg} J \left(\frac{p_g}{p_0}\right)^\kappa \frac{R_d}{p_g}\frac{\partial \theta}{\partial p_g} = 0 \tag{6.223}$$

These equations are very similar to the quasigeostrophic equations except that they are defined in semigeostrophic space. However, the vertical advection term in the thermodynamic equation is different which makes it more complicated to find analytical solutions. A simple solution can be found by assuming **zero potential vorticity**. To do this we must first see how the potential vorticity is defined in geostrophic coordinates. This is indeed simply proportional to $J\partial\theta/\partial p_g$ since

$$J\frac{\partial \theta}{\partial p_g} = -\frac{1}{f_0}\frac{\partial v'_g}{\partial p}\frac{\partial \theta}{\partial x} + \frac{\zeta_g}{f_0}\frac{\partial \theta}{\partial p}$$

$$= -\frac{\zeta_g \cdot \left(\frac{\partial \theta}{\partial x}\mathbf{i} + \frac{\partial \theta}{\partial z_p}\mathbf{k}\right)}{g f_0 \rho_p} = -\frac{P_{\theta p}}{f_0 g} \tag{6.224}$$

where $P_{\theta p}$ is the potential vorticity of the geostrophic flow in pseudo-height coordinates (cf. Section 2.4.3). It can be shown that the geostrophic momentum equations conserve $P_{\theta p}$ under adiabatic and frictionless conditions (see Hoskins and Bretherton, 1972).

With (6.224) we can write the thermodynamic equation as

$$\frac{\partial}{\partial t_g}\left(\frac{\partial \phi_g}{\partial p_g}\right) - F_1 x_g \frac{\partial}{\partial x_g}\left(\frac{\partial \phi_g}{\partial p_g}\right) + w_{a,sg}\left(\frac{p_g}{p_0}\right)^{\kappa}\frac{R_d}{gf_0 p_g}P_{\theta p} = 0 \quad (6.225)$$

Simple solutions as in the previous subsection are not possible because of the height-dependent factor in front of $P_{\theta p}$. Hoskins and Bretherton (1972) circumvented this problem by making a Boussinesq approximation which, however, is only valid for fronts with small vertical extent compared to the troposphere height. The simple special case $\partial\theta/\partial p_g = 0$ leads to zero potential vorticity. Then, the semigeostrophic equations are identical to the quasigeostrophic ones for $\Xi = 0$ except that they are defined in semigeostrophic space. In this special case the potential temperature anomaly θ' does not depend on p_g. A corresponding specific volume anomaly describing a developing front as in the previous subsection becomes

$$v' = \hat{v}\left(\frac{p_g}{p_0}\right)^{\kappa-1}\sum_{j=0}^{\frac{n-1}{2}}\binom{n}{j}\frac{1}{n-2j}\sin(k_j x_g) \quad (6.226)$$

and solves Eq. (6.223) in the case $P_{\theta p} = 0$. The semigeostrophic geopotential is obtained by vertical integration of (6.210)

$$\phi_g = \left[\phi_c - \frac{\hat{v}p_0}{\kappa}\left(\frac{p_g}{p_0}\right)^{\kappa}\right]\sum_{j=0}^{\frac{n-1}{2}}\binom{n}{j}\frac{1}{n-2j}\sin(k_j x_g) \quad (6.227)$$

where ϕ_c denotes a constant of integration. With the geopotential we can calculate the geostrophic meridional wind anomaly

$$v'_g = \frac{1}{f_0}\frac{\partial \phi_g}{\partial x_g} = \frac{1}{f_0}k_0 e^{F_1 t_g}$$

$$\left[\phi_c - \frac{\hat{v}p_0}{\kappa}\left(\frac{p_g}{p_0}\right)^{\kappa}\right]\sum_{j=0}^{\frac{n-1}{2}}\binom{n}{j}\cos(k_j x_g) \quad (6.228)$$

To calculate the transformed overturning circulation, we solve the meridional momentum equation (6.213) for $u_{a,sg}$:

$$u_{a,sg} = -\frac{1}{f_0}\left(F_1 v'_g + \frac{\partial v'_g}{\partial t_g} - F_1 x_g \frac{\partial v'_g}{\partial x_g}\right) = -2\frac{F_1}{f_0}v'_g \quad (6.229)$$

where the second identity results from (6.228). The stream function of this overturning circulation can be calculated by vertical integration of (6.229) which gives

$$\psi_{a,sg} = 2\frac{F_1 k_0 e^{F_1 t_g}}{f_0^2} \left[(p_g - p_c)\phi_c - \frac{\hat{v} p_0 p_g}{\kappa^2 + \kappa} \left(\frac{p_g}{p_0}\right)^{\kappa} \right]$$

$$\times \sum_{j=0}^{\frac{n-1}{2}} \binom{n}{j} \cos(k_j x_g) \tag{6.230}$$

where p_c is another constant of integration. The constants ϕ_c and p_c can be determined by prescribing the boundary condition $\psi_{a,sg} = 0$ at $p_g = p_b$ and $p_g = p_t$ which is left as an exercise for the reader. Equations (6.226), (6.227), (6.228) and (6.230) fully describe the semigeostrophic solution which is identical to the quasigeostrophic one if the coordinates were physical. Of interest is also to determine the semigeostrophic solution in physical space. For this purpose, we must replace the geostrophic by the physical coordinates using the transformation rules (6.201) and evaluating the relations

$$\phi' = \phi_g - \frac{1}{2}v_g'^2, \quad u_a = u_{a,sg} - \frac{J\omega_{a,sg}}{f_0}\frac{\partial v_g'}{\partial p_g}, \quad \omega = J\omega_{a,sg} \tag{6.231}$$

First, we get J in geostrophic coordinates by applying $\partial/\partial x_g$ to v_g'

$$\frac{\partial v_g'}{\partial x_g} = \frac{1}{J}\frac{\partial v_g'}{\partial x} = \frac{f_0(J-1)}{J} \leftrightarrow J = \frac{f_0}{f_0 - \frac{\partial v_g'}{\partial x_g}} \tag{6.232}$$

The next issue is to express x_g in terms of physical coordinates. This is, however, not possible because

$$x_g(x,p,t) = x + \frac{v_g'[x_g(x,p,t),p,t]}{f_0} \tag{6.233}$$

cannot be solved for x_g in our case. Therefore, the transformation must be conducted by numerical interpolation.

Figure 6.21 shows the evolution of the semigeostrophic solution at several time points in semigeostrophic space (left panel) and physical space (right panel). Note that the solution in semigeostrophic space corresponds to the quasigeostrophic solution of the same problem.

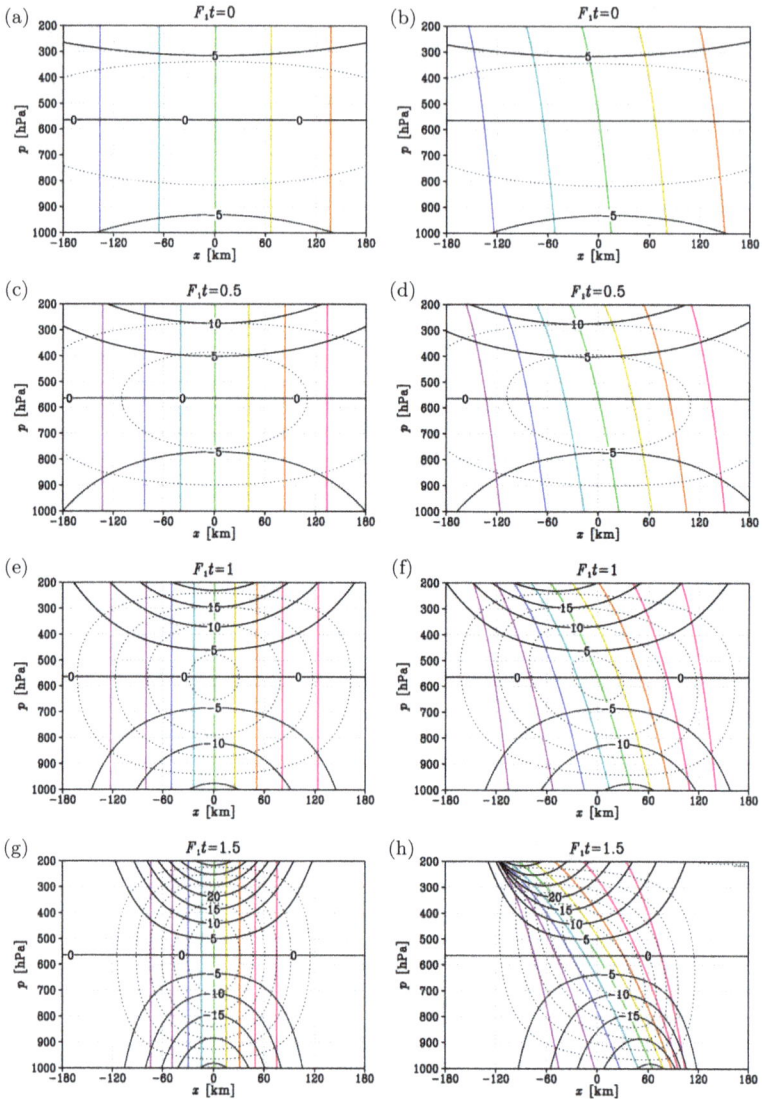

Fig. 6.21. Potential temperature anomaly θ' (colored isolines, contour interval 1 K), geostrophic meridional wind anomaly v'_g (black isolines, contour interval 5 m/s) and stream function ψ_a (dotted isolines, contour interval 10,000 Pa m/s) of the semigeostrophic frontogenesis solution at different time points. The left panels (a,c,e,g) show the fields in semigeostrophic space and the right panels (b,d,f,h) the corresponding fields in physical space. The solution parameters are $f_0 = 10^{-4} \text{s}^{-1}$, $\Xi = 0$, $k_0 = 0.675 \times 10^{-7} \text{m}^{-1}$ and $F_1 = 1.15 \times 10^{-5} \text{s}^{-1}$. The parameter $\hat{v}$ has been calculated such that the θ' the difference across the front becomes 10 K.

The isolines of potential temperature run in the vertical direction in the quasigeostrophic case due to the vanishing static stability parameter Ξ. The converging movement of the isentropes is accompanied with jet formation at the upper and lower boundaries. The upper jet is somewhat stronger than the lower jet because the mean density decreases with height. The semigeostrophic solution in physical space clearly reveals the impact of advection by the ageostrophic flow. The isolines of potential temperature slant so that warm air moves above cold air. Therefore, the static stability becomes increasingly positive in the semigeostrophic solution. The upper and lower jets move west- and eastward, respectively, which is understandable because of the x-coordinate transformation in (6.201). In the last snap-shot in Fig. 6.21, zonal gradients of specific volume and meridional wind take extremely large values at the upper boundary. Somewhat later, a **frontal collapse** occurs, which is characterized by the occurrence of an infinite gradient. Afterward, the solution turns out to be unphysical because a unique transformation from semigeostrophic to physical space is not possible anymore. A frontal collapse appears when

$$\frac{\partial v'_g}{\partial x_g} = f_0 \qquad (6.234)$$

is fulfilled somewhere in the fluid. Then, the vorticity becomes infinite as seen by inspection of (6.232). In a real fluid, this cannot take place because mixing by turbulence or molecular diffusion prevents the formation of a discontinuity. However, it is very complicated to include such processes in the semigeostrophic model and, therefore, we did not consider them. A frontal collapse can also happen at the lower boundary but at a later timepoint because of the slight vertical asymmetry.

The semigeostrophic equations without prescribed y–independence are obtained from the transformation with the following geostrophic coordinates:

$$x_g = x + \frac{v_g}{f_0}, \quad y_g = y - \frac{u_g}{f_0}, \quad p_g = p, \quad t_g = t \qquad (6.235)$$

Obviously, they fulfill the relations

$$\frac{Dx_g}{Dt_g} = u + \frac{1}{f_0}\frac{Dv_g}{Dt} = u_g, \quad \frac{Dy_g}{Dt_g} = v - \frac{1}{f_0}\frac{Du_g}{Dt} = v_g \qquad (6.236)$$

With the modified geopotential

$$\phi_g \equiv \phi' + \frac{1}{2}(u_g{}^2 + v_g{}^2) \tag{6.237}$$

it can be shown that

$$\frac{\partial \phi_g}{\partial x_g} = \frac{\partial \phi'}{\partial x}, \quad \frac{\partial \phi_g}{\partial y_g} = \frac{\partial \phi'}{\partial y}, \quad \frac{\partial \phi_g}{\partial p_g} = \frac{\partial \phi'}{\partial p} \tag{6.238}$$

For the proof of these relations and a full derivation of the semi-geostrophic equations, we refer to Hoskins (1975) who made use of the Boussinesq approximation. The resulting equations turn out to be very similar to the quasigeostrophic equations in the uniform potential vorticity case except for an additional nonlinear term in the definition of PV.

The advantage of the semigeostrophic equations is that a nearly quasigeostrophic flow results and the transformation to physical space reveals the effect of ageostrophic advection. Figure 6.22 illustrates the impact of this transformation on a simulated midlatitude cyclone. We see that by the transformation

- the area of lows shrinks and that of highs increases,
- warm- and cold fronts become sharper,
- geostrophic streamlines attain a more pronounced kink in an anti-clockwise direction at the fronts (high cyclonic vorticity),
- the warm sector shrinks (occlusion process).

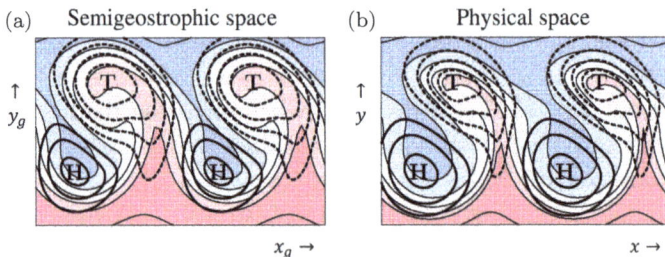

Fig. 6.22. Illustration of the impact of the transformation from (a) semi-geostrophic space to (b) physical space for the example of extratropical highs and lows. The colored shadings and isolines show temperature and geopotential at the lower boundary, respectively. The markers H and T indicate the centers of the highs and lows, respectively.

Therefore, the transformation introduces features that are well-known to synoptic weather analysts and are not reproducible in a quasigeostrophic model. However, not all aspects of a cyclone-anticyclone asymmetry are revealed by the transformation. The central pressure of cyclones and anticyclones remains unaffected because at the centers geostrophic wind vanishes (cf. Eq. (6.237)). Therefore, the semigeostrophic model cannot explain the development of deeper lows as seen in a primitive equation model (cf. Fig. 5.9). Snyder *et al.* (1991) revealed several deficiencies of the semigeostrophic equations and Rotunno *et al.* (2000) introduced a next order correction of the quasigeostrophic equations (QG^{+1}) which compares much better with the primitive equations.

6.5.3 *Stationary front models*

Margules (1906) found a steady-state solution for a sloped atmospheric temperature front. It fulfills the Euler equations for an f-plane geometry and hydrostatic as well as geostrophic balance are assumed. In general, these assumptions yield the thermal wind relation (4.86). A steady-state solution only results if the wind is directed parallel to the isotherms. Therefore, the flow must be parallel and this means that the wind direction is uniform. Furthermore, Margules assumed a temperature discontinuity at the front that draws the angle α with the horizontal surface plane (see Fig. 6.23). At the

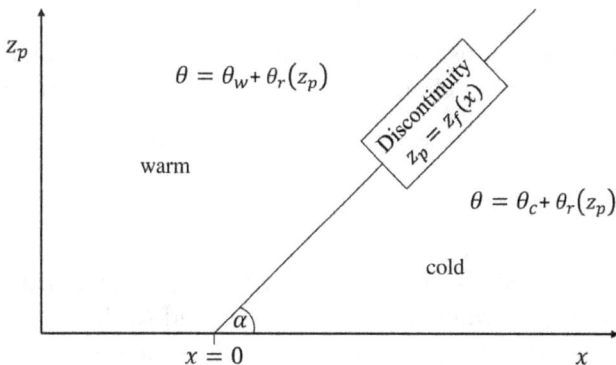

Fig. 6.23. Margules model of an atmospheric front. Two layers are separated by a temperature discontinuity that forms a sloped plane. The potential temperature θ changes abruptly by $\Delta\theta = \theta_w - \theta_c$ at the discontinuity.

same pseudoheight level z_p the potential temperature in the upper layer is higher than in the lower layer by the amount $\Delta\theta = \theta_w - \theta_c$. In each layer, the potential temperature is uniform except for a height-dependent reference potential temperature $\theta_r(p)$. The geostrophic and hydrostatic balance equations become for a y-independent front

$$v = \frac{1}{f_0}\frac{\partial\phi}{\partial x_g}, \quad g\frac{\theta}{\theta_0} = \frac{\partial\phi}{\partial z_p} \tag{6.239}$$

where for the vertical coordinate the pseudo-height z_p has been chosen (cf. Section 2.4.3). By vertical integration of the hydrostatic balance equation we get for $x > 0$:

$$\phi = \begin{cases} \phi_s(x) + g\dfrac{\theta_c}{\theta_0}z_p + \dfrac{g}{\theta_0}\displaystyle\int_0^{z_p}\theta_r dz_p' & \text{for } z_p \leq z_f \\[3mm] \phi_s(x) + g\dfrac{\theta_c}{\theta_0}z_f + g\dfrac{\theta_w}{\theta_0}[z_p - z_f] + \dfrac{g}{\theta_0}\displaystyle\int_0^{z_p}\theta_r dz_p' & \text{for } z_p > z_f \end{cases} \tag{6.240}$$

where $z_f = \tan(\alpha)\,x$ is the height of the temperature discontinuity. With this result the meridional wind becomes

$$v = \begin{cases} \dfrac{1}{f_0}\dfrac{\partial\phi_s}{\partial x} & \text{for } z_p \leq z_f \\[3mm] \dfrac{1}{f_0}\dfrac{\partial\phi_s}{\partial x} - g\dfrac{\Delta\theta}{f_0\theta_0}\tan(\alpha) & \text{for } z_p > z_f \end{cases} \tag{6.241}$$

Obviously, there is also a discontinuity in the meridional wind at $z_p = z_f$ and the wind change $\Delta v = v_w - v_c$ is related to the frontal slope $\tan(\alpha)$ by

$$\tan(\alpha) = -\frac{f_0\theta_0}{g}\frac{\Delta v}{\Delta\theta} \tag{6.242}$$

When the warm air is above the cold ($\alpha > 0$), the meridional wind decreases abruptly when moving from the cold to the warm air mass. Then, a cyclonic wind jump ($\Delta v/\Delta x > 0$) occurs at the front. This is the usual case since the warm air moves above the cold air during frontogenesis (see previous subsection). Moreover, a front with cold air above warm air is convectively unstable which would lead to vigorous mixing and disorganization of the front. Typical values of $\tan(\alpha)$ range from 1:50 to 1:300 (Kurz, 1990).

The front model by Margules cannot describe a jet stream that often appears at fronts. The assumption of a discontinuity that divides the atmosphere into two layers must be exaggerated and instead, a continuous temperature field with large gradients at the frontal zone should be prescribed in a more realistic front model.[g] With the potential temperature field

$$\theta = -\frac{\Delta\theta}{\pi} \tan^{-1}\left[\frac{2}{L}(x - \cot\alpha\, z_p)\right] + \theta_r(z_p) \tag{6.243}$$

we obtain a **slanted frontal zone** in which the gradients are high but finite. Indeed, it approaches the potential temperature field of Margules' front model in the limit $L \to 0$. By the thermal wind relation, we get for the vertical gradient of meridional wind

$$\frac{\partial v}{\partial z_p} = \frac{g}{f_0\theta_0}\frac{\partial\theta}{\partial x} = -\frac{g\Delta\theta}{\pi f_0\theta_0}\frac{2/L}{1 + \frac{4}{L^2}(x - \cot\alpha\, z_p)^2} \tag{6.244}$$

and vertical integration yields

$$
\begin{aligned}
v &= v_s(x) - \int_0^{z_p} \frac{g\Delta\theta}{\pi f_0\theta_0}\frac{2/L}{1 + \frac{4}{L^2}(x - \cot\alpha\, z_p')^2}\,dz_p' \\
&= v_s(x) + \frac{g}{f_0\theta_0}\frac{\Delta\theta}{\pi\cot\alpha}\left\{\tan^{-1}\left[\frac{2}{L}(x - \cot\alpha\, z_p)\right] - \tan^{-1}\left[\frac{2}{L}x\right]\right\}
\end{aligned}
\tag{6.245}
$$

The wind solution is determined except for the integration constant $v_s(x)$ that represents the surface wind. Figure 6.24 shows several cases for the choice of $v_s(x)$. With $v_s(x) = 0$ the surface wind vanishes and a northerly jet stream maximum appears at the top of the model layer (Fig. 6.24(a)). However, it is also possible to find a solution in which the wind at the top of the model ($z_p = H$) vanishes. Then, a southerly jet stream maximum arises at the surface

---

[g]Note that discontinuities only evolve at the upper and lower boundaries in the semigeostrophic model described in the previous subsection and this is not in disagreement with the development of jet streams.

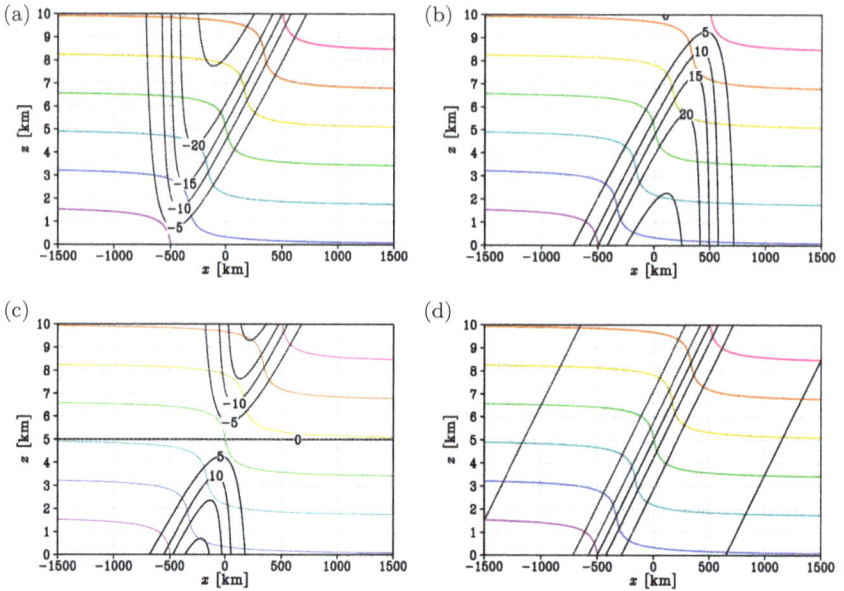

Fig. 6.24.   Potential temperature (colored isolines, contour interval 5 K) and meridional wind (black isolines, contour interval 5 m/s) of the stationary front solution for (a) $v(x,0) = 0$, (b) $v(x,H) = 0$, (c) $v(x,H/2) = 0$ and (d) $v \propto \tan^{-1}[2/L(x-\cot\alpha\, z_p)]$. Negative isolines are dashed. The solution parameters are $f_0 = 10^{-4}\mathrm{s}^{-1}$, $\Delta\theta = 5\,\mathrm{K}$, $\theta_0 = 300\,\mathrm{K}$, $\tan\alpha = 0.01$, $L = 300\,\mathrm{km}$ and $H = 10\,\mathrm{km}$.

(Fig. 6.24(b)). If the wind disappears at mid-level, two jet streams with opposite signs appear, one at the model top and the other at the surface (Fig. 6.24(c)). Finally, Fig. 6.24(d) displays the case for $v_s(x) = g\Delta\theta/(f_0\theta_0\,\pi\cot\alpha)\tan^{-1}(2x/L)$. Then, no jet stream occurs anymore and the meridional wind has only enhanced vertical and cyclonic shear in the slanted frontal zone. The cross-section through an observed jet stream at the polar front (Fig. 2.5(b)) reveals a picture that agrees most with the case of zero surface wind (Fig. 6.24(a)). On the other hand, frontogenesis as described in the previous subsection yields a flow that corresponds to the case with two opposed jet streams (Fig. 6.24(c)). The lower boundary jet stream appears because surface friction is absent in this model. Inclusion of an underlying Ekman layer would highly reduce the amplitude of this jet and would yield a picture that lies between Figs. 6.24(a) and 6.24(c). The cases in Figs. 6.24(b) and 6.24(d) are unrealistic and do not appear in observations.

6.6   Tropical Cyclones

Tropical cyclones are very intense weather systems and form over tropical oceans. They are nearly axisymmetric and have a warm core with a cloud-free **eye** in the center. Figure 6.25 shows a satellite picture of a mature tropical cyclone. We see clearly the eye in the center and the surrounding air reveals a dense and nearly axisymmetric cloud. Spiral bands appear at a larger distance from the center. The typical diameter of a tropical cyclone is about 800 km (Dean *et al.*, 2009). The scale appears to fall into the synoptic range but the quasigeostrophic approximation is invalid because of the large Rossby number. Furthermore, the energetic inner core of a tropical cyclone has a much smaller scale. Before making theoretical considerations, we describe some empirical facts gathered from observations. Then, a suitable approximation of the governing equations is introduced which will be used to analyze **tropical cyclogenesis** within idealized linear and nonlinear models. Finally, a steady state solution is derived to see how much intensity tropical cyclones can attain.

Fig. 6.25.   Satellite picture of hurricane Dennis (2005) near peak intensity on July 10, 2005.
Source: https://en.wikipedia.org/wiki/File:Dennis_2005-07-10_0410Z.jpg.

6.6.1 *Empirical facts*

Tropical cyclones are referred to as **hurricanes** in the Atlantic and northeastern Pacific and **typhoons** in the northwestern Pacific. The designation is simply tropical cyclone in the Southern Hemisphere and the Northern Indian Ocean. A tropical cyclone must have a maximum near-surface wind speed of more than 33 m/s. For smaller wind speeds the designations **tropical storm** (18–33 m/s) and **tropical depression** (less than 18 m/s) are used. About 85 tropical storms develop per year while only about 50 of them become eventually tropical cyclones (Webster *et al.*, 2005). Tropical cyclones are quite rare compared to extratropical cyclones, of which about 150 develop per month (Hodges *et al.*, 2011).

Tropical cyclones migrate with the larger scale flow but their tracks are also affected by the β-effect causing a poleward **beta drift**. They decay after making **landfall** where they are responsible for fatalities and a lot of damages due to the high wind speed and flooding. Sometimes, a tropical cyclone does not meet land and moves to higher latitudes. It transforms into an extratropical cyclone when they reach the polar front. This process is called **extratropical transition**.

Gray (1979) introduced the **seasonal genesis parameter** (SGP) that is a seasonal index for the frequency of tropical cyclones in a certain region. It takes the form

$$
SGP = C_G |f| \left(\frac{\xi_{950}}{10^{-6}\text{s}^{-1}} \frac{f}{|f|} + 5 \right) \left(|\mathbf{v}_{h200} - \mathbf{v}_{h950}| + 3\frac{\text{m}}{\text{s}} \right)^{-1}
$$

$$
\times (T_{ml} - 299.15K)(\theta_{e1000} - \theta_{e500} + 5K) \frac{\mathcal{H}_{700} + \mathcal{H}_{500} - 0.8}{0.6}
$$

$$
(6.246)
$$

where T_{ml} is the mean temperature of the upper 60 m of the ocean (**mixed layer**), $C_G = 4000 \text{ m/K}^2$ a scaling factor and the number in the subindex indicates the pressure level of the respective field quantity. This parameter should yield the number of tropical cyclogenesis events of 20 consecutive seasons within a 5° longitude-latitude box and Gray (1979) found a remarkable agreement for the July-August-September seasons 1958–1977. The SGP parameter is the product of six factors which already hint at the cyclogenesis mechanism. The SGP is set to zero whenever one factor becomes negative. Therefore, tropical cyclogenesis becomes impossible if

1. the Coriolis parameter vanishes,
2. the relative vorticity at 950 hPa is negative (positive) and less (larger) than $-5 \cdot 10^{-6} \mathrm{s}^{-1} (5 \cdot 10^{-6} \mathrm{s}^{-1})$ in the northern (southern) hemisphere,
3. the mean mixed layer temperature is below $26°\mathrm{C}$,
4. the equivalent potential temperature difference between 1000 hPa and 500 hPa is below $-5\,\mathrm{K}$,
5. the mean relative humidity $(\mathcal{H}_{700} + \mathcal{H}_{500})/2$ is below 40%.

Criterion 4 indicates that the atmospheric stability with respect to deep moist convection is low or absent. Indeed, tropical cyclones always originate from **convective cloud clusters**. The atmosphere must be sufficiently moist according to criterion 5 which states that convection must produce a minimum amount of precipitation and latent heating following the concept of the Kuo scheme (cf. Section 6.4.4.1). Criterion 1 states that tropical cyclogenesis is impossible at the equator. It appears that the Coriolis force is essential for the formation by inducing cyclonic rotation in converging flows. Criterion 2 hints at a supporting role of Ekman pumping that lifts low-level air and could trigger deep convection. Furthermore, additional rotation due to positive vorticity may help to form the tropical storm. Criterion 3 indicates that heat fluxes from the ocean are important for tropical cyclogenesis. Finally, the thermal wind $\mathbf{v}_{h200} - \mathbf{v}_{h950}$ reduces the number of tropical cyclones being generated. The shear tilts the vertical axis of the tropical disturbance and this tends to obstruct the intensification of the low. This is in contrast to the formation of extratropical cyclones for which a significant unidirectional thermal wind is essential. Figure 6.26 shows the SGP for the seasons July–August–September 2005 (JAS 2005) and January–February–March 2005 (JFM 2005) together with the significant tracks of tropical lows observed during these seasons. We see that nearly all tracks originate at a location where the SGP is nonzero. The typical track starts at a low latitude with westward migration. Later, the cyclone makes a poleward turn. Afterwards, it makes landfall, dissipates or it reaches the mid-latitude where the movement becomes eastward and an extratropical transition takes place. Most of the tropical cyclones appear in the late summer season (JAS for the Northern Hemisphere and JFM for the Southern

(a)
JAS 2005

☐ Tropical Depression   ☐ Tropical Storm   ☐ Tropical Cyclone

(b)
JFM 2005

☐ Tropical Depression   ☐ Tropical Storm   ☐ Tropical Cyclone

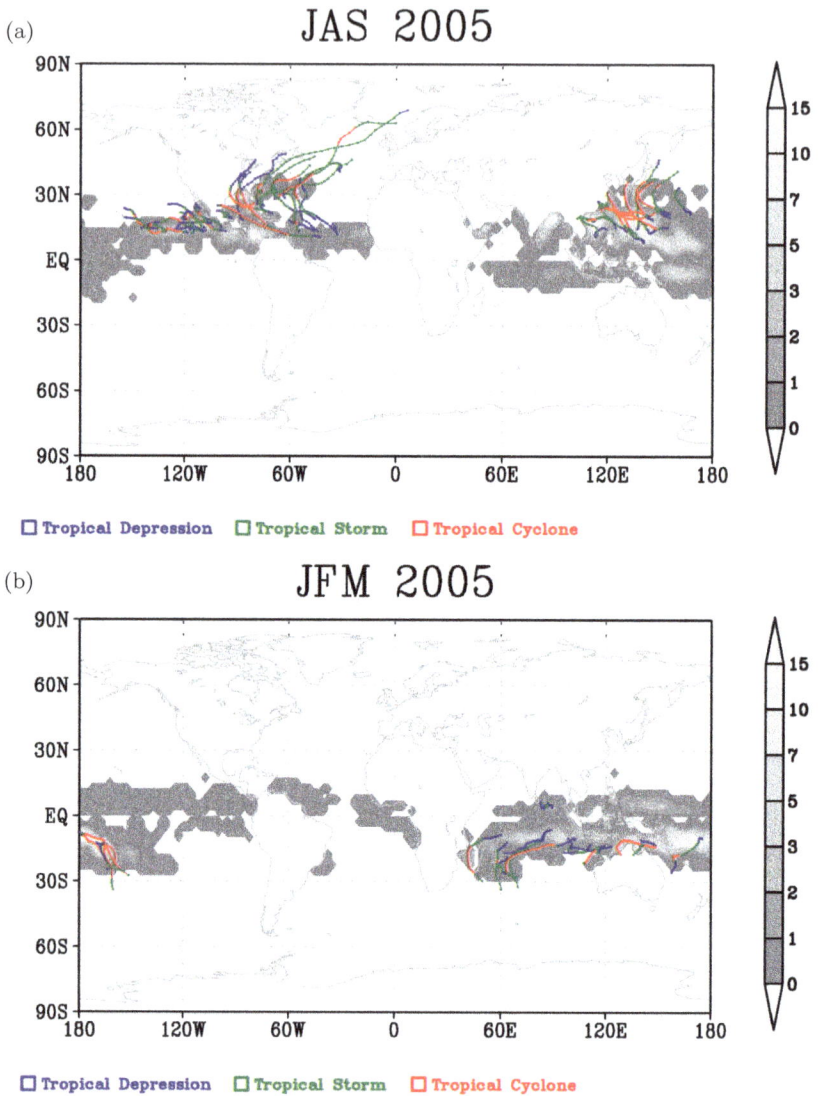

Fig. 6.26.   Seasonal genesis parameter (grey shadings) deduced from NCEP-reanalysis and tropical cyclone tracks for the seasons (a) JAS 2005 and (b) JFM 2005. The cyclone track data from the West Pacific, Northern Indian Ocean and Southern Hemisphere were provided by the Joint Typhoon Warning Center (https://www.metoc.navy.mil/jtwc/jtwc.html?best-tracks) and those from the North Atlantic and East Pacific stem from the National Hurricane Center (https://www.nhc.noaa.gov/data/).

Hemisphere). JAS 2005 was a very active season with 27 tropical cyclones. In this season, 15 lows just reach tropical storm intensity while only three remain tropical depressions. JFM 2005 reveals 10 tropical cyclones and nine tropical storms in the Southern Hemisphere. However, one tropical storm and one tropical depression developed in the Northern Indian Ocean although it is the winter season. In the Southern Hemisphere fewer tropical cyclones are observed because of the lower sea surface temperature. No tropical cyclones develop in the South Atlantic at all.

The SGP cannot tell us where tropical cyclogenesis actually takes place. Tropical cyclones form usually in the troughs of larger scale tropical waves (**easterly waves** in the North-Atlantic and **monsoon troughs** in the North West Pacific). In these troughs, the circulation is already cyclonic and the deformation is low. Figure 6.27 shows snapshots temperature at 400 hPa, streamlines at 825 hPa and precipitation during the formation of hurricane Dennis (2005). On the 2nd July, there is a trough at 45 W which has a closed cyclonic circulation. It is an easterly wave trough that originated in the **African easterly jet**. Such a wave is generated by internal jet instability according to the criterion of Charney and Stern (Burpee, 1972). At this stage, the perturbation is nearly barotropic and no organized convection is visible. Two days later the core of the trough warmed and precipitation starts. Shortly thereafter, a tropical depression develops. After another two days it becomes a tropical storm for the first time. The core has a higher temperature and the streamlines reveal a closed cyclonic circulation. Finally, on 8th July, the cyclone becomes a fully developed hurricane with near-surface wind speeds up to 120 knots. Note that hurricane Cindy (2005) and tropical storm Dora (2005) formed during this period south of Yucatán Peninsula and in the North East Pacific at 15N, respectively. They also develop from pre-existing lows with little precipitation and no warm core. The fields shown stem from ERA-5 reanalysis data (Hersbach *et al.*, 2020) which have a higher resolution compared to NCEP reanalysis but which are still too coarse to resolve the high wind speed core of a tropical cyclone.

6.6.2 *Balance approximation*

It is reasonable to use cylindrical coordinates for the theoretical analysis of tropical cyclones because the inner core is nearly circular

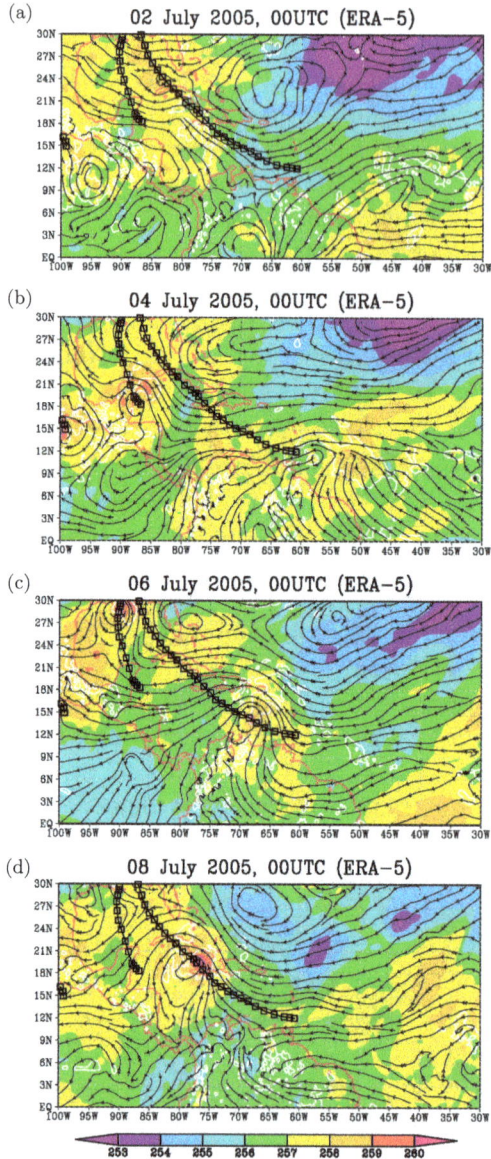

Fig. 6.27.   Temperature at 400 hPa (colored shadings, K), precipitation (the white isoline mark the 1 mm/h contour) and streamlines at 825 hPa on (a) July 2, 2005, (b) July 4, 2005, (c) July 6, 2005 and (d) July 8, 2005. The data are derived from ERA-5 reanalysis. The connected black squares show the cyclone tracks of Hurricane Dennis (2005), Hurricane Cindy (2005) and Tropical Storm Dora (2005) (from right to left).

symmetric. As an idealization we assume that no basic flow and no β-effect exist since they break the circular symmetry. To derive a suitable approximation, we perform a scale analysis following Shapiro and Willoughby (1982) and assume that the tangential wind has a larger magnitude than the radial wind. This assumption is valid for the tropical cyclone flow above the boundary layer. However, in the boundary layer the radial wind can also be very strong due to the frictional inflow. For the horizontal wind vector, we will use the notation $\mathbf{v}_h = u\mathbf{e}_r + v\mathbf{e}_\alpha$. With this notation, the primitive equations on an f-plane become in cylindrical and isobaric coordinates

$$\frac{\partial u}{\partial t} + u\frac{\partial u}{\partial r} + \frac{v}{r}\frac{\partial u}{\partial \alpha} + \omega\frac{\partial u}{\partial p} - \left(f_0 + \frac{v}{r}\right)v = -\frac{\partial \phi'}{\partial r} + f_{T,r} \qquad (6.247)$$

$$\frac{\partial v}{\partial t} + u\frac{\partial v}{\partial r} + \frac{v}{r}\frac{\partial v}{\partial \alpha} + \omega\frac{\partial v}{\partial p} + \left(f_0 + \frac{v}{r}\right)u = -\frac{1}{r}\frac{\partial \phi'}{\partial \alpha} + f_{T,\alpha} \qquad (6.248)$$

$$\frac{\partial \phi'}{\partial p} = -\frac{R_d}{p}\Pi\theta' \qquad (6.249)$$

$$\frac{1}{r}\frac{\partial}{\partial r}(ru) + \frac{1}{r}\frac{\partial v}{\partial \alpha} + \frac{\partial \omega}{\partial p} = 0 \qquad (6.250)$$

$$\frac{\partial \theta'}{\partial t} + u\frac{\partial \theta'}{\partial r} + \frac{v}{r}\frac{\partial \theta'}{\partial \alpha} + \omega\frac{\partial \theta'}{\partial p} + \frac{\partial \theta_r}{\partial p}\omega = \frac{Q}{c_p\Pi} \qquad (6.251)$$

For the scale analysis, we dimensionalize the variables as follows:

$$\tilde{r} = \frac{r}{L}, \quad \tilde{p} = \frac{p}{\Delta p}, \quad \tilde{u} = \frac{u}{U}, \quad \tilde{v} = \frac{v}{V}, \quad \tilde{\omega} = \frac{\omega}{\Delta pU/L},$$

$$\tilde{f}_{T,r} = \frac{f_{T,r}}{U^2/L}, \quad \tilde{f}_{T,\alpha} = \frac{f_{T,\alpha}}{UV/L}$$

$$\tilde{t} = \frac{t}{L/U}, \quad \tilde{\phi}' = \frac{\phi'}{V^2}, \quad \tilde{\theta}' = \frac{\theta'}{V^2/R_d}, \quad \tilde{Q} = \frac{R_dL}{UV^2}\frac{Q}{c_p} \qquad (6.252)$$

and we assume that the ratio

$$\epsilon = \frac{U}{V} \qquad (6.253)$$

is a small number while the Rossby number Ro $= \mathrm{V}/(f_0\mathrm{L})$ is of the order of unity. Using an expansion as in (6.16) we get for the zeroth order

$$\left(\frac{1}{\mathrm{Ro}} + \frac{v^{(0)}}{\tilde{r}}\right) v^{(0)} = \frac{\partial \phi'^{(0)}}{\partial \tilde{r}} \tag{6.254}$$

$$\frac{v^{(0)}}{\tilde{r}} \frac{\partial v^{(0)}}{\partial \alpha} = -\frac{1}{\tilde{r}} \frac{\partial \phi'^{(0)}}{\partial \alpha} \tag{6.255}$$

$$\frac{\partial \phi'^{(0)}}{\partial \tilde{p}} = -\Pi \frac{\theta'^{(0)}}{\tilde{p}} \tag{6.256}$$

$$\frac{1}{\tilde{r}} \frac{\partial v^{(0)}}{\partial \alpha} = 0 \tag{6.257}$$

Consequently, the zeroth order tangential wind, geopotential and temperature fields are axisymmetric. The first-order equations are

$$\left(\frac{\partial}{\partial \tilde{t}} + u^{(0)}\frac{\partial}{\partial \tilde{r}} + w^{(0)}\frac{\partial}{\partial \tilde{p}}\right) v^{(0)} + \left(\frac{1}{\mathrm{Ro}} + \frac{v^{(0)}}{\tilde{r}}\right) u^{(0)}$$

$$= -\frac{1}{\tilde{r}} \frac{\partial \phi'^{(1)}}{\partial \alpha} + \tilde{f}_{T,\alpha} \tag{6.258}$$

$$\frac{1}{\tilde{r}} \frac{\partial}{\partial r}(\tilde{r} u^{(0)}) + \frac{\partial w^{(0)}}{\partial \tilde{p}} = 0 \tag{6.259}$$

$$\left(\frac{\partial}{\partial \tilde{t}} + u^{(0)}\frac{\partial}{\partial \tilde{r}} + w^{(0)}\frac{\partial}{\partial \tilde{p}}\right) \theta'^{(0)} + \frac{v^{(0)}}{r}\frac{\partial \theta'^{(1)}}{\partial \alpha} + \frac{\partial \tilde{\theta}_r}{\partial \tilde{p}} w^{(0)} = \frac{\tilde{Q}}{\Pi}$$

$$\tag{6.260}$$

These equations may include some fields which are not axisymmetric. However, we can eliminate them by taking the azimuthal average

$$[\cdot\cdot]_\alpha \equiv \frac{1}{2\pi} \int_0^{2\pi} \cdot\cdot \, d\alpha \tag{6.261}$$

since these fields do not appear in the form of nonlinear products. With the new notations

$$u = U[u^{(0)}]_\alpha, \quad v_{gr} = Vv^{(0)}, \quad \omega = \frac{\Delta p}{L} U[\omega^{(0)}]_\alpha, \quad \phi' = V^2 \phi'^{(0)},$$

$$\theta' = V^2/R_d\, \theta'^{(0)}, \quad Q = c_p \frac{UV^2}{R_d L} [\tilde{Q}]_\alpha \qquad (6.262)$$

we obtain the **axisymmetric balance equations**

$$\left(f_0 + \frac{v_{gr}}{r}\right) v_{gr} = \frac{\partial \phi'}{\partial r} \qquad (6.263)$$

$$\frac{\partial v_{gr}}{\partial t} + u\frac{\partial v_{gr}}{\partial r} + \omega\frac{\partial v_{gr}}{\partial p} + \left(f_0 + \frac{v_{gr}}{r}\right) u = f_{T,\alpha} \qquad (6.264)$$

$$\frac{\partial \phi'}{\partial p} = -\Pi \frac{R_d\theta'}{p} \qquad (6.265)$$

$$\frac{1}{r}\frac{\partial}{\partial r}(ru) + \frac{\partial \omega}{\partial p} = 0 \qquad (6.266)$$

$$\frac{\partial \theta'}{\partial t} + u\frac{\partial \theta'}{\partial r} + \omega\frac{\partial \theta'}{\partial p} + \frac{\partial \theta_r}{\partial p}\omega = \frac{Q}{c_p\Pi} \qquad (6.267)$$

For the derivation, we have assumed that $\mathrm{Ro} = o(1)$ and $\partial\tilde{\theta}_r/\partial\tilde{p} = o(1)$. However, both of them have larger magnitudes in tropical cyclones. Considering a large Ro would lead to the omission of the Coriolis force which on the other hand has importance in the outer part of the cyclone where $\mathrm{Ro} \leq o(1)$. A large $\partial\tilde{\theta}_r/\partial\tilde{p}$ would yield a zeroth order balance between adiabatic cooling and latent heat release like in (6.44). In the inner part of the tropical cyclone this is reasonable because clouds are present. Then, adding zeroth and first-order contributions yields an equation which is identical to (6.267). With the same reasoning, we can treat the budget equation for water vapor (2.74). After collecting first-order terms it becomes

$$\frac{\partial m_v}{\partial t} + u\frac{\partial m_v}{\partial r} + \omega\frac{\partial m_v}{\partial p} = S_{l,v} + D_{m_v} \qquad (6.268)$$

where m_v is here the zeroth order and axisymmetric part of the specific humidity field and the terms on the right-hand side are azimuthally averaged sources and sinks.

The approximation resembles Eqs. (6.194)–(6.197) resulting from the geostrophic momentum approximation (cf. Section 6.5.2). The main difference is that in the latter the geostrophic wind balance is assumed while in the former the **gradient wind balance** equation (6.263) is evaluated. It describes the force balance between Coriolis, centrifugal and pressure gradient forces. From this balance, we get for the **gradient wind**

$$v_{gr} = -\frac{f_0 r}{2} \pm \sqrt{\frac{f_0^2 r^2}{4} + r\frac{\partial \phi'}{\partial r}} = -\frac{f_0 r}{2} \pm \sqrt{\frac{f_0^2 r^2}{4} + f_0 r v_g} \quad (6.269)$$

where v_g denotes the geostrophic tangential wind.

The solutions are sketched in Fig. 6.28, which show the following results for gradient wind balanced flows: (i) There are two solutions for a low of which only the one with the plus sign in front of the square root is realistic and, therefore, named regular (Fig. 6.28(a)). The other solution is anomalous which, if at all, rarely occurs as it describes a low-pressure regime with an anticyclonic vortex in which the wind speed becomes very high because the centrifugal

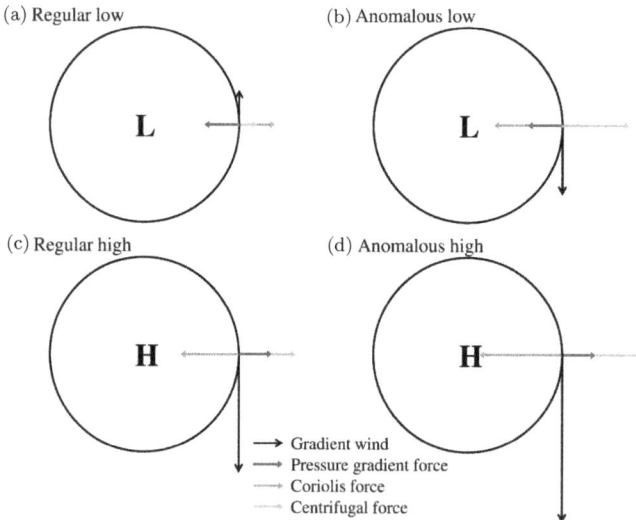

Fig. 6.28.   Sketch of the gradient wind balance in (a) a regular low, (b) anomalous low, (c) regular high and (d) anomalous high.

force must balance the pressure gradient and Coriolis forces (see Fig. 6.28(b)). (ii) A high-pressure system (see Fig. 6.28(c) and 6.28(d)) also has two gradient wind solutions both of which describe an anticyclonic circulation. Highs cannot be in gradient wind balance if the geostrophic wind speed exceeds $f_0 r/4$, because then the square root in (6.269) becomes imaginary. That is, in a vortex of 50 km radius a maximum geostrophic wind of 0.625 m/s is allowed for $f_0 = 0.5 \times 10^{-4} \text{s}^{-1}$. The condition $v_g > -f_0 r/4$ may play a role in the inner warm core of a tropical cyclone (100 km diameter). It has no significant high-pressure system at the top due to the impossible gradient wind balance. Thus, the maximum gradient wind occurs at the surface, which is one reason why tropical cyclones have so much destructive power. At large radii the regular gradient wind solution (Fig. 6.28(c)) approaches the geostrophic tangential wind v_g and the balance approximation becomes equivalent to the geostrophic momentum approximation. The anomalous high solution (Fig. 6.28(d)) does not approach the geostrophic wind at large radii and is usually inertially unstable. Therefore, it does not appear in reality.

6.6.3 *Conditional instability of the second kind*

To date, there is no accepted concise theory for tropical cyclogenesis. On the other hand, it is possible to simulate the complete life cycle of a tropical cyclone in much detail with state-of-the-art numerical models. Indeed, it could be true that the genesis process can be modeled but cannot be described adequately with a concise theory. Nevertheless, we will examine in the following an "old" simple theory that has some weaknesses. The discussion of these weaknesses hints at some additional processes which are essential during tropical cyclogenesis and are considered in the subsequent subsection.

The theory of the **conditional instability of the second kind (CISK)** was introduced by Charney and Eliassen (1964). The idea was that convection organizes into a large-scale cloud cluster which induces large-scale instability by the collective effect of the cumulus clouds. The boundary layer plays an important role because it provides the necessary moisture convergence and Ekman pumping in the developing cyclone. The heating by the released convection leads in turn to a surface pressure fall and a subsequent increase of moisture convergence which closes the feedback loop.

To describe this instability in a model we linearize the balance equations (6.263)–(6.267) with respect to an atmosphere at rest. Furthermore, we neglect turbulent exchange and make the two-level approximation as in Section 4.6.1. Then, we obtain the following linear equation set:

$$v'_{gr1} = \frac{1}{f_0}\frac{\partial \phi'_1}{\partial r}, \quad v'_{gr3} = \frac{1}{f_0}\frac{\partial \phi'_3}{\partial r} \tag{6.270}$$

$$\frac{\partial v'_{gr1}}{\partial t} + f_0 u'_1 = 0, \quad \frac{\partial v'_{gr3}}{\partial t} + f_0 u'_3 = 0 \tag{6.271}$$

$$\frac{1}{r}\frac{\partial}{\partial r}(ru'_1) + \frac{\omega'_2}{\Delta p} = 0, \quad \frac{1}{r}\frac{\partial}{\partial r}(ru'_3) + \frac{\omega'_4 - \omega'_2}{\Delta p} = 0 \tag{6.272}$$

$$\frac{\partial}{\partial t}\left(\frac{\phi'_1 - \phi'_3}{\Delta p}\right) = \Xi_2\omega'_2 + \frac{R_d Q_2}{p_2 c_p} \tag{6.273}$$

The index denotes the considered equidistant pressure levels as indicated in Fig. 4.9, but the pressure values are different because (i) a boundary layer is introduced below level 4 and (ii) the upper boundary is the tropopause where vanishing vertical velocity is assumed. Typical values for the pressure levels could be 100, 300, 500, 700 and 900 hPa. We see that the equations are quasigeostrophic because all nonquasigeostrophic terms vanished by linearization. Therefore, they are at best valid in the first growth stage of a tropical cyclone. The vertical velocity ω'_4 is nonzero since we couple the model with an underlying boundary layer. The linearized momentum equations in this layer are

$$\frac{\partial u'_b}{\partial t} - f_0(v'_b - v'_{gr3}) = -\frac{u'_b}{\tau_b}, \quad \frac{\partial v'_b}{\partial t} + f_0 u'_b = -\frac{v'_b}{\tau_b} \tag{6.274}$$

where τ_b is the timescale for Rayleigh friction that describes the effect of surface drag in a simple way. In these equations, we have assumed that the pressure gradient is identical to that at level 3. Eliminating the boundary tangential wind yields

$$\left[\left(\frac{\partial}{\partial t} + \frac{1}{\tau_b}\right)^2 + f_0^2\right] u'_b = -f_0\left(\frac{\partial}{\partial t} + \frac{1}{\tau_b}\right) v'_{gr3}, \tag{6.275}$$

The boundary layer does not include the balance approximation because here it is inappropriate. However, in the original CISK model, it is assumed and time tendencies are neglected. Then, $v_b' = v_{gr3}'$ and

$$u_b' = -\frac{v_{gr3}'}{f_0 \tau_b} \tag{6.276}$$

hold. At the upper boundary of the boundary layer, we get approximately for the vertical velocity

$$\omega_4' = \frac{\Delta p_b}{r} \frac{\partial}{\partial r} (r u_b') \tag{6.277}$$

where Δp_b denotes the pressure difference between the bottom and top of the boundary layer.

Finally, we have to express the latent heating Q_2 in terms of other variables. The model cannot resolve individual convection cells and, therefore, convection must be parameterized. Here, we adopt the Kuo scheme as this conforms to the CISK mechanism. Indeed, it can be suspected that Kuo was inspired by the study of Charney and Eliassen (1964). On the other hand, no instability would result in the model with the Betts–Miller scheme because convective exchange does not depend on moisture convergence. The moisture convergence I is approximated by

$$
\begin{aligned}
I &= -\int_0^{p_s} \nabla_h \cdot (\mathbf{v}_h m_v) \frac{dp}{g} \\
&\approx -\overline{m}_{vb} \frac{\Delta p_b}{gr} \frac{\partial}{\partial r} (r u_b') - \overline{m}_{v2} \frac{2 \Delta p}{gr} \frac{\partial}{\partial r} \left(r \frac{u_1' + u_3'}{2} \right) \\
&= -(\overline{m}_{vb} - \overline{m}_{v2}) \frac{\omega_4'}{g}
\end{aligned}
\tag{6.278}
$$

It is positive in the case of upward motion because usually $\overline{m}_{vb} > \overline{m}_{v2}$ holds. An appropriate level approximation for (6.162) is

$$
\begin{aligned}
\delta M_v &= \int_{p_t}^{p_l} (\delta m_{v1} + \delta m_{v2}) \frac{dp}{g} \\
&\approx \left[\frac{c_p}{L_v} (\overline{T}_{p2} - \overline{T}_2) + (\overline{m}_{vs2} - \overline{m}_{v2}) \right] \frac{\Delta p}{g}
\end{aligned}
\tag{6.279}
$$

All variables with an overbar are treated as constants in the linearized model. Eventually, we get from (6.166) in the case $I > 0$ for the latent heating rate

$$\frac{Q_2}{c_p} = -\frac{\overline{m}_{vb} - \overline{m}_{v2}}{\frac{c_p}{L_v}(\overline{T}_p - \overline{T}_2) + \overline{m}_{vs2} - \overline{m}_{v2}} \frac{\overline{T}_{p2} - \overline{T}_2}{\Delta p} \omega_4'$$

$$\equiv -\Xi_2 \eta \frac{p_2}{R_d} \omega_4' \tag{6.280}$$

where η is a constant. Therefore, the thermodynamic equation takes the form

$$\frac{\partial}{\partial t}\left(\frac{\phi_1' - \phi_3'}{\Delta p}\right) = \begin{cases} \Xi_2 \omega_2' - \Xi_2 \eta \omega_4' & \text{for } I > 0 \\ \Xi_2 \omega_2' & \text{for } I \leq 0 \end{cases} \tag{6.281}$$

With the solution ansatz

$$b' = \hat{b}(r)e^{\varsigma t} \tag{6.282}$$

we can derive from the equations (6.270) to (6.272), (6.275), (6.277) and (6.281) an equation for the radial velocity profile $\hat{u}_3$

$$\frac{2 + F_\varsigma(\varsigma)}{2L_R^2}\hat{u}_3 = \begin{cases} [F_\varsigma(\varsigma) + 1 - \eta F_\varsigma(\varsigma)]\dfrac{d}{dr}\left[\dfrac{1}{r}\dfrac{d}{dr}(r\hat{u}_3)\right] & \text{for } I > 0 \\[12pt] [F_\varsigma(\varsigma) + 1]\dfrac{d}{dr}\left[\dfrac{1}{r}\dfrac{d}{dr}(r\hat{u}_3)\right] & \text{for } I \leq 0 \end{cases} \tag{6.283}$$

where

$$F_\varsigma(\varsigma) = \begin{cases} \dfrac{f_0^2(\varsigma\tau_b + 1)}{f_0^2 + \left(\varsigma + \frac{1}{\tau_b}\right)^2}\dfrac{1}{\varsigma\tau_b}\dfrac{\Delta p_b}{\Delta p} & \text{unbalanced boundary layer} \\[16pt] \dfrac{1}{\varsigma\tau_b}\dfrac{\Delta p_b}{\Delta p} & \text{balanced boundary layer} \end{cases} \tag{6.284}$$

and $L_R = \sqrt{\Xi_2}\Delta p/\left(\sqrt{2}f_0\right)$ denotes the baroclinic Rossby radius. To solve this equation, it is constructive to use eigenfunctions of the operator $d/dr[1/rd/dr(\cdot\cdot)]$. Suitable functions are the **Bessel functions of the first kind**:

$$J_n(kr) = \sum_{j=0}^{\infty} \frac{(-1)^j \left(\frac{kr}{2}\right)^{n+2j}}{j!(n+j)!} \qquad (6.285)$$

and the associated **MacDonald function (modified Bessel function of the second kind)**:

$$K_n(kr) = \frac{\pi}{2} \frac{i^n J_{-n}(ikr) - i^{-n} J_n(ikr)}{\sin(n\pi)} \qquad (6.286)$$

where the index n denotes the order of the function. These functions are displayed in Fig. 6.29 for zero and first order. For the first-order function, we have the relations

$$\frac{d}{dr}\left\{\frac{1}{r}\frac{d}{dr}[rJ_1(kr)]\right\} = -k^2 J_1(kr) \qquad (6.287)$$

$$\frac{d}{dr}\left\{\frac{1}{r}\frac{d}{dr}[rK_1(kr)]\right\} = k^2 K_1(kr) \qquad (6.288)$$

The functions $J_1(kr)$ and $K_1(kr)$ are suitable for the inner region with $I > 0$ and the outer region with $I \leq 0$, respectively because

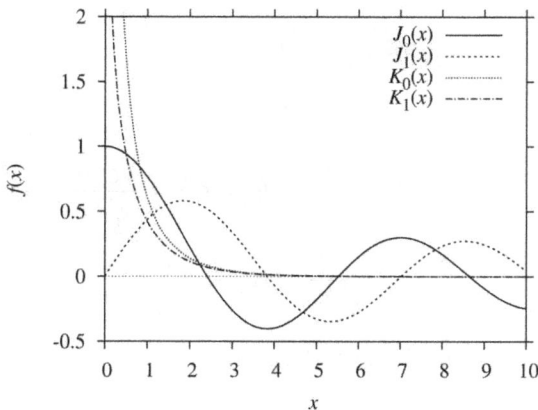

Fig. 6.29. Bessel functions of the first kind $J_n(x)$ and McDonald functions $K_n(x)$ for the orders $n = 0$ and $n = 1$.

$J_1(0) = 0$ and $\lim_{r \to \infty} K_1(kr) = 0$ Therefore, we set

$$\hat{u}_3(r) = \begin{cases} \hat{u}_I J_1(k_I r) & \text{for } I > 0 \\ \hat{u}_O K_1(k_O r) & \text{for } I \leq 0 \end{cases} \qquad (6.289)$$

Inserting these solutions yields the dispersion relations

$$k_I{}^2 = \frac{2 + F_\varsigma(\varsigma)}{2L_R{}^2[(\eta - 1)F_\varsigma(\varsigma) - 1]} \qquad (6.290)$$

$$k_O{}^2 = \frac{2 + F_\varsigma(\varsigma)}{2L_R{}^2(F_\varsigma(\varsigma) + 1)} \qquad (6.291)$$

The kinematic boundary condition of a continuous radial velocity must be satisfied at the radius r_b between the inner and outer regions, i.e.

$$\hat{u}_I J_1(k_I r_b) = \hat{u}_O K_1(k_O r_b) \qquad (6.292)$$

In addition to that, we have to fulfill the dynamic boundary condition of a steady geopotential transition at $r = r_b$. Because the temperature equation (6.281) leads to

$$\varsigma \frac{\hat{\phi}_1 - \hat{\phi}_3}{f_0{}^2} = \begin{cases} -\dfrac{F_\varsigma(\varsigma) + 1}{k_I^2 r} \dfrac{d}{dr}(r\hat{u}_3) & \text{for } I > 0 \\ \dfrac{F_\varsigma(\varsigma) + 1}{k_O^2 r} \dfrac{d}{dr}(r\hat{u}_3) & \text{for } I \leq 0 \end{cases} \qquad (6.293)$$

we must demand at $r = r_b$:

$$-\frac{\hat{u}_I}{k_I^2} \frac{d}{dr}[r J_1(k_I r_b)] = -\frac{\hat{u}_I}{k_I} r_b J_0(k_I r_b) = \frac{\hat{u}_O}{k_O^2} \frac{d}{dr}[r K_1(k_I r_b)]$$

$$= -\frac{\hat{u}_O}{k_O} r_b K_0(k_I r_b) \qquad (6.294)$$

Elimination of $\hat{u}_I/\hat{u}_O$ using (6.292) yields

$$\frac{J_0(k_I r_b)}{J_1(k_I r_b)} = \frac{k_I K_0(k_O r_b)}{k_O K_1(k_O r_b)} \qquad (6.295)$$

This equation constitutes the eigenvalue equation that relates the growth rate ς to the prescribed radius r_b. Before showing the results, we derive the solution for the simpler case in which the thermodynamic equation (6.281) for $I > 0$ holds at all radii with a coupling to a balanced boundary layer. Then, heating is unconditional and (6.290) yields the growth rate

$$\varsigma = \frac{1}{\tau_b} \frac{\Delta p_b}{\Delta p} \frac{\eta - 1 - \frac{1}{2}\left(\frac{r_b}{2.4048 L_R}\right)^2}{\left(\frac{r_b}{2.4048 L_R}\right)^2 + 1} \tag{6.296}$$

where we set the radius of the cloud region to $r_b = 2.4048/k_I$ (first zero of the Bessel function $J_0(k_I r)$ that describes the radial profile of vertical wind). With this result, we see that $\eta > 1$ is a necessary condition for instability. Therefore, the boundary layer must be moist enough for instability. The existence of nonzero CAPE is not sufficient for instability when the adiabatic cooling outside of the cumulus clouds is higher than the latent heating inside of the clouds. Instability turns into stability when the cloud region exceeds the radius $r_b = 2.4048\sqrt{2(\eta - 1)}L_R$. Maximum growth happens with the smallest possible diameter of the perturbation and the growth rate function resembles that of ordinary inviscid convection (see Fig. 6.1). Hence, there is no preferred scale of the perturbation and no short-wave cutoff takes place as for baroclinic instability (cf. Section 4.6). It appears in a model with a more precise vertical resolution when the atmosphere becomes very moist (see Charney and Eliassen, 1964). Then, the growth rate becomes infinite at a certain radius and the instability vanishes for perturbations with a smaller cloud region. This case with infinite growth at a finite cloud region radius was inspected by Fraedrich and McBride (1995) in more detail. They suggested that this case describes a viable mechanism for tropical cyclone formation. However, they also showed that the associated mode vanishes when the balance approximation is omitted. Therefore, it is not clear whether the large growth rate is rather an error due to the balance approximation. Figure 6.30 shows the growth rate as a function of r_b for $\eta = 2$ and three different solutions. Indeed, all solutions reveal qualitatively similar results. The solution including an unbalanced boundary layer yields a larger growth rate than the solution with a balanced boundary layer. The radius r_b where

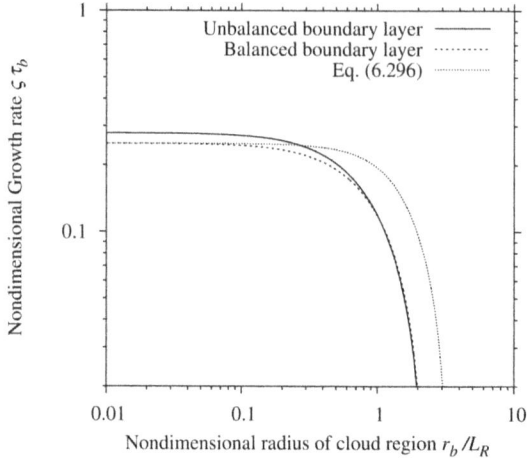

Fig. 6.30. Nondimensional growth rate $\varsigma\tau_b$ as a function of nondimensional radius of the cloud region r_b/L_R in the CISK model for the solution with an unbalanced boundary layer (solid line), balanced boundary layer (dashed line) and unconditional heating with a balanced boundary layer (dotted line).

instability turns into stability is smaller with conditional heating than with unconditional heating. Consequently, the stability properties hold also in the more complicated cases with conditional heating and an unbalanced boundary layer.

It is hard to get $\eta > 1$ for typical values observed in the tropics even if the relative humidity is very large. Kuo (1974) stated that under such conditions a very low amount of produced cloud water remains in the atmosphere. Therefore, he suggested an improved formula for his scheme. Then, latent heating becomes

$$\frac{Q_2}{c_p} = -f_Q \frac{L_v(\overline{m}_{vb} - \overline{m}_{v2})}{c_p \Delta p} \omega_4' \equiv -\Xi_2 \eta \frac{p_2}{R_d} \omega_4' \qquad (6.297)$$

where f_Q is a parameter that is slightly smaller than 1. In the general circulation model Planet Simulator this factor is a strictly monotonic increasing function of mean relative humidity (Lunkeit *et al.*, 2011). For the sounding displayed in Fig. 6.14, we get $\eta = 2.39$ using $f_Q = 1$. A typical tropical sounding was compiled by Jordan (1958). For this sounding the parameter becomes $\eta = 1.86$. Therefore, both atmospheric states are unstable with respect to CISK with the modified Kuo scheme even if the factor f_Q is slightly smaller than one.

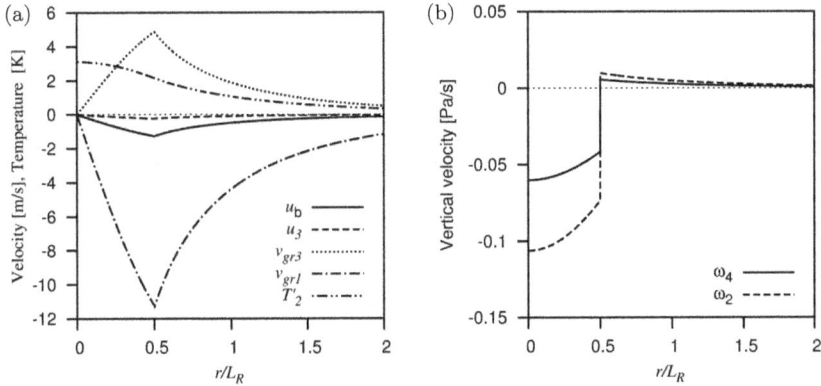

Fig. 6.31. Radial profiles of the growing normal mode solution for $\eta = 2$ and $r_b = 0.5L_R$: (a) Radial boundary layer wind (solid line), radial wind at level 3 (dashed line), gradient wind at level 3 (dotted line), gradient wind at level 1 (dot-dashed line), temperature anomaly at level 2 (double dot-dashed line), (b) vertical wind at level 4 (solid line) and vertical wind at level 2 (dashed line).

Figure 6.31 shows radial profiles of the growing solution for $\eta = 2$ and $r_b = 0.5L_R$. The solution comprises a cyclone and anticyclone in the lower and upper layers of the atmosphere, respectively. A warm-core results at the center due to the thermal wind relation. The maximum wind appears at the cloud region boundary $r = r_b$ where upwelling turns into downwelling. The inflow in the boundary layer is stronger than in the overlying free atmosphere layer. The upper-layer anticyclone is more intense than the lower-layer cyclone because the upper-layer outflow must compensate for the inflows of both lower layers.

The CISK theory for tropical cyclogenesis has some shortcomings. First, the growing mode has an anticyclone in the upper layer but in the previous section, we learned that anticyclones cannot be in gradient wind balance even for quite a small perturbation amplitude. Therefore, the linear stability analysis based on geostrophic balance seems questionable because the perturbation would rapidly reach a state where the linearization is invalid. Secondly, the convection stops when the convective stability measure $\overline{T}_2 - \overline{T}_{p2}$ (known as **lifted index**) becomes positive. The Jordan sounding yields a lifted index of about -4.5 K. Therefore, convection would stop at the center when the vortex has a maximum cyclonic wind of less than 10 m/s under such conditions. Therefore, CISK cannot explain the development of a fully developed tropical cyclone. Thirdly, the CISK mode predicts a vertical wind maximum at the center while in developed tropical cyclones it appears close to the **radius of maximum wind** (RMW) where the **eyewall** forms. Furthermore, the RMW of a developed tropical cyclone is much smaller than that of the initial perturbation. Possibly, a nonmodal perturbation (cf. Section 4.6.3) that describes the shrinking of the scale during the development could provide a more realistic solution. Anyhow, it appears inevitable to consider a nonlinear model for a complete understanding of tropical cyclogenesis.

6.6.4 *Nonlinear axisymmetric cyclogenesis*

One of the first attempts to simulate tropical cyclogenesis numerically in a nonlinear axisymmetric model was performed by Ogura (1964). He also investigated CISK in a linearized version of his model. However, the model has some deficiencies which were already mentioned by Ogura. The static stability is prescribed by a constant, the boundary layer model is based on the balanced approximation, and latent heat fluxes from the sea surface are ignored. Here, we describe a similar nonlinear axisymmetric model but without these deficiencies. To derive the governing model equations, it is more convenient to work with **specific angular momentum** given by

$$m = vr + \frac{1}{2} f_0 r^2 \tag{6.298}$$

Strictly speaking, it is the vertical component of the specific angular momentum vector $\mathbf{v} \times \mathbf{r}$ for which the origin lies in the center of the cyclone and the velocity of the rotating f-plane $(f_0 r/2)$ is included in $\mathbf{v}$. The material time derivative of specific angular momentum becomes

$$\frac{Dm}{Dt} = r\frac{Dv}{Dt} + uv + f_0 ru \tag{6.299}$$

In the axisymmetric model, we get due to Eq. (6.248)

$$\frac{Dm}{Dt} = r f_{T,\alpha} \tag{6.300}$$

Hence, angular momentum is conserved in the inviscid case ($f_{T,\alpha} = 0$). However, the horizontal turbulent exchange must be included in the balance model because otherwise, a frontal collapse takes place at a finite time due to latent heating which was shown by Frisius (2005). We will assume a constant exchange coefficient and parameterize $f_{T,\alpha}$ by

$$f_{T,\alpha} = \mathbf{e}_\alpha \cdot \mathbf{f}_T = K_M \mathbf{e}_\alpha \cdot \nabla_h^2 \mathbf{v}_h = K_M \left(\frac{\partial^2 v}{\partial r^2} + \frac{1}{r} \frac{\partial v}{\partial r} - \frac{v}{r^2} \right)$$

$$= K_M \frac{\partial}{\partial r} \left[\frac{1}{r} \frac{\partial}{\partial r} (vr) \right] = K_M \frac{\partial}{\partial r} \left(\frac{1}{r} \frac{\partial m}{\partial r} \right) \tag{6.301}$$

The derivation of the third identity is left to the reader (cf. Section 6.4.2). Hence, the angular momentum equation takes the form

$$\frac{\partial m}{\partial t} + u \frac{\partial m}{\partial r} + \omega \frac{\partial m}{\partial p} = r K_M \frac{\partial}{\partial r} \left(\frac{1}{r} \frac{\partial m}{\partial r} \right) \tag{6.302}$$

By assuming gradient wind balance we obtain the relation

$$\frac{m^2}{r^3} = \frac{\partial \phi}{\partial r} + \frac{1}{4} f_0^2 r \tag{6.303}$$

Another interesting relation is

$$\frac{1}{r} \frac{\partial m}{\partial r} = \zeta_z \tag{6.304}$$

Therefore, the radial gradient of angular momentum is proportional to the vertical component of absolute vorticity. This has the consequence that angular momentum must increase radially in an inertially stable vortex (cf. Section 3.11.1) which is usually fulfilled.

We use the two-level approximation as in the previous subsection and in the model by Ogura (1964). With this approximation, the governing equations of the free atmosphere can be written as

$$\frac{\partial m_1^2}{\partial t} + u_1 \frac{\partial m_1^2}{\partial r} + \omega_2 \frac{m_3^2 - m_1^2}{2\Delta p} = 2r m_1 f_{T,\alpha 1} \tag{6.305}$$

$$\frac{\partial m_3^2}{\partial t} + u_3 \frac{\partial m_3^2}{\partial r} + \omega_2 \frac{m_3^2 - m_1^2}{2\Delta p} + \omega_4 \frac{m_b^2 - m_3^2}{\Delta p} = 2r m_3 f_{T,\alpha 3} \tag{6.306}$$

$$\frac{m_1^2}{r^3} = \frac{\partial \phi_1}{\partial r} + \frac{1}{4} f_0^{\,2} r, \qquad \frac{m_3^2}{r^3} = \frac{\partial \phi_3}{\partial r} + \frac{1}{4} f_0^2 r \qquad (6.307)$$

$$\frac{\phi_1 - \phi_3}{\Delta p} = \frac{R_d T_2}{p_2} \qquad (6.308)$$

$$\frac{1}{r} \frac{\partial}{\partial r}(r u_1) + \frac{\omega_2}{\Delta p} = 0, \qquad \frac{1}{r} \frac{\partial}{\partial r}(r u_3) + \frac{\omega_4 - \omega_2}{\Delta p} = 0 \qquad (6.309)$$

$$\frac{\partial T_2}{\partial t} + \frac{u_1 + u_3}{2} \frac{\partial T_2}{\partial r} + \omega_2 \frac{\Pi_2 T_b - T_2}{\Delta p} = \frac{Q_2}{c_p} \qquad (6.310)$$

$$\omega_4 = \frac{\Delta p_b}{r} \frac{\partial}{\partial r}(r u_b) \qquad (6.311)$$

where we have approximated the vertical gradient of potential temperature by $(\partial \theta / \partial p)_2 \approx (T_b - T_2/\Pi_2)/\Delta p$ since temperature is only evaluated at level 2 and in the boundary layer.

To solve these equations, it is convenient to have a diagnostic equation for the toroidal overturning circulation. For this purpose, we first derive the thermal wind balance equation by combining (6.307) and (6.308) so that we get

$$\frac{m_1^2}{r^3} - \frac{m_3^2}{r^3} = \Delta p \frac{R_d}{p_2} \frac{\partial T_2}{\partial r} \qquad (6.312)$$

By taking the time derivative of this equation and inserting 6.305), (6.306) and (6.310) we obtain

$$-\frac{u_1}{r^3} \frac{\partial m_1^2}{\partial r} + \frac{u_3}{r^3} \frac{\partial m_3^2}{\partial r} + \frac{\omega_4}{r^3} \frac{m_b^2 - m^2}{\Delta p} + \frac{2}{r^2}(m_1 f_{T,a1} - m_3 f_{T,a3})$$

$$= \Delta p \frac{\partial}{\partial r}(\Xi_2 \omega_2) + \Delta p \frac{R_d}{p_2} \frac{\partial}{\partial r}\left(\frac{Q_2}{c_p} - \frac{u_1 + u_3}{2} \frac{\partial T_2}{\partial r}\right) \qquad (6.313)$$

where $\Xi_2 = -R_d/p_2(\Pi_2 T_b - T_2)/\Delta p$. The vertical velocity is related to the mass stream function Ψ by

$$\omega = \frac{1}{r} \frac{\partial \Psi}{\partial r} \rightarrow \omega_2 = \frac{1}{r} \frac{\partial \Psi_2}{\partial r} \quad \text{and} \quad \omega_4 = \frac{1}{r} \frac{\partial \Psi_4}{\partial r} \qquad (6.314)$$

The stream function is also related to the radial velocities due to the continuity equations (6.309):

$$u_1 = -\frac{\Psi_2}{r\Delta p}, \quad u_3 = \frac{\Psi_2 - \Psi_4}{r\Delta p} \tag{6.315}$$

These relations can be used to eliminate u_1, u_3 and ω_2 in (6.313). Then, we get the following diagnostic differential equation for Ψ_2:

$$\frac{\partial}{\partial r}\left(\frac{\Xi_2}{r}\frac{\partial \Psi_2}{\partial r}\right) - \Sigma_2\Psi_2 = -\frac{1}{\Delta p^2 r^4}\frac{\partial m_3^2}{\partial r}\Psi_4 + \frac{\omega_4}{r^3}\frac{m_b^2 - m_3^2}{\Delta p^2}$$

$$-\frac{R_d}{p_2}\frac{\partial}{\partial r}\left(\frac{Q_2}{c_p} + \frac{\Psi_4}{2r\Delta p}\frac{\partial T_2}{\partial r}\right)$$

$$+2\frac{m_1 f_{T,\alpha 1} - m_3 f_{T,\alpha 3}}{\Delta p r^2} \tag{6.316}$$

where

$$\Sigma_2 = \frac{1}{\Delta p^2 r^4}\frac{\partial}{\partial r}(m_1^2 + m_3^2) \tag{6.317}$$

is a parameter that measures inertial stability. Equation (6.317) is the two-level approximation of the **Sawyer-Eliassen equation**[h] applied to an axisymmetric vortex. For wavelike Bessel functions the left-hand side is proportional to $-\Psi_2$. Therefore, the toroidal overturning circulation results from the source terms on the right-hand side of this equation. These appear due to horizontal and vertical advection of angular momentum at the lower boundary, latent heat release, horizontal temperature advection and turbulent exchange. Some of them do not appear in the vertically continuous equation because they result from the two-level approximation. Latent heat release is the most important source term and it yields upward motion at the location where the heating takes place. The overturning circulation decreases obviously with increasing static stability Ξ_2 and increasing inertial stability Σ_2.

---

[h]This equation is based on the works by Sawyer (1956) and Eliassen (1951). For a derivation, see Shapiro and Willoughby (1982).

The equations for the free atmosphere must be coupled to a boundary layer to complete the model. The momentum equations in the boundary layer are given by

$$\frac{\partial u_b}{\partial t} + u_b \frac{\partial u_b}{\partial r} + \frac{\omega_4 + |\omega_4|}{2\Delta p_b}(u_3 - u_b) - \left(\frac{v_b}{r} + f_0\right) v_b$$

$$= -\frac{\partial \phi_3}{\partial r} + K_M \frac{\partial}{\partial r}\left[\frac{1}{r}\frac{\partial}{\partial r}(u_b r)\right] - \frac{C_D}{h_b}|\mathbf{v}_{10m}|u_b \qquad (6.318)$$

$$\frac{\partial v_b}{\partial t} + u_b \frac{\partial v_b}{\partial r} + \frac{\omega_4 + |\omega_4|}{2\Delta p_b}(v_{gr3} - v_b) + \left(\frac{v_b}{r} + f_0\right) u_b$$

$$= K_M \frac{\partial}{\partial r}\left[\frac{1}{r}\frac{\partial}{\partial r}(v_b r)\right] - \frac{C_D}{h_b}|\mathbf{v}_{10m}|v_b \qquad (6.319)$$

where h_b denotes the height of the boundary layer and $|\mathbf{v}_{10m}|$ the horizontal wind speed at 10 m height. The special formulation of vertical advection guarantees that the boundary layer is affected by the free atmosphere only in the case of downward motion. The introduction of a realistic surface drag as described in Section 1.13 yields nonlinear friction that is incompatible with the CISK model. Indeed, CISK would not appear with the boundary layer equations (6.318) and (6.319). To circumvent this issue one can set the 10 m wind speed $|\mathbf{v}_{10m}|$ to a constant or can introduce a "gust velocity" v_G such that $|\mathbf{v}_{10m}| = |\mathbf{v}_b| + v_G$. This gust velocity could also crudely account for the usual migration of the tropical cyclone relative to the sea surface. Note that the 10 m wind is usually smaller than the depth-averaged boundary layer wind but this error can approximately be corrected by choosing a smaller C_D.

To determine convection properly it is also necessary to introduce boundary layer equations for thermodynamic variables. For simplicity, we assume that the temperature in the boundary layer is constant and identical to that at the sea surface. Then, we only need an equation for specific humidity which is introduced by

$$\frac{\partial m_{vb}}{\partial t} + u_b \frac{\partial m_{vb}}{\partial r} + \frac{\omega_4 + |\omega_4|}{2\Delta p_b}(m_{v3} - m_{vb})$$

$$= K_{mv}\frac{1}{r}\frac{\partial}{\partial r}\left(r\frac{\partial m_{vb}}{\partial r}\right) + \frac{C_H}{h_b}|\mathbf{v}_{10m}|(m_{vs} - m_{vb}) \qquad (6.320)$$

where m_{vs} is the specific humidity at the sea surface. The second term on the right-hand side describes the evaporation from the sea surface where m_{vs} is mainly a function of the sea surface temperature.

Finally, the convective parameterization must be described. Convection only takes place when the lifted index $T_2 - T_{p2}$ is negative. The mid-level specific humidity m_{v2} is constant because we assume for simplicity that all produced cloud water is removed by precipitation ($f_Q = 1$). Thus, the moisture convergence I becomes

$$I = -(m_{vb} - m_{v2})\frac{\omega_4}{g} - \frac{\Delta p_b u_b}{g}\frac{\partial m_{vb}}{\partial r} \qquad (6.321)$$

and for latent heat release, we obtain

$$\frac{Q_L}{c_p} = -\frac{L_v}{c_p \Delta p}\left[(m_{vb} - m_{v2})\omega_4 + \Delta p_b u_b \frac{\partial m_{vb}}{\partial r}\right] \qquad (6.322)$$

in the case of a negative lifted index and positive I. This together with horizontal diffusion yields the heating

$$\frac{Q_2}{c_p} = \frac{Q_L}{c_p} + K_T \frac{1}{r}\frac{\partial}{\partial r}\left(r\frac{\partial T_2}{\partial r}\right) \qquad (6.323)$$

We initialize the model with a normal mode of the linearized equation to see how CISK develops in the nonlinear model. For this purpose, we set $|\mathbf{v}_{10m}| = v_G$ and ignore tendencies in boundary layer moisture. Then, the linearized model is compatible with the CISK theory. However, we cannot use the analytical solution because additional horizontal diffusion is included in the nonlinear model. Therefore, the normal mode is obtained by integrating the linearized model numerically until it is established. The model simulates a cylindrical domain with a radius of 2000 km and the radial grid point distance is 500 m. For the various model parameters we select: $f_0 = 0.5 \times 10^{-4}\text{s}^{-1}, C_D = 0.0015, v_G = 10\,\text{m/s}, K_M = K_T = K_{m_v} = 5000\,\text{m}^2/\text{s}, T_b = 300.15\,\text{K}, \overline{T}_2 = 263.75\,\text{K}, \overline{m}_{vb} = 20\,\text{g/kg}, \overline{m}_{v2} = 2.6\,\text{g/kg}$. The model is initialized with the normal mode such that the maximum tangential wind at level 3 is $1\,\text{m/s}$. The initial normal mode is shown in Fig. 6.32. The profiles are — in contrast to the analytical CISK mode — smooth at the boundary of the cloud region (cf. Fig. 6.31). The radius of maximum gradient wind at 190 km is now within this region. The numerical procedure selects automatically the fastest growing mode and,

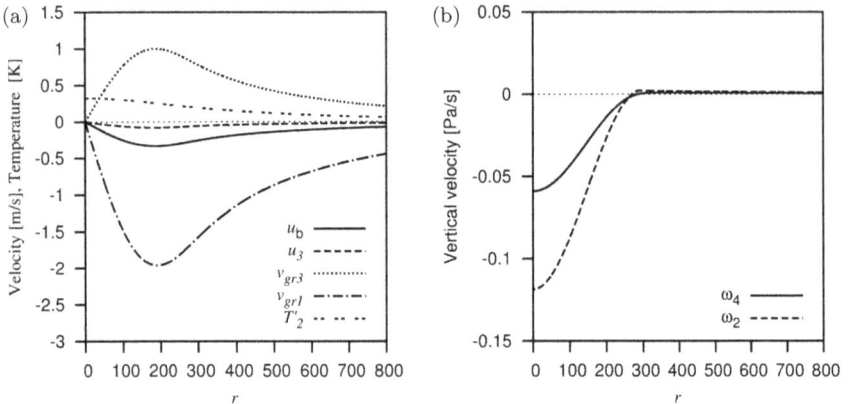

Fig. 6.32.   (a), (b) As in Fig. 6.31 but for the fastest-growing normal mode of the linearized tropical cyclone model.

obviously, its scale corresponds to that of a tropical cyclone. In the three experiments we assume (i) $|\mathbf{v}_{10m}| = v_G$, (ii) $|\mathbf{v}_{10m}| = |\mathbf{v}_b| + v_G$ and (iii) $|\mathbf{v}_{10m}| = |\mathbf{v}_b|$. Note that the last experiment does not support CISK instability because surface drag is nonlinear.

Figure 6.33 shows the time evolution of maximum tangential wind at level 3 for the three experiments. The wind increases exponentially with time in the first two experiments until growth accelerates on the second day. Growth stops when a fully developed tropical cyclone is established. Afterwards, the wind speed remains approximately constant. The development is faster in case (ii) in which the boundary layer wind speed contributes to the surface drag. Obviously, nonlinearities increase growth in the tropical cyclone model which is in contrast to baroclinic instability where these terms dampen the development. No tropical cyclone develops in the third case because the CISK mechanism is absent. However, tropical cyclogenesis might take place when the model is initialized with a more intense vortex.

To understand the acceleration of growth, we have a look at the time development of radial profiles as shown in Fig. 6.34. After 48 hours the radius of maximum gradient wind (RMGW) at the lower level moved inward while the anticyclonic gradient wind maximum at the upper level occurs at a larger radius (see Fig. 6.34(a)). These changes result from the radial angular momentum advection by the toroidal overturning circulation. Heating and temperature have still their maximum at the vortex axis and the negative lifted index

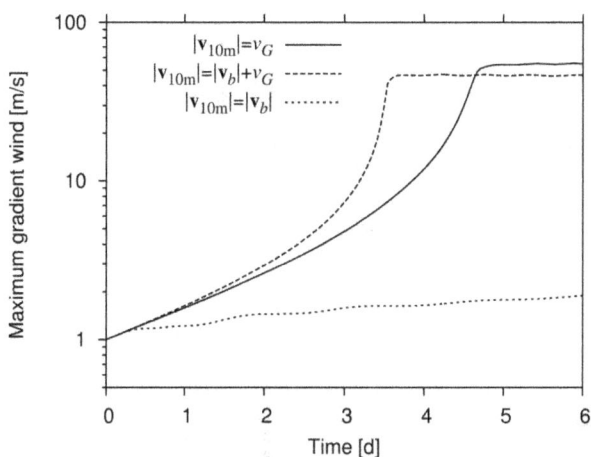

Fig. 6.33. Maximum gradient wind as a function of time for the three simulations assuming (i) $|\mathbf{v}_{10m}| = v_G$ (solid line), (ii) $|\mathbf{v}_{10m}| = |\mathbf{v}_b| + v_G$ (dashed line) and (iii) $|\mathbf{v}_{10m}| = |\mathbf{v}_b|$ dotted line. Note the logarithmic scaling of the ordinate.

indicates convective instability at all radii (Fig. 6.34(b)). A significant modification of the perturbation pattern can be noted at 72 h (Figs. 6.34(c) and 6.34(d)). Now, the maximum heating takes place at a finite radius of about 80 km and cyclonic gradient winds occur at the upper level near to this radius. The temperature amplitude is too high for establishing gradient wind balance with anticyclonic upper-level winds with the same amplitude as the cyclonic lower-level winds. This is represented by the dashed line showing the minimum possible gradient wind of $-f_0 r/3$. Consequently, radial outflow and vertical momentum advection increase leading to a cyclonic upper-level gradient wind. Furthermore, the more intense gradient wind in the boundary layer enhances the moisture convergence locally so that the vertical motion maximizes at a finite radius to form a convective ring. At 96 h (Figs. 6.34(e) and 6.34(f)) the convective ring becomes narrower and has moved inward while the maximum gradient wind has doubled and the cyclonic shear on the inner side of the gradient wind maximum has increased dramatically. This indicates strong frontogenesis due to radial momentum advection. Indeed, the vorticity is very high at this location and it can be expected that a frontal collapse would take place if horizontal diffusion were absent. The temperature T_2 and the lifted index have a negative gradient indicating the slow

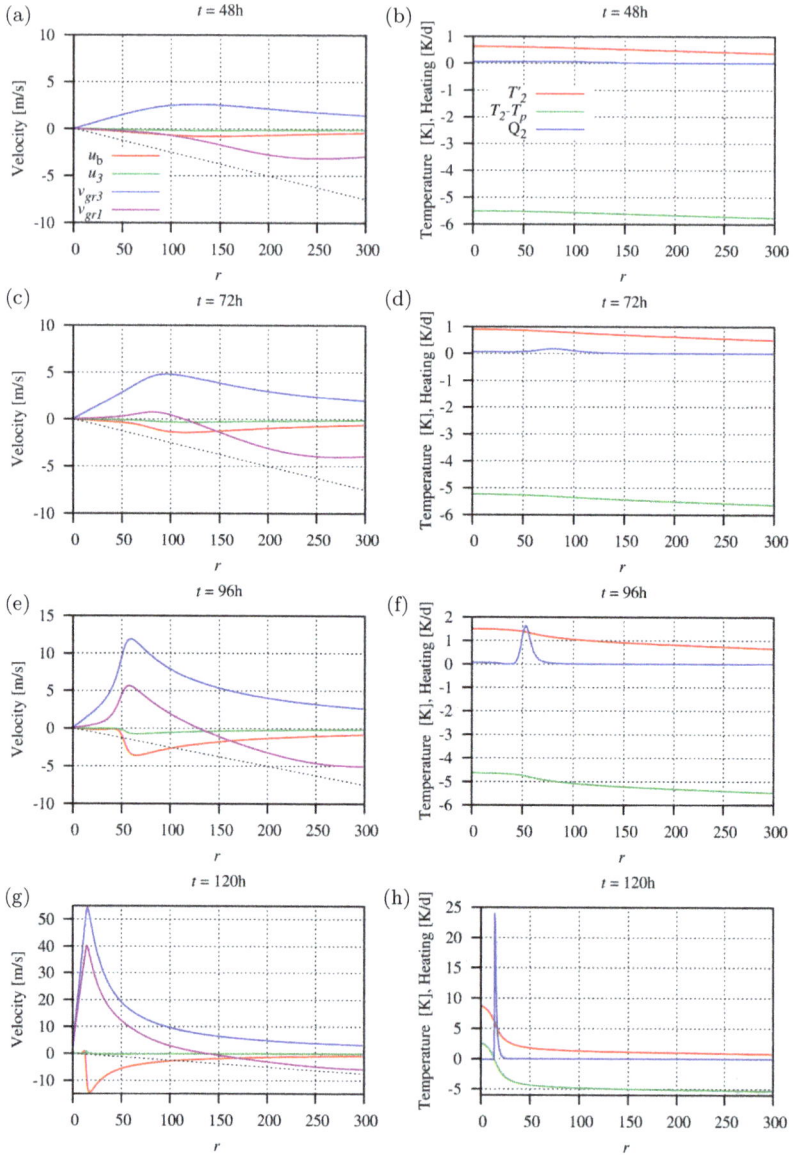

Fig. 6.34.   Radial profiles of radial boundary layer wind (red line), radial wind at level 3 (green line), gradient wind at level 3 (blue line), gradient wind at level 1 (magenta line) at several time points (left panels a, c, e, g) and temperature anomaly at level 2 (red line), lifted index (green line) and heating (blue line) at the same time points (right panels b, d, f, h). The black dashed line in the left panels shows the minimum possible gradient wind.

stabilization of the inner core. At 120 h (Figs. 6.34(g) and 6.37(h)) the vortex has developed into a mature tropical cyclone. The RMGW is only 17 km and intense latent heat release takes place in a narrow ring close to this radius. This ring forms the eyewall surrounding the eye where the lifted index indicates moist convective stability. The maximum temperature T_2 is nearly 9 K warmer than that of the environment and is found at the center of the vortex. This high-temperature excess mainly results from subsidence in the eye. Convective instability cannot support such a high value because the lifted index is everywhere above -6 K.

The **rapid intensification** between 96 h and 120 h can be better understood by considering the prognostic equation for the gradient wind at level 3:

$$\frac{\partial v_{gr3}}{\partial t} = -\frac{u_3}{r}\frac{\partial m_3}{\partial r} - \left(\omega\frac{\partial m}{\partial p}\right)_3 + f_{T,\alpha3} \qquad (6.324)$$

The first term on the right-hand side of this equation is responsible for intensification. It brings higher angular momentum to the inner core of the cyclone. The radial velocity u_3 must be negative in this case which is fulfilled because the latent heat release induces a toroidal overturning circulation that establishes gradient wind balance. The Sawyer–Eliassen equation yields such a result as shown by Shapiro and Willoughby (1982). Here, we obtain u_3 from the two-level analog (6.316).

Figure 6.35 shows a radius-time diagram of the radial advection term in (6.324) together with gradient wind at level 3 and the lifted index during the rapid intensification period. Obviously, the maximum gradient wind increase is found inside the RMGW. Therefore, the RMGW moves inward, and the gradient wind maximum increases since the angular momentum is approximately conserved by inward advection. Inward migration and intensification stop immediately after the eye has formed which can be identified by the region having a positive lifted index. This is understandable because no convection can be released inside of the eye so that it acts as a barrier for the convective ring which eventually forms the eyewall of the fully developed tropical cyclone. The RMGW is approximately constant and the radial advection term becomes negative after the eye has formed. This happens because vertical advection of **supergradient** boundary layer winds ($v_b > v_{gr3}$) cause outward motion

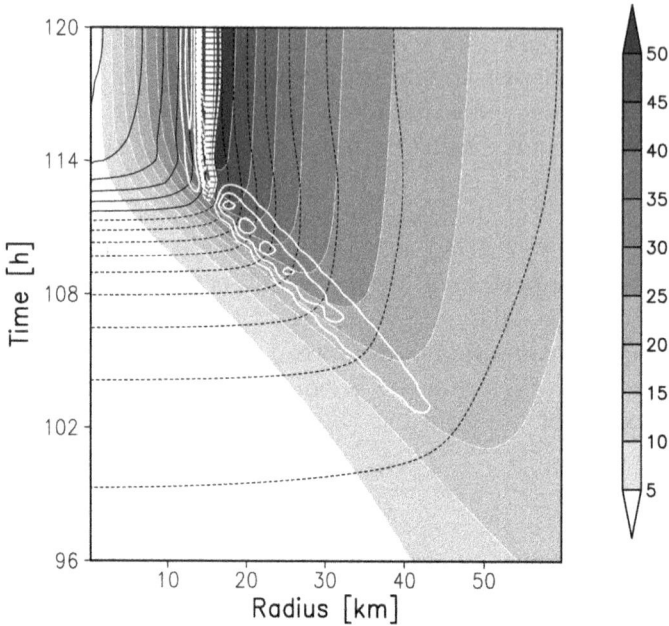

Fig. 6.35. Gradient wind at level 3 (shadings, m/s), gradient wind tendency due to radial angular momentum advection at level 3 (white isolines, contour interval $2\,\text{m/s/h}$) and lifted index (black isolines, contour interval $0.5\,\text{K}$) as a function of radius and time. The zero isoline has been omitted and negative isolines are dashed.

at level 3 to reestablish gradient wind balance. The mechanism of rapid intensification is sketched in Fig. 6.36. The pressure gradient induces a boundary layer inflow. It overshoots the RMGW inward due to the inertia. The associated deceleration causes flow convergence, upward motion and release of cumulus convection. The latent heat release warms the air and the adjustment to gradient wind balance causes an inward flow at the lower levels of the free atmosphere. The inward transport of angular momentum leads to a higher maximum of gradient wind and a smaller RMGW. The intensified gradient wind yields a stronger and more inward-directed boundary layer inflow that causes further intensification and inward migration. The gradient wind adjustment generates a downward flow inside of the convective ring which eventually forms the eye.

The described mechanism of rapid intensification has some commonalities with CISK but there are also significant discrepancies.

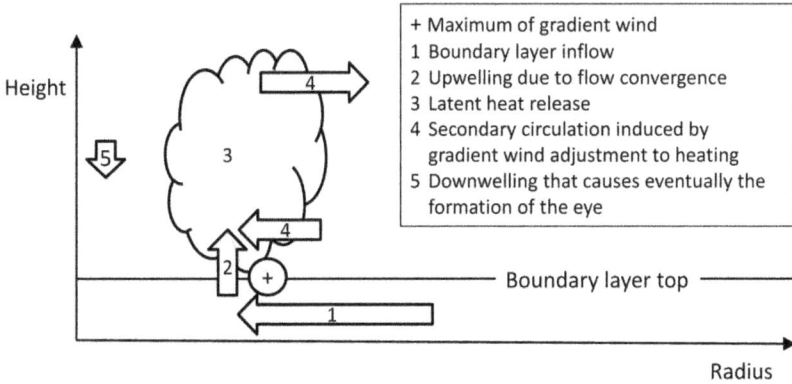

Fig. 6.36. Sketch of the rapid intensification mechanism in the nonlinear tropical cyclone model (see text).

Both mechanisms describe a feedback loop between moisture convergence by boundary layer inflow, latent heat release and gradient wind adjustment. However, there is only a narrow convective ring at a finite radius in the phase of rapid intensification while CISK incorporates a wide region with many convective clouds where the maximum vertical velocity appears at the vortex axis. The small scale of the convective ring makes the convection parameterization scheme questionable. Therefore, it should be rather viewed as a parameterization for a single convective cell. Nevertheless, the contracting convective ring appears in observations indeed and can cause rapid intensification as found by Willoughby (1990). Furthermore, convection-resolving models also reveal such a behavior (e.g. Stern *et al.*, 2015). Another difference is that the RMW decreases considerably during rapid intensification in the nonlinear model while the spatial perturbation pattern of the CISK mode is time-invariant.

We have to consider the equation for boundary layer humidity (6.320) for a more realistic simulation. Then, the humidity in the boundary layer decreases by entrainment of mid-level air due to downwelling and increases by evaporation from the sea surface. The initial vortex must have a finite amplitude to obtain tropical cyclogenesis because the model does not support CISK with nonlinear surface drag. Furthermore, the upper anticyclone is replaced by a cyclone due to the impossibility of anticyclonic gradient wind balance for the chosen vortex with a gradient wind maximum of $4\,\mathrm{m/s}$. The

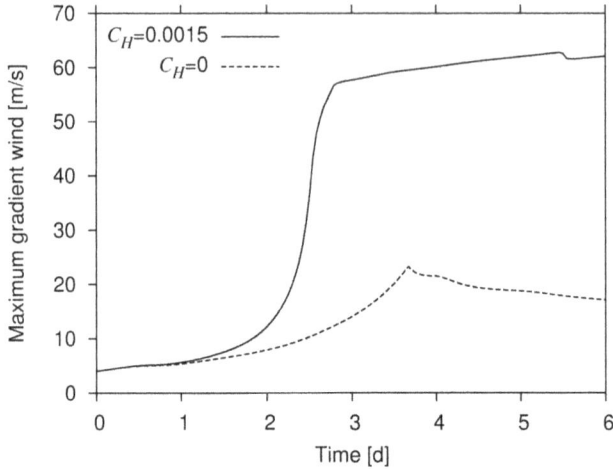

Fig. 6.37.   Maximum gradient wind as a function of time for the experiments with time-dependent boundary layer humidity assuming $C_H = 0.0015$ (solid line) and $C_H = 0.0$ (dashed line).

surface transfer coefficient C_H has been set to 0.0015 and 0. The comparison of the experiments with these values reveals the role of evaporation from the sea surface. Figure 6.37 shows the time evolution of maximum gradient wind for these experiments. Rapid intensification happens in the experiment with $C_H = 0.0015$ after two days with slow development and the final gradient wind amplitude is close to 60 m/s. Inspection of the fields reveals the same intensification mechanism as found in the previous experiments. However, the experiment without surface evaporation ($C_H = 0$) does not show such a strong intensification and the vortex never reaches the strength of a tropical cyclone. The reason is that the boundary layer inflow entrains dry air from the free atmosphere and this decreases the latent heat release. Eventually, the atmosphere becomes stable with respect to convection so that the vortex decays. Evaporation at the sea surface is obviously essential for tropical cyclogenesis and this effect is called **wind-induced surface heat exchange (WISHE)**.[i]

---

[i]Emanuel (1986,1997) suggested that an "air sea interaction instability" can result by WISHE and can explain tropical cyclogenesis with marginal CAPE. However, this theory is still debatable and has been challenged by Montgomery *et al.* (2015) and Lee and Frisius (2018).

The nonlinear model introduced in this section is still quite simple. It has a very coarse vertical resolution and ignores deviations from axisymmetry. It can be expected that tropical cyclogenesis rarely takes place with an axisymmetric convective ring. Instead, convection could possibly be organized in a spiral band which curls up. The dynamics of such a flow is obviously more complicated but in the azimuthal mean the picture of contracting convection is possibly still visible. However, it remains unclear whether nonaxisymmetric processes have a significant impact on the evolution dynamics. Furthermore, the model ignores convective downdrafts into the boundary layer although this process leads to a dramatic obstruction of tropical cyclogenesis (see Frisius and Hasselbeck, 2009). Finally, the use of a convective parameterization scheme is questionable due to the small scale of the developing inner core. Therefore, it is more appropriate to use a model with resolved convection for simulating tropical cyclones realistically.

6.6.5 *Steady-state solution of a tropical cyclone*

Emanuel (1986) derived an analytical model for a steady-state tropical cyclone which gained much popularity since it provides an estimation of the **potential intensity** of tropical cyclones. In this subsection, we introduce the more recent model by Emanuel and Rotunno (2011). Figure 6.38 shows a sketch of the steady-state model. The outward-tilted streamlines have certain values of specific

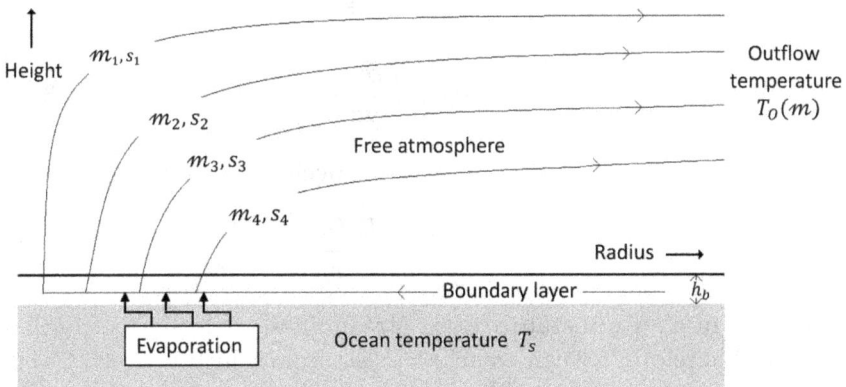

Fig. 6.38. Sketch of the steady-state tropical cyclone model.

entropy s and specific angular momentum m. The **outflow temperature** is observed at a large radius where the streamlines are nearly horizontal. Below the free atmosphere, a boundary layer is added in which most of the inflow takes place.

First, we treat the free atmosphere in which turbulent diffusion is neglected. Furthermore, water vapor saturation is assumed and all processes are moist adiabatic (cf. Section 1.10). Then, the axisymmetric balanced equations can be written as

$$\frac{m^2}{r^3} = \frac{\partial \phi}{\partial r} + \frac{1}{4} f_0^2 r \tag{6.325}$$

$$\frac{Dm}{Dt} = 0 \tag{6.326}$$

$$\frac{\partial \phi}{\partial p} = -v \tag{6.327}$$

$$\frac{1}{r}\frac{\partial}{\partial r}(ru) + \frac{\partial w}{\partial p} = 0 \tag{6.328}$$

$$\frac{Ds}{Dt} = 0 \tag{6.329}$$

These equations lead to an interesting consequence for potential vorticity. We find that

$$\begin{aligned}\zeta &= -\frac{\partial v}{\partial z}\mathbf{e}_r + \left(\frac{\partial u}{\partial z} - \frac{\partial w}{\partial r}\right)\mathbf{e}_\alpha + \left[\frac{1}{r}\frac{\partial}{\partial r}(rv) + f_0\right]\mathbf{e}_z \\ &= -\frac{1}{r}\frac{\partial m}{\partial z}\mathbf{e}_r + \frac{1}{r}\frac{\partial m}{\partial r}\mathbf{e}_z + \left(\frac{\partial u}{\partial z} - \frac{\partial w}{\partial r}\right)\mathbf{e}_\alpha\end{aligned} \tag{6.330}$$

Therefore, potential vorticity based on specific entropy becomes

$$P_s = \frac{\zeta \cdot \nabla s}{\rho} = \frac{J_{rz}(m,s)}{r\rho} \tag{6.331}$$

However, in a steady-state vortex, streamlines are isolines of both s and m. Therefore, we can write $s = s(m)$ from which we get $P_s = 0$. Hence, the potential vorticity P_b of a steady state inviscid axisymmetric flow is zero if the scalar property b is conserved.

First, we derive the shape of the streamlines. From (6.325) and (6.327) we get the following thermal wind balance equation:

$$\frac{1}{r^3}\frac{\partial m^2}{\partial p} = -\frac{\partial v}{\partial r} \tag{6.332}$$

By using the chain rule for differentiation, we can write this equation in the form[j]

$$\frac{1}{r^3}\frac{\partial m^2}{\partial p} = -\frac{\partial v}{\partial s}\bigg|_p \frac{\partial s}{\partial r} \tag{6.333}$$

This equation can be further manipulated by considering Gibbs's fundamental equation (1.114) that can be written as

$$dh = T ds + v dp + a_{l,v} dm_{vs} \tag{6.334}$$

where $dm_d = 0$ is assumed. The affinity $a_{l,v}$ vanishes in thermodynamic equilibrium which is suitable for cloud air. Therefore, we get

$$\frac{\partial T}{\partial p}\bigg|_s = \frac{\partial v}{\partial s}\bigg|_p \tag{6.335}$$

since (6.334) is an exact differential. Substituting this identity in (6.333) yields

$$\frac{1}{r^3}\frac{\partial m^2}{\partial p} = -\frac{\partial T}{\partial p}\bigg|_s \frac{\partial s}{\partial r} \tag{6.336}$$

Since $s = s(m)$ we can write

$$\frac{1}{r^3}\frac{\partial m^2}{\partial p} = -\frac{\partial T}{\partial p}\bigg|_s \frac{ds}{dm}\frac{\partial m}{\partial r} \tag{6.337}$$

The slope of a streamline in the $r - p$ diagram results from the ratio $-(\partial m/\partial r)/(\partial m/\partial p)$ because m is conserved. Therefore,

$$\frac{1}{r^3}\frac{\partial r}{\partial p}\bigg|_m = \frac{1}{2m}\frac{\partial T}{\partial p}\bigg|_s \frac{ds}{dm} \tag{6.338}$$

---

[j]The notation $\frac{\partial v}{\partial s}|_p$ means that the variable p is kept constant during differentiation.

This equation can be integrated along the streamline on which m and s are constant. Since ds/dm is also constant we get by integration

$$\left(\frac{1}{r^2} - \frac{1}{r_O{}^2}\right) = -\frac{T - T_O}{m}\frac{ds}{dm} \qquad (6.339)$$

where the index O indicates the outflow radius at which the gradient wind vanishes for the first time. Given that temperature decreases with height we find an outward slope of the streamline in the warm core case $ds/dm < 0$. With (6.339) and the definition of angular momentum (6.298), we can deduce the gradient wind because

$$\left(\frac{m}{r^2} - \frac{m}{r_O^2}\right) = \left(\frac{v_{gr}r + \frac{f_0}{2}r^2}{r^2} - \frac{\frac{f_0}{2}r_O^2}{r_O^2}\right) = \frac{v_{gr}}{r} \qquad (6.340)$$

Therefore,

$$\frac{v_{gr}}{r} = -(T - T_O)\frac{ds}{dm} \qquad (6.341)$$

We demand that the gradient wind coincides with the tangential boundary layer wind at $z = h_b$. Hence,

$$\frac{v_b}{r} = -(T_b - T_O)\frac{ds}{dm} \text{ at } z = h_b \qquad (6.342)$$

where T_b is the temperature at the top of the boundary layer. Consequently, gradient wind imbalance is neglected in the model. For a model with an unbalanced boundary layer, we refer to Frisius *et al.* (2013). Furthermore, the boundary layer entropy coincides with that of the free atmosphere so that

$$s_b = s \quad \text{at } z = h_b \qquad (6.343)$$

Therefore, CAPE vanishes in the model. Instead, the vortex is neutral with respect to **slantwise convection**. This means that the displacement in the horizontal and vertical are repelled by inertial and static stability, respectively, but that motion along the outflow streamline does not experience a repelling force. This can be understood by the circumstance that such a motion does not modify the pressure and wind fields due to the conservation of entropy and angular momentum. The nonexistence of CAPE is an assumption which is usually not fulfilled. Frisius and Schönemann (2012) included CAPE in the model and found a higher intensity in agreement with numerical model simulations.

A simple balanced slab-boundary model is used for coupling with the free atmosphere. The equations for tangential wind and specific entropy in the boundary layer are as follows:

$$u_b \frac{dv_b}{dr} + \left(\frac{v_b}{r} + f_0 \right) u_b = \frac{u_b}{r} \frac{dm_b}{dr} = -\frac{C_D}{h_b} |v_b| v_b \qquad (6.344)$$

$$u_b \frac{ds_b}{dr} = \frac{C_H}{h_b} |v_b| (s_s - s_b) \qquad (6.345)$$

where m_b is the specific angular momentum in the boundary layer and s_s the specific entropy at the sea surface. Eliminating radial wind yields

$$-v_b r \left(\frac{dm_b}{dr} \right)^{-1} \frac{ds_b}{dr} = \frac{C_H}{C_D} (s_s - s_b) \qquad (6.346)$$

Choosing m_b as the radial coordinate leads to the simplification

$$-v_b r \frac{ds_b}{dm_b} = \frac{C_H}{C_D} (s_s - s_b) \qquad (6.347)$$

In the core of a tropical cyclone, we have $m_b \approx v_b r$. Emanuel and Rotunno (2011) showed that this approximation introduces only a small error and, therefore, we will use it here. Then, (6.347) becomes

$$-m_b \frac{ds_b}{dm_b} = \frac{C_H}{C_D} (s_s - s_b) \qquad (6.348)$$

Assuming constant s_s as done by Emanuel and Rotunno (2011)[k] yields the solution

$$s_b(m_b) = s_s - (s_s - s_0) \left(\frac{m_b}{m_{b0}} \right)^{\frac{C_H}{C_D}} \qquad (6.349)$$

---

[k] The entropy at the sea surface also depends on surface pressure (cf. Eq. (1.142)). Indeed, this dependence leads to substantially higher intensities for high sea surface temperatures. For simplicity, we ignore this fact and refer for more details of this effect to Emanuel (1986).

where $s_0 = s(m_{b0})$ denotes the reference entropy. Now, we couple the free atmosphere with the boundary layer. With the assumptions (6.342) and (6.343) we can eliminate ds/dm by combining (6.342) and (6.347) such that

$$v_b{}^2 = \frac{C_H}{C_D}(T_b - T_O)(s_s - s_b) \tag{6.350}$$

The outflow temperature T_O is still an unknown in the solution. Emanuel (1986) set T_O to a constant but this would produce a very unrealistic wind profile. So, he fixed this problem with additional assumptions for the boundary layer entropy. However, Emanuel and Rotunno (2011) realized that T_O is not constant because different streamlines approach different heights at large radii. Instead of using a constant, they proposed that the Richardson number of the outflow is constant and found a vertical dependency of the outflow temperature. They justified this assumption by the strong vertical mixing in the outflow when the Richardson number is below a critical value. According to Emanuel and Rotunno (2011), the Richardson number in cloud air is given by

$$\text{Ri} = \frac{c_p \frac{\Gamma_m}{\theta_{es}} \frac{\partial \theta_{es}}{\partial z}}{\left(\frac{\partial u}{\partial z}\right)^2 + \left(\frac{\partial v}{\partial z}\right)^2} = \frac{\Gamma_m \frac{\partial s}{\partial z}}{\left(\frac{\partial u}{\partial z}\right)^2 + \left(\frac{\partial v}{\partial z}\right)^2} \approx \frac{\Gamma_m \frac{\partial s}{\partial z}}{\left(\frac{\partial v}{\partial z}\right)^2} \tag{6.351}$$

It can be expected that the pseudo-adiabatic lapse rate Γ_m is close to that of the outflow temperature profile. Therefore,

$$\text{Ri}_O \approx -\frac{\frac{\partial T_O}{\partial z} \frac{\partial s}{\partial z}}{\left(\frac{\partial v}{\partial z}\right)^2} = -r_O^2 \frac{\frac{\partial T_O}{\partial z} \frac{\partial s}{\partial z}}{\left(\frac{\partial m}{\partial z}\right)^2} = -r_O^2 \frac{\partial T_O}{\partial m}\bigg|_{r=r_O} \frac{ds}{dm} \tag{6.352}$$

The outflow adjusts to a critical value Ri_c so that

$$\frac{\partial T_O}{\partial m}\bigg|_{r=r_O} = -\frac{\text{Ri}_c}{r_O^2}\left(\frac{ds}{dm}\right)^{-1} \tag{6.353}$$

We can insert the solution (6.349) because of the coupling with the boundary layer ($s = s_b, m = m_b$). This yields for the streamline that intersects the RMGW at $z = h_b$

$$\frac{\partial T_O}{\partial m}\bigg|_{r=r_t} = \frac{C_D}{C_H}\frac{\text{Ri}_c}{r_t^2}\frac{m}{(s_s - s_m)}\left(\frac{m_m}{m}\right)^{\frac{C_H}{C_D}} \qquad (6.354)$$

where the index m refers to this streamline and r_t denotes the corresponding outflow radius. Integration of this equation leads to

$$T_O = T_t + \frac{C_D/C_H}{2 - C_H/C_D}\frac{\text{Ri}_c}{r_t^2}\frac{m_m^{\frac{C_H}{C_D}}}{s_s - s_m}\left(m^{2-\frac{C_H}{C_D}} - m_m^{2-\frac{C_H}{C_D}}\right) \qquad (6.355)$$

where T_t is the outflow temperature of the streamline intersecting the RMGW. We can eliminate the temperatures T_O and T_t by using (6.342) and the approximation $m = m_b \approx v_b r$. The result is

$$\left(\frac{m_b}{m_m}\right)^{2-\frac{C_H}{C_D}} = \frac{\frac{r^2}{r_m^2} + \frac{1}{2-C_H/C_D}\frac{\text{Ri}_c}{r_t^2}r^2}{1 + \frac{1}{2-C_H/C_D}\frac{\text{Ri}_c}{r_t^2}r^2} \qquad (6.356)$$

where r_m is the RMGW. The maximum of the corresponding gradient wind profile appears at

$$r_{\max} = \sqrt{\frac{C_H}{C_D}\frac{1}{\text{Ri}_c}}r_t \qquad (6.357)$$

Since $r_{\max} = r_m$ must hold we can eliminate r_t^2/Ri_c in (6.356). Then, the solution for boundary layer angular momentum becomes

$$\left(\frac{m_b}{m_m}\right)^{2-\frac{C_H}{C_D}} = \frac{2r^2/r_m^2}{2 - C_H/C_D(1 - r^2/r_m^2)} \qquad (6.358)$$

We still need to find out the maximum gradient wind since m_m has not been determined so far. Using the solution (6.349) we find approximately

$$s_O = s_s - (s_s - s_m)\left(\frac{f_0 r_{bO}^2}{2v_{\max}r_m}\right)^{\frac{C_H}{C_D}} \qquad (6.359)$$

Here, s_O denotes the specific boundary layer entropy at the radius r_{bO} where v_b vanishes and $v_{\max}$ is the maximum gradient wind. On the other hand, inserting the angular momentum at the radius r_{bO} in (6.358) yields

$$\left(\frac{f_0 r_{bO}^2}{2v_{\max} r_m}\right)^{2-\frac{C_H}{C_D}} = \frac{2r_{bO}^2/r_m^2}{2 - C_H/C_D\left(1 - r_{bO}^2/r_m^2\right)} \approx 2\frac{C_D}{C_H} \quad (6.360)$$

because $r_{bO}^2/r_m^2 \gg 1$. Combination of both equations leads to

$$s_s - s_m = \left(\frac{1}{2}\frac{C_H}{C_D}\right)^{\frac{1}{2C_D/C_H-1}} (s_s - s_O) \quad (6.361)$$

Applying (6.350) at the gradient wind maximum and inserting (6.361) gives the desired relation

$$v_{\max} = \sqrt{(T_b - T_t)\frac{C_H}{C_D}\left(\frac{1}{2}\frac{C_H}{C_D}\right)^{\frac{1}{2C_D/C_H-1}} (s_s - s_O)} \quad (6.362)$$

This formula can be used to estimate the potential intensity as a function of observed quantities. The temperature T_b can be chosen to be the observed temperature at the top of the boundary layer. The specific entropies s_s and s_O can be calculated by Eq. (1.147) where observed near-surface values could be used for s_O. Finally, the outflow temperature T_t could be roughly estimated by the tropopause temperature. However, the formula yields higher potential intensity for a dry boundary layer (low s_O) than for a moist boundary layer (high s_O) which does not appear to be plausible. This happens because the outflow stratification is determined by the ambient boundary layer moisture in the model but likely not in reality. To circumvent this issue, it is better to prescribe a certain relative humidity that has been established at the outer radius r_{bO}. Figure 6.39 shows potential intensity for the seasons JAS 2005 and JFM 2005 using NCEP reanalysis data. We can clearly see the enhanced potential intensity in the tropical latitudes. However, the area in which the potential intensity exceeds the minimum tropical cyclone strength is much larger than the area where the SGP predicts a nonzero probability of tropical cyclogenesis. Indeed, a tropical cyclone can be maintained in regions where tropical cyclogenesis is impossible, e.g. at the equator. On the other hand, this is not true in regions where the shear is too large for the maintenance of the storm as it is the case in the mid-latitudes.

(a)

JAS 2005

(b)

JFM 2005

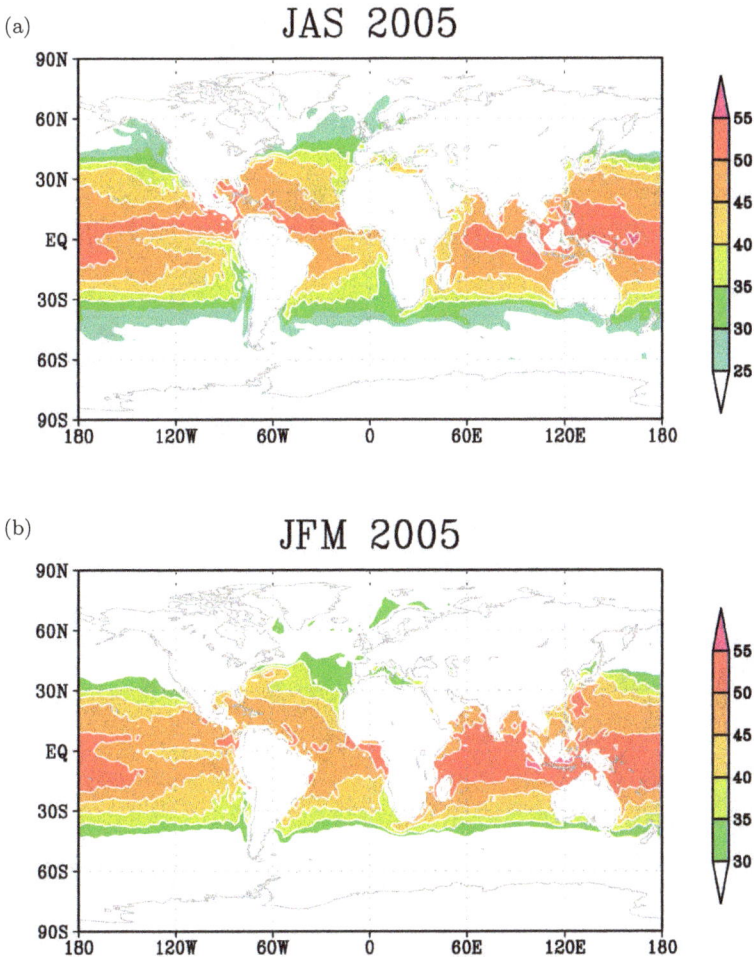

Fig. 6.39. Potential intensity (m/s) deduced from Eq. (6.362) using NCEP reanalysis data for the seasons (a) JAS 2005 and (b) JFM 2005. The ratio $C_H/C_D = 1$ and a boundary layer relative humidity of 70% at the outer radius have been assumed.

It is notable that significant potential intensity arises in high latitudes during the JFM season, such as the Labrador Sea, Norwegian Sea and Barents Sea. Indeed, in these regions, **polar lows** are frequently observed (Rasmussen and Turner, 2003). Polar lows are intense mesoscale cyclones which share many properties with tropical cyclones. However, they form in contrast to tropical cyclones due to a large temperature difference between the air and the sea surface. Emanuel and Rotunno (1989) applied their Carnot cycle theory for tropical cyclones to polar lows and found corresponding results which led them to the term **arctic hurricane** for such a cyclone.

To determine the tangential boundary layer wind profile, we insert the definition of specific angular momentum in Eq. (6.358) and make use of the approximation $m_m \approx v_{\max} r_m$ so that we obtain

$$v_b = v_{\max} \frac{r_m}{r} \left[\frac{2r^2/r_m^2}{2 - C_H/C_D(1 - r^2/r_m^2)} \right]^{\frac{1}{2C_H/C_D - 1}} - \frac{f_0 r}{2} \quad (6.363)$$

The radial boundary layer wind results from the slab-boundary layer equation (6.344)

$$u_b = -r \frac{C_D |v_b| v_b}{h_b \frac{dm_b}{dr}} \quad (6.364)$$

With this result, we get for the vertical wind at the top of the boundary layer

$$w_b = -\frac{h_b}{r} \frac{d}{dr}(r u_b) = \frac{1}{r} \frac{d}{dr} \left[C_D |v_b| v_b r^2 \left(\frac{dm_b}{dr} \right)^{-1} \right] \quad (6.365)$$

where an incompressible boundary layer flow has been assumed. Finally, the specific entropy profile can be deduced by inserting (6.358) in Eq. (6.349).

Figure 6.40 shows the radial profiles of the steady-state solution for various ratios C_H/C_D. These profiles should be compared with Figs. 6.34(g) and 6.34(h) displaying profiles for the quasi-steady state tropical cyclone simulated by the numerical two-level model. Obviously, the gradient wind of the analytical model has a smoother shape in the vicinity of the maximum. Furthermore, the maximum radial boundary layer inflow appears at a much larger radius than in the numerical two-level model. This has the consequence that the upward motion has a smooth maximum and is distributed over a wide radial range. These differences can be attributed to the simple balanced boundary layer model of the analytical model. Inflow characteristics being more similar to the numerical model arise when an unbalanced boundary layer is implemented (see Frisius *et al.*, 2013). We see a substantial sensitivity to C_H/C_D, i.e. the maximum gradient wind and

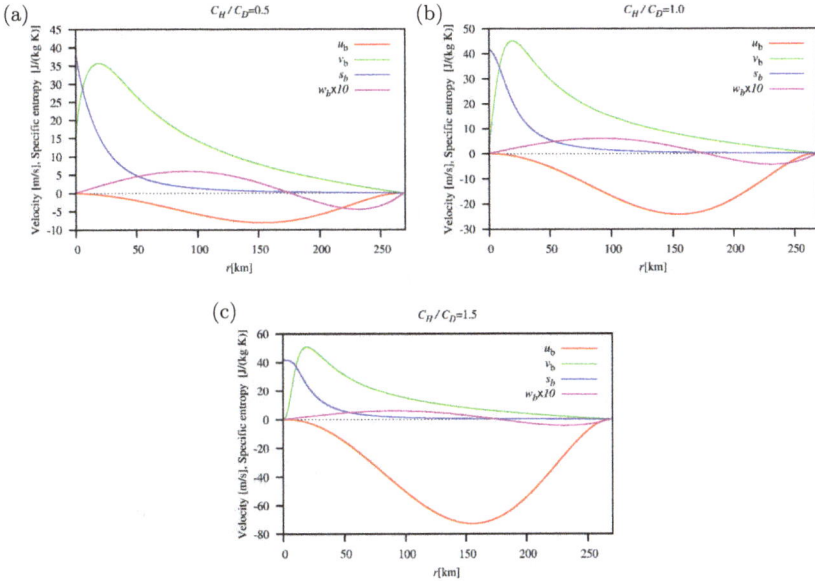

Fig. 6.40. Radial profiles of radial boundary layer wind u_b (red lines), gradient wind v_b (green lines), boundary layer entropy s_b (blue lines) and vertical wind at the top of the boundary layer (blue lines) of the analytical tropical cyclone solution for (a) $C_H/C_D = 0.5$, (b) $C_H/C_D = 1.0$ and (c) $C_H/C_D = 1.5$. The chosen solution parameters are $r_m = 20$ km, $C_D = 0.0015$, $T_s = 301.15$ K, $T_B - T_t = 100$ K, $h_b = 1500$ m and $f_0 = 0.5 \times 10^{-4}$s^{-1}. A boundary layer relative humidity of 70% at the outer radius has been assumed for the specification of s_O.

the boundary layer inflow increase with this ratio. For $C_H/C_D = 1.5$ a maximum radial wind speed of about 70 m/s occurs which is very unrealistic since it exceeds the gradient wind. The reason for this high value is the vorticity of the balanced flow that is close to zero in the outer part of the vortex. We see from (6.304) and (6.364) that the radial boundary layer wind approaches infinity in the limit $\zeta_z \to 0$. Obviously, the balanced slab-boundary model is inappropriate in this case. The structure of the gradient wind profile also changes with C_H/C_D. For $C_H/C_D = 0.5$ a sharp increase of wind is noted at the vortex axis leading to extremely large vorticity. On the other hand, for $C_H/C_D = 1.5$ the wind is rather calm close to the vortex axis. The analytical solution has no eye since the free atmosphere is saturated everywhere and no downwelling appears in the vortex center.

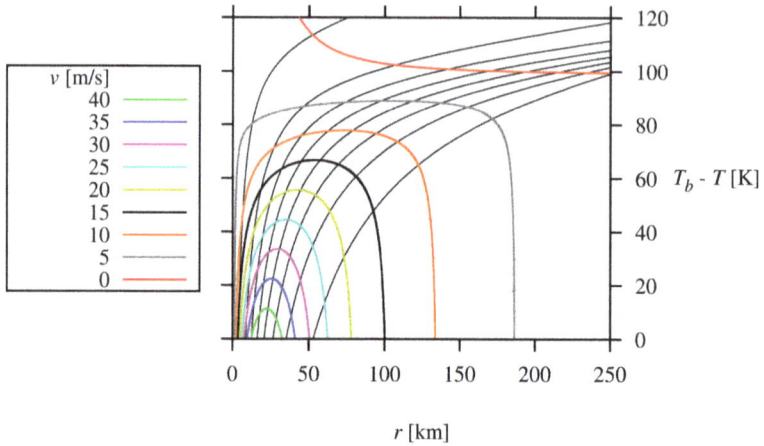

Fig. 6.41. Specific entropy (black isolines) and gradient wind (colored isolines) of the analytical tropical cyclone solution as a function of radius and height coordinate $T_b - T$ for $C_H/C_D = 1$. The other solution parameters are as in Fig. 6.40.

Figure 6.41 shows the gradient wind and specific entropy fields above the boundary layer for $C_H/C_D = 1$ where $T_b - T$ is used as the vertical coordinate in the diagram. The isentropes slope outwards as in the sketch displayed in Fig. 6.38 while the gradient wind decreases with height which is a consequence of the specific entropy declining in radial direction. The gradient wind becomes anticyclonic at large heights in the outer part of the vortex. At large radii, the atmosphere has a nearly isentropic stratification and is, therefore, neutral to moist convective instability. It is questionable if the ambient atmosphere provides such stratification. It is at least impossible with this solution to describe the transition to a convectively stable environment. Furthermore, the downward motion at large radii requires that the air has enough cloud water or ice for keeping the descending air parcel saturated until it reaches the top of the boundary layer. Such a condition is unrealistic. However, the analytical tropical cyclone solution by Emanuel and Rotunno (2011) is, despite some shortcomings, very insightful since it provides a theory for the maintenance of tropical cyclones due to the thermodynamical air-sea disequilibrium and the resulting formula for potential intensity is a useful diagnostic for weather and climate analyses.

Appendix A

Introduction to Some Useful Mathematical Tools

To understand the physics of weather dynamics, it is necessary to use mathematical tools. Therefore, we introduce some mathematical tools which are required for the chapters of the book.

A.1 Vectors

Physics happens in a three-dimensional space which we can assume in our context to be an **Euclidean space**. The location of a point in space is determined by the three **coordinates** x, y and z. The location can be visualized graphically in a **Cartesian coordinate system** by a **vector** that points from the origin to the respective location (see Fig. A.1). We can interpret the three coordinates as **components** of the **location vector**

$$\mathbf{r} = x\mathbf{e}_x + y\mathbf{e}_y + z\mathbf{e}_z \qquad (A.1)$$

where $\mathbf{e}_x$, $\mathbf{e}_y$ and $\mathbf{e}_z$ denote the unit vectors of the coordinate system. They point to the directions of the respective coordinate axes. Usually, vectors are highlighted with a bold letter while scalars and components are written in italic. The shorter notations $\mathbf{r} = (x, y, z)$ (**row vector**) or $\mathbf{r} = \begin{pmatrix} x \\ y \\ z \end{pmatrix}$ (**column vector**) are also used to codify a vector. The length of a vector is in Euclidean space the distance to

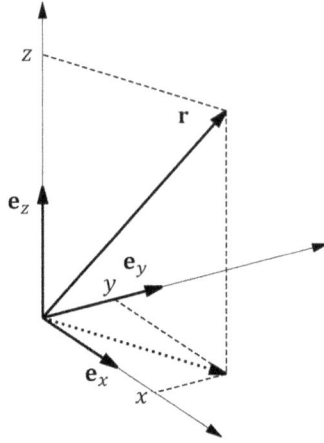

Fig. A.1. Representation of the location vector in the Cartesian coordinate system. The coordinates x, y and z result from a projection of the vector on the three coordinate axes.

the origin which results from the equation

$$|\mathbf{r}| = \sqrt{x^2 + y^2 + z^2} \tag{A.2}$$

This equation constitutes the **Euclidean norm** of a vector.

The unit vectors have by definition the length of one unit. The vector equation

$$\mathbf{a} = \mathbf{b} \tag{A.3}$$

always implies that all components have to be identical. Consequently, addition rules also apply to vectors. Therefore, we add two vectors to obtain a third vector:

$$\mathbf{c} = \mathbf{a} + \mathbf{b} \tag{A.4}$$

Equation (A.4) corresponds to three equations for the components

$$c_x = a_x + b_x, \quad c_y = a_y + b_y, \quad c_z = a_z + b_z \tag{A.5}$$

The vector addition can also be done graphically as shown in Fig. A.2. Obviously, the additions $\mathbf{a} + \mathbf{b}$ and $\mathbf{b} + \mathbf{a}$ lead to the same result. Therefore, vector addition is commutative.

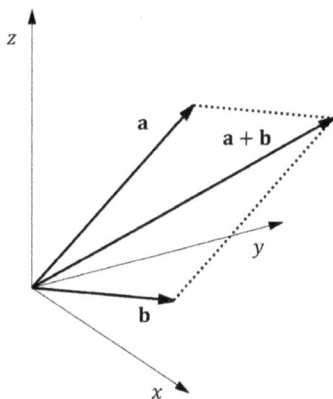

Fig. A.2. Addition of two vectors by drawing a parallelogram.

Furthermore, we find the following calculation rules:

$$a|\mathbf{a}| = |a\mathbf{a}|, \quad a(\mathbf{a} + \mathbf{b}) = a\mathbf{a} + a\mathbf{b}, \quad (a + b)\mathbf{a} = a\mathbf{a} + b\mathbf{a} \quad (A.6)$$

The **scalar product** is an important tool in vector algebra. It is defined by

$$\mathbf{a} \cdot \mathbf{b} = a_x b_x + a_y b_y + a_z b_z \quad (A.7)$$

Obviously, we see that

$$\mathbf{a} \cdot \mathbf{a} = |\mathbf{a}|^2 \quad (A.8)$$

Therefore, the scalar product between two identical vectors yields the Euclidean norm squared of this vector. For the scalar product, the calculation rules

$$\mathbf{a} \cdot \mathbf{b} = \mathbf{b} \cdot \mathbf{a}, \quad \mathbf{a} \cdot (\mathbf{b} + \mathbf{c}) = \mathbf{a} \cdot \mathbf{b} + \mathbf{a} \cdot \mathbf{c}, \quad a\mathbf{a} \cdot \mathbf{b} = (a\mathbf{a}) \cdot \mathbf{b} = \mathbf{a} \cdot (a\mathbf{b})$$
$$(A.9)$$

hold. The scalar products between the unit vectors become

$$\mathbf{e}_x \cdot \mathbf{e}_x = \mathbf{e}_y \cdot \mathbf{e}_y = \mathbf{e}_z \cdot \mathbf{e}_z = 1, \quad \mathbf{e}_x \cdot \mathbf{e}_y = \mathbf{e}_x \cdot \mathbf{e}_z = \mathbf{e}_y \cdot \mathbf{e}_z = 0$$
$$(A.10)$$

A set of vectors with these properties form an **orthogonal basis** of Euclidean space. The directions of orthogonal basis vectors are not unique and the transformation from one orthogonal basis to another is referred to as an **orthogonal transformation**.

The scalar product can also be geometrically interpreted in Euclidean space by the relation

$$\mathbf{a} \cdot \mathbf{b} = |\mathbf{a}||\mathbf{b}| \cos \alpha \qquad (A.11)$$

where α denotes the angle between the two vectors. The geometrical determination of the scalar product is sketched in Fig. A.3. First, we draw perpendicularly a line from vector $\mathbf{a}$ onto $\mathbf{b}$. The resulting distance yields $|\mathbf{a}| \cos \alpha$. Therefore, the scalar product is identical to the area of the rectangle with the edge lengths $|\mathbf{a}| \cos \alpha$ and $|\mathbf{b}|$. For parallel vectors, the result is just the product of their Euclidean norms. The same holds for antiparallel vectors, the only difference being that the product is negative. The scalar product vanishes in the case of perpendicular vectors because $\cos \pi/2 = \cos 3\pi/2 = 0$.

The **vector product** is another important tool in vector algebra. It can also be defined on the basis of geometrical considerations in Euclidean space. If

$$\mathbf{c} = \mathbf{a} \times \mathbf{b} \qquad (A.12)$$

defines the vector product between vector $\mathbf{a}$ and $\mathbf{b}$, then

1. $\mathbf{c}$ is perpendicular to both $\mathbf{a}$ and $\mathbf{b}$.
2. $\mathbf{a}, \mathbf{b}$ and $\mathbf{c}$ constitute a right-handed vector basis (right-hand rule).
3. The Euclidean norm of $\mathbf{c}$ is identical to the parallelogram area spanned by $\mathbf{a}$ and $\mathbf{b}$. Therefore, the relation $|\mathbf{c}| = |\mathbf{a} \times \mathbf{b}| = |\mathbf{a}||\mathbf{b}| \sin \alpha$ holds.

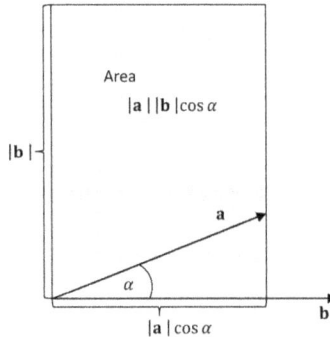

Fig. A.3.   Sketch of the scalar product. The scalar product is identical to the area of the shaded rectangle.

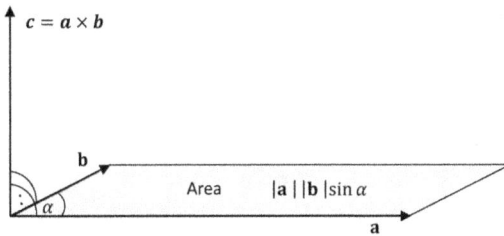

Fig. A.4. Sketch of the geometrical determination of the vector product.

Figure A.4 sketches the vector product. The resulting vector is, because of 1, perpendicular to the parallelogram area spanned by **a** and **b**. The right-hand rules state that **a** is directed rightward with respect to **b** when vector $\mathbf{c} = \mathbf{a} \times \mathbf{b}$ is assumed to be the vertical direction. Finally, the vector product is uniquely determined by assigning the Euclidean norm of **c** with the parallelogram area.

For the vector product, the following calculation rules apply:

$$\mathbf{a} \times \mathbf{b} = -\mathbf{b} \times \mathbf{a}, \quad \mathbf{a} \times (\mathbf{b} + \mathbf{c}) = \mathbf{a} \times \mathbf{b} + \mathbf{a} \times \mathbf{c},$$

$$a\mathbf{a} \times \mathbf{b} = (a\mathbf{a}) \times \mathbf{b} = \mathbf{a} \times (a\mathbf{b}) \quad \text{(A.13)}$$

From the geometrical considerations, we find the following relations for the unit vectors:

$$\mathbf{e}_x \times \mathbf{e}_x = \mathbf{e}_y \times \mathbf{e}_y = \mathbf{e}_z \times \mathbf{e}_z = 0,$$

$$\mathbf{e}_x \times \mathbf{e}_y = \mathbf{e}_z, \quad \mathbf{e}_y \times \mathbf{e}_z = \mathbf{e}_x, \quad \mathbf{e}_z \times \mathbf{e}_x = \mathbf{e}_y \quad \text{(A.14)}$$

Since every vector can be represented as in (A.1), it can be shown with (A.13) and (A.14) that

$$\mathbf{a} \times \mathbf{b} = (a_y b_z - a_z b_y)\mathbf{e}_x + (a_z b_x - a_x b_z)\mathbf{e}_y + (a_x b_y - a_y b_x)\mathbf{e}_z \quad \text{(A.15)}$$

The scalar product can be combined with the vector product to form the so-called **triple product**:

$$V = \mathbf{a} \cdot (\mathbf{b} \times \mathbf{c}) \quad \text{(A.16)}$$

The result is identical to the volume of the parallelepiped shown in Fig. A.5. One can easily verify this in the following way. First, the

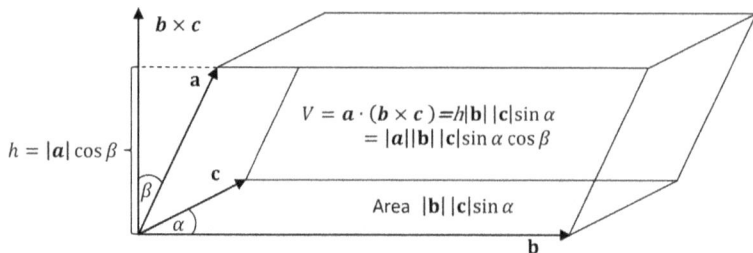

Fig. A.5.   Sketch of a parallelepiped and its relation to the triple product.

Euclidean norm of the vector product yields the area of the bottom parallelogram on which the result vector stands perpendicular. Second, the projection of vector $\mathbf{a}$ on the unit vector $\mathbf{b} \times \mathbf{c}/|\mathbf{b} \times \mathbf{c}|$ can be identified with the height h of the parallelepiped. Therefore, the relation $h = \mathbf{a} \cdot \mathbf{b} \times \mathbf{c}/|\mathbf{b} \times \mathbf{c}|$ holds which proves the identity (A.16)

The triple product remains identical by a circular shift of its factors. Hence, we have

$$\mathbf{a} \cdot (\mathbf{b} \times \mathbf{c}) = \mathbf{c} \cdot (\mathbf{a} \times \mathbf{b}) = \mathbf{b} \cdot (\mathbf{c} \times \mathbf{a}) \qquad (A.17)$$

Using the scalar and vector product, one can find the following useful algebraic vector identities:

Graßmann identity:

$$\mathbf{a} \times (\mathbf{b} \times \mathbf{c}) = (\mathbf{a} \cdot \mathbf{c})\mathbf{b} - (\mathbf{a} \cdot \mathbf{b})\mathbf{c} \qquad (A.18)$$

Jacoby identity:

$$\mathbf{a} \times (\mathbf{b} \times \mathbf{c}) + \mathbf{b} \times (\mathbf{c} \times \mathbf{a}) + \mathbf{c} \times (\mathbf{a} \times \mathbf{b}) = 0 \qquad (A.19)$$

Lagrange identity:

$$(\mathbf{a} \times \mathbf{b}) \cdot (\mathbf{c} \times \mathbf{d}) = (\mathbf{a} \cdot \mathbf{c})(\mathbf{b} \cdot \mathbf{d}) - (\mathbf{b} \cdot \mathbf{c})(\mathbf{a} \cdot \mathbf{d}) \qquad (A.20)$$

Vector product of two vector products:

$$(\mathbf{a} \times \mathbf{b}) \times (\mathbf{c} \times \mathbf{d}) = \mathbf{b}[\mathbf{a} \cdot (\mathbf{c} \times \mathbf{d})] - \mathbf{a}[\mathbf{b} \cdot (\mathbf{c} \times \mathbf{d})] \qquad (A.21)$$

These identities can be verified by evaluating them in component form.

A.2 Tensors[a]

A **dyadic tensor** results from the **tensor product** between two vectors. It is written as

$$\mathbf{D} = \mathbf{a} \otimes \mathbf{b} \tag{A.22}$$

Often, the shorter notation $\mathbf{D} = \mathbf{ab}$ is used. Then, a tensor product applies if no symbol appears between two vectors. We will do it likewise. A dyadic tensor can be multiplied with vectors in the following way:

$$\mathbf{c} \cdot \mathbf{D} = \mathbf{c} \cdot (\mathbf{ab}) = (\mathbf{c} \cdot \mathbf{a})\mathbf{b} \tag{A.23}$$

$$\mathbf{D} \cdot \mathbf{c} = (\mathbf{ab}) \cdot \mathbf{c} = (\mathbf{b} \cdot \mathbf{c})\mathbf{a} \tag{A.24}$$

Hence, the dot product of a vector and a dyadic tensor yields a vector. Dyadic tensor can also be added

$$\mathbf{D} = \mathbf{ab} + \mathbf{cd} + \mathbf{ef} + \cdots \tag{A.25}$$

and the following rules apply for the combination of a sum with the dot product:

$$\mathbf{c} \cdot (\mathbf{D}_1 + \mathbf{D}_2) = \mathbf{c} \cdot \mathbf{D}_1 + \mathbf{c} \cdot \mathbf{D}_2 \tag{A.26}$$

$$(\mathbf{D}_1 + \mathbf{D}_2) \cdot \mathbf{c} = \mathbf{D}_1 \cdot \mathbf{c} + \mathbf{D}_2 \cdot \mathbf{c} \tag{A.27}$$

Using these rules, an arbitrary tensor can be represented in component form by

$$\begin{aligned}
\mathbf{D} = &D_{xx}\mathbf{e}_x\mathbf{e}_x + D_{xy}\mathbf{e}_x\mathbf{e}_y + D_{xz}\mathbf{e}_x\mathbf{e}_z \\
&+ D_{yx}\mathbf{e}_y\mathbf{e}_x + D_{yy}\mathbf{e}_y\mathbf{e}_y +_{yz}\mathbf{e}_y\mathbf{e}_z \\
&+ D_{zx}\mathbf{e}_z\mathbf{e}_x + D_{zy}\mathbf{e}_z\mathbf{e}_y + D_{zz}\mathbf{e}_z\mathbf{e}_z
\end{aligned} \tag{A.28}$$

This is not necessarily a dyadic tensor since it has nine components which can be independent while two vectors have six components.

---

[a]Here, we restrict to tensors of second order. In general, a tensor can also have a higher order or one which is simply a vector.

In this way, the tensor can be treated like a matrix and the products with vectors can be done in the usual way

$$\mathbf{D} \cdot \mathbf{a} = \begin{pmatrix} D_{xx} & D_{xy} & D_{xz} \\ D_{yx} & D_{yy} & D_{yz} \\ D_{zx} & D_{zy} & D_{zz} \end{pmatrix} \cdot \begin{pmatrix} a_x \\ a_y \\ a_z \end{pmatrix} = \begin{pmatrix} D_{xx}a_x + D_{xy}a_y + D_{xz}a_z \\ D_{yx}a_x + D_{yy}a_y + D_{yz}a_z \\ D_{zx}a_x + D_{zy}a_y + D_{zz}a_z \end{pmatrix}$$

$$(A.29)$$

or

$$\mathbf{a} \cdot \mathbf{D} = (a_x, a_y, a_z) \cdot \begin{pmatrix} D_{xx} & D_{xy} & D_{xz} \\ D_{yx} & D_{yy} & D_{yz} \\ D_{zx} & D_{zy} & D_{zz} \end{pmatrix}$$

$$= \begin{pmatrix} D_{xx}a_x + D_{yx}a_y + D_{zx}a_z \\ D_{xy}a_x + D_{yy}a_y + D_{zy}a_z \\ D_{xz}a_x + D_{yz}a_y + D_{zz}a_z \end{pmatrix} \qquad (A.30)$$

where the shorter notation for vectors has been used. Tensors can be transposed like a matrix. Transposition implies a permutation of row and line indices. Therefore, the **transpose** of tensor $\mathbf{D}$ becomes:

$$\mathbf{D}^{\mathrm{T}} = \begin{pmatrix} D_{xx} & D_{yx} & D_{zx} \\ D_{xy} & D_{yy} & D_{zy} \\ D_{xz} & D_{yz} & D_{zz} \end{pmatrix} \qquad (A.31)$$

The **dot product** and the **double dot product** can be applied to dyadic tensors. The first one is defined by

$$\mathbf{ab} \cdot \mathbf{cd} = \mathbf{a}(\mathbf{b} \cdot \mathbf{c})\mathbf{d} = (\mathbf{b} \cdot \mathbf{c})\mathbf{ad} \qquad (A.32)$$

Consequently, the dot product between two tensors yields another dyadic tensor. (It corresponds to the multiplication of two matrices.) In contrast, the double dot product leads to a scalar. It is defined by

$$\mathbf{ab} : \mathbf{cd} = (\mathbf{a} \cdot \mathbf{c})(\mathbf{b} \cdot \mathbf{d}) \qquad (A.33)$$

Obviously, the double dot product is commutative so that $\mathbf{ab} : \mathbf{cd} = \mathbf{cd} : \mathbf{ab}$. With the decomposition (A.28), we can also define the double dot product for general tensors, i.e. $\mathbf{C} : \mathbf{D}$ using the rule $(\mathbf{ab} + \mathbf{cd}) : \mathbf{ef} = \mathbf{ab} : \mathbf{ef} + \mathbf{cd} : \mathbf{ef}$. Therefore, the double dot product just results by adding all products of the same matrix elements of both tensors.

A.3 Nabla Operator

The laws of atmospheric dynamics are based on the notion that the air constitutes a continuous fluid. Then, the state of the atmosphere can be described by continuous **field functions** of the form

$$b = b(x, y, z) \equiv b(\mathbf{r}) \tag{A.34}$$

Often, it is necessary to differentiate these field functions by a **partial derivative**. The partial derivative with respect x to results from the limiting value of the following expression:

$$\frac{\partial b}{\partial x} \equiv \lim_{\Delta x \to 0} \frac{b(x + \Delta x, y, z) - b(x, y, z)}{\Delta x} \tag{A.35}$$

The partial derivative with respect to y and z reads likewise. A field function can be visualized by isosurfaces in space as shown in Fig. A.6(a). Then, the partial derivative can be approximately deduced by dividing the difference of the function value of two isosurfaces by the distance along the respective coordinate direction.

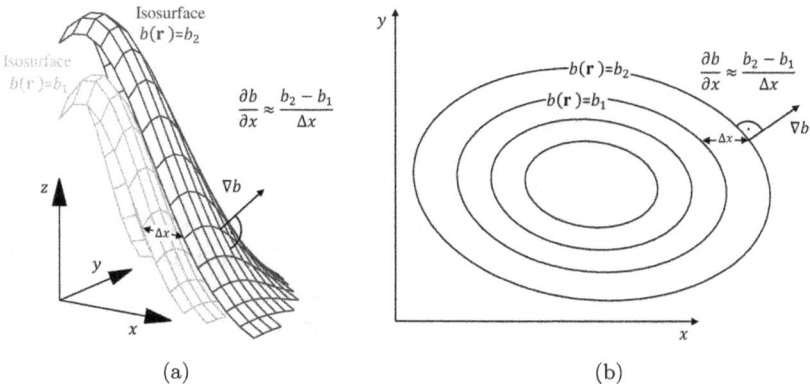

(a)   (b)

Fig. A.6. Graphical representation of a field function by (a) isosurfaces and (b) isolines. The partial derivative $\partial b/\partial x$ approximately results by taking the difference $b_2 - b_1$ and dividing the result by Δx. The gradient of a field function is always perpendicular to an isosurface or isoline.

Partial derivatives are commutable and, therefore, identities like

$$\frac{\partial}{\partial x}\left(\frac{\partial b}{\partial y}\right) = \frac{\partial}{\partial y}\left(\frac{\partial b}{\partial x}\right) \tag{A.36}$$

hold. For a fixed z the field function only depends on the two coordinates x and y. Then, it is convenient to visualize the field function by isolines as sketched in Fig. A.6(b). This is indeed common practice for the operational weather forecast where meteorologically relevant fields are plotted on weather maps in the form of isolines.

The **Nabla operator** ∇ represents an important differential vector operator that can be applied to field functions. It is given by

$$\nabla \equiv \mathbf{e}_x\frac{\partial}{\partial x} + \mathbf{e}_y\frac{\partial}{\partial y} + \mathbf{e}_z\frac{\partial}{\partial z} \tag{A.37}$$

We can operate with the Nabla operator in a similar way as with a vector. Gradient, divergence and rotation constitute the most important three applications of the Nabla operator. They are described in the following.

A.3.1   *Gradient*

The gradient is applied to a scalar field function to give

$$\nabla b \equiv \frac{\partial b}{\partial x}\mathbf{e}_x + \frac{\partial b}{\partial y}\mathbf{e}_y + \frac{\partial b}{\partial z}\mathbf{e}_z \tag{A.38}$$

Therefore, the gradient generates a **vector field** which means that a vector is defined at every location in space. The gradient can be interpreted geometrically since it is perpendicular to the isosurface on which the origin of gradient lies (see Fig. A.6(a)). We can prove this result by considering the difference in the field function between two neighboring points $\mathbf{r}$ and $\mathbf{r} + \Delta\mathbf{r}$. For the difference, we obtain approximately

$$\Delta b = b(\mathbf{r} + \Delta\mathbf{r}) - b(\mathbf{r}) \approx \frac{\partial b}{\partial x}\Delta x + \frac{\partial b}{\partial y}\Delta y + \frac{\partial b}{\partial z}\Delta z = \nabla b \cdot \Delta\mathbf{r}$$

$$\tag{A.39}$$

This approximation turns into an identity in the limit $\Delta \mathbf{r} \to \mathbf{0}$. Then, the difference Δb becomes infinitesimally small and one obtains the **total differential**

$$db \equiv \lim_{\Delta \mathbf{r} \to \mathbf{0}} \nabla b \cdot \Delta \mathbf{r} \equiv \nabla b \cdot d\mathbf{r} \tag{A.40}$$

We get $db = 0$ if $d\mathbf{r}$ is tangential to the isosurface since the b value does not change in this case.[b] Consequently, ∇b is perpendicular to the isosurface.

We can use the gradient to determine the **directional derivative** that specifies the change of a field function per segment at $\mathbf{r} = \mathbf{r}_0$ in a certain direction. Let l be a coordinate in the direction $\mathbf{e}_l$ and $b_l(l)$ the function which coincides with the field function b along the line $\mathbf{r}(l) = \mathbf{r}_0 + l\mathbf{e}_l$. Consequently, we have formally the representation $b_l(l) = b[\mathbf{r}(l)]$. With the chain rule for differentiation, we find for the directional derivative

$$\frac{db_l}{dl} = \frac{\partial b}{\partial x}\frac{dx}{dl} + \frac{\partial b}{\partial y}\frac{dy}{dl} + \frac{\partial b}{\partial z}\frac{dz}{dl} = \nabla b \cdot \frac{d\mathbf{r}}{dl} = \nabla b \cdot \frac{d(l\mathbf{e}_l)}{dl} = \nabla b \cdot \mathbf{e}_l \tag{A.41}$$

Obviously, the directional derivative maximizes when $\mathbf{e}_l$ directs parallel to the gradient ∇b. Therefore, the gradient specifies the direction of maximum positive change per segment.

Applying the rules for differentiation, one can easily verify the following identities

$$\nabla(a + b) = \nabla a + \nabla b, \quad \nabla(ab) = a\nabla b + b\nabla a \tag{A.42}$$

A.3.2 *Divergence*

It is also possible to apply the Nabla operator to a vector field. One operation is the scalar product between ∇ and a vector field $\mathbf{a}(\mathbf{r})$.

---

[b]The fact that the limit $\Delta \mathbf{r} \to \mathbf{0}$ leads to vanishing differences Δb may create confusion. Therefore, it is better for the understanding to divide the differential db by another quantity that vanishes in the limit $\Delta \mathbf{r} \to \mathbf{0}$. With the directional derivative treated below, one can circumvent the problem in this way.

This operation constitutes the **divergence** of $\mathbf{a}(\mathbf{r})$. We simply get by evaluating the dot product:

$$\nabla \cdot \mathbf{a} = \left(\mathbf{e}_x \frac{\partial}{\partial x} + \mathbf{e}_y \frac{\partial}{\partial y} + \mathbf{e}_z \frac{\partial}{\partial z} \right) \cdot (a_x \mathbf{e}_x + a_y \mathbf{e}_y + a_z \mathbf{e}_z)$$

$$= \frac{\partial a_x}{\partial x} + \frac{\partial a_y}{\partial y} + \frac{\partial a_z}{\partial z} \tag{A.43}$$

Thus, the divergence applies to a vector field and yields a scalar field function. Combining divergence with gradient gives

$$\nabla \cdot \nabla b = \frac{\partial^2 b}{\partial x^2} + \frac{\partial^2 b}{\partial y^2} + \frac{\partial^2 b}{\partial z^2} \tag{A.44}$$

The associated operator $\nabla \cdot \nabla \equiv \nabla^2$ is known as the **Laplace operator**. For the divergence, the following calculation rule applies:

$$\nabla \cdot (b\mathbf{a}) = \nabla b \cdot \mathbf{a} + b \nabla \cdot \mathbf{a} \tag{A.45}$$

A.3.3 Curl

With the **curl**, we conduct the vector product of the Nabla operator and a vector field. We get

$$\nabla \times \mathbf{a} = \left(\mathbf{e}_x \frac{\partial}{\partial x} + \mathbf{e}_y \frac{\partial}{\partial y} + \mathbf{e}_z \frac{\partial}{\partial z} \right) \times (a_x \mathbf{e}_x + a_y \mathbf{e}_y + a_z \mathbf{e}_z)$$

$$= \left(\frac{\partial a_z}{\partial y} - \frac{\partial a_y}{\partial z} \right) \mathbf{e}_x + \left(\frac{\partial a_x}{\partial z} - \frac{\partial a_z}{\partial x} \right) \mathbf{e}_y + \left(\frac{\partial a_y}{\partial x} - \frac{\partial a_x}{\partial y} \right) \mathbf{e}_z$$

$$\tag{A.46}$$

Thus, the curl applies to a vector field and yields, in contrast to the divergence, a vector field. Combining the curl with the gradient yields

$$\nabla \times \nabla b = \left(\frac{\partial^2 b}{\partial y \partial z} - \frac{\partial^2 b}{\partial z \partial y} \right) \mathbf{e}_x + \left(\frac{\partial^2 b}{\partial z \partial x} - \frac{\partial^2 b}{\partial x \partial z} \right) \mathbf{e}_y$$

$$+ \left(\frac{\partial^2 b}{\partial x \partial y} - \frac{\partial^2 b}{\partial y \partial x} \right) \mathbf{e}_z = 0 \tag{A.47}$$

because partial derivatives are commutable. Therefore, the curl of a gradient always vanishes. That is, gradient fields are **curl-free**.

Furthermore, we find

$$\nabla \cdot (\nabla \times \mathbf{a}) = \frac{\partial}{\partial x}\left(\frac{\partial a_z}{\partial y} - \frac{\partial a_y}{\partial z}\right) + \frac{\partial}{\partial y}\left(\frac{\partial a_x}{\partial z} - \frac{\partial a_z}{\partial x}\right)$$

$$+ \frac{\partial}{\partial z}\left(\frac{\partial a_y}{\partial x} - \frac{\partial a_x}{\partial y}\right) = 0 \tag{A.48}$$

Consequently, the divergence of the vector curl also vanishes. It is said that curl fields are **source-free**. A substantiation of this term will follow below. For the curl, we have the following calculation rules:

$$\nabla \times (b\mathbf{a}) = \nabla b \times \mathbf{a} + b\nabla \times \mathbf{a} \tag{A.49}$$

$$\nabla \cdot (\mathbf{a} \times \mathbf{b}) = \mathbf{b} \cdot \nabla \times \mathbf{a} - \mathbf{a} \cdot \nabla \times \mathbf{b} \tag{A.50}$$

$$\nabla \times (\nabla \times \mathbf{a}) = \nabla(\nabla \cdot \mathbf{a}) - \nabla^2 \mathbf{a} \tag{A.51}$$

These identities can be verified by evaluating the operations in component form.

A.3.4 *Helmholtz theorem*

The **Helmholtz theorem** states that every three-dimensional vector field can be decomposed into a curl-free and source-free contribution. Then, the vector field $\mathbf{a}$ can be expressed by the negative gradient of a **scalar potential** $\chi(\mathbf{r})$ and the curl of a **vector potential** $\boldsymbol{\Psi}(\mathbf{r})$ so that

$$\mathbf{a} = \nabla\chi + \nabla \times \boldsymbol{\Psi} \tag{A.52}$$

The decomposition is not unique because a part of a vector field could be both curl- and source-free. The Helmholtz theorem also applies in two dimensions:

$$\mathbf{a} = \frac{\partial\chi}{\partial x}\mathbf{e}_x + \frac{\partial\chi}{\partial y}\mathbf{e}_y + \left(\frac{\partial\psi_z}{\partial y} - \frac{\partial\psi_y}{\partial z}\right)\mathbf{e}_x + \left(\frac{\partial\psi_x}{\partial z} - \frac{\partial\psi_z}{\partial x}\right)\mathbf{e}_y$$

$$\tag{A.53}$$

Without loss of generality, we can assume vertical independence for two-dimensional vector fields so that $\partial\psi_x/\partial z = \partial\psi_y/\partial z = 0$.

Then, by introducing the **stream function** $\psi \equiv -\psi_z$ we obtain the decomposition

$$\mathbf{a} = \frac{\partial \chi}{\partial x}\mathbf{e}_x + \frac{\partial \chi}{\partial y}\mathbf{e}_y - \frac{\partial \psi}{\partial y}\mathbf{e}_x + \frac{\partial \psi}{\partial x}\mathbf{e}_y = \nabla_h \chi + \mathbf{e}_z \times \nabla_h \psi \quad \text{(A.54)}$$

where ∇_h denotes the two-dimensional Nabla operator applying only to the horizontal coordinates x and y.

A.3.5 *Nabla in connection with tensors*

The Nabla operator can also be combined with a vector in the form of a tensor product. Such a product results in

$$\begin{aligned} \nabla \mathbf{a} = &\frac{\partial a_x}{\partial x}\mathbf{e}_x\mathbf{e}_x + \frac{\partial a_y}{\partial x}\mathbf{e}_x\mathbf{e}_y + \frac{\partial a_z}{\partial x}\mathbf{e}_x\mathbf{e}_z \\ &+\frac{\partial a_x}{\partial y}\mathbf{e}_y\mathbf{e}_x + \frac{\partial a_y}{\partial y}\mathbf{e}_y\mathbf{e}_y + \frac{\partial a_z}{\partial y}\mathbf{e}_y\mathbf{e}_z \\ &+\frac{\partial a_x}{\partial z}\mathbf{e}_z\mathbf{e}_x + \frac{\partial a_y}{\partial z}\mathbf{e}_z\mathbf{e}_y + \frac{\partial a_z}{\partial z}\mathbf{e}_z\mathbf{e}_z \quad \text{(A.55)} \end{aligned}$$

On the other hand, the Nabla operator can also be multiplied with a dyadic tensor by the dot product:

$$\begin{aligned} \nabla \cdot (\mathbf{ab}) = &\frac{\partial}{\partial x}(a_x b_x)\mathbf{e}_x + \frac{\partial}{\partial x}(a_x b_y)\mathbf{e}_y + \frac{\partial}{\partial x}(a_x b_z)\mathbf{e}_z \\ &+\frac{\partial}{\partial y}(a_y b_x)\mathbf{e}_x + \frac{\partial}{\partial y}(a_y b_y)\mathbf{e}_y + \frac{\partial}{\partial y}(a_y b_z)\mathbf{e}_z \\ &+\frac{\partial}{\partial z}(a_z b_x)\mathbf{e} + \frac{\partial}{\partial z}(a_z b_y)\mathbf{e}_y + \frac{\partial}{\partial z}(a_z b_z)\mathbf{e}_z \\ = &(\nabla \cdot \mathbf{a})\mathbf{b} + \mathbf{a} \cdot \nabla \mathbf{b} \quad \text{(A.56)} \end{aligned}$$

Other Nabla operations that include tensors are as follows:

$$\nabla(\mathbf{a} \cdot \mathbf{b}) = \mathbf{a} \cdot \nabla \mathbf{b} + \mathbf{b} \cdot \nabla \mathbf{a} + \mathbf{a} \times (\nabla \times \mathbf{b}) + \mathbf{b} \times (\nabla \times \mathbf{a}) \quad \text{(A.57)}$$

$$\nabla \times (\mathbf{a} \times \mathbf{b}) = \mathbf{a}\nabla \cdot \mathbf{b} - \mathbf{b}\nabla \cdot \mathbf{a} + \mathbf{b} \cdot \nabla \mathbf{a} - \mathbf{a} \cdot \nabla \mathbf{b} \quad \text{(A.58)}$$

$$\mathbf{a} \cdot \nabla \mathbf{a} = \frac{1}{2}\nabla(\mathbf{a} \cdot \mathbf{a}) - \mathbf{a} \times (\nabla \times \mathbf{a}) \quad \text{(A.59)}$$

$$\nabla \cdot (\mathbf{D} \cdot \mathbf{a}) = \mathbf{a} \cdot (\nabla \cdot \mathbf{D}) + \mathbf{D} : \nabla \mathbf{a} \quad \text{(A.60)}$$

In the last equation, $\mathbf{D}$ denotes a tensor. These identities can be proven by evaluation in component form.

A.3.6 *Nabla in orthogonal curvilinear coordinates*

Coordinate systems of atmospheric models have often curvilinear coordinate lines. For example, the spherical geometry of Earth suggests the use of spherical coordinates. Consequently, the Nabla operators must be transformed from a Cartesian to a **curvilinear coordinate system**. Let $q_1(\mathbf{r})$, $q_2(\mathbf{r})$ and $q_3(\mathbf{r})$ be the curvilinear coordinates. Then, we can compute the gradient by the chain rule of differentiation

$$\nabla b = \nabla q_1 \frac{\partial b}{\partial q_1} + \nabla q_2 \frac{\partial b}{\partial q_2} + \nabla q_3 \frac{\partial b}{\partial q_3} \tag{A.61}$$

The gradient of curvilinear coordinates is parallel to the coordinate lines in the special case of orthogonality. The gradient of one coordinate is perpendicular to the directions of the other two coordinates lines since these must be parallel to the isosurface of the one coordinate. Consequently, the gradient of a coordinate is parallel to the respective coordinate line. This is sketched in Fig. A.7. We find for the unit vectors of the orthogonal curvilinear coordinate system

$$\mathbf{e}_{q_1} = \frac{\nabla q_1}{|\nabla q_1|}, \quad \mathbf{e}_{q_2} = \frac{\nabla q_2}{|\nabla q_2|}, \quad \mathbf{e}_{q_3} = \frac{\nabla q_3}{|\nabla q_3|} \tag{A.62}$$

The Euclidean norms in the denominators define the **Lamé coefficients** which are given by

$$h_{q_1} = 1/|\nabla q_1|, \quad h_{q_2} = 1/|\nabla q_2|, \quad h_{q_3} = 1/|\nabla q_3| \tag{A.63}$$

With these definitions, we can write the gradient as

$$\nabla b = \frac{1}{h_{q_1}} \frac{\partial b}{\partial q_1} \mathbf{e}_{q_1} + \frac{1}{h_{q_2}} \frac{\partial b}{\partial q_2} \mathbf{e}_{q_2} + \frac{1}{h_{q_3}} \frac{\partial b}{\partial q_3} \mathbf{e}_{q_3} \tag{A.64}$$

To derive the divergence in the orthogonal curvilinear coordinates, we make use of the rules (A.14) for a right-handed orthogonal vector

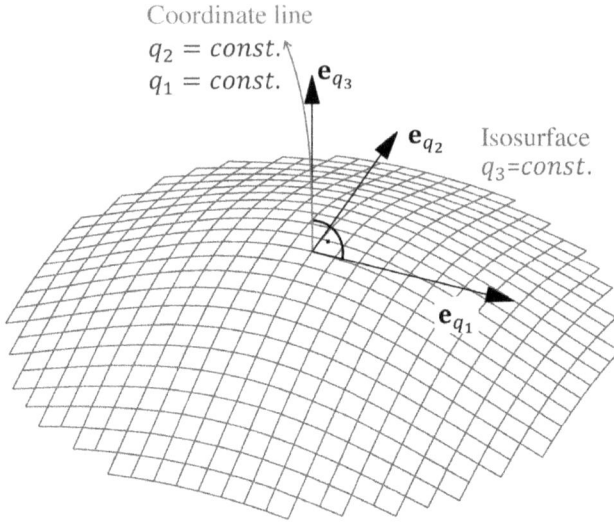

Fig. A.7. Orientation of the unit vectors in an orthogonal curvilinear coordinate system.

basis: Then, we get for the divergence of one vector component

$$\nabla \cdot (a_{q_1} \mathbf{e}_{q_1}) = \nabla \cdot (a_{q_1} \mathbf{e}_{q_2} \times \mathbf{e}_{q_3}) = \nabla \cdot (a_{q_1} h_{q_2} h_{q_3} \nabla q_2 \times \nabla q_3)$$

$$= a_{q_1} h_{q_2} h_{q_3} \nabla \cdot (\nabla q_2 \times \nabla q_3) + \frac{\mathbf{e}_{q_1}}{h_{q_2} h_{q_3}} \cdot \nabla (h_{q_2} h_{q_3} a_{q_1})$$

$$= \frac{1}{h_{q_1} h_{q_2} h_{q_3}} \frac{\partial}{\partial q_1} (h_{q_2} h_{q_3} a_{q_1}) \tag{A.65}$$

Equations (A.47), (A.50) and (A.64) have been used for the last conversion. Applying the procedure to the other two vector components yields the divergence in orthogonal curvilinear coordinates:

$$\nabla \cdot \mathbf{a} = \frac{1}{h_{q_1} h_{q_2} h_{q_3}}$$

$$\times \left[\frac{\partial}{\partial q_1} (h_{q_2} h_{q_3} a_{q_1}) + \frac{\partial}{\partial q_2} (h_{q_1} h_{q_3} a_{q_2}) + \frac{\partial}{\partial q_3} (h_{q_1} h_{q_2} a_{q_3}) \right] \tag{A.66}$$

For determining the rotation, we also apply the operator to one component first:

$$\nabla \times (a_{q_1} \mathbf{e}_{q_1}) = \nabla \times (a_{q_1} h_{q_1} \nabla q_1)$$

$$= a_{q_1} h_{q_1} \nabla \times \nabla q_1 + \nabla (a_{q_1} h_{q_1}) \times \nabla q_1$$

$$= \frac{1}{h_{q_1}} \nabla (a_{q_1} h_{q_1}) \times \mathbf{e}_{q_1} \qquad \text{(A.67)}$$

Evaluating the result in component form yields

$$\nabla \times (a_{q_1} \mathbf{e}_{q_1}) = \frac{1}{h_{q_1} h_{q_3}} \frac{\partial}{\partial q_3} (a_{q_1} h_{q_1}) \mathbf{e}_{q_2} - \frac{1}{h_{q_1} h_{q_2}} \frac{\partial}{\partial q_2} (a_{q_1} h_{q_1}) \mathbf{e}_{q_3}$$

$$\text{(A.68)}$$

Repeating this procedure for the other two components leads to the desired result

$$\nabla \times \mathbf{a} = \frac{1}{h_{q_2} h_{q_3}} \left[\frac{\partial}{\partial q_2} (a_{q_3} h_{q_3}) - \frac{\partial}{\partial q_3} (a_{q_2} h_{q_2}) \right] \mathbf{e}_{q_1}$$

$$+ \frac{1}{h_{q_1} h_{q_3}} \left[\frac{\partial}{\partial q_3} (a_{q_1} h_{q_1}) - \frac{\partial}{\partial q_1} (a_{q_3} h_{q_3}) \right] \mathbf{e}_{q_2}$$

$$+ \frac{1}{h_{q_1} h_{q_2}} \left[\frac{\partial}{\partial q_1} (a_{q_2} h_{q_2}) - \frac{\partial}{\partial q_2} (a_{q_1} h_{q_1}) \right] \mathbf{e}_{q_3} \qquad \text{(A.69)}$$

Conversion of other Nabla operations in component form requires tedious calculations which will not be conducted here.

A.4 Integral Theorems

Integral theorems are very useful tools in hydrodynamics. In this section, **Gauss's divergence theorem** and **Stokes' theorem** are introduced.

A.4.1 *Gauss's divergence theorem*

The divergence theorem by Gauss relates a volume integral to the integral over the surface of the same volume. We can write the theorem as follows:

$$\oiint_{\partial V} \mathbf{a} \cdot \mathbf{n} \, da = \iiint_V \nabla \cdot \mathbf{a} \, dV \qquad (A.70)$$

In this equation, V denotes the volume, ∂V the closed bounding surface of the volume, da the differential area element, dV the differential volume element and $\mathbf{n}$ the **surface normal** which is the local unit vector directing outward and perpendicular to the differential surface element. The theorem indicates that the volume integral of the divergence of a vector field coincides with the surface integral of the scalar product between the vector field and the surface normal. The latter constitutes the flux of the vector field out of the volume. Therefore, the flux must have a "source" inside of the volume. This is the reason why non-divergent vector fields are called source-free.

We illustrate the validity of the divergence theorem for the example of a cuboid-shaped volume that is sketched in Fig. A.8. The

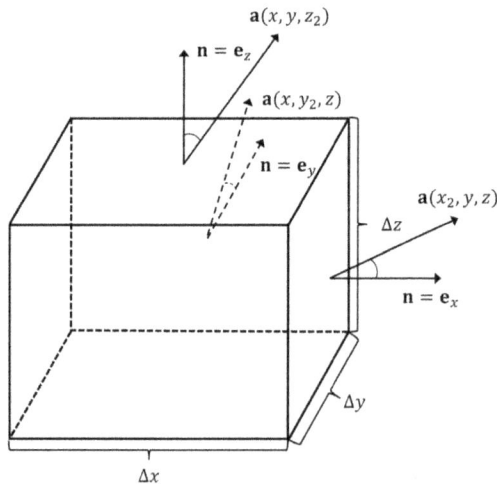

Fig. A.8. Orientation of the surface normal $\mathbf{n}$ and the vector $\mathbf{a}$ on a cuboid-shaped volume. The integral over the surface of the cuboid yields the volume integral of the divergence of $\mathbf{a}$ as predicted by Gauss's divergence theorem.

volume integral for the sketched cuboid becomes:

$$\iiint_V \nabla \cdot \mathbf{a} dV = \int_{z_1}^{z_2} \int_{y_1}^{y_2} \int_{x_1}^{x_2} \left(\frac{\partial a_x}{\partial x} + \frac{\partial a_y}{\partial y} + \frac{\partial a_z}{\partial z} \right) dxdydz$$

$$(A.71)$$

The volume integral on the right-hand side of Eq. (A.71) can be split into three solvable integrals. The first of these turns into

$$\int_{z_1}^{z_2} \int_{y_1}^{y_2} \int_{x_1}^{x_2} \frac{\partial a_x}{\partial x} dxdydz = \int_{z_1}^{z_2} \int_{y_1}^{y_2} [a_x]|_{x_1}^{x_2} dydz$$

$$= \int_{z_1}^{z_2} \int_{y_1}^{y_2} a_x(x_2, y, z) dydz - \int_{z_1}^{z_2} \int_{y_1}^{y_2} a_x(x_1, y, z) dydz$$

$$= \iint_{a_{x_2}} \mathbf{a} \cdot \mathbf{n} da + \iint_{a_{x_1}} \mathbf{a} \cdot \mathbf{n} da \qquad (A.72)$$

where a_{x_2} and a_{x_1} denote the left and right face of the cuboid, respectively (see Fig. A.8). The other two integrals can be solved likewise which proves the validity of the theorem for a cuboid-shaped volume. Gauss's divergence theorem applies to arbitrarily shaped volumes. For a proof, we refer to the mathematical literature (Spiegel *et al.*, 2009).

A.4.2 *Stokes' theorem*

Stokes' theorem relates a surface integral with a line integral along the boundary of the surface. The theorem reads

$$\iint_a (\nabla \times \mathbf{a}) \cdot \mathbf{n} da = \oint_{\partial a} \mathbf{a} \cdot \mathbf{t} dl \qquad (A.73)$$

where a is the area, ∂a the closed boundary of a, dl the differential line element and $\mathbf{t}$ the tangent unit vector directing anticlockwise along the boundary line.

Stokes' theorem can be illustrated for a surface that takes the simple form of a rectangle as sketched in Fig. A.9. The area integral

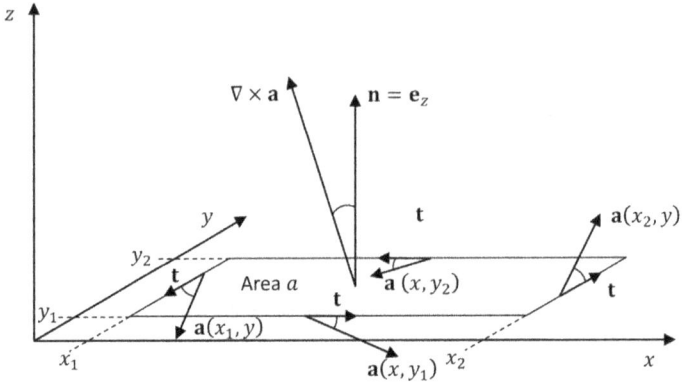

Fig. A.9. Orientation of the tangent unit vector **t** and the vector **a** along the boundary line of a rectangle. The integral along this line yields the surface integral of the curl of **a** as predicted by Stokes' theorem.

for the sketched rectangle becomes:

$$\iint_a (\nabla \times \mathbf{a}) \cdot \mathbf{n} da = \int_{y_1}^{y_2} \int_{x_1}^{x_2} \frac{\partial a_y}{\partial x} - \frac{\partial a_x}{\partial y} dx dy \qquad (A.74)$$

Again, the right-hand side can be split into solvable integrals. We get

$$\int_{y_1}^{y_2} \int_{x_1}^{x_2} \frac{\partial a_y}{\partial x} - \frac{\partial a_x}{\partial y} dx dy = \int_{y_1}^{y_2} [a_y]|_{x_1}^{x_2} dy - \int_{x_1}^{x_2} [a_x]|_{y_1}^{y_2} dx$$

$$= \int_{y_1}^{y_2} a_y(x_2, y, 0) dy - \int_{x_1}^{x_2} a_x(x, y_2, 0) dx$$

$$- \int_{y_1}^{y_2} a_y(x_1, y, 0) dy + \int_{x_1}^{x_2} a_x(x, y_1, 0) dx$$

$$= \oint_{\partial a} \mathbf{a} \cdot \mathbf{t} dl \qquad (A.75)$$

Stokes' theorem is valid for any closed boundary line and also for nonplanar surfaces.

Bibliography

Achatz, U., Schmitz, G. and Greisiger, K. (1995). Principal interaction patterns in baroclinic wave life cycles, *Journal of Atmospheric Sciences*, 52(18), pp. 3201–3213, doi: 10.1175/1520-0469(1995)052<3201:PIPIBW>2.0.CO;2.

Agee, E. M., Chen, T. S. and Dowell, K. E. (1973). A review of mesoscale cellular convection, *Bulletin of the American Meteorological Society*, 54(10), pp. 1004–1012, doi: 10.1175/1520-0477(1973)054<1004:AROMCC>2.0.CO;2.

Bénard, H. (1900). Étude expérimentale des courants de convection dans une nappe liquide. Régime permanent: Tourbillons cellulaires, *Journal of Physics: Theories and Applications*, 9(1), pp. 513–524, doi: 10.1051/jphystap: 019000090051300.

Betts, A. K. (1986). A new convective adjustment scheme. Part I: Observational and theoretical basis, *Quarterly Journal of Royal Meteorological Society*, 112, pp. 677–691, doi: 10.1002/qj.49711247307.

Betts, A. K. and Miller, M. J. (1986). A new convective adjustment scheme. Part II: Single column tests using GATE wave, BOMEX, ATEX and arctic air-mass data sets, *Quarterly Journal of Royal Meteorological Society*, 112, pp. 693–709. doi: 10.1002/qj.49711247308.

Bjerknes, J. (1898). Über einen hydrodynamischen Fundamentalsatz und seine Anwendung besonders auf die Mechanik der Atmosphäre und des Weltmeeres, *Kongl Sven. Vetensk. Akad. Handlingar*, 31, pp. 1–35.

Bjerknes, J. and Solberg, H. (1922). Life cycle of cyclones and the polar front theory of atmospheric circulation, *Geophysics Publications*, 3(1), pp. 3–18.

Blackadar, A. K. (1957). Boundary layer wind maxima and their significance for the growth of nocturnal inversions, *Bulletin of the American Meteorological Society*, 38(5), pp. 283–290, doi: 10.1175/1520-0477-38.5.283.

Blackadar, A. K. (1962). The vertical distribution of wind and turbulent exchange in a neutral atmosphere, *Journal of Geophysics Research*, 67(8), pp. 3095–3102, doi: 10.1029/JZ067i008p03095.

Blanchard, D. O., Cotton, W. R. and Brown, J. M. (1998). Mesoscale circulation growth under conditions of weak inertial instability, *Monthly Weather Review*, 126(1), pp. 118–140, doi: 10.1175/1520-0493(1998)126<0118: MCGUCO>2.0.CO;2.

Bronshtein, I. N., Semendyayev, K. A. Musiol, G. and Mühlig, H. (2015). *Handbook of mathematics*, sixth edition, Springer-Verlag, Berlin, Heidelberg, 1207 p.

Brunt, D. (1927), The period of simple vertical oscillations in the atmosphere, *Quarterly Journal of Royal Meteorological Society*, 53, pp. 30–32. doi: 10.1002/qj.49705322103.

Burpee, R. W. (1972). The origin and structure of easterly waves in the lower troposphere of North Africa, *Journal of Atmospheric Sciences*, 29(1), pp. 77–90, doi: 10.1175/1520-0469(1972)029<0077:TOASOE>2.0.CO;2.

Businger, J. A., Wyngaard, J. C., Izumi, Y. and Bradley, E. F. (1971). Flux-profile relationships in the atmospheric surface layer, *Journal of Atmospheric Sciences*, 28(2), pp. 181–189. doi: 10.1175/1520-0469(1971)028<0181: FPRITA>2.0.CO;2.

Charney, J. G. (1947). The dynamics of long waves in a baroclinic westerly current, *Journal of Atmospheric Sciences*, 4(5), pp. 136–162. doi: 10.1175/1520-0469(1947)004<0136:TDOLWI>2.0.CO;2.

Charney, J. G. (1948). On the scale of atmospheric motions, *Geophysics Publications Oslo*, 17(2), pp. 1–17.

Charney, J. G. and DeVore, J. G. (1979). Multiple flow equilibria in the atmosphere and blocking, *Journal of Atmospheric Sciences*, 36(7), pp. 1205–1216, doi: 10.1175/1520-0469(1979)036<1205:MFEITA>2.0.CO;2.

Charney, J. G. and Eliassen, A. (1949). A numerical method for predicting the perturbations of the middle latitude westerlies, *Tellus*, 1(2), pp. 38–54, doi: 10.3402/tellusa.v1i2.8500.

Charney, J. G. and Eliassen, A. (1964). On the growth of the hurricane depression, *Journal of Atmospheric Sciences*, 21(1), pp. 68–75, doi: 10.1175/1520-0469(1964)021<0068:OTGOTH>2.0.CO;2.

Charney, J. G., Fjörtoft, R. and von Neumann, J. (1950). Numerical integration of the barotropic vorticity equation, *Tellus*, 2, pp. 237–254, doi: 10.3402/tellusa.v2i4.8607.

Charney, J. G. and Stern, M. E. (1962). On the stability of internal baroclinic jets in a rotating atmosphere, *Journal of Atmospheric Sciences*, 19(2), pp. 159–172, doi: 10.1175/1520-0469(1962)019<0159:OTSOIB>2.0.CO;2.

Dalcher, A. and Kalnay, E. (1987). Error growth and predictability in operational ECMWF forecasts, *Tellus A*, 39A, pp. 474–491, doi: 10.3402/tellusa.v39i5.11774.

Dean, L., Emanuel, K. A. and Chavas, D. R. (2009). On the size distribution of Atlantic tropical cyclones, *Geophysics Research Letters*, 36, L14803, doi: 10.1029/2009GL039051.

Dijkstra, H. A. (2005). *Nonlinear physical oceanography: A dynamical systems approach to the large scale ocean circulation and El Niño*, 2nd revised and enlarged edition, Springer, Dordrecht, The Netherlands, 534 p. doi: 10.1007/1-4020-2263-8.

Drazin, P. G. and Reid, W. H. 1981. *Hydrodynamic stability*, Cambridge University Press, Cambridge, 527 p.

Eady, E. T. (1949). Long waves and cyclone waves, *Tellus*, 1(3), pp. 33–52, doi: 10.3402/tellusa.v1i3.8507.

Edmon, H. J., Hoskins, B. J. and McIntyre, M. E. (1980). Eliassen-Palm cross sections for the troposphere, *Journal of Atmospheric Sciences*, 37(12), pp. 2600–2616, doi: 10.1175/1520-0469(1980)037<2600:EPCSFT>2.0.CO;2.

Ekman, V. W. (1905). On the influence of the earth's rotation on ocean currents, *Arkiv for Matematik, Astronomi och Fysik*, 2(11), pp. 1–52.

Eliassen, A. (1951). Slow thermally or frictionally controlled meridional circulation in a circular vortex, *Astrophysica Norvegica*, 5(2), pp. 19–60.

Eliassen, A. and Palm, E. (1961). On the transfer of energy in stationary mountain waves, *Geophysics Publications*, 22, pp. 1–23.

Emanuel, K. A. (1986). An air-sea interaction theory for tropical cyclones. Part I: Steady-state maintenance, *Journal of Atmospheric Sciences*, 43(6), pp. 585–605, doi: 10.1175/1520-0469(1986)043<0585:AASITF>2.0.CO;2.

Emanuel, K. A. and Rotunno, R. (1989). Polar lows as arctic hurricanes, *Tellus A: Dynamic Meteorology and Oceanography*, 41(1), pp. 1–17, doi: 10.3402/tellusa.v41i1.11817.

Emanuel, K. A. (1997). Some aspects of hurricane inner-core dynamics and energetics, *Journal of the Atmospheric Sciences*, 54(8), pp. 1014–1026, doi: 10.1175/1520-0469(1997)054%3C1014:SAOHIC%3E2.0.CO;2.

Emanuel, K. and Rotunno, R. (2011). Self-stratification of tropical cyclone outflow. Part I: Implications for storm structure, *Journal of the Atmospheric Sciences*, 68(10), pp. 2236–2249, doi: 10.1175/JAS-D-10-05024.1.

Ertel, H. (1942). Ein neuer hydrodynamischer Erhaltungssatz, *Naturwissenschaften*, 30, pp. 543–544, doi: 10.1007/BF01475602.

Exner, F.M. (1925). *Dynamische meteorologie*, second edition, Springer-Verlag Wien, Leipzig, 422 p., doi: 10.1007/978-3-642-52603-9.

Farrell, B. (1984). Modal and non-modal baroclinic waves, *Journal of Atmospheric Sciences*, 41(4), pp. 668–673, doi: 10.1175/1520-0469(1984)041<0668:MANMBW>2.0.CO;2.

Fraedrich, K. and Böttger, H. (1978). A wavenumber-frequency analysis of the 500 mb geopotential at 50°N, *Journal of Atmospheric Sciences*, 35(4), pp. 745–750, doi: 10.1175/1520-0469(1978)035<0745:AWFAOT>2.0.CO;2.

Fraedrich, K. and Frisius, T. (2001), Two-level primitive-equation baroclinic instability on an f-plane, *Quarterly Journal of Royal Meteorological Society*, 127, pp. 2053–2068, doi: 10.1002/qj.49712757611.

Fraedrich, K. and McBride, J. L. (1995). Large-scale convective instability revisited, *Journal of Atmospheric Sciences*, 52(11), pp. 1914–1923, doi: 10.1175/1520-0469(1995)052<1914:LSCIR>2.0.CO;2.

Fraedrich, K., Jansen, H., Kirk, E., Luksch, U. and Lunkeit, F. (2005). The planet simulator: Towards a user friendly model, *Meteorologische Zeitschrift*, 14(3), pp. 299–304, doi: 10.1127/0941-2948/2005/0043.

Fraedrich, K. and Ziehmann, C. (1995). Praktische Vorhersagbarkeit: Persistenz in rotem Rauschen, *Meteorologische Zeitschrift*, 4(4), pp. 139–149, doi: 10.1127/metz/4/1995/139.

Frierson, D. M. W. (2007). The dynamics of idealized convection schemes and their effect on the zonally averaged tropical circulation, *Journal of the Atmospheric Sciences*, 64(6), pp. 1959–1976, doi: 10.1175/JAS3935.1.

Frisius, T. (1998). A mechanism for the barotropic equilibration of baroclinic waves, *Journal of the Atmospheric Sciences*, 55(18), pp. 2918–2936, doi: 10.1175/1520-0469(1998)055<2918:AMFTBE>2.0.CO;2.

Frisius, T. (2003). The development of a cyclone–anticyclone asymmetry within a growing baroclinic wave, *Journal of the Atmospheric Sciences*, 60(23), pp. 2887–2906. doi: 10.1175/1520-0469(2003)060<2887:TDOACA>2.0.CO;2.

Frisius, T. (2005). An atmospheric balanced model of an axisymmetric vortex with zero potential vorticity, *Tellus A: Dynamic Meteorology and Oceanography*, 57(1), pp. 55–64, doi: 10.3402/tellusa.v57i1.14605.

Frisius, T. and Hasselbeck, T. (2009). The effect of latent cooling processes in tropical cyclone simulations, *Quarterly Journal of Royal Meteorological Society*, 135, pp. 1732–1749, doi: 10.1002/qj.495.

Frisius, T. and Wacker, U. (2007). Das massenkonsistente axialsymmetrische Wolkenmodell HURMOD, Deutscher Wetterdienst, Arbeitsergebnisse 85, 42 p.

Frisius, T., Schönemann, D. and Vigh, J. (2013). The impact of gradient wind imbalance on potential intensity of tropical cyclones in an unbalanced slab boundary layer model, *Journal of the Atmospheric Sciences*, 70(7), pp. 1874–1890, doi: 10.1175/JAS-D-12-0160.1.

Frisius, T. and Schönemann, D. (2012). An extended model for the potential intensity of tropical cyclones, *Journal of the Atmospheric Sciences*, 69(2), pp. 641–661, doi: 10.1175/JAS-D-11-064.1.

Gray, W. M. (1979). Hurricanes: Their formation, structure and likely role in the tropical circulation, *Meteorology over the Tropical Oceans*, D. B. Shaw, Ed., Royal Meteorological Society, Bracknell, pp. 155–218.

Hart, J. E. (1985). A laboratory study of baroclinic chaos on the f-plane, *Tellus A: Dynamic Meteorology and Oceanography*, 37(3), pp. 286–296, doi: 10.3402/tellusa.v37i3.11673.

Hasselmann, K. (1988). PIPs and POPs: The reduction of complex dynamical systems using principal interaction and oscillation patterns, *Journal of Geophysics Research*, 93(D9), pp. 11015–11021, doi: 10.1029/JD093iD09p11015.

Haurwitz, B. (1940). The motion of atmospheric disturbances on the spherical earth, *Journal of Marine Research*, 3(5), pp. 254–267.

Helmholtz, H. (1858). Über Integrale der hydrodynamischen Gleichungen, welche den Wirbelbewegungen entsprechen, *Angew. Mathematics*, 55, pp. 25–55.

Hersbach, H., Bell, B., Berrisford, P., *et al.* (2020). The ERA5 global reanalysis, *Quarterly Journal of Royal Meteorological Society*, 146, pp. 1999–2049, doi: 10.1002/qj.3803.

Hodges, K. I., Lee, R. W. and Bengtsson, L. (2011). A comparison of extratropical cyclones in recent reanalyses ERA-Interim, NASA MERRA, NCEP CFSR, and JRA-25, *Journal of Climate*, 24(18), pp. 4888–4906, doi: 10.1175/2011JCLI4097.1.

Hoskins, B. J. (1973). Stability of the Rossby-Haurwitz wave, *Quarterly Journal of Royal Meteorological Society*, 99, pp. 723–745, doi: 10.1002/qj.49709942213.

Hoskins, B. J. (1975). The geostrophic momentum approximation and the semi-geostrophic equations, *Journal of Atmospheric Sciences*, 32(2), pp. 233–242, doi: 10.1175/1520-0469(1975)032<0233:TGMAAT>2.0.CO;2.

Hoskins, B. J. and Bretherton, F. P. (1972). Atmospheric frontogenesis models: Mathematical formulation and solution, *Journal of Atmospheric Sciences*, 29(1), pp. 11–37, doi: 10.1175/1520-0469(1972)029<0011:AFMMFA>2.0.CO;2.

Hoskins, B. J., Draghici, I. and Davies, H. C. (1978). A new look at the ω-equation, *Quarterly Journal of Royal Meteorological Society*, 104, pp. 31–38, doi: 10.1002/qj.49710443903.

Hoskins, B. J. and Simmons, A. J. (1975). A multi-layer spectral model and the semi-implicit method, *Quarterly Journal of Royal Meteorological Society*, 101, pp. 637–655, doi: 10.1002/qj.49710142918.

Hoskins, B. J., McIntyre, M. E. and Robertson, A. W. (1985). On the use and significance of isentropic potential vorticity maps, *Quarterly Journal of Royal Meteorological Society*, 111, 877–946, doi: 10.1002/qj.49711147002.

Iyer, K. P., Scheel, J. D., Schumacher, J. and Sreenivasan, K. R. (2020). Classical 1/3 scaling of convection holds up to Ra = 1015, *Proceedings of the National Academy of Sciences*, 117(14), pp. 7594–7598, doi: 10.1073/pnas.1922794117.

James, I. N. (1987). Suppression of baroclinic instability in horizontally sheared flows, *Journal of Atmospheric Sciences*, 44(24), pp. 3710–3720, doi: 10.1175/1520-0469(1987)044<3710:SOBIIH>2.0.CO;2.

James, I. N. (1994). *Introduction to circulating atmospheres*, Cambridge University Press, Cambridge, 422 p. doi: 10.1017/CBO9780511622977.

Jordan, C. L. (1958). Mean soundings for the West Indies area, *Journal of Atmospheric Sciences*, 15(1), pp. 91–97, doi: 10.1175/1520-0469(1958)015<0091:MSFTWI>2.0.CO;2.

Kalnay, E. (2003). Historical overview of numerical weather prediction.*Handbook of Weather, Climate, and Water: Dynamics, Climate, Physical Meteorology, Weather Systems, and Measurements*, pp. 95–115.

Kalnay, E., Kanamitsu, M., Kistler, R., Collins, W., Deaven, D., Gandin, L., Iredell, M., Saha, S., White, G., Woollen, J., Zhu, Y., Chelliah, M., Ebisuzaki, W., Higgins, W., Janowiak, J., Mo, K. C., Ropelewski, C., Wang, J., Leetmaa, A., Reynolds, R., Jenne, R. and Joseph, D. (1996). The NCEP/NCAR 40-year reanalysis project, *Bulletin of the American Meteorological Society*, 77(3), pp. 437–472, doi: 10.1175/1520-0477(1996)077<0437:TNYRP>2.0.CO;2.

Kessler, E. (1969). On the distribution and continuity of water substance in atmospheric circulations, *Meteorological Monograph*, 10(32), 88 p.

Kuettner, J. P. (1971). Cloud bands in the earth's atmosphere: Observations and theory, *Tellus*, 23(4–5), pp. 404–426, doi: 10.3402/tellusa.v23i4-5.10519.

Kuo, H. L. (1949). Dynamic instability of two-dimensional nondivergent flow in a barotropic atmosphere, *Journal of Atmospheric Sciences*, 6(2), pp. 105–122, doi: 10.1175/1520-0469(1949)006<0105:DIOTDN>2.0.CO;2.

Kuo, H. L. (1965). On formation and intensification of tropical cyclones through latent heat release by cumulus convection, *Journal of Atmospheric Sciences*, 22(1), pp. 40–63, doi: 10.1175/1520-0469(1965)022<0040:OFAIOT>2.0.CO;2.

Kuo, H. L. (1974). Further studies of the parameterization of the influence of cumulus convection on large-scale flow, *Journal of Atmospheric Sciences*, 31(5), pp. 1232–1240, doi: 10.1175/1520-0469(1974)031<1232:FSOTPO>2.0.CO;2.

Kurz, M. (1990). Synoptische Meteorologie. Leitfäden für die Ausbildung im Deutschen Wetterdienst Nr. 8. Offenbach am Main, Selbstverlag des DWD, 197 p.

Lee, M. and Frisius, T. (2018). On the role of convective available potential energy (CAPE) in tropical cyclone intensification, *Tellus A: Dynamic Meteorology and Oceanography*, 70(1), pp. 1–18, doi: 10.1080/16000870.2018.1433433.

Lamb, H. (1932). *Hydrodynamics*, sixth edition, Cambridge University Press, Cambridge, pp. 548–549.

LeMone, M. A. (1973). The structure and dynamics of horizontal roll vortices in the planetary boundary layer, *Journal of Atmospheric Sciences*, 30(6), pp. 1077–1091, doi: 10.1175/1520-0469(1973)030<1077:TSADOH>2.0.CO;2.

Lilly, D. K. (1962). On the numerical simulation of buoyant convection, *Tellus*, 14, pp. 148–172, doi: 10.1111/j.2153-3490.1962.tb00128.x.

Longuet-Higgins, M. S. (1968). The eigenfunctions of Laplace's tidal equation over a sphere, *Philosophical Transactions of the Royal Society of London*, Series A, Mathematical and Physical Sciences, 262, pp. 511–607, doi: 10.1098/rsta.1968.0003.

Lorenz, E. N. (1955). Available potential energy and the maintenance of the general circulation, *Tellus*, 7(2), pp. 157–167, doi: 10.3402/tellusa.v7i2.8796.

Lorenz, E. N. (1963). Deterministic nonperiodic flow, *Journal of Atmospheric Sciences*, 20(2), pp. 130–141, doi: 10.1175/1520-0469(1963)020<0130:DNF>2.0.CO;2.

Lorenz, E. N. (1969). The predictability of a flow which possesses many scales of motion, *Tellus*, 21(3), pp. 289–307, doi: 10.3402/tellusa.v21i3.10086.

Lorenz, E. (1972). Predictability: Does the flap of a butterfly's wing in Brazil set off a tornado in Texas? Paper presented at the annual meeting of the American Association for the Advancement of Science, Washington, DC.

Lunkeit, F., Borth, H., Böttinger, M., Fraedrich, K., Jansen, H., Kirk, E., Kleidon, A., Luksch, U., Paiewonsky, P., Schubert, S., Sielmann, S. and Wan, H. (2011). Planet simulator reference manual version 16, Technical report, Hamburg, Germany, 85 p.

Margules, M. (1893). Luftbewegungen in einer rotirenden Sphäroidschale (II. Teil), *Sitz-Ber. kaiserl. Akad. Wissensch. Wien*, 102, pp. 11–56.

Margules, M. (1905). Über die Energie der Stürme, *Jahrbücher d. k. k. Zentralanstalt für Meteorologie und Erdmagnetismus*, 40, pp. 1–26.

Margules, M., (1906). Über Temperaturschichtung in stationär bewegter und in ruhender Luft, *Meteorol. Z.*, Hann-Volume, pp. 243–254, doi: 10.1127/metz/2016/0833.

Matsuno, T. (1966). Quasi-Geostrophic motions in the equatorial area, *Journal of the Meteorological Society of Japan Series II*, 44(1), pp. 25–43, doi: 10.2151/jmsj1965.44.1_25.

McCormmach, R. (1970). H. A. Lorentz and the Electromagnetic view of nature, *Isis*, 61(4), pp. 459–497, doi: 10.1086/350681.

Montgomery, M. T., Persing, J. and Smith, R. K. (2015). Putting to rest WISHEful misconceptions for tropical cyclone intensification, *Journal of Advanced Modelling Earth Systems*, 7, pp. 92–109, doi: 10.1002/2014MS000362.

Montgomery, R. B. (1937). A suggested method for representing gradient flow in isentropic surfaces, *Bulletin of the American Meteorological Society*, 18, pp. 210–212, doi: 10.1175/1520-0477-18.6-7.210.

Müller, G. and Chlond, A. (1996). Three-dimensional numerical study of cell broadening during cold-air outbreaks, *Boundary-Layer Meteorology* 81, pp. 289–323, doi: 10.1007/BF02430333.

Nakamura, N. (1993). An illustrative model of instabilities in meridionally and vertically sheared flows, *Journal of Atmospheric Sciences*, 50(3), pp. 357–376, doi: 10.1175/1520-0469(1993)050<0357:AIMOII>2.0.CO;2.

Ogura, Y. (1964). Frictionally controlled, thermally driven circulations in a circular vortex with application to tropical cyclones, *Journal of Atmospheric Sciences*, 21(6), pp. 610–621, doi: 10.1175/1520-0469(1964)021<0610:FCTDCI>2.0.CO;2.

Ogura, Y. and Phillips, N. A. (1962). Scale analysis of deep and shallow convection in the atmosphere, *Journal of Atmospheric Sciences*, 19(2), pp. 173–179, doi: 10.1175/1520-0469(1962)019<0173:SAODAS>2.0.CO;2.

Orlanski, I. (1968). Instability of frontal waves, *Journal of Atmospheric Sciences*, 25(2), pp. 178–200, doi: 10.1175/1520-0469(1968)025<0178:IOFW>2.0.CO;2.

Orlanski, I. (1975). A rational subdivision of scales for atmospheric processes, *Bulletin of the American Meteorological Society*, 56(5), pp. 527–530.

Pedlosky, J. (1983). The growth and decay of finite-amplitude baroclinic waves, *Journal of Atmospheric Sciences*, 40(8), pp. 1863–1876, doi: 10.1175/1520-0469(1983)040<1863:TGADOF>2.0.CO;2.

Peixoto, J. P. and Oort, A. H. (1992). *Physics of climate*, American Institute of Physics, New York, 520 p.

Petterssen, S. and Smebye, S. J. (1971), On the development of extratropical cyclones, *Quarterly Journal of Royal Meteorological Society*, 97, pp. 457–482, doi: 10.1002/qj.49709741407.

Phillips, N. A. (1957). A coordinate system having some special advantages for numerical forecasting, *Journal of Atmospheric Sciences*, 14(2), pp. 184–185, doi: 10.1175/1520-0469(1957)014<0184:ACSHSS>2.0.CO;2.

Polvani, L. M., Scott, R. K. and Thomas, S. J. (2004). Numerically converged solutions of the global primitive equations for testing the dynamical core of atmospheric GCMs, *Monthly Weather Review*, 132(11), pp. 2539–2552, doi: 10.1175/MWR2788.1.

Pomeau, Y., Manneville, P. (1980). Intermittent transition to turbulence in dissipative dynamical systems, *Communications Mathematics & Physics*, 74, pp. 189–197, doi: 10.1007/BF01197757.

Prandtl, L. (1942). *Führer durch die Strömungslehre*, Vieweg & Sohn Braunschweig, 321 p.

Rasmussen, E. A. and Turner, J. (2003). *Polar lows: Mesoscale weather systems in the polar regions*, Cambridge University Press, Cambridge, 626 p.

Rayleigh, L. (1880). On the stability, or instability, of certain fluid motions, *Proceedings of the London Mathematical Society*, 11, pp. 25–43.

Rayleigh, L. (1916). On convection currents in a horizontal layer of fluid, when the higher temperature is on the under side, *The London, Edinburgh, and Dublin Philosophical Magazine and Journal of Science*, 32(192), pp. 529–546, doi: 10.1080/14786441608635602.

Reinhold, B. B. and Pierrehumbert, R. T. (1982). Dynamics of weather regimes: Quasi-stationary waves and blocking, *Monthly Weather Review*, 110(9), pp. 1105–1145, doi: 10.1175/1520-0493(1982)110<1105:DOWRQS>2.0.CO;2.

Richtmyer, R. D. (1957). *Difference methods for initial-value problems*, Interscience Publishers, Inc., New York, 238 p.

Riemann-Campe, K., Fraedrich, K. and Lunkeit, K. (2009). Global climatology of convective available potential energy (CAPE) and convective inhibition (CIN) in ERA-40 reanalysis, *Atmospheric Research*, 93, pp. 534–545, doi: 10.1016/j.atmosres.2008.09.037.

Roeckner, E., Bäuml, G., Bonaventura, L., Brokopf, R., Esch, M., Giorgetta, M., *et al.* (2003). The atmospheric general circulation model ECHAM 5. PART I: Model description, *Report / Max-Planck-Institut für Meteorologie*, 349, 127 p.

Rossby, C. G. (1936). Dynamics of steady ocean currents in the light of experimental fluid mechanics, *Papers Physics Oceanographic Meteorology*, 5, pp. 1–43.

Rossby, C. G. (1939). Relation between variations in the intensity of the zonal circulation of the atmosphere and the displacements of the semi-permanent centers of action, *Journal of Marine Research*, 2, pp. 38–55.

Rotunno, R., Muraki, D. J. and Snyder, C. (2000). Unstable baroclinic waves beyond quasigeostrophic theory, *Journal of the Atmospheric Sciences*, 57(19), pp. 3285–3295, doi: 10.1175/1520-0469(2000)057<3285:UBWBQT>2.0.CO;2.

Saltzman, B. (1962). Finite amplitude free convection as an initial value problem-I, *Journal of Atmospheric Sciences*, 19(4), pp. 329–341, doi: 10.1175/1520-0469(1962)019<0329:FAFCAA>2.0.CO;2.

Sawyer, J. S. (1956). The vertical circulation at meteorological fronts and its relation to frontogenesis, *Proceedings of the Royal Society London*, A234, pp. 346–362, doi: 10.1098/rspa.1956.0039.

Schultz, D. M., Keyser, D. and Bosart, L. F. (1998). The effect of large-scale flow on low-level frontal structure and evolution in midlatitude cyclones, *Monthly Weather Review*, 126(7), pp. 1767–1791, doi: 10.1175/1520-0493(1998)126<1767:TEOLSF>2.0.CO;2.

Sela, J. (2010). The derivation of sigma pressure hybrid coordinate semi-Lagrangian model equations for the GFS. NCEP Office Note #462, 31 p.

Shapiro, M. A. and Keyser, D. (1990). Fronts, jet streams and the tropopause, *Extratropical Cyclones, The Erik Palmén Memorial Volume*, C. W. Newton and E. O. Holopainen, Eds., American Meteorological Society, Boston, MA, pp. 167–191, doi: 10.1007/978-1-944970-33-8_10.

Shapiro, L. J. and Willoughby, H. E. (1982). The response of balanced hurricanes to local sources of heat and momentum, *Journal of Atmospheric Sciences*, 39(2), pp. 378–394, doi: 10.1175/1520-0469(1982)039<0378: TROBHT>2.0.CO;2.

Shuman, F. G. and Hovermale, J. B. (1968). An operational six-layer primitive equation model, *Journal of Applied Meteorology and Climatology*, 7(4), pp. 525–547. doi: 10.1175/1520-0450(1968)007<0525:AOSLPE>2.0.CO;2.

Simmons, A. J. and Burridge, D. M. (1981). An energy and angular-momentum conserving vertical finite-difference scheme and hybrid vertical coordinates, *Monthly Weather Review*, 109(4), pp. 758–766, doi: 10.1175/1520-0493 (1981)109<0758:AEAAMC>2.0.CO;2.

Simmons, A. J., Burridge, D. M., Jarraud, M. *et al.* (1989). The ECMWF medium-range prediction models development of the numerical formulations and the impact of increased resolution, *Meteorological Atmosphere, Phys.* 40, pp. 28–60, doi: 10.1007/BF01027467.

Snyder, C., Skamarock, W. C. and Rotunno, R. (1991). A comparison of primitive-equation and semigeostrophic simulations of baroclinic waves, *Journal of Atmospheric Sciences*, 48(19), pp. 2179–2194, doi: 10.1175/1520-0469(1991) 048<2179:ACOPEA>2.0.CO;2.

Sparrow, E., Goldstein, R. and Jonsson, V. (1964). Thermal instability in a horizontal fluid layer: Effect of boundary conditions and non-linear temperature profile, *Journal of Fluid Mechanics,* 18(4), pp. 513–528, doi: 10.1017/ S0022112064000386.

Spiegel, M. R., Lipschutz, S. and Spellman, D. (2009). *Schaum's outlines. Vector analysis*, second edition, McGraw-Hill, New York, 238 p.

Stern, D. P., Vigh, J. L., Nolan, D. S. and Zhang, F. (2015). Revisiting the relationship between eyewall contraction and intensification, *Journal of the Atmospheric Sciences*, 72(4), pp. 1283–1306, doi: 10.1175/JAS-D-14-0261.1.

Stull, R. B. (1988). *An introduction to boundary layer meteorology*, Springer, Dordrecht, 666 p., doi: 10.1007/978-94-009-3027-8.

Taylor, K. E. (1979). Formulas for calculating available potential energy over uneven topography, *Tellus*, 31(3), pp. 236–245, doi: 10.1111/j.2153-3490. 1979.tb00902.x.

Thomson, W., (1867). On vortex atoms, *Proceedings of the Royal Society Edinburgh*, 6, pp. 94–105.

von Neumann, J. (1955). *Mathematical foundations of quantum mechanics*, Princeton University Press, Princeton, NJ,0020445 p.

Webster, P. J., Holland, G. J., Curry, J. A. and Chang, H.-R. (2005). Changes in tropical cyclone number, duration, and intensity in a warming environment, *Science*, 309(5742), pp. 1844–1846, doi: 10.1126/science.1116448.

Willoughby, H. E. (1990). Temporal changes of the primary circulation in tropical cyclones, *Journal of Atmospheric Sciences*, 47(2), pp. 242–264, doi: 10.1175/1520-0469(1990)047<0242:TCOTPC>2.0.CO;2.

Wolf, A., Swift, J. B., Swinney, H. L. and Vastano, J. A. (1985). Determining Lyapunov exponents from a time series, *Physica D: Nonlinear Phenomena*, 16(3), pp. 285–317, doi: 10.1016/0167-2789(85)90011-9.

Yanai, M. and Maruyama T. (1966). Stratospheric wave disturbances propagating over the equatorial Pacific, *Journal of the Meteorological Society of Japan. Series II*, 44(5), pp. 291–294, doi: 10.2151/jmsj1965.44.5_291.

Zhang, F., Sun, Y. Q., Magnusson, L., Buizza, R., Lin, S., Chen, J. and Emanuel, K. (2019). What is the predictability limit of midlatitude weather?, *Journal of the Atmospheric Sciences*, 76(4), 1077–1091, doi: 10.1175/JAS-D-18-0269.1.

Index

www.ingramcontent.com/pod-product-compliance
Lightning Source LLC
Chambersburg PA
CBHW052115230326
41598CB00079B/3700